BIOFEEDBACK
and SELF-CONTROL 1972

EDITORS

DAVID SHAPIRO
Senior Associate
Department of Psychiatry
Massachusetts Mental Health Center
Harvard Medical School
Boston, Massachusetts

T. X. BARBER
Director of Psychological Research
Medfield State Hospital
Harding, Massachusetts

LEO V. DiCARA
Professor
Department of Psychiatry
Ann Arbor, Michigan

JOE KAMIYA
Director, Psychophysiology Lab
The Langley Porter Institute
University of California Medical Center
San Francisco, California

NEAL E. MILLER
Professor
Laboratory of Physiological Psychology
The Rockefeller University
New York, New York

JOHANN STOYVA
Associate Professor
Department of Psychiatry
University of Colorado Medical Center
Denver, Colorado

BIOFEEDBACK and SELF-CONTROL 1972

an Aldine Annual on the regulation of bodily processes and consciousness

ALDINE PUBLISHING COMPANY
Chicago

Copyright © 1973 by Aldine Publishing Company

All rights reserved.
No part of this publication may be reproduced
or transmitted in any form or by any means, electronic
or mechanical, including photocopy, recording, or
any information storage and retrieval system, without
permission in writing from the publisher.

First published 1973 by
Aldine Publishing Company
529 South Wabash Avenue
Chicago, Illinois 60605

Library of Congress Catalog Card Number 73-75702
ISBN 0-202-25107-1

Printed in the United States of America
Second Printing 1974

PREFACE

Nineteen seventy-two was a year in which biofeedback continued to inspire productive research and significant applications by behavioral scientists and clinicians, and also continued to excite the imagination of journalists and science writers. Not only did the topic show up again in newsweeklies and the Sunday supplements, but it was brought to other readers. A full-page advertisement in the *New York Times* announced a two-part "Profile on Visceral Learning" in the *New Yorker* magazine, the ad framed around a pensive portrait of Neal Miller. The profile was not of a highly publicized popularizer of behavioral science but rather of the psychologists' psychologist who has dedicated almost four decades to the unraveling of basic mechanisms of motivation and learning. The article paid tribute to the many elegant, more recent demonstrations of learned regulation of visceral processes in the rat in the experiments by Miller, Leo V. DiCara, and their colleagues. It was also a tribute to the many independent achievements of other innovators in the field, particularly of the human researchers during the past 10-15 years whose curiosity about physiological self-regulation in man, and presumably about themselves, had brought them to perform similar experiments and draw similar conclusions. *Playboy* magazine also carried a piece late in the year, playing up the more sensational and somewhat freakish feats of voluntary control and pointing to the potentials of "the whole business of bugging your body" for creativity, varied states of consciousness, and psychosomatic good health.

But the outstanding achievements of the year are not truly reflected in these public images. Rather they are shown in the many extensions of fact, theory, and application described in the excellent papers reprinted in this volume. This year's collection attests to the growing theoretical and methodological sophistication of biofeedback research. Empirical studies appear to be consolidating around a number of critical issues, and the excitement of initial discovery and demonstration has given way to hard-nosed queries into the significance and interpretation of the earlier experimental data. Sharper, more precise questions are being posed about the relative contributions to voluntary control of different types and amounts of feedback using varying instructions (articles 16, 17, 19, and 23); attention to and awareness of one's own internal responses (articles 6 and 16); choice and scheduling of rewards (articles 13 and 24); cognitive and somatomotor mechanisms (articles 9, 13, 14, 15, 25, and 26). How can the facts of visceral self-control be best integrated into the mainstreams of psychological and physiological theory and research, and how do the phenomena relate to accumulated knowledge about learning, motivation, and other varieties of skilled performance, especially in human behavior (part II). Can the techniques be effectively applied to treatment and research in psychosomatic medicine, psychiatry, and clinical psychology (part X)? What does the research tell us about cognitive influences on physiological functioning and behavior (part IX)? What new dimensions of consciousness, creativity, and other human potentials can be unfolded (part III)?

Johann Stoyva, in his preface to the 1971 *Annual,* remarked that some of the authors that year had sounded a cheerful, confident note, while others had voiced gloomier sentiments and had emphasized that little was known for sure, especially on the human level. Similar contrasting positions are taken this year, although the prevailing tone is more a mix of optimism and skepticism, which is in my opinion the hallmark of a productive scientific attitude.

PROBLEM OF REPLICATION

A major crisis of confidence about the future of biofeedback research has preoccupied the attention and concern of all. Surprisingly, it centers around certain difficulties encountered by Neal Miller and his fellow researchers at Rockefeller in attempting to duplicate the results of earlier experiments on learned control of heart rate in rats that are paralyzed by neuromuscular blocking agents. In the early studies, the percent learned changes in heart rate, averaged over increase and decrease groups, varied between 15 to somewhat over 20 percent of baseline. In more recent studies, this change declined to between 5 and 10 percent. These facts and associated issues are discussed by Miller in article 34. However, a brighter note was sounded in the studies by Hothersall and Brener and by Slaughter, Hahn, and Rinaldi. In these independent laboratories, replications have been reported that are considered to be statistically highly reliable, and the average change is about 10 percent of baseline.

It is perhaps ironic that it is the animal rather than the human studies that are called into question; the latter are usually challenged on the grounds of inadequate controls for respiration and other voluntary somatomotor processes or because of effects of too small magnitude. The Miller experiments in the curarized rat, historically speaking, had presumably laid a firm foundation for research on learned autonomic control, and had bolstered the human research, a great deal of which had already been published. The effects in the animal research had been dramatic in terms of magnitude of changes and in terms of apparent specificity. The fact that given visceral responses could be selectively reinforced, while others did not change differentially, strengthened the instrumental conditioning model as applied to these behaviors. From a clinical standpoint, of course, one is prepared to utilize whatever means, whether somatomotor maneuvers or particular thoughts or images, that the patient can contrive to alter his own responses in a given direction or to bring unwanted behaviors under control. The critical practical problem is to demonstrate effective learned control of a desired magnitude, persistence of the change over time, and hopefully transfer of the acquired learning to real-life conditions.

THE MEDIATION ISSUE REVISITED

While the effects obtained in curarized animals may be smaller than originally believed, the accumulated data definitely support the conclusion that *peripheral* somatomotor responses are not necessary to learned changed in autonomic responses. However, the theoretical significance of such data has been called into question. Abe Black (article 9) argues that the attempt to rule out mediation in operant neural conditioning has the purpose of showing that the results of such conditioning are novel, rather than an inadvertent outcome of the easily-obtained conditioning of some observable response. Mediation data, he argues, "could provide information about the neural and behavioral systems of which the reinforced pattern might be a component [and] about the neural and behavioral systems that are activated when a given reinforced pattern occurs, and, therefore, about the mechanisms that are involved in performing a given operantly conditioned neural response. They could provide information on the specificity of conditioning; that is, on whether a response is conditioned as part of a global behavioral pattern, or as an individual unit over which we have precise and specific control. These issues are, I think, more important than the demonstration that one has conditioned a novel response."

In regard to operant autonomic conditioning, article 25 by Goesling and Brener also deals with the mediation issue. These authors take the stance that cardiovascular and somatomotor activities, rather than being autonomously controlled functions, represent two components of a general response process. This has also been proposed by Paul Obrist and his associates, in addition to Black and other investigators. Goesling and Brener conditioned rats either to run or to stay immobile and found that the pre-training effects were more significant determinants of heart rate change when the animals were subsequently curarized and punished for either low or high heart rates. As these authors put it, "In other words, the use of curare . . . did not prevent somatomotor activity from occurring—it simply prevented the experimenter from observing it by blocking its most peripheral manifestations." In effect, since the curarized preparation does not affect the central linkage between cardiovascular and somatomotor activities, it cannot rule out the influence of central nervous system somatomotor activities on the learning process.

Goesling and Brener point out, however, that DiCara and Miller in an earlier paper in 1969 presented data indicating that transfer of instrumentally-learned heart-rate changes from curarized to noncurarized state is initially accompanied by related somatomotor activities. Further training in the noncurarized state, however, indicates that the cardiovascular effects can be magnified and the somatomotor effects diminished, suggesting that the two processes can be dissociated by learning techniques. These dynamic processes of dissociation, particularly between somatomotor and autonomically-mediated behaviors, or between autonomic and central nervous processes, are classical problems in psychophysiology and psychosomatic medicine, and the methods of biofeedback and operant conditioning may provide new knowledge about such relationships. Article 27 contributes significantly to some of the basic neurophysiological processes and theoretical issues. Research on the curarized animal has helped clarify many of the relationships involved, and further work along these lines using surgical lesions and biochemical agents is needed. Studies of visceral and neural control in humans with muscular and neural disorders is also indicated.

Questions of response interaction during the conditioning of neural and visceral functions are taken up in different ways in almost all of the papers in this volume. As has long been known, somatomotor and autonomic processes are integrated in varying degrees at all levels of the nervous system. Bernard Engel (article 13) states simply, "It is my personal view that there must be some degree of integration among cardiovascular behavior, somatic behavior, and central nervous system behavior. Specificity is biologically useful since it supports adaptation. However, specificity to the point of physiological disintegration would be biologically disastrous." Engel, in his presidential address to the

Society for Psychophysiological Research, offers a particularly down-to-earth and enlightening discussion of the multiple questions of mediation and inter-response correlation during conditioning. He deals with mechanisms of cardiac control in conditioning experiments, analyzed in terms of cardiodynamics, hemodynamics, neural and psychological regulations. One of his major conclusions is that the learning mechanisms in early stages of training differ significantly from those of late phases, and he suggests that the analysis of control mechanisms should probably be derived from chronic studies. Engel's long-term studies of cardiac control in normal human subjects, in patients with cardiac arrhythmias, and in monkeys have clarified these mechanisms and indicate the need for further solid contributions to research along these lines.

A Model of Visceral Learning

In regard to the issues of mediation and specificity, Gary Schwartz in article 14 makes an outstanding contribution to our understanding of the data showing that individual functions can be brought under voluntary control and indicating specificity of learning similar to that observed in somatomotor responses. One of the earliest demonstrations of specificity in human conditioning was made in our laboratories at the Massachusetts Mental Health Center (Shapiro, Crider, and Tursky, *Psychonomic Science,* 1964), in which it was shown that subjects could learn relatively to increase or decrease the rate of occurrence of spontaneous skin potential responses independently of presumably associated functions such as respiration rate, heart rate, and basal skin potential level. Later work indicated that gross bodily movement was not related to these differences. These results were considered to be examples of specificity because it was assumed on several grounds that the other functions should co-vary with electrodermal response frequency. However, if the other functions were in fact independent of the electrodermal activity in natural or nonlearning conditions, the dissociation itself would not be surprising, but only the fact that the same reward could be used either to increase or decrease the particular response in question.

Schwartz started with observations about blood pressure and heart rate, which should covary, according to available evidence. It had been found in our research that human subjects rewarded for increases or decreases in systolic blood pressure showed relative changes in pressure in the appropriate direction without corresponding changes in heart rate. Similarly, subjects could learn to increase or decrease their heart rate without corresponding changes in systolic blood pressure. Schwartz reasoned that if in fact two functions such as these are highly correlated, when one is reinforced and shows learning the other should as well. If they are uncorrelated, then reinforcing one would result in learned changes in that function but not the other. In effect, when you condition one function in this manner, the degree to which other functions also show learning is informative about the natural interrelationships of the functions to begin with.

Schwartz developed a behavioral and biological model that explains specificity of learning in the autonomic nervous system, which he tested out in an ingenious fashion. He developed an on-line procedure for assessing the simultaneous pattern of systolic blood pressure and heart rate (or other functions) and for providing feedback and rewards to subjects for integrating these functions (increase or decrease both jointly) or for differentiating them (raise one and simultaneously lower the other). He obtained positive experimental evidence for both integration and differentiation, although the latter was not as clear cut. Where specificity had previously been observed, now the two functions could be associated or deliberately dissociated. Schwartz concluded that the model clarified the earlier specificity findings, and could provide a framework for research and theory on the control of "multiautonomic" functions. He also speculated that it may be possible to apply these techniques to the treatment of specific clinical disorders. For example, conditioning decreases in blood pressure and heart rate may help reduce the pain of angina pectoris.

Article 15 offers additional data on the analysis of multiple responses in the conditioning of visceral responses. Studies such as these depend at the very least on complex logic and programming apparatus and can be more effectively carried out using laboratory computers which can detect and analyze patterns of several simultaneous responses. More importantly, computer technology in future research will be needed more and more to provide on-line control of experiments, particularly as a means of perfecting dynamic feedback displays and techniques of modifying response criteria such as in continuous shaping procedures.

These new developments in examining psychological and physiological mechanisms of learned visceral and neural control are significant not only for what they may reveal about the self-control phenomena themselves but also for our understanding of basic organization and interrelationships of visceral, somatomotor, and central nervous processes.

Voluntary Control: New or Old?

The first paper in this *Annual* was published earlier than 1972, almost 90 years earlier, and it is included along with articles 2-5 to indicate that unusual instances of voluntary control of physiological functions such as heart rate have long been observed and noted in the scientific literature. These are beautiful papers that are full of keen observations and the gamut of questions that have also preoccupied modern researchers—questions about cognitive, pharmacological, somatomotor, central nervous system, and physiological mechanisms concerning the nature of voluntary control.

Any biofeedback researcher worth his salt has made observations of individuals who have great physiological skills, with or without the aid of feedback. The modern and fantastic exploits of the Swami Rama in Elmer Green's laboratory at the Menninger Foundation are by now well known (article 11). Article 18 by Stephens *et al.* describes large magnitude changes in heart rate observed in selected subjects. In our laboratories at the Massachusetts Mental Health Center, in earlier experiments on the operant control of skin potential response out of 150 or more subjects, there were about a dozen who were reactive in their electrodermal activity and who seemed to be able to turn it on and to some extent to turn it off at will, although the latter was more difficult. Some of these subjects were able to evolve idiosyncratic mental strategies to meet the demands of different operant schedules used in the experiments. Some felt unique sensations in their body that accompanied electrodermal responding. In regard to cardiovascular functions, I have also observed many instances of unusual control. One subject who was particularly outstanding in heart rate control could literally make the cardiotachometer move up and down "at will" for periods of minutes at a time. When I asked, "How do you do it?", he replied, "How do you move your arm?" His behavior matched the description by Lindsley and Sassman (article 3) of a subject who had voluntary control of the pilomotors. "The question arises as to how the subject is able to control the erection of the body hairs. So far as he is aware, the process is essentially similar to that of initiating contractions in one of his skeletal muscles. He does not call up an image of a painful or fearful experience. . . . There is no straining or tensing of skeletal muscles and there are no observable movements of any sort . . . he is able to inhibit the inflex erection of the hairs and the appearance of 'gooseflesh' normally induced by stepping out of a hot shower into a cold draft."

It was fascinating to read in the Lindsley-Sassaman article that this exploit of voluntary control was first discovered, as in some of the other cases that have been reported in the literature, during childhood, either during "severe fatigue, a critical illness, or some other unusual condition." In McClure's case (article 5) of willful cardiac arrest for periods of several seconds at a time, the individual stated that he had rheumatic fever at the age of seven and was bedridden for a long time. In the case of heart control observed in my lab, this person had also been hospitalized for a long period with a severe illness during his childhood. There is so much to learn from such cases about physiological self-regulation and control. What is the significance of illness in these cases, if any? Does medical interest and attention in and of itself play a role? Does persistent concentration on one's own organs, as Tarchanoff suggested in 1885, change the function of those organs, even though direct awareness of the change is rather undefined? Is the control brought about by conditioning processes, operant or classical, which are more potent at an early age?

Ogden and Shock (article 4) wrote in 1939 that only 15 cases of voluntary acceleration of pulse rate had been reported. A later paper in 1947, not included in this volume, states that the number was 21 at that time. Is such voluntary control so rare in fact or have we been blinded by bias and disbelief? One function of biofeedback research, perhaps, is to uncover cases of prior learned visceral control. Another, of course, is to use the techniques to perfect such control, where desirable, in people who have not already acquired it. One key to further understanding of physiological self-control may lie in the intensive study of these apparent oddities of human nature.

On Consciousness and Cognitive Processes

This year's collection of papers includes some excellent contributions to the psychophysiology of consciousness and cognitive processes in parts II, III, and IX. What Hefferline and Bruno (article 6) have to say about the psychophysiology of private events touches on this topic, as well as on all the other topics of the book, as they trip from clinic to laboratory and from laboratory to clinic. Hefferline traces the origins of his own thinking that led to the development of an "electrophysiological analysis of private events." His earlier clinical work with Fritz Perls and Paul Goodman, described in *Gestalt Therapy* in 1951, concerned "Mobilizing the Self" which dealt with the problem of "somatic residuals"—aches, pains, and tensions occurring in the "here and now" but persisting from earlier conflicts and learning, and with procedures to "cultivate awareness of just how one produces the physical hang-up, experiment with it, even try to make it momentarily more severe, while remaining alert to possible anxiety and bits of fearful fantasy of what might happen if one let go." Budzynski's (article 10) observations and speculations about biofeedback-produced twilight states in relation to reverie, hynogogic imagery, sleep learning, and the delightful states of drowsiness associated with complete relaxation offer new insights into the study of subjective experience. The articles on hypnosis by Barber and others in part IX, particularly the theory of hypnotic induction procedures (article 29), continue to advance our knowledge on the critical variables involved in hypnosis, trances, and suggestion.

The reader will be interested in the ingenious uses of suggestion in the improvement of visual acuity (article 31) and the control of peripheral skin temperature (article 32). In this last article, Maslach, Marshall, and Zimbardo suggest that therapeutic control of psychosomatic symptoms "may thus be best achieved by combining the precision of reinforcement contingencies with the power of a more pervasive cognitive approach to dealing with such mind-body interactions." Finally, Wallace and Benson (article 28) trace out in superb detail the effects on oxygen consumption, blood lactate, and other critical physiological measures of the practice of meditation, which appears to be a powerful and relatively simple means of producing a low arousal state that is particularly antithetical to stress reactions. Meditation and other related yogic and Zen experiences seem to offer one antidote to the stresses of contemporary life. Budzynski and Stoyva (article 33) also discuss the various uses of biofeedback in modifying the "defense-alarm" reaction which seems to be implicated in various psychiatric and psychosomatic disorders.

On Clinical Applications

The bread and butter of biofeedback research is in its application to problems of health and illness. And the contributions in part X should be of great interest to clinicians and researchers concerned with mental health, psychosomatic and behavioral disorders, from the viewpoints of both etiology and treatment. Other sections in the *Annual* also contain papers of practical interest. See, for example, articles 6, 10, 11, 12 as well as part VI on electromyographic control. This last section contains some remarkable contributions by Basmajian and Marinacci on the control of muscles and on the alleviation of muscular dysfunctions, by having patients listen to electromyographic recordings of their muscles. Marinacci characterizes it, "This method is similar to that of driving an automobile and learning how one can control the motor by the amount of pressure exerted on the accelerator." Feedback procedures can be used as a means of restoring voluntary control in weak, injured, or paralytic muscles by reestablishing functionally certain neural pathways. According to Marinacci, the most important factors in neuromuscular reeducation are an increase in the voltage, duration, frequency, and promptness of the motor unit's response to voluntary efforts. These changes are facilitated by feedback and achieved through the development of physiological compensatory factors seen in the formation of *giant* motor units. The increases in voltage and duration appear to be due to a hypertrophy of the muscles from overuse, and the acquisition of additional muscle fibers by the surviving motor units. This addition of muscle fibers is the result of sprouting or budding of the terminals of functional axons which assume control of adjoining denervated elements. As a result, voluntary power is increased.

Budzynski and Stoyva (article 33) discuss novel applications of feedback in combination with autogenic, progressive relaxation, and systematic desensitization techniques. In psychosomatics, the direction of clinical work currently in progress concerns such disorders as essential hypertension, cardiac arrhythmias, tension headaches, Raynaud's disease, and migraine headaches (see article 35). Essential hypertension is a disorder that is believed to affect more than 20 million Americans. It is a serious public health problem because high blood pressure increases the risk of stroke and heart attack. Biofeedback and other behavioral approaches to hypertension, either in treatment or in prevention, in addition to other therapies already in practice, are therefore of enormous significance. Comprehensive clinical research is needed to examine the value and cost of these procedures.

As investigators move directly into clinical applications, they are confronted with the critical issues of patient motivation and management, choice and use of incentives, medical suitability, and drugs and placebo effects. Case studies and more systematic clinical data are beginning to emerge, but a great deal remains to be done in establishing the effectiveness and therapeutic utility of the techniques. While it is expected that clinical applications will proceed at a rapid pace, basic studies on the processes and mechanisms of physiological self-regulation are vital to provide answers to fundamental questions. On the basis of physiological and pharmacological evidence, in any given disorder, what specific physiological functions should be fed back to subjects? For example, the studies on premature ventricular contractions by Engel and Weiss indicate that sympathetic or parasympathetic mechanisms are involved in different patients. In a given patient, it is desirable that clinical goals be thoughtfully devised in relation to thorough knowledge of the physiological bases of the disorder and the patient's history.

What kinds of feedback displays are most effective? Does the sensory modality make a difference? Or the amount of feedback? What techniques can be used to increase the individual's direct awareness of his own physiological processes, and does increased awareness facilitate control? What is the role of thoughts and imagery? How can suggestions, attitudes, and beliefs be most effectively utilized in improving voluntary control? How can feedback be effectively integrated with techniques of muscular relaxation and control, yogic and other exercises, autogenic procedures, and meditation? What combinations of procedures are best in achieving desired permanent modification of visceral functions and effective self-control? We are looking forward to reprinting studies on these and related issues in future *Annuals*.

This volume could not have been published without the cooperation of various organizations, professional societies, journals, publishers, and authors who kindly granted permission to reprint their articles. Each article was photographed and printed exactly as it originally appeared in order to complete the *Annual* and distribute it as rapidly as possible. This year's *Annual* includes a selection of earlier papers that were neglected in previous *Annuals,* and we will continue to make up for such omissions, as we become aware of them, in future *Annuals*. Because of space limitations, we regret that a number of excellent papers published in 1972 could not be included. One innovation in the *Annual* this year was to include abstracts of relevant papers presented at the meetings of the Society for Psychophysiological Research and the Biofeedback Research Society. This will help familiarize our readers with work currently in progress but not at the stage of final publication.

A *Reader* containing papers published before 1970 and *Annuals* for the years 1970 and 1971 have been published. In June of each coming year we plan to bring out a new *Annual* covering the major works published in the previous year. The Board of Editors will welcome any comments on these publications and any suggestions for improvement.

DAVID SHAPIRO
For The Editors

CONTENTS

I. VOLUNTARY CONTROL IN HISTORICAL PERSPECTIVE

1. *Voluntary Acceleration of the Heart Beat in Man*
   J. R. Tarchanoff, *translated by* David A. Blizard ... 3
2. *An Instance of Voluntary Acceleration of the Pulse*
   J. T. King, Jr. ... 21
3. *Autonomic Activity and Brain Potentials Associated with "Voluntary" Control of the Pilomotors*
   Donald B. Lindsley *and* William H. Sassaman ... 26
4. *Voluntary Hypercirculation*
   Eric Ogden *and* Nathan W. Shock ... 34
5. *Cardiac Arrest through Volition*
   C. M. McClure ... 49

II. OVERVIEW

6. *The Psychophysiology of Private Events*
   Ralph F. Hefferline *and* Louis J. J. Bruno ... 53
7. *Learning Mechanisms*
   Leo V. DiCara ... 81
8. *Learning of Glandular and Visceral Responses*
   Neal E. Miller ... 90
9. *The Operant Conditioning of Central Nervous System Electrical Activity*
   A. H. Black ... 96

III. CONSCIOUSNESS AND CREATIVITY

10. *Some Applications of Biofeedback Produced Twilight States*
    Thomas H. Budzynski ... 145
11. *Biofeedback for Mind-Body Self-Regulation: Healing and Creativity*
    Elmer Green ... 152
12. *Alpha and the Development of Human Potential*
    Robert M. Nideffer ... 167

IV. CARDIOVASCULAR CONTROL

13. *Operant Conditioning of Cardiac Function: A Status Report*
    Bernard T. Engel — 191
14. *Voluntary Control of Human Cardiovascular Integration and Differentiation through Feedback and Reward*
    Gary E. Schwartz — 209
15. *Control of Diastolic Blood Pressure in Man by Feedback and Reinforcement*
    David Shapiro, Gary E. Schwartz, *and* Bernard Tursky — 217
16. *Sources of Information which Affect Training and Raising of Heart Rate*
    Joel S. Bergman *and* Harold J. Johnson — 227
17. *Some Experiments on Instrumental Modification of Automatic Responses*
    Keiichi Hamano and Tsunetaka Okita — 238
18. *Large Magnitude Heart Rate Changes in Subjects Instructed to Change their Heart Rates and Given Exteroceptive Feedback*
    Joseph H. Stephens, Alan H. Harris, *and* Joseph V. Brady — 247

V. ELECTROENCEPHALOGRAPHIC CONTROL

19. *Similar Effects of Feedback Signals and Instructional Information on EEG Activity*
    Jackson Beatty — 253
20. *Localized EEG Alpha Feedback Training: A Possible Technique for Mapping Subjective, Conscious, and Behavioral Experiences*
    Erik Peper — 262

VI. ELECTROMYOGRAPHIC CONTROL

21. *Electromyography Comes of Age*
    John V. Basmajian — 273
22. *The Basic Principles Underlying Neuromusclar Re-education*
    Alberto A. Marinacci — 286

VII. ELECTRODERMAL CONTROL

23. *Effects of Exteroceptive Feedback and Instructions on Control of Spontaneous Galvanic Skin Response*
    Valerie Klinge — 297
24. *Timing Characteristics of Operant Electrodermal Modification: Fixed-Interval Effects*
    David Shapiro *and* Takami Watanabe — 311

VIII. ANIMAL STUDIES

25. *Effects of Activity and Immobility Conditioning Upon Subsequent Heart-Rate Conditioning in Curarized Rats*
    Wendall J. Goesling *and* Jasper Brener   321
26. *Discriminative Shock Avoidance Learning of an Autonomic Response Under Curare*
    Ali Banuazizi   328
27. *Sequential Representation of Voluntary Movement in Cortical Macro-Potentials: Direct Control of Behavior by Operant Conditioning of Wave Amplitude*
    Joel P. Rosenfeld *and* Stephen S. Fox   339

IX. MEDITATION AND HYPNOSIS

28. *The Physiology of Meditation*
    Robert K. Wallace *and* Herbert Benson   353
29. *A Theory of Hypnotic Induction Procedures*
    T. X. Barber *and* W. D. DeMoor   365
30. *Suggested ("Hypnotic") Behavior: Trance Paradigm versus an Alternative Paradigm*
    T. X. Barber   386
31. *The Effect of Suggestion on Visual Acuity*
    Charles Graham *and* Herschel W. Liebowitz   412
32. *Hypnotic Control of Peripheral Skin Temperature*
    Christina Maslach, Gary Marshall, *and* Philip Zimbardo   429

X. CLINICAL APPLICATIONS

33. *Biofeedback Techniques in Behavior Therapy*
    Thomas H. Budzynski *and* Johann Stoyva   437
34. *Interactions Between Learned and Physical Factors in Mental Illness*
    Neal E. Miller   460
35. *Biofeedback and Visceral Learning: Clinical Applications*
    David Shapiro *and* Gary E. Schwartz   477

XI. SELECTED ABSTRACTS

36. *Abstracts of Papers Presented at the Eleventh Annual Meeting of The Society for Psychophysiological Research*   495
37. *Abstracts of Papers Presented at the Biofeedback Research Society Annual Meeting, 1972*   500

Name Index   519
Subject Index   524

BIOFEEDBACK
and SELF-CONTROL 1972

I

VOLUNTARY CONTROL
IN HISTORICAL PERSPECTIVE

Voluntary Acceleration of the Heart Beat in Man

J. R. Tarchanoff, translated by David A. Blizard

Translator's note.

The name of Tarchanoff has hitherto been mainly associated with the early history of electrodermal research. Specifically, he has been credited with the independent discovery of the electrodermal phenomenon known as skin potential response(Landis, 1932), although Neumann and Blunton(1970) provide evidence to suggest that Tarchanoff was aware of Fere's prior contribution to this field of research <u>before</u> the publication of his own papers.

Somewhat before his contribution to the analysis of skin potential responses were published, Tarchanoff described his work in another field in Pflugers Archives(1885). This paper was concerned with the voluntary control of heart-rate and is certainly one of the earliest contributions to this field of inquiry. The investigations are presented in some detail and many of the topics that are discussed are currently occupying researchers in this field. Examples of each issue are the role of individual differences in producing a given level of autonomic change, the importance of somatic activation in the aetiology of autonomic control and the nature of autonomic response profiles associated with heart-rate control.

Methodological flaws do exist in the experiments which he reports. Despite this, the early date of the investigation and the discussion of research issues in the article which are central to this area of inquiry today make Tarchanoff's paper of considerable topical interest.

*Supported by Grants MH 13189 and 19183 from National Institute of Mental Health to N.E. Miller.

This paper is a translation of "Uber die Willkurliche Acceleration der Herzschlage beim Menschen," *Pflugers Archives*, 1885, Vol. 35, 109-135. It is published here with permission of the original publisher and the translator. Please note that we were unable to reproduce the figures that appeared in the 1885 version.

Although the function of the heart is not directly subject to the influence of the mind ordinarily, we have today at our disposal a considerable number of examples which indicate the intimate dependence of the heart beat (as far as rate is concerned) on the psychic life of man.

Corresponding to the two types of regulatory nervous pathways which connect the heart to the brain, the slowing and acceleratory nerves, the various psychological states influence the heart sometimes by slowing and sometimes by accelerating the pulse.

The strict dependence of heart action on psychical activity has been pointed out more than once with particular persuasive power by Kurschner[1], Carpenter[2], and Claude Bernard[3]. It is easy to put the existence of such a relationship out of all doubt by the enumeration of some relevant facts in the follwoing investigations.

Everyone knows how clearly the feelings and emotions influence the rhythm of the heart: pleasant and joyful feelings and emotions accelerate, unpleasant and sad emotions depress or slow the heart beat.

In this respect, a case of Professor Botkin's appears especially instructive, concerning a patient who was suffering from progressive muscle atrophy and who possessed the rare characteristic of being able to slow his heart beat very strikingly at any time by choice and to disturb its regular rhythm. He had only to imagine his sad condition to produce this demonstration. However, joyful thoughts did not bring forth the reverse effects.[4]

The highest level of violent and abrupt sensations of joy or sadness are able to bring about a momentary cessation of the heart. These events are too well known to detain us any longer.

More interesting are the cases that indicate the activity of thought processes on heart beat.

Millner Fothergill[5] wrote me recently of the following interesting facts: By the use of a stethoscope placed over the heart of a nervous young man he noticed that the consideration of each difficult question directed to him immediately caused a slowing of the heart beat, during which the pulse became very irregular; the questions which did not make demands on thought processes had no effect on the heart beat.

It is also well known that persistent, concentrated observation of the heart can change the function of this organ. In this way, from reading about diseases of the heart, Frank[6] began to direct his attention to the beating of his own heart; there occurred as a consequence of this a most irregular, intermittent pulse. Only a diversionary journey could free him from the condition he had developed.

An observation of Morgagni's is in complete harmony with this. He freed a Bolognese professor from an irregular pulse by advising him never to count his pulse himself.[7]

In view of the manifest influence of different moods on heart beat, it might be conceded that the will also might regulate the heart beat in one way or another, and we shall present the following experiments in favour of this hypothesis.

Tuke mentions a member of the Royal Society of London. At 79 years of age he proved to the author by experiment his ability to accelerate his heart rate by 10 or 20 beats per minute.[8] Furthermore, there was mentioned in the literature a certain Lieutenant Townsend who possessed the noteworthy ability to stop his heart and respiration at will and even to fall into a deathlike state; his body would begin to cool down and become stiff, his eyes became fixed, and finally consciousness was lost. Within several hours he came to himself again. Finally, he died after a similar experiment that he had continued, before many spectators, into the evening of the same day. The autopsy revealed no organ degeneration with the exception of the right kidney.[9]

The same ability to stop the heart is possessed by a well-known American psychologist.[10]

In both cases the cessation of the heart was brought about without any pressure on the vagus; that is to say, without any manipulation by the help of which Professor Czermak of Prague could slow his heart and bring it to a stop.

However, cautious reference should be made to the above cases of apparently voluntary acceleration of the heart, since the proof has been offered, especially by Weber, that changes in respiration rhythm directly influence pulse frequency. Infrequent respiration interrupted by deep inspiration brings about a striking slowing of the heart beat, in one way by changing intrathoracic pressure and in another by facilitating the accumulation of CO_2 whereby the neural centres in the medulla oblongata for slowing the heart are stimulated; frequent, shallow breathing leads to the opposite effect. Weber has demonstrated that by holding respiration in the inspiration phase and the energetic contraction of chest muscles by closing the glottis(that is to say, by hindering expiration), complete and deep unsciousness can be produced with stoppage of the heart.

In just this way Weber explained the mechanism by which Lieutenant Townsend apparently brough about the stoppage of his heart by an act of the will.[11]

In evaluating Weber's experiments under the conditions described, Donders[12] obtains a disappearance of the pulse, heart beat, and heart sounds in many subjects.

On the basis of the above data, the cases of apparent voluntary acceleration or slowing of heart rate must rouse doubt every time on account of the in mode of origin. That is to say, whether or not the effects influencing heart regulation stem from, for example, voluntarily evoked ideas of this or that type, or from voluntary changes of respiratory rhythm, rather than from the direct influence of the will on the regulation of the heart. Such an analysis has not been employed in any of the known cases, using as an aid the graphical method; on this account they possess no power of proof for our questions.

By the discovery of a striking case of voluntary acceleration of the heart, I was naturally concerned to investigate as much as possible all the causes linked with this, in order to clarify the mechanism of the phenomenon. With the aid of suitable apparatus, I followed the change in frequency and character of the pulse, the change in volume of the respiration, and the change in volume of the extremities, both during the acceleration period and before and after it. At the same time I studied arterial blood pressure and the variation in skin temperature. Such an investigation provided us with interesting and purely objective data for the analysis of our case. Only one subject factor which might <u>a priori</u> effect the acceleration of the heart evaded the investigation. We refer to the voluntary evocation of any ideas or images which can accelerate the heart. I had necessarily to trust the subject for the expression of these; we had reason to believe his word, for he was an earnest, educated young man who know the seriousness of the investigation and was himself very interested in the correct explanation for his gift of being able to accelerate his heart.

The individual of interest to us, a student from the senior class of the Medical-Surgical Academy, Eugene Salome, is tall, rather thin, and somewhat nervous and sensitive. Between his tenth and fifteenth years, he suffered from heart palpatations that occurred without any apparent cause. In the course of time, these attacks diminished almost to disappearance under the influence of planned treatment and other favourable measures, and he might have almost completely forgotten them if his fellow students had not noticed his special ability to change his pulse rhythm strikingly by relatively insignificant outward means. Concentrating his attention on the arousability of his heart, Salome noticed once by chance that it sufficed only to desire to accelerate his heart and devote the relevant effort of will to this purpose in order to achieve the desired effect.

From our first meeting he afforded striking proof of his power to do this and accelerated his heart beat from 70-105, b.p.m. i.e., 35 beats per minute. He repeated this experiment with the same consequences, although the degree of acceleration decreased with each repetition.

Convinced of the reliability of the phenomenon, I made use of Salome's worthy offer and undertook a series of experiments and observations with him in order to elicit the true nature of the apparent voluntary acceleration of his heart.

The first assumption was that Salome, in order to bring about the acceleration of the pulse, evoked some pleasant or joyful thoughts normally associated with an increase in pulse rate. If this were the case, the increase might offer a close analogy concerning its mode of origin with the appearance of, for example, goose pimples and salivary secretion, etc. by voluntary means. In these cases, the effects are indeed brought about by voluntary production of suitable fantasy images, in the first case of cold, in the second of tasty or sour things, and it appears that the decisive determinant of goose pimples or salavation is not the voluntary impulse but the mediating idea.

We are not authorized even so to concede such an analogy in our case, as Salome categorically maintains that he does not evoke his heart-rate acceleration by means of ideas or imagination but lightly concentrates his attention on the beating of the heart and directs his conscious willpower to the aim of accelerating his heart. This willpower resembles completely in quality the perceptible sensation of the voluntary contraction of certain muscle groups. During the session he has, to be sure, an uncertain feeling of a contraction or tension of neck muscles and also in the region of his heart, but this is transient and by no means constantly present. The feeling of tensing is not accompanied, however, by objective signs of contraction of the neck or chest muscles.

It is important to remark here that, according to Salome, mere concentration of attention with the aim of acceleration is not usually sufficient; an effort of will is necessary. This reinforces almost the same impression as exists in the sphere of voluntary movement of striated muscles.

It is known that tensing or contracting the muscles to a certain degres usually influences the heart in the direction of acceleration. On this account, we had to investigate whether the subject stimulated his muscles to bring about an increase in heart rate in different positions: lying down, sitting, standing. There were no noticeable contractions or changes in tension in the muscles of the trunk or the extremeties. Usually the somatic system appeared in a state of complete rest. During the session, only a noticeable blushing of the face, and a slight tension of the neck muscles which did not always appear, were evidence of some exertion in the acceleration phase. In passing it should be noticed that he could turn his neck quite normally and no kind of tumour was in evidence.

In raising the question of the skeletal system of our individual, it does not appear without interest that he controls it to perfection; he was especially capable of contracting certain muscles which are not usually under voluntary control; thus, he could move his ears, contract the platysma myoid muscle on both sides combined or each one isolated, flex voluntarily each selected third finger joint, and contract the different muscle groups of the hip on their own, etc.

The heart muscle subjected to voluntary control appears to be a special case of this quite extraordinary neuromusclar organization.

From the cited works of Weber and of Donders et al., it is known that factors such as rhythm of respiratory movements clearly play a role in the frequency of heart beat. It had, therefore, to be shown whether the acceleration of the heart was the result of voluntary changes in respiratory movements in our case. To this end, experiments were arranged in the following manner.

> Four traces were drawn at the same time on smoked paper on a horizontal, steadily moving, cylinder of Balzar-type construction.
>
> The first line was drawn with a Marcy-type chronograph, each deflection equalling one second. The second was applied from the pen of the Depres electrical instrument and represented in the form of a simple white, line the state of rest of the subject; the time of the **acceleration** period was represented by a broad white stripe. The latter was arranged so that the subject pressed lightly with the index finger on the end of a horizontal lever coincident with the beginning of the voluntary impulse to accelerate heart beat so that during the elevation of the lever the access to the electrical signal was connected to a series of induction oscillations (of the usual DuBois induction apparatus) which set the metal pens of the signals into operation; the fast oscillations of these pens were not individually written out because of the relative slowness of the cylinder's

rotation, but they fuse into each other and appear in the form of a white stripe whose width corresponds to the oscillations of the indicator pens of the machine. With the cessation of the voluntary effort directed towards acceleration of the heart beat, the subject ceased to press on the lever; by the depression of another lever, the current from the Depres machine was simultaneously cut off and the pen came to rest and began to draw a white line again.

The third trace was drawn with Marcy's pneumograph which was fastened on the subject's chest across the nipples. The undulations of this tracing gave the precise magnitude and rhythm of respiratory movements.

The fourth trace was written with the aid of a Mosso-Frankish plethysmograph (easily modified), in which a foot was placed, and two kinds of oscillations were available: first that of the pulse, secondly, the more pronounced changes in organ circumference. The quite clearly indicated pulse deviations require no commentary; the diminution in the circumference of the organ expresses itself by abating the whole trace, the enlargement by a rise of the curve.13

The investigation was planned in such a way as to allow one to ascertain clearly how large the obtained increases were, how soon these followed the evoked voluntary impulse, with what consequences these were linked and disappeared, and whether the increase was accompanied by changes of breathing movements and oscillations of the extremities.

Figure 1 gives only one example out of the whole series; one glance is enough to see that the increase of the heart beat cannot be completely linked with changes in respiratory movements.

Thus, we have before the beginning of the acceleration period, in the course of 20 seconds 32 heart beats in 6 breaths. During the acceleration period of the heart, we have 41 beats in 7 breaths in the course of the same time; that is, by counting by the minute we get 96 beats in 18 breaths in the first case, and in the second 123 beats in 21 breaths.

It is indicated that such a striking increase in heart rate (about 27 beats per minute) could be in no way dependent on such an ambiguous change in breathing movements (an increase of about three breaths in a minute). Other data traces in my possession completely support this conclusion. By consideration of the respiration trace, one notices easily that the character of respiratory movements changes a little; they become a little uneven, the height of the respiratory undulations is augmented a little versus the norm and, because of this, appears inconstant; the peaks of these curves become sharper, indicating an abbreviation of the breathing pause, which agrees totally with the slight increase in respiratory movements during the period of heart rate increase. In passing, this characteristic does not appear constant after observation of a large number of traces, and can hardly have any meaning because of its insignificance in the appearance of the increase in heart rate.

As proof of our thoughts, we permit ourselves to cite data drawn from other representative traces in our possession, as follows.

20-second period

No. of Expt.		Breaths	Heart Rate
1	rest	7.25	31
	voluntary acceleration	6.5	40
	continuation of acceleration	6	36
	6 sec after onset of rest	5.75	31
	rest	7.75	31
2	voluntary acceleration	7	38
3	repeated voluntary acceleration	6	40
	continued	6.5	39
	rest	6	31

In order to convince myself further on the subject of the independence of the increase in heart rate from every arbitrary change in breathing, I examined the influence of the relative frequency and depth of extreme and striking respiratory movements on the rhythm of the heart beat by the use of the same graphical method. Here are some of the results.

20-second period

Breaths	Character of breath	Heart Rate
4	deep	37
22	very shallow	31
8.5	medium depth	33
9	medium depth	32
8	medium depth	33
0	hold breath at inspiration	32
0	hold breath at expiration	33

As these extraordinarily clear respiratory movements were only weakly reflected in the heart rhythm of our subject, one can hardly doubt that the increase in heart beat dependent on will that was observed did not in any way depend on those produced by slight changes in respiratory movements; these latter changes appear as if an incidental complication similar to those observed in the sphere of breathing during the intensified voluntary contraction of certain muscle groups.

On the plethysmographic curve(Fig. 1, trace 4) one can easily acknowledge that the act of increasing heart rate is accompanied some time after the beginning of the trace by a clear reduction in the circumference of the extremities, a reduction which(because of the restriction and complete quiescence of the extremities in the apparatus)might depend a <u>priori</u> either on the contraction of blood vessels in the extremity or by a reduced or restricted filling of peripheral blood vessels caused by changed condition of the heart. The decision as to which of these propositions was true was made possible by the measurement of arterial blood pressure in the extremities throughout the period of heart-rate increase and also before and after it. For this I used, in collaboration with Dr. Schumova, the Basch sphygmomanometer with which the measurement of arterial blood pressure can be ascertained in the radial artery. The following results were obtained.

Experiment 1. While the subject was in a lying position and completely relaxed, blood pressure varied in the course of 3 minues between 105 and 100 mm Hg; the pulse for the whole time averaged 76 beats per minute.

During the increase period, when the pulse increased from 76 to 100 in one minute, blood pressure rose gradually from 110 mm Hg to 112, 115, 118, and 120 mm; after the cessation of the voluntary impulse toward acceleration, when the pulse began to return to normal, blood pressure reached its maximum and, after the pulse had sunk to its normal frequency(i.e., 76 beats per minute), the blood pressure wavered between 118 and 120 mm Hg. It was not until 5 minutes after the heart-rate increase that blood pressure returned to its normal level, i.e., 105 to 108 mm Hg.

Experiment 2. In this experiment, pulse and blood pressure were ascertained every 15 seconds for the whole period.

State of Subject	Time(seconds)	Pulse	Blood Pressure
Rest	15	20	110-112
Rest	15	21	110-112
Rest	15	20	110-112
Increase	15	15	118-120
Increase	15	22	120,125-128
Increase	15	25	130-132
Increase	15	26	130
Increase	15	24	132
Increase	15	23	132
Increase	15	22	125-122
Increase	15	22	128
Increase	15	23	122-120
Increase	15	21	132
Increase	15	24	130-128
Increase	15	24	
Increase	15	22	120-125
Increase	15	19	128

Towards the end the subject became tired and was no longer able to produce the intense effort needed for heart-rate acceleration.

As can be seen, the pulse rate became as low as the normal rate during rest towards the end of the acceleration phase. The whole duration of the period when willpower was directed towards increasing heart rate lasted 12 minutes; of that I used altogether only 3.75 minutes for the measurement of pulse and blood pressure. Fifty-five minutes after the beginning of complete rest, the subject gave the following results in consecutive periods of 15 seconds each:

Pulse	Blood Perssure
20	120-122
19	118-120
19	125
18	125

Experiment 3. This experiment was undertaken after a half-hour's rest after the first two experiments. The subject certainly felt tired and could not bring about such an energetic acceleration as in the first two attempts. However, this experiment provided the advantage that pulse and blood pressure were registered after each 15-second period in unbroken sequence for the whole time, and the independence of the increase in heart rate from the rise in blood pressure was thrown into sharp prominence.

State of Subject	Time(seconds)	Pulse	Blood Pressure
Rest	15	18	122
Rest	15	17	125
Rest	15	18	122
Voluntary acceleration	15	20	132-135
Voluntary acceleration	15	21	135
Voluntary acceleration	15	23	138-140
Transition to state of rest			
Rest	15	19	145
Rest	15	19	138-140
Rest	15	18	142-140
Rest	15	20	140-142
Rest	15	19	
Rest	15	21	140-138
Rest	15	19	
Rest	15	19	135
Rest	15	18	135-132
Rest	15	18	135

All three experiments show unanimously that with the beginning of heart-rate acceleration, sooner or later an increase in blood pressure sets in and lasts longer than the increase in pulse rate; thus, pulse rate might return completely to normal while blood pressure continued at relatively high levels.

This consequence becomes especially plain if we present graphically the data from the last experiment.

From the graphical representation(Figure 2), it becomes clear that with the beginning of voluntary heart-rate acceleration blood pressure begins to rise step-wise and reaches its maximum only after the subject has ceased the effort of directing his will towards acceleration of heart rate and the pulse has already shown a clear deceleration. The rise in blood pressure lasts a longer time, shows a slight tendency to decrease and, toward the end of the experiment when the pulse has already completely returned to normal (during rest), blood pressure remains about 13 mm Hg above its normal level. In this respect the results of the blood pressure measurements coincide with the indications of the plethysmographic curve(Figure 1, trace 4) where we noticed that the voluntary increase in heart rate was accompained by a diminution in the periphery, a diminution which lasted only as long as the period of the increase in heart rate.

The sphygmomanometer records, which have traces almost parallel to those of the plethysmograph, afford the key to the clarification of the changes.

It is evident that the vasomotor appearance of blood vessel constriction accompanying and outlasting the voluntary increase in heart rate causes the diminution in the extremities, as the lateral pressure in the larger arteries clearly rises with it. Our presupposition above, that the diminution in the periphery during the increase phase might depend on insufficient flow of arterial blood as a consequence of diminished heart beat, loses all foundation as a consequence of the above experiments with blood pressure, for we should have obtained, rather than an increase, a fast drop in blood pressure--which did not happen in reality. If the reduction in circumference of the extremity were an indirect consequence of the decreased work of the heart supplying blood in insufficient quantities, then, with the immediate return of normal heart rhythm after the increase phase, the recovery of the extremities must also occur. This recovery is only completed in an approximately parallel sequence to the stepwise drop of the blood pressure to normal levels.

It is doubtless that the act of increasing the heart rate is accompanied by simultaneous arousal of vasomotor controls in the periphery and that the effect of this arousal lasts some time beyond the voluntary act of increasing heart rate. As, within certain limits, elevated arterial and intracordial blood pressure can serve as the origin of heart-rate increase, so the question can be shown to be not an idle one, whether or not the increase in heart rate is dependent on the elevated arterial blood pressure. In other words, our case might be presented in the following way: the subject puts the peripheral arteries into a state of contraction by means of a display of will, raises his blood pressure by this means, and this in turn causes an acceleration in heart rate. The increase in heart rate would appear then as a related secondary phenomenon, without direct connection of voluntary action.

It is easy to see, however, that such a supposition cannot stand up to criticism. From the comparison of the blood presure curve with that of the pulse rate, it is clear that the increase in blood pressure does not precede the increase in heart rate, but accompanies it and outlasts it by a fairly considerable period of time.

The subject was able to suspend his efforts aimed at increasing his heart rate, but the arterial pressure continued to rise and remained at a fairly high level for some time after the pulse rate had already returned to normal. It is now clear that there is no possibility in our case to link the rise in blood pressure and the increase of heart rate as cause and effect.

Touching on the vasomotor phenomenon with regard to this subject, I do not consider it without interest to add that his hands felt cold during and after the phase of voluntary acceleration, corresponding to the previously described peripheral vasoconstriction; skin temperature of the hands measured with a thermometer fell by about 1-2° C, while the skin temperature of the face, forehead, and cheek rose by about 0.5° C and more. Here we meet the same series of vasomotor phenomena that accompanies any severe arousal of the nerves, namely, a constriction of peripheral vessels together with congestion in the head.

At this point, a phenomenon can be described that gives evidence of the high arousability of the vasomotor system of the subject.

In the room where the investigation on blood pressure was made, there were some doctors who now and then directed questions to each other in whispers. Despite these precautions, the subject's blood pressure increased in the radial artery by 5-15 mm Hg and only returned to normal when silence was resumed. This phenomenon recalled to me the well-known plethysmographic investigations of Mossa, in which every little sharp or sudden arousal of the nervous system was accompanied by a diminution of a peripheral circumference; that is, a contraction of peripheral vessels. In our case, the vasomotor apparatus was responsive to even the weakest and smallest of external auditory stimuli.

The more than trifling arousal of the subject's heart is linked with the high arousal of the vasomotor system. During the slowing of his heart rate(before the increase phase), he closed his eyes and eliminated all light. By this means he succeeded in achieving his goal; often the pulse beat became noticeably more regular and sometimes somewhat slower. Opening the eyes gave the opposite result. Similarly quietening effects on the heart were factors such as concentrating on some boring treatise. He did not seldom have recourse to these tricks to bring peace to his excited heart.

After we had explored all existing considerations that had reference to our case of voluntary increase of the heart rate, we inevitably reached the conclusion that each acceleration was not the product of imaginations and ideas(which were brought forth by the subject), nor the result of certain muscle movements, changed respiratory movements, or changed blood pressure, but probably was the direct consequence of voluntary effects on regulatory nerve centres and nerve pathways of the heart.

It may be asked now upon which regulatory mechanism the mind worked in our case. The acceleration of the heart beat via the influence of the will might occur through heart-slowing centres in the medulla oblongata being depressed or paralysed by the voluntary impulses of our individual, or through heart-rate increase centres being aroused in the upper part of the spinal cord in the neck region. Which of these two explanations is then more probable? For the solution of this question we must first clarify the path of the acceleration in heart rate directly after the beginning of the effort of will directed at increasing, and also the path of the disappearance of acceleration after the cessation of the same voluntary impulses. Secondly, we must direct our attention to the changes of the pulse curve brought about by the voluntary effort, and, thirdly, consider the changes in the circumference of the extremities.

With regard to the first point, one glance at the pulse curve during acceleration of heart beat(Figure 2) is enough to understand that, after an effort of will, pulse rate rises only gradually and may reach its maximum between 1/2 and 3/4 of a minute after the beginning of the voluntary attempt. On the other hand, pulse rate goes down, after its arousal by this acceleratory voluntary impulse, gradually in the same way, taking a few minutes to reach normal level, during which time some secondary fluctuations,up or down, are evident. This gradual arousal of the heart-rate increasing mechanism, i.e., the centrifugal fibres(nn, accelerantes) and their centre; it is known that, in this case also, acceleration will not be reached at once but will grow gradually to a certain limit. The removal of stimulation causes a similarly gradual disappearance of the acceleration until the normal level is reached once more. Quite a different matter is the paralysis of the slowing nerves of the heart, i.e., by destruction of the heart's slowing centres in the medulla or by severance of the vagus. Here, the pulse accelerates almost immediately in a striking manner, and reaches its maximum quickly. If the pulse is influenced by other methods, e.g., by arousal of the heart-slowing mechanism, and this arousal is strong enough to bring forth an exceedingly quick deceleration of the heart beat, the preceding increase disappears quickly. By consideration of what has been said, we could hardly err if we say that in our case voluntary acceleration of the heart was achieved by the voluntary arousal of the acceleratory nerves.

We now come to a consideration of the pulse complex. From the plethysmographic curve of pulse waves (Figure 1, curve 4), one can clearly see that the increase in rate is accompanied by gradual but striking weakening of the pulse waves, in which, in the period of the most important increase (which can be noticed by the restrained behavior of the subject), the height is one-third of normal height and can sink to an even deeper level. As it was valuable to study these changes in the pulse waves more exactly, we recorded them with the Knollish as well as with the Mareyish sphygmograph. As the latter afforded us better and more successful curves, we shall consider them.

From the appearance of the pulse curve taken thus, one can clearly see that the acceleratory voluntary impulse changes the character as well as the frequency of the pulse. The systolic rise of the wave gets gradually less after the development of the increase in heart beat, and drops to approximately one-third of its normal height during the rest of the period. Also, the steepness of the systolic rise in the acceleration period usually gives a less striking impression. Both these circumstances are naturally related to a less energetic heart; we can see here a striking phenomenon that is produced by clever arousal of the heart-rate increase mechanism: those pulsations that are increased in frequency are diminished in strength. Against these explanations it could be countered that the changes in the pulse-wave curve are not the consequence of changed output by the heart but of the change in blood pressure brought about by constriction of peripheral blood vessels, a constriction which we have seen accompained our subject's increase in heart rate. Such a hypothesis might be based on this fact because the higher the mean blood pressure (which places on a single pulse a unit of increase in pressure), the weaker the pulse height becomes. One cannot agree because of the following three proofs: first, the degree of increase of arterial blood pressure in the period of heart-rate increase is so very insignificant (20-25 mm Hg) that the striking changes obtained in the pulse rate are not explained by it; second, the less steep rise (more easily achieved) of the systolic pulse wave during the acceleration of the heart rate would remain inexplicable in the increase period from this point of view; and third; the changes in the character of the pulse do not go hand-in-hand with the undulations of blood pressure, which is not evident after the cessation of the acceleratory voluntary impulses; when the pulse wave returned to normal, the blood pressure remained at a relatively high level. These data are enough to permit the conclusion that the change in character of the pulse in the increase period depends not on the undulations of blood pressure but on the change in heart activity.

In regard to the decreasing portion of the pulse wave, corresponding to the diastolic decline, it is plain that the steepness of the decline becomes more trifling during the period of increase in heart rate; i.e., the gradient of the decrease becomes more gradual, and the secondary and tertiary waves (the dichrotic, trichrotic, and polychrotic waves) stand out more clearly from the pulse wave. Both these phenomena can be clarified in the rising portion of the pulse wave by the constricted, taut nature of the peripheral vessels that accompanies the act of heart-rate increase as we have seen. In fact, as a consequence of the constriction of the vessels, the arteries unavoidably empty the blood more slowly, resulting in a more oblique curve of diastolic decrease; otherwise, the counstricted state of the arterial walls must cause the formation of secondary waves of vibrations and favour the appearance of polykrotismus. The changes in character of the pulse indicate (in the increase phase of heart rate) that our subject did not achieve his goal of arousal by means of depressing the slowing mechanism but by arousal of the acceleratory mechanism via voluntary influences.

Indirect proof indicates the same thesis. It is well known that the change in heart beat elicited by arousal of the acceleratory heart-rate control centre resembles most closely the change obtained by warming the same organ. Therefore, it was very interesting for me to compare the changes in pulse wave brought about by keeping a man in a hot Russian bathroom with those that our subject produced by effort of will. In the hot room the body temperature of a man rises by about .5-2°C and the heart reacts to the effect. In the work published by Kostjurm on the physiological effect of the Russian bathroom on the human organism are to be found completely satisfactory methods to change the frequency and character of the pulse under the influence of increased bathroom temperature (a mean of 57°C.[14] The comparison of the sphygmographic curve obtained by him with one obtained from our subject during the period of heart-rate increase gives almost identical results. Similar to the changes in pulse in our individual during the transition from rest to voluntary acceleration are the changes in pulse during the transition from normal room temperature to the high temperature of the bathroom. In addition to the striking acceleration of the pulse, the character of the changes of the pulse wave are also similar; the systolic rise and the steepness of the diastolic decline clearly become less substantial, and the dichrotismus and diastolic decreases are clearly augmented. Briefly, the result obtained from our subject during the period of heart-rate increase by voluntary influences is repeated almost word for word. What was caused in our case by mere effort of will is made possible in other men by exposing them to the effect of heart on the organism and therefore on the heart.

The almost complete identity of the changes brought about, as demonstrated above, by warming the heart and those which can be obtained by direct stimulation of the acceleratory nerves in animals are in favour of the opinion that the subject aroused his acceleratory mechanism by willpower and that it was in this way that he increased his heart rate.

This is further supported by the analogy between the changes in the pulse wave and the voluntary acceleration of heart rate and the changes induced in healthy men by a high bathroom temperature.

The peripheral changes in the circumference of the extremities in our subject lead one to the same conclusion. From that curve we have seen that the transition to voluntary acceleration of the pulse is accompanied by a diminution in the periphery. This immediately prevents one from explaining the increase in our subject's heart rate by a voluntary depression of the mechanism that slows the heart rate, because the opposite effect-- i.e., a striking peripheral increase as a consequence of the increased blood flow in the arteries, would be inevitable on account of the strong coincident increase in blood pressure. In fact, a striking constriction was shown in the extremities together with only a trifling increase in blood pressure.

On the basis of the above data and experiences, I consider it completely confirmed that Salome possesses the noteworthy and rare gift of voluntarily stimulating the acceleratory mechanism of his heart. There must naturally be at the basis of this ability a special organization of the nervous system. The simplest explanation was the possibility of the existence of direct nervous connections in his central nervous system between the highest voluntary centres of the cerebral hemispheres and the centres that accelerate the heart(situated in the upper part of the spinal cord). Perhaps such a bond exists in the nervous system of man, with most people not able to call upon it voluntarily, but subject in our case for reasons unknown to arousal by will.

Our subject controlled his heart specifically in only one direction, i.e., toward acceleration; he was not able to slow his heart rate in the slightest.

We come now to the factors influencing our subject's voluntary acceleration of the heart.

He usually achieved pulse acceleration most easily and with the most significant consequences in the morning after a night spent peacefully, when he felt completely wide awake and calm, and before taking coffee, tea, or other warming drinks.

It is noteworthy that any fatigue or exhaustion of the nervous system, especially influential on the heart, weakened our subject's capability to accelerate his heart; e.g., straining of muscles, intense mental work, hot drinks, sexual excess, a prolonged time in a hot bathroom, a sleepless night, increased tobacco smoking. Under such unfavourable conditions, it cost him a colossal effort of will to increase his pulse by 10-18 beats per minute. In the same manner, successive acts of acceleration, interrupted by only short pauses, were unfavourably affected; with each successive attempt the acceleration became less significant and was coupled with ever greater strain. Finally, the subject became so tired that he was completely incapable of bringing about any further acceleration.

We can point out two pharmacological agents that influenced the ability of our subject to produce voluntary acceleration: arsenic and nitrous oxide.

I received the information on the first drug from Salome himself. He had noticed one day, on taking a solution of Fowler's arsenic, that he could bring about much more easily the acceleration of his heart, and that the acceleration was more striking than on days when arsenic was not taken. The effect was so constant and reliable that he took small doses of arsenic to strengthen it on occasions when he gave a demonstration of his ability to increase his heart rate. I do not take it on myself to explain how arsenic favours the elicitation of voluntary acceleration of the heart, whether it is by heightening the ability to arouse the acceleratory mechanism or by increasing the general well-being and augmenting the energy of the organism. But the observation makes it worthwhile to investigate the effect of arsenic on the functioning of the acceleratory mechanism.

The experiences on the influence of nitrous oxide were most obligingly sent to me by the late head of Professor Botkin's clinic, Dr. Kilkowitsch. He subjected Salome to inhalations of a gas mixture of 4 vol nitrous oxide and 1 vol CO_2, i.e., a mixture that, by restricting the necessary amount of breathing so as not to affect his consciousness, yet offered the subject the possibility of increasing the heart rate.

The pulse rate was counted before the inhalation and then its acceleration was ascertained over the same length of time under the influence of the voluntary impulses. Under normal circumstances, our subject could increase the pulse by 20-30 beats per minute. As it was already known from the work of Goldstein[15] that inhalation of nitrous oxide accelerates the heart by weakening the slowing function of the vagus, Klikowitsch ascertained at first the height of our subject's acceleration under the influence of 6-10 inhalations of the gas mixture described above. The pulse rate increased by only 6-8 beats a minute without any acceleratory direction of the subject's will. After a recovery period of 15 minutes, the subject again took 6-10 inhalations of the gas mixture and, at Klikowitsch's request, attempted acceleration of the heart. The attempt remained quite without consequence; the pulse remained completely unchanged, and the subject's powerlessness continued, despite his full consciousness, until after the respiration had completely recovered from the influence of the laughing gas. Then the usual voluntary impulse was enough to increase his heart rate by 24-30 beats per minute as before. The subject compared his powerlessness after inhalation of the laughing gas to the vain attempt to impart a voluntary motor impulse to a crippled limb. The experiment was repeated four times in the same form and the same result always obtained; the ability to increase his heart beat voluntarily was consistently denied under the nitrous oxide.

Finally, it appears necessary to bring out, for the sake of greater completeness, the results of clinical studies of Salome undertaken by Professor Botkin. Because of its great interest, the skilled clinician's analysis is set out word for word:[16] "Subject is quite tall, of average constitution and nutrition, with slight sluggishness of the thyroid. Diffuse heart beats,* i.e., between the 3-4, 4-5, and at its clearest between the fifth and sixth ribs; perfect dullness of the heart begins beneath the third rib and ends at the sixth; transverse diameter to the right almost to the left median+ to the left almost to the left mammillary. In

*Refers to point of location of maximum impulse.
+A line perpendicular to and bisecting the clavicle.

the left parasternum a slight dullness is noticeable, which begins almost under the clavicle and extends to the right to the left median and to the left is separated by about three fingers' breadth from the edge of the sternum.

"The liver begins at the sixth rib and apparently extend to the curve of the ribs. The spleen is percussible from the eighth rib, it is not palpable. In the lungs, vesicular breath sounds, in the left parasternum, in the region of the dullness noted, a somewhat weaker and shorter respiratory murmur than in the right. In the apex of the heart, two sounds, in the aorta and pulmonary artery; accent on the second tone; in the left parasternum under the third rib, a systolic murmur, made stronger by pressure with the stethoscope. Both sounds in the carotids. Pulse 84 in the first moment of the investigation, raised quickly to 96; fairly full, soft, and even. Most noticeably, abnormal irritability of the heart, for a single investigation was enough to bring out a pulse acceleration of 12-16 beats a minute.

"After Salome had calmed down to some degree, the repetition of the objective examination showed nothing abnormal.

"During the act of will directed at increasing the heart, the heart beat is weaker and remains perceptible only between the fifth and sixth ribs, thus nearer to the left mammillary; the longitudinal axis of the heart begins at the third and ends at the sixth rib; transverse axis has enlarged, almost to the left median; the systolic murmur in the left parasternum is less clear; in the region of the dullness, however, appear small murmurs in which the dulness itself becomes clearer. The character of the pulse changes in frequency and strength; from 94-96, increases to 116, 118, and 120 beats; the pulse becomes softer and weaker, easily displaced; the single pulse wave incomparably full. After a rest, results were obtained that had already been found with regard to percussion, auscultation, and palpation before the voluntary acceleration."

This clinical investigation confirms completely, first the results of the physiological investigation of the pulse that was undertaken, secondly it leads to the same conclusion as proof of the weakening of the heart beat during the acceleratory voluntary impulse that we had reached by another route. The acceleratory phase of the heart is accompanied by diminishing of the heart's output. Thirdly, the same proof, which is important, that during the voluntary acceleration such striking changes of output occur as are capable of showing themselves by the transverse diagonal of the heart. This final circumstance stopped the long series of experiments and observations that we had had in mind to carry out with Salome. He had quite voluntarily proposed the investigation of his heart, but we advised him to give up further excercises in this regard as protection against the final outcome of heart failure.

Appendix

Thanks to the goodwill of doctors with whom I am acquainted, I have succeeded in making observations on yet another very noteworthy case of voluntary acceleration of heart rate. A very nervous and sensitive young man demontrated in the first examination which I had with him an increase in his pulse rate from 85 bpm to 130, or an increase of 45 bpm at will. Unfortunately, during this examination of his ability to increase his pulse rate, an involuntary trembling of his whole body set in, and not only that but also a faster and more energetic mode of respiration. However, it was not possible to explain this very considerable increase in pulse rate by the trembling or by the change in character of breathing, as the voluntary repeat of these tremblings and of the energetic and frequent respiratory movements alone were not enough to bring about an acceleration of the heart rate; for this, a quite special voluntary effort of the subject was absolutely indispensable. Quite special implications struck me; especially as from the first in this individual the muscle system was subject to voluntary control in much greater latitude than in the majority of men. Just like Herr Salome, this individual was capable of contracting his ear muscles quite freely, also the third joint of the finger, and various muscle groups of the extremities and of the neck throat, etc., etc. Consequently, it occurred to me in this subject that dependence of the heart action on certain acceleratory voluntary impulses to be a special function of his general neuromuscular organization, which has already been shown to be subject to quite special voluntary impulses.

Thus, I had noticed that both subjects which I had investigated who were capable of accelerating their heart rate exhibited a quite special striking capability to contract muscle groups which, in the majority of men, are not subject to voluntary control. It occurred to me to make use of this characteristic immediately in further searches for other subjects capable of accelerating their hearts. The results have strengthened my expectations to some extent. I met a Dr. S. who moved his ear muscles and the third joint of his finger and was of nervous constitution, so I asked him if he also was capable of increasing his heart rate voluntarily. As he answered that he had not yet tried, I asked him to accelerate his heart at once; immediately, at the first attempt, he increased it by more than 20 beats per minute. After he had practiced it somewhat, for about a month, he improved it so much that he increased it from 85 to 160 beats per minute during one of my public lectures, in the presence of a large group of people.

After I had observed several other persons who had more or less striking ability to control muscles which are not usually under voluntary control, I found two more people-- one, a young Russian poet, the other a young professor--who voluntarily were capable of increasing the pulse rate by 15-20 bpm. Until now, however, I have not succeeded in finding people with a perfectly normal neuromuscular organization who are nevertheless able to accelerate the heart voluntarily. On these groups it appears to me that very probably that such people capable of accelerating the heart may very often be found in the ranks of members of the theatre world, who usually control their muscle systems more adeptly and with incomparably greater skill and art than people of other professions. It would be very worthwhile on this account if people and especially doctors who have especially free access to the theatrical world would make this investigation.

I have this firm conviction from these personal observations that cases capable of voluntary acceleration of the heart exist much more frequently than has hitherto been suspected and that they have remained and remain unnoticed because it does not occur to individuals capable of such acceleration as well as the people with whom they have dealings to direct their attention to it.

From a theoretical standpoint the fact of the dependence of the heart on certain voluntary impulses now appears quite according to expectations as recently the investigations of Wedenski have shown that pure motor fibres run in the vagus nerve to the heart.

[1] Kurschner, Herz und Herzthatigkeit; Wagner's Handworterbuch d. Physiologie, 1844. Bd. II, p. 82-84.

[2] Carpenter, Principles of Human Physiology, 1864, p. 735.

[3] Claude-Bernard, Lecons sur les proprietes des tissues vivants, , p. 465.

[4] Klinische Wochenschrift. Herausgegeben von Botkin. 1881. No. 10 (russisch: Eschenedelnaja Klinitscheskaja Gazeta).

[5] Millner Fothergill, Gaillard's Medical Journal, Feb. 1880, p. 161.

[6] Joseph Frank, Praxae medicae universae praecepta, Lipsiae, Th. II. Bd. II Abth. II, p. 373.

[7] Wagner's Handworterbuch d. Physiol. 1844, Bd. II, p. 82.

[8] Daniel Hack Tuke, Illustrations of the Influence of the Mind upon the Body in Health and Disease, designed to elucidate the Action of the Imagination; London 1872.

[9] Symond's; Miscellanies, 1871, p. 160 und Carpenter, Human Physiology, 1853, p. 1103.

[10] Millner Fothergill 1. c. p. 160.

[11] Weber, Ueber ein Verfahren, den Kreislauf des Blutes und die Function des Herzens willkurlich zu unterbrechen, Arch. F. Anat. und Physiol und wissenuschaftl. Mediein, von J. Miiller, 1851, p. 88 u.f.

[12] Donders, Weitere Beitrage zur Physiologie der Respiration und Circulation. Zeitschr. f. ration. Medic. 1854, p. 241 f.

[13] Translator. This device, coupled with blood pressure and heart-rate readings, was intended to provide an estimate of vasoconstriction/vasodilatation by measuring the circumference of the limbs.

[14] Meschdunarodnaja Klinika(Russian) June 1883, p. 32.

[15] Dies Archiv 1878, 17 Bd, 7. u. 8. Heft, S.331.

[16] Dlinische Wochenschrift, Herausgeg. von Botkin. 1881. No. 10 (russich: Eschenedelnaja Klinitscheskaja Gazeta).

An Instance of Voluntary Acceleration of the Pulse

J. T. King, Jr.

It is a matter of universal knowledge that acceleration of the pulse-rate commonly attends such psychic states as fear, pain and anger. Whatever may be the ultimate source of the stimuli which arouse the pace-maker of the heart to greater activity in such conditions, it is now well established that the immediate control of the heart's rate lies in a delicate balance of influence between two opposing nerve mechanisms--the vagus cardio-inhibitory nerves and the sympathetic cardioaccelerator nerves. Such a balance of influence might be disturbed by certain psychic states through either a reflex inhibition of the vagi, particularly of the right vagus.

In contrast to the frequent instances of acceleration of the pulse-rate by emotional states, in which the acceleration is reflex and involuntary, voluntary increase of the pulse-rate is very rare. It is possible that emotional states affect the cardiac nerve mechanism reflexly through their influence upon the body as a whole, rather than purely through the discharge of nerve stimuli from the higher to the lower centers. At any rate, the pulse-rate is strikingly immune from control by the will except in a few isolated cases.

West and Savage* have recently described the fifteenth case of recorded voluntary acceleration of the heart beat. The case reported by these writers was that of a young man whose heart was normal on physical examination and electrocardiography; the subject could voluntarily accelerate his pulse, and the pulse-rate increased by 25 beats per minute following the injection of 0.02 gm. of atropin sulphate. During the strain of voluntary acceleration this subject showed dilatation of the pupils, which has been described in connection with previous cases, and elevation of the blood-pressure. It is not easy, even in a case so carefully studied as that of West and Savage, to state the mechanism by which the acceleration was accomplished: dilatation of the pupils suggests activity of the sympathetic nerves, while marked acceleration from the atropin suggests that the vagus inhibition might be readily lifted.

*H.F. West and W.F. Savage: Voluntary Acceleration of the Heart Beat. Arch. Int. Med., 1918, XXII, 290.

Reprinted with permission of the publisher from the *Johns Hopkins Hospital Bulletin*, 1920, 303-305.

An instance of voluntary acceleration of the heart beat has recently come to my attention. The subject is a white man, aged 26, single. His general health has always been good, though he has been subject to naso-pharyngeal infections, with symphatic hypertrophy and glandular enlargement. He has always been an unusually "high strung" individual, excelling in games which require speed and quick muscular control. For many years he has been subject to attacks of uncontrollable thumping of the heart and tachycardia when under any marked emotion. A few years ago he gave up school for several months and rested, by the advise of physicians, on account of what was called valvular heart disease.

When the recent war broke out, he made three unsuccessful attempts to enlist, but was refused enlistment on account of his heart. He was summoned early in the war before his local draft board, from which he was referred to a medical advisory board, where he came to my attention. Examination revealed a normal heart, with marked tachycardia, and cardio-respiratory murmurs. He was a "high strung" man, and, as the tachycardia was apparently due to excitement, he was accepted. He went through the training without trouble and saw active service, including the Argonne Forest drive. After the armistice, he reported to a medical officer for a minor complaint, and a diagnosis of effort syndrome was made. He was returned to this country, trained in camp, and discharged as well.

He consulted me in October, 1919, not for any particular complaint, but because he was concerned about the diagnosis of effort syndrome, which had been made in the army. He was rather nervous, and having given up exercise on discharge from the army, he was somewhat out of "condition."

Examination showed a man organically sound. The reflexes were moderately exaggerated throughout. He could voluntarily move his ears and posterior scalp muscles. There was marked tachycardia during examination and a sinus arrhythmia. On counting the pulse, I noticed that there was a rapid rise in rate. The patient then gave the voluntary information that he could accelerate the pulse-rate by "thinking about it."

This power to accelerate the pulse voluntarily was readily verified. Almost immediately on command to accelerate, the rise in rate sets in sharply. The pupils dilate slightly, the respirations become somewhat irregular and seem a little more shallow. The subject says that he creates no emotional state whatever, and that the rise in pulse is due to a calm voluntary effort. He is conscious of his heart beat when it is accelerated, and he knows without feeling the pulse whether or not his efforts are strikingly successful. As to the mechanism of the acceleration, his only conviction is that his respirations are not perfectly natural during the effort to accelerate. It will be seen below, however, by pulse tracings, that the acceleration was equally marked when the subject held his breath as it was under the usual conditions.

Several pulse tracings were made by Dr. Donald R. Hooker. The tracings were made at the brachial artery with the Erlanger sphygmomanometer and cuff. The time markings indicate intervals of one second. During the early readings, the subject was rather excited, and the excitement tachycardia made the voluntary acceleration less striking.

Voluntary Acceleration of the Pulse

There was never, however, a failure to accelerate the pulse promptly when the effort was made. The least acceleration per minute was one rise of 20, and the greatest rise was of 40 beats per minute.

Retardation of the pulse-rate seems in all instances to have been a passive one. It was always possible for the subject to approach the initial rate after his effort to retard, but he was not able to retard below the initial rate. In one experiment the patient was ordered to accelerate, and the high speed was maintained for 17 1/2 seconds. In this case the pulse "got beyond control," because the high rate was held for so long a period, and it fell very gradually when the effort to retard was made. The pulse-rate at the beginning of this experiment was 108; it rose with effort to 131, was sustained for 17 1/2 minutes, then fell to 117 after 20 seconds. After this, the subject suggested that he be allowed to indicate when he would make the efforts to accelerate and retard: this was done, and the results were somewhat more prompt. A characteristic tracing is shown in Fig. 1.

Electrocardiographic tracings were made by Dr. E.P. Carter. The mechanism of the heart-beat was normal throughout.

The results were as follows:

At rest; sitting. Leads 1, 2 and 3: rate 88, rhythm regular; the P-R interval measures 0.12 second.

No. 2: The R-R interval measures 0.56 second(actual rate 107). Following the signal to accelerate, the R-R interval shortens to 0.48 second(rate 125). Following the signal to retard the R-R slows to 0.60 second(rate 100).

No. 4: At rest the R-R interval measures 0.72 second (rate 83); following the signal to accelerate, it shortens to 0.60 second(rate 100) and then to 0.56(rate 107). Following the signal to slow, the R-R interval drops to 0.72 second(rate 83).

No. 5: At rest the R-R interval measures 0.54 second (rate 111). Inspiration, R-R interval measures 0.64(rate 93). Acceleration, R-R interval shortens to 0.50 second (rate 120). Retardation, R-R interval lengthens to 0.64 second(rate 93). Expiration, sinus arrhythmia.

No. 6: Right vagus pressure. Before, $-$ interval measures 0.68 second(rate 88); afterwards, R-R interval measures 0.64 second(rate 93).

No. 7: Left vagus pressure. No conspicuous change.

Comment.--The changes in pulse-rate are not so striking as are those seen in the pulse tracings. The greatest change in pulse-rate by the usual procedure was in No. 4, in which the heart-beats were accelerated by 24 per minute. In No. 5, the effort to accelerate was made while the breath was held in inspiration. The pulse-rate was retarded by 18 beats per minute following inspiration. On effort, with the breath held, the acceleration amounted to 37 per minute, the retardation to 27 per minute. It is apparent, therefore, that acceleration of the pulse was more marked while the breath was held in inspiration than it was under usual conditions.

Pharmacologic tests were not made, owing to the subject's unwillingness to submit to them. He stated that he was very susceptible to the effect of any drug, especially

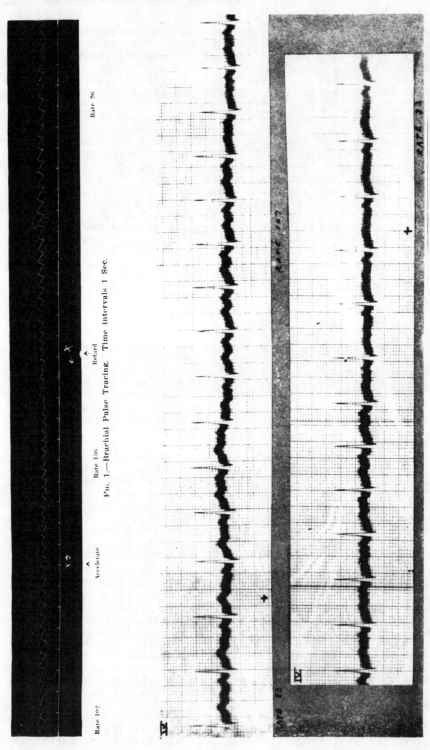

Rate 102   Accelerate   Rate 136   Retard   Rate 96

Fig. 1.—Brachial Pulse Tracing. Time intervals 1 Sec.

Fig. 2.—Electrocardiogram (Second Lead). The signals are not reproduced in this print, but the first mark indicates the signal to accelerate, the second indicates the signal to retard.

to that of belladonna. Some years ago he was prostrated by taking capsules containing a small amount of belladonna. He is uncertain as to his exact symptoms at the time, but thinks that he suffered from palpitation and a weak pulse.

The explanation of the phenomenon with which we are dealing is not easy and must be purely hypothetical. The evidence at hand is as follows:

1. "High strung" individual, with highly developed voluntary muscle control.
2. Marked emotional tachycardia.
3. History of sensitiveness to belladonna.
4. Statement of patient that the respiration is in some way involved in the process of acceleration.
5. Marked acceleration with the breath held.
6. Sinus arrhythmia.

The last four of these items rather suggest that the acceleration in the individual may be brought about because of an unstable vagus control of the heart beat, allowing the acceleration influences to gain control at times. It is known that sinus arrhythmia is due to rhythmical changes in vagus activity occurring during respiration. Our subject showed marked slowing of the pulse following his holding the breath in inspiration. However, when the effort to accelerate was made, with the breath still held, there was a sharp accelerator response; slowing took place at will with the breath still held. We might interpret these phenomena as signifying: (1) Slowing from vagus following deep inspiration; (2) acceleration from accelerator (sympathetic) activity with the vagus quiescent; (3) slowing from release of accelerator activity; (4) sinus arrhythmia from rhythmic activity of the vagus as the subject resumes breathing. The patient's feeling that his respiratory movements are linked in some way with the process of acceleration and the observed irregularity of breathing suggests that vagus activity may be minimized by such procedure. The patient's history of sensitiveness to belladonna, if reliable, is strong evidence in favor of his having a vagus system which may be readily neutralized, thus allowing accelerator stimuli to gain control of the beat's rhythm.

West and Savage were of the opinion that in their case the voluntary acceleration of the heart-beat was brought about through a primary decrease in vagus inhibition, probably augmented at times by accelerator influences. Although it is difficult to draw conclusions from our case, there is nothing which could conflict with this hypothesis of West and Savage, and all the evidence would tend to substantiate such a conception of this unusual phenomenon.

Autonomic Activity and Brain Potentials Associated with "Voluntary" Control of the Pilomotors

Donald B. Lindsley and William H. Sassaman

THE CONTROL of most of the functions of the autonomic nervous system is generally considered to be "involuntary" in character. A few cases have been described, however, in which one or another of the autonomic activities is under "voluntary" control. Most of these deal with some aspect of the visual mechanism, such as, "voluntary" control of accommodation,[2,3,16,17] pupillary dilatation[14] and constriction.[11] The case to be described here is of particular interest since it involves "voluntary" control of the *arrectores pilorum* muscles which are innervated only by the sympathetic division of the autonomic nervous system.

The subject is a middle-aged male, who since the age of 10 years has been aware of the ability to control the erection of hairs over the entire surface of his body. Experimental study of the subject has revealed that the erection of hairs is accompanied by a number of other autonomic phenomena of which he was not aware. These consist of an increase in heart rate, an increase in the rate and depth of respiration, dilatation of the pupils, an increase in the electrical potentials of the skin over regions rich in sweat glands and characteristic changes in the electrical potentials over the premotor area of the brain.

METHODS

Sixteen millimeter motion picture records, at a measured speed of 18 frames per second, were made of the erection of the hair on the dorsal surface of the forearm, the lateral surface of the thigh, and the anterior surface of the leg just below the knee. Similar records of the dilatation of the pupil, which occurs during "voluntary" erection of the hair, were made while the eye was illuminated by a 500 watt flood lamp at an angle of approximately 45° to the anterior surface of the cornea. Fixation of the eyes was always maintained on a point 15 feet away. A quick movement of the experimenter's hand before the lens of the camera marked the film and served as a signal for the subject to begin the erection of the hair; the same movement repeated meant to lower the hair. The movement fell on the periphery of the visual field and did not intercept the line of regard. The signals used in other aspects of the study consisted of a barely audible "click" in a headphone and the "flash" of a dim light. Control experiments in every case indicated that none of the effects observed was due to the stimulus value of the signals themselves.

Simultaneous recording of electrical phenomena, such as the electrocardiogram and skin potentials, or brain potentials from two regions on the surface of the head, was accomplished by means of two independent amplifying and recording systems. The amplifiers are of the resistance-capacity coupled type, with pre-amplifiers employing balanced input circuits. The time-constant of the amplifiers may be made appropriate to the particular type of phenomena to be recorded by changing the inter-stage coupling capacity (0.1, 0.5, or 2.0 mfd.). A type-PA Westinghouse oscillograph with matched elements served as the recording unit. The mechanical devices for recording respiration and changes in blood pressure are incorporated in the oscillograph case so that all phenomena may be recorded simultaneously on the same record if desired. Continuous changes in blood pressure were recorded with the pressure in the cuff on the arm maintained at 70 to 80 mm. of Hg for short periods of time. Systolic and diastolic readings by the auscultatory method were also made before and during the erection of the hair.

Electrodes for recording brain and skin potentials consisted of small gold discs, 8 mm. in diameter, each sunk in a small bakelite block. The cup of the block was filled with an electrode jelly which served as the conducting medium. When brain potentials were recorded the electrodes were attached to the surface of the scalp, one to two inches apart, and held in place by bandages. For the recording of skin potentials the electrodes were attached to the palmar surface of the hand. Small silver plates and electrode jelly applied to the left wrist and ankle furnished leads for the electrocardiogram. Electrograms from individual arrectores pilorum muscles were obtained by means of needle electrodes inserted through the skin at the base of a hair on the side toward which it leaned. The needle electrodes consisted of a fine gauge hypodermic needle with either one or two fine insulated wires cemented in its lumen. When the first type was used the exposed tip of the single inner wire led to grid and the sheath of the needle to ground; in the latter case the exposed tips of the two wires led to the balanced input of a pre-amplifier and the sheath of the needle was grounded.

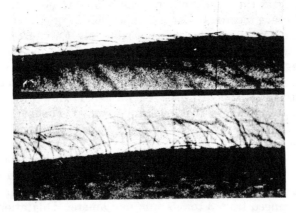

FIG. 1. "Voluntary" erection of the body hairs. Sections from a motion picture record of the hair on the lateral surface of the thigh, before the signal to erect the hairs and about 10 seconds after the signal showing the hairs erect and the "gooseflesh" appearance of the skin.

RESULTS

Several series of experiments have been made in which the erection of the hair on the arms and legs was photographed. Fig. 1 shows two frames from a motion picture record of the erection of the hair on the thigh. The first shows the hair in its normal position close to the surface of the skin before the signal to erect the hairs. The second, taken approximately 10 secs. after the signal, shows the hairs almost fully erected and the "gooseflesh" appearance of the surface of the skin.

Fig. 2 shows three frames from a motion picture record of the dilatation of the pupil which accompanies the "voluntary" erection of the hairs. The first is from a section of the record before the signal to erect the hairs and shows the pupil well constricted. The second shows the increased diameter of the pupil while the hairs are erect; the third shows the decreased diameter

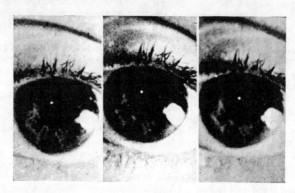

FIG. 2. Dilatation of the pupil during "voluntary" erection of the body hairs. Sections from a motion picture record before, during and after the erection of the hairs.

a few seconds after the signal to lower the hairs. Frame by frame analysis of the motion picture records under magnification in an Edinger projector revealed that the hairs began to rise approximately 1 sec. after the signal was given. By plotting the distance of the tip of any hair from the surface of the skin in successive frames (see Fig. 3) it was found that the erection of a hair proceeds in a step-wise fashion. Maximum elevation of a number of hairs studied was attained in about 7 secs. After a signal to lower the hairs was given 5 to 6 secs. elapsed before actual lowering occurred. The lowering of the hairs also proceeded by steps and was completely accomplished between 15 to 20 secs. after the signal to lower them.

The degree of illumination necessary for photographing the pupils caused considerable constriction which competed with the dilatation accompanying the "voluntary" erection of the hairs. The pupils at the start of an experiment usually ranged from 3 to 3.5 mm. in diameter. Dilatation of the pupils (see Fig. 3) began 0.3 to 0.5 of a sec. after the signal to raise the hairs. Maximum enlargement of the diameter under these conditions of high illumination ranged from 0.5 to 1.5 mm. and was usually attained in 0.5 of a sec. Enlargement of the pupils was maintained throughout the duration of the erection of the hairs at a level well above the starting diameter, although there were frequent fluctuations due to the competing tendencies for dilatation and constriction. Five-tenths of a sec. after the signal to lower the hairs the pupils began to constrict in a step-wise manner. Constriction always proceeded to a level below that of the original diameter of the pupils but slowly returned to it.

Simultaneous records of heart rate (electrocardiogram), respiration and skin potentials from the palm of the hand (see Fig. 4) show that changes in

"Voluntary" Control of the Pilomotors 29

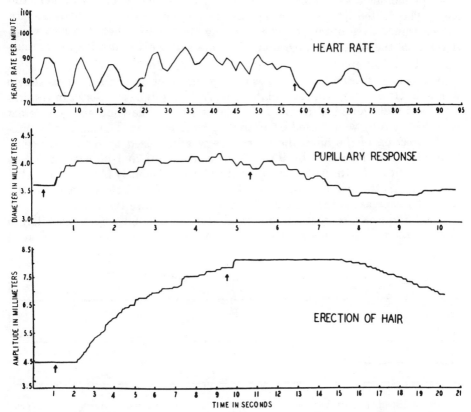

Fig. 3. Graphs showing changes in the heart rate, diameter of the pupil and the distance of a hair from the surface of the skin following a signal (first arrow) to erect the hairs and a signal (second arrow) to lower them.

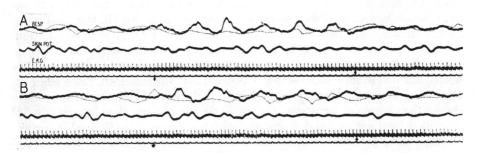

Fig. 4. Simultaneous records of heart rate (E.K.G.), respiration and skin potentials from the palm of the hand showing distinct changes during the "voluntary" erection of the body hairs. The dotted line is a continuous record of changes in heart rate; an increasing rate is shown by a descent of the line. The first arrow represents the signal to erect the hairs, the second to lower them. Time is shown in seconds at the bottom of each record.

these phenomena occurred approximately 1 sec. after the signal to erect the hairs. The dotted line in these records gives moment to moment heart rate values which were obtained by measuring the interval between successive R-waves of the electrocardiogram with proportional dividers adjusted to multiply by 10 and erecting an ordinate over the midpoint of each interval. A typical curve of heart rate changes during the erection of the hairs is shown in Fig. 3. Heart rate, which oscillates widely during the two phases of respiration, was maintained at a definitely higher level while the hairs were erect than before they were raised or after they were lowered. During the "voluntary" erection of the hairs the heart rate was increased by 6 to 8 per cent; respiration was increased by 9 to 25 per cent in rate and by approximately 100 per cent in depth. Anticipation of the signal frequently increased the skin potentials in the palm of the hand but the degree of activity was usually distinctly increased about 1 sec. after the signal to raise the hairs.

Fig. 5 shows electrograms obtained from individual arrectores pilorum muscles by means of needle electrodes during the "voluntary" erection of

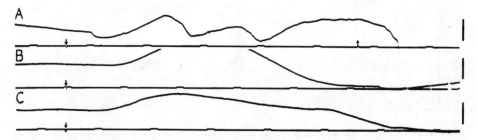

FIG. 5. Electrograms from individual arrectores pilorum muscles during "voluntary" elevation of the body hairs. The first arrow indicates the signal to erect the hairs, the second to lower them. Time is shown in seconds at the bottom of each record. Calibrations at right of each record equal 150 microvolts.

the hairs. The electric response occurs from 0.5 to 0.7 of a sec. after the signal to raise the hairs and thus precedes the mechanical movement of the hairs which has a latency of about 1 sec. As shown in Fig. 5, the electric response may consist of rhythmic oscillations (A), or long, slow undulations (B and C). Occasionally small, quick, spike-like components are superimposed. Both the rhythmic and the more prolonged electric responses have been described in smooth muscle by Lambert and Rosenblueth,[13] Rosenblueth, Davis and Rempel,[15] and by Eccles and Magladery.[6,7] The electric responses shown here seem to correspond in time relations to the slow component III described by Lambert and Rosenblueth,[13] but there is no "initial complex" corresponding to their components I and II. Since, for the purposes of this study, latency is the main concern, further analysis of the response will not be attempted here.

Continuous records of the relative changes in blood pressure have shown that there is a very slight increase during the "voluntary" erection of the hairs. Repeated readings by the auscultatory method have shown that there are also very slight elevations of the diastolic and systolic pressures. Although the changes were always in the direction of an increase they have for the most part been too small to be considered significant.

Electrical potentials of the brain were recorded simultaneously from over two regions of the same hemisphere during "voluntary" erection of the hairs, usually from the premotor and occipital, premotor and temporal, premotor and parietal, or the premotor and the anterior part of the frontal region. Characteristic responses consisting of large, slow waves were found only over the premotor region (see Fig. 6). The latency of the electric response of the premotor region ranges from 0.23 to 0.35 of a sec. There is sometimes an "off response" 0.2 to 0.3 of a sec. after the signal to lower the hairs. The general pattern of the premotor response is essentially the same on repetition of the experiment. The largest response is obtained when the electrodes are over a region just anterior to the central sulcus, in a line parallel with, but about 1 inch off the mid-line. According to careful estimate by Chiene's

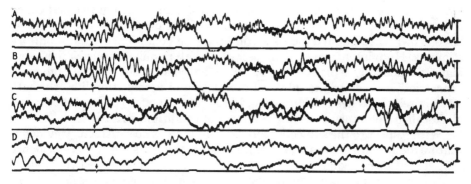

Fig. 6. Electrical potentials of the brain simultaneously recorded from over the premotor (middle line of each record) and occipital (upper line) areas during "voluntary" erection of the hairs. In A, B, and C the premotor electrodes were over the upper part of 6aα; in D they were over the upper part of 6aβ. The first arrow indicates the signal to erect the hairs, the second to lower them. Time is given in seconds at the bottom of each record. Calibrations at the right of each record equal 20 microvolts.

method[5] for determining the relationship between cranio-cerebral topography, the area between the electrodes probably included, mainly, the upper part of area 6aα and parts of areas 4 and 6aβ. Responses were obtained from over the outlying regions of 6aα and 6aβ but they were never as prominent as those from the upper part of 6aα.

The nature and direction of the various changes in autonomic activity associated with the "voluntary" erection of the hairs, together with the general concurrence of these changes suggest that there is a generalized sympathetic discharge. The fact that marked changes in the brain potentials occur only over the premotor area and precede the peripheral autonomic phenomena further suggests that the premotor cortex plays a part in the generalized sympathetic discharge associated with the "voluntary" erection of the body hairs.

Discussion

The question arises as to how the subject is able to control the erection of the body hairs. So far as he is aware the process is essentially similar to that of initiating contraction in one of his skeletal muscles. He does not call up an image of a painful or fearful experience as has been reported in connection with one case of so-called "voluntary" control of pupillary dilatation.[14] There is no straining or tensing of skeletal muscles and there are no observable movements of any sort. The subject is able to erect the hairs when in any position and while carrying on other activities. Of special importance is the fact that he is able to inhibit the reflex erection of the hairs and the appearance of "gooseflesh" normally induced by stepping out of a hot shower into a cold draft.

In this case, as in some of the others which have been described, "voluntary" control of an autonomic function was first discovered during childhood; either during severe fatigue, a critical illness or some other unusual condition. It may be that, unknown to these subjects, some form of conditioning took place when, under unusual circumstances, their attention was first called to the reflex manifestation of an autonomic function. Cason[4] and Hudgins[11] have shown that pupillary constriction may be conditioned to an auditory stimulus. Hudgins has even demonstrated that pupillary constriction originally induced by a light stimulus may be conditioned to the command "constrict" or even to the subvocal production of verbal stimuli by the subject himself. This has led Hunter and Hudgins[12] to offer the hypothesis that all so-called voluntary behavior is a form of conditioned response.

With respect to the subject described here there is no satisfactory evidence that he did not always have the ability to erect the hairs nor is there evidence that some form of conditioning took place when he first noticed the ability.

No matter how the "voluntary" control of the pilomotors is brought about in this case the familiar tendency of the sympathetic system to respond as a whole is apparent in the accompanying phenomena. The latencies of these various responses are of such an order as to indicate that they are a part of a generalized discharge of the sympathetic system. The finding of characteristic brain potential changes over the premotor area suggests that they are precursors of the peripheral autonomic changes. This corresponds with the results from experiments on animals[8,9,10,18] and human subjects[1] which have indicated that there is an autonomic representation in the cortex, particularly in the premotor area.

Summary

A subject has been described who has "voluntary" control of the pilomotors (*arrectores pilorum* muscles). Experimental study of the subject has revealed that the "voluntary" erection of the body hairs is accompanied by an increase in heart rate, an increase in the rate and depth of respiration, dilatation of the pupils, an increase in the electrical potentials of the skin over regions rich in sweat glands, and a very slight increase in blood pressure. The nature and direction of these changes together with their latencies indicate a tendency for a generalized sympathetic discharge.

Simultaneous records of electrical potentials from two regions of the brain, the premotor and one of the other regions, during the "voluntary" erection of the body hairs have shown characteristic changes only over the premotor area. These premotor responses precede and appear to be associated with the peripheral autonomic changes. This has been interpreted as further evidence of the representation of the autonomic nervous system in the premotor area of the cortex.

The possibility that the subject's ability to control the erection of the hairs is the result of a conditioning process has been discussed, but no satisfactory evidence for or against this hypothesis is at hand.

REFERENCES

1. Bucy, P. C., and Case, T. J. Cortical innervation of respiratory movements. I. Slowing of respiratory movements by cerebral stimulation. *J. nerv. ment. Dis.*, 1936, *84*: 156–168.
2. Carr, H. Apparent control of the position of the visual field. *Psychol. Rev.*, 1907, *14*: 357–382.
3. Carr, H. Voluntary control of the distance location of the visual field. *Psychol. Rev.* 1908, *15*: 139–149.
4. Cason, H. Conditioned pupillary reactions. *J. exp. Psychol.*, 1922, *5*: 108–146.
5. Cunningham, *Textbook of Anatomy*. 6th ed. London: Oxford University Press, 1931 (pp. 1343–1348).
6. Eccles, J. C., and Magladery, J. W. The excitation and response of smooth muscle. *J. Physiol.*, 1937, *90*: 31–67.
7. Eccles, J. C., and Magladery, J. W. Rhythmic responses of smooth muscle. *J. Physiol.*, 1937, *90*: 68–99.
8. Fulton, J. F., Kennard, Margaret A., and Watts, J. W. Autonomic representation in the cerebral cortex. *Amer. J. Physiol.*, 1934, *109*: 37.
9. Green, H. D., and Hoff, E. C. Effects of faradic stimulation of the cerebral cortex on limb and renal volumes in the cat and monkey. *Amer. J. Physiol.*, 1937, *118*: 641–658.
10. Hoff, E. C., and Green, H. D. Cardiovascular reactions induced by electrical stimulation of the cerebral cortex. *Amer. J. Physiol.*, 1936, *117*: 411–422.
11. Hudgins, C. V. Conditioning and the voluntary control of the pupillary light reflex. *J. gen. Psychol.*, 1933, *8*: 3–51.
12. Hunter, W. S., and Hudgins, C. V. Voluntary activity from the standpoint of behaviorism. *J. gen. Psychol.*, 1934, *10*: 198–204.
13. Lambert, E. F., and Rosenblueth, A. A further study of the electric responses of smooth muscle. *Amer. J. Physiol.*, 1935, *114*: 147–159.
14. Petrovic, A., and Tschemolossow, A. Zur Frage über "willkurliche" Pupillenerweiterung. *Klin. Mbl. Augenheilk.*, 1931, *85*: 23–34.
15. Rosenblueth, A., Davis, H., and Rempel, B. The physiological significance of the electric response of smooth muscle. *Amer. J. Physiol.*, 1936, *116*: 387–407.
16. Sisson, E. D. A case of voluntary control of accommodation. *J. gen. Psychol.*, 1937, *17*: 170–174.
17. Sisson, E. D. Voluntary control of accommodation. *J. gen. Psychol.*, 1938, *18*: 195–198.
18. Smith, W. K. The representation of respiratory movements in the cerebral cortex. *J. Neurophysiol.*, 1938, *1*: 55–68.

Voluntary Hypercirculation 4
Eric Ogden and Nathan W. Shock

At least 9 reports[3,7,9,11,15-19] have been published describing 15 persons who have been able to increase their pulse rate on demand. The present report deals with 2 who were able to increase their pulse rate and also to raise their blood pressure at will. This investigation was undertaken to throw some light upon the relationships between the various physiologic adjustments which accompany these phenomena.

Subjects. The one subject, "A. L.," was not aware of his ability to control his circulation, until he was attending a course of instruction in human physiology, when he announced his ability to increase his pulse rate. When a preliminary examination had verified this, he was questioned as to the mental processes involved, but was unable to elucidate them other than by comparison with some ordinary somatic movements such as raising an arm. This may be compared with the report of Hoskins' introspection.[3] "It is *not* induced by thinking of appropriately stimulating conditions or situations. It is probably no more—and no less—mysterious than is, for example, the flexing of a biceps muscle. Introspectively the two procedures seem similar. One simply 'wills' the change, and it takes place." In contrast, Cameron[16] reports introspectively the processes by which he trained himself to associate a spontaneous physiologic response to a fear stimulus with the will to accelerate his heart.

The other subject, "A. B.," stated that he had been able to increase his heart rate for several years before he studied physiology. His procedure was as follows: he would make himself conscious of his heart, either by feeling the pulse or the apex beat, by leaning his ear on a pillow or by any other of the common means familiar to

Reprinted with permission of Lea & Febiger and the author from the *American Journal of the Medical Sciences*, 1939, Vol. 198, 329-342.

us all. He would then attempt to anticipate each heart beat, or, as he said, "to pace his heart," and thus by what he termed "a sustained mental effort" applied before each beat, he was able to maintain an increased heart rate. In neither case was the subject aware of any emotional component either in the production of acceleration or in the process by which he learned to produce it.

For the more detailed investigation of this phenomenon, the following experiment was repeated 4 times, that is, twice for each subject.

Experimental Procedure. The subjects reported to the laboratory under basal conditions for the experiments. After 30 minutes' rest in the supine posture, 3 8-minute determinations were made of the basal metabolism. A half mask was used and the determination made by the Tissot open-circuit method with gas analyses by the Boothby-Sandiford[2] modification of the Haldane technique. At the beginning of the rest period a blood pressure cuff was fastened to the left upper arm. With this the technician made 5 auscultatory readings of systolic and diastolic blood pressures during the rest period and in between the basal determinations, and a pair of readings every half minute during the subsequent procedure. After the basal data had been collected, a second blood pressure cuff was attached to the left ankle. This was connected to an ink-writing tambour and inflated to 50 mm. Hg to record the pulse on a polygraph. The expiratory hose from the mask was then connected to one of a pair of 9-liter Tissot spirometers, so connected that when one filled it operated a relay-driven valve switching the expired air into the other spirometer. In this way a *continuous collection* of expired air was made. The duration of collection of each 9 liters of expired air was automatically recorded on the polygraph. A duplicate determination of the CO_2 and O_2 content was made on a sample from each of these collections of expired air. In some instances it was found convenient to pool several 9-liter collections, mix them and then obtain one sample. Respiratory frequency was determined from the polygraph record of a tambour actuated by the slight pressure changes within the mask.

Two to four 9-liter collections were made and the subject was then instructed to accelerate his heart. After a further 27 liters had been expired, he was asked to end the acceleration. Collection was continued for a further 10 minutes to observe the recovery period,* and immediately following this the subject was instructed to tense his arm and leg muscles. After tensing his muscles for a period corresponding to 27 liters' expiration, a final 10-minute period of observations was made and the experiment concluded.

The tensing of muscles was used as a control partly as a simple method of inducing a moderate metabolic increase and partly to test the suggestion of Carpenter, Hoskins, and Hitchcock[3] that the tensing of skeletal muscles may be an essential part of the process of voluntary acceleration of the heart.

Blood Pressure and Pulse Rate. Figures 1, 2, 3, and 4 show the data from these experiments plotted against time. In the first place, it will be noted that in the subject (A. L., Figs. 1 and 2) to

---

* One of the experiments was concluded at this point.

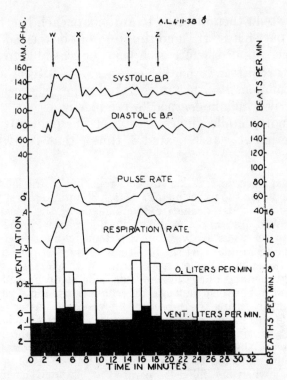

Fig. 1.—Increase of circulatory, respiratory, and metabolic functions on command (Subject "A. L.," 4/11/38). The upper 4 lines (systolic and diastolic blood pressure, pulse rate, and respiration rate) and the lower black histogram (respiration volume) are plots of the actual values measured. The upper (oxygen consumption) histogram is computed from the ventilation volumes and analyses of the expired air. The time blocks in the histograms indicate the duration of the collections of expired air. The pulse rate was counted for 6 seconds every half minute from a continuous polygraph record and was plotted in the middle of the interval so counted; the respiration rate, counted for the whole of each half minute, was plotted similarly. The blood pressure readings were made at approximately 30-second intervals and plotted at times at which they were taken as indicated by a foot-operated recording signal. The unavoidable difference in the methods of plotting results in a slight discrepancy in the actual time relations in the graph, but we do not believe that this affects the conclusion discussed in the text. W and X indicate the time at which the subject was commanded to accelerate and to stop accelerating the pulse rate. Y and Z indicate the time at which he was commanded to stop and start the muscle tensing. The "ventilation in liters per minute" and "oxygen consumption in liters per minute" are both plotted as histograms, and hence no temporal confusion results. The same method of plotting cannot be followed for the systolic and diastolic blood pressures or for the respiration rate. Since our readings are not instantaneous, but extend over a period of time, it is customary to plot data of this sort at the middle of the temporal interval rather than at the end. Hence, when the point just preceding the command "W" is connected with the measurement made immediately after the command is given, the impression is given that the rise precedes the command, which is not the case. The question as to whether the rise in respiratory rate began before, simultaneously with, or after the blood pressure rise, can be answered only partially by critical examination of the plot. Our experimental method provided us with continuous measurements of pulse rate and respiration rate, but our measurements of blood pressure are necessarily made at specified temporal intervals. Our first blood pressure measurement after the command, occurred between 20 and 30 seconds after the order was given. Furthermore, irregularities in respiration make it extremely hazardous to draw conclusions about the speed at which the respiratory rate may change.

whom the procedure appeared effortless, the new frequency was well maintained, and the return to normal was abrupt in both experiments. In the other subject (A.B., Figs. 3 and 4) the building up of a high frequency was somewhat delayed; he was not completely successful in maintaining the high rate, and at the end of the effort the normal rate reappeared more gradually. In each case the systolic blood pressure also increased considerably. As to the diastolic pressure, the subjects differ in that "A.L." induces a clear increase, whereas "A.B." shows no change in one experiment (Fig. 4) and a barely recognizable rise in the other. In either case, however, the marked increase in pulse rate without corresponding drop in pulse pressure would be interpreted by many investigators (Read and Barnett;[13] Liljestrand and Zander[10]) as evidence for an increased cardiac output. As to whether these subjects really showed an increased cardiac output, we have at present no direct evidence.

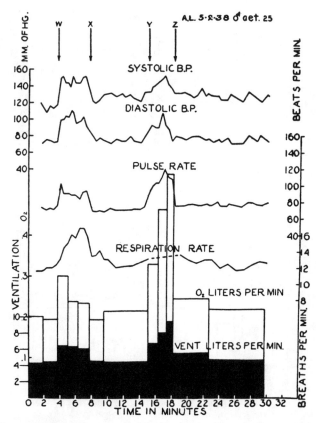

Fig. 2.—Increase of circulatory, respiratory, and metabolic functions on command (Subject "A. L.," 5/2/38). (See legend to Fig. 1.) Note the increase in circulatory functions on tensing muscles and the very great metabolic cost of this.

Toward the end of the experiment in each case the subjects' hands broke out into a cold sweat, and although we were unable to recognize any change with a skin resistance thermometer, this feeling of coldness suggests that in our subjects the blood flow through the limbs was diminished, as was shown to be the case by Taylor and Cameron[16] using calorimetry, and by Tarchanoff[15] and Pease[11] using plethysmographs. In this connection it may be noted that the slight drop in level of the pulse record which appeared every time (see Fig. 5) may be regarded as an indication of diminished limb volume and vasoconstriction.

Increased Metabolism and Respiration. Our observations may be divided into 7 periods, each separated by rest or recovery periods. Four of these (*i. e.*, 1 on each of 2 days for each subject) were periods during which the subject was attempting to accelerate his heart. These are the "experimental" periods. The 3 remaining or "control" periods, 2 for "A.L." and 1 for "A.B." were those in which the sub-

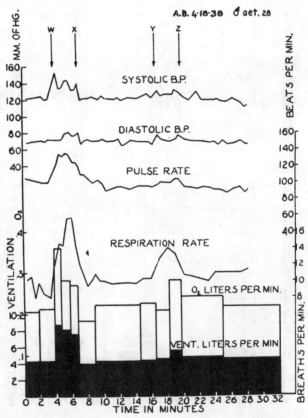

Fig. 3.—Increase of circulatory, respiratory, and metabolic functions on command (Subject "A. B.," 4/18/38). (See legend to Fig. 1.)

jects were tensing their skeletal muscles and not making any attempt at cardiac acceleration.

In each of the 7 periods, experimental and control alike, we find that there is an increase in respiratory frequency and volume of air breathed per minute, and a slight increase in oxygen consumption. In 6 of these periods the gas analyses (of which Table 1 is a sample) were characteristic of mild exercise. That is to say, the increase was found in both oxygen intake (Column 4, Table 1) and carbon dioxide elimination (Column 6, Table 1) with the respiratory quotient* (Column 7, Table 1) rising towards unity (in 1 case, "A.B.," 4/4/38, to 1.09). That a true increase in metabolism, probably due in part to increased work, was taking place is further shown by the

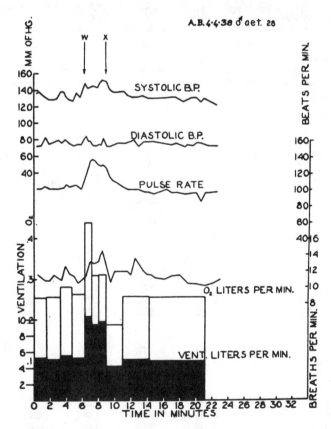

Fig. 4.—Increase of circulatory, respiratory, and metabolic functions on command (Subject "A. B.," 4/4/38). (See legend to Fig. 1.) This experiment was concluded without any muscle tensing control.

* The term "respiratory quotient" is here used, as it was originally, to signify the ratio of carbon dioxide to oxygen in the expired air. It is not intended to include in the term any interpretation as to chemical processes occurring in the blood stream or in the tissues.

40 Historical Perspective

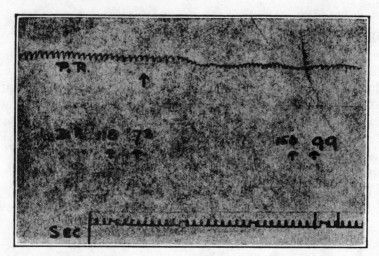

FIG. 5.—Part of polygraph record, showing the rapidity of onset of increase of pulse rate and blood pressure (Subject "A. L.," 5/2/38). Top line = record of changes in limb volume from cuff on ankle inflated to 50 mm. Hg. Arrow indicates command to increase heart rate. Second line = systolic and diastolic blood pressure. Arrows indicate moments of reading. Third line = time in seconds.

TABLE 1.—EXPIRED AIR ANALYSIS.

Experiment No. 2. A. L.   April 11, 1938·

Time for collection of 9.19 liters expired air, minutes. (1)	Expired air.		Apparent oxygen consumption.		CO_2 elimination, liters per minute. (6)	R. Q. (7)
	% O_2. (2)	% CO_2. (3)	Liters per minute. (4)	% of basal. (5)		
Resting.						
1.85	16.88	3.03	.194	91	.136	.702
1.82	16.91	3.17	.193	91	.144	.746
Voluntary Acceleration.						
1.28	16.48	3.61	.304	143	.235	.773
1.25	17.45	3.41	.232	109	.226	.974
1.36	17.63	3.17	.204	96	.194	.951
Resting.						
1.91	16.95	3.53	.178	84	.153	.860
5.18†	16.83	3.37	.206	97	.162	.786
Muscle Tensing.						
1.37	16.85	3.31	.261	123	.201	.770
1.23	16.55	3.44	.313	148	.232	.741
1.55	16.46	3.52	.253	119	.189	.747
Resting.						
5.24†	16.54	3.54	.219	103	.168	.767
5.33†	17.22	3.15	.180	85	.147	.817

† 25.57 liters.

fact that for each of the seven instances there is a clear-cut excess oxygen utilization if the oxygen intake for the periods of increase and the subsequent recovery periods be added and compared with the rate for the immediately preceding period (Table 2).

Hyperventilation. Whenever the voluntary acceleration was induced, we found evidence of hyperventilation (Taylor and Cameron;[16] Carpenter, Hoskins and Hitchcock[3]) in addition to the increased work. The evidence for this appears in the gas analyses (Table 1), where the carbon dioxide content (Column 3) and the oxygen content (Column 2) of the expired air are found to be high during the period of voluntary acceleration. This change is responsible to some extent for the R. Q. (Table 1, Column 7; and Table 2) which is noticeably higher during acceleration than during muscle tensing. The falling carbon dioxide content of expired air during the latter two-thirds of the voluntary acceleration (from 3.6% to 3.17%), in contrast to the rising concentration of carbon dioxide in the expired air during muscle tensing (3.31% to 3.52%) is further evidence that hyperventilation occurs in the former case. It may be seen clearly from the figures that the increase in ventilation has an abrupt onset and is maintained with very little change up to the end of the experimental period, whereas the increase in oxygen retention is much greater in the first than in subsequent periods. With hyperventilation the composition of the air in the chest tends to approximate more closely that of atmospheric air and the oxygen involved in this change gives a spurious appearance of a greater increase in

TABLE 2.—EXCESS METABOLISM DURING ACCELERATION.

	Apparent O_2 consumption for whole period (liters).		R. Q. (peak value).
	Measured.	Excess.	
A. L. 4/11/38			
*Rest	0.71	..	0.746
Acceleration	2.37	0.24	0.974
Muscle tensing	3.24	0.39	0.817
A. L. 5/2/38			
*Rest	0.74	..	0.719
Acceleration	2.55	0.32	0.901
Muscle tensing	3.99	1.11	0.805
A. B. 4/18/38			
*Rest	0.80	..	0.756
Acceleration	2.98	0.33	0.989
Muscle tensing	3.63	0.33	0.780

* The preliminary rest periods only. The periods labeled "acceleration" and "muscle tensing" include the following period of recovery (see figures).

The first experiment on A. B. (4/4/38) is omitted from this table. The interpretation of this experiment is similar to that of the others, but the somewhat different arrangement of the timing makes the figures confusing and not strictly comparable to those in this table. Values in Column 2 computed from measurements in Column 1 and duration of periods. (See Table 1, Column 1; and Figs. 1–4).

oxygen consumption during the first periods than is real. At the end of the experiment the respiration abruptly returns to its previous level and the reverse change takes place. During this time the excess oxygen in the lungs is absorbed for metabolic use without appearing as oxygen consumption in the analyses. Accordingly, the observed oxygen removal from the room shows a spurious drop during the early part of the recovery, a drop which has been used in Table 2 to neutralize the effect of spurious excess oxygen absorption. This drop is a further index of the existence of hyperventilation. These variations in apparent oxygen consumption have been considered in detail in experimental hyperventilation in persons with "effort syndrome" (Soley and Shock[14]).

Although we have definite evidence that hyperventilation was present in all the experiments and in none of the controls, we are inclined to agree with Taylor and Cameron[16] and with Carpenter, Hoskins and Hitchcock[3] that it is merely incidental and not an essential part of the mechanism for acceleration, since in a series of experiments (in preparation for publication) in which subjects were made to hyperventilate (by 3- or 4-fold) no increase of pulse rate or systolic blood pressure was found. Soley and Shock[14] found that hyperventilation *per se* appeared to increase the rate of oxygen consumption and also to produce cutaneous vasoconstriction and sweating. The hyperventilation, then, may contribute to an uncertain degree to these phenomena in our experiments, but almost certainly is not the cause of the changes in blood pressure and pulse rate.

Muscle Tensing. In view of the experiments in which Hoskins[3] accelerated his heart by deliberate tensing of muscles, it becomes imperative to consider the possibility that muscle tensing may be the underlying factor responsible for the cardiac acceleration in the cases reported here. Considering first the case of "A.L." as observed on April 11, 1938 (Fig. 1), it may be seen that in the experiment and the control during acceleration and muscle tensing respectively the apparent increase in oxygen consumption is of the same order of magnitude. The circulatory functions, on the other hand, are decidedly more increased during the experimental than during the control period. In fact, the real excess oxygen utilization (Table 2) was greater in the control period than in the other and the pulse acceleration was less, while the increase in blood pressure during the muscle tensing period was negligible. In the other experiment (Fig. 2) on this subject, it is true that all the circulatory measurements of the experimental period were closely imitated during the muscle tensing period, but this was at very much greater metabolic cost. On the evidence of these two experiments, then, it seems that the circulatory changes were determined by something more than the necessity for satisfying a metabolic need created by muscle tensing. Just as the subject has been shown to be hyperventilating, so we may speak of him as independently "hypercirculating."

Cardiac Output. In view of the 60% and 40% increases in pulse rate and the simultaneous 20% and 10% increases in pulse pressure, it is very likely that there was a sudden increase in cardiac output. The evidences of peripheral vasoconstriction in such cases—cold, pale extremities—and in the instances where measurements have been made, diminution in limb volume (Tarchanoff[15]), and limb blood flow (Taylor and Cameron[16]), together with the dilatation of the pupil observed in our subjects suggests the possibility that the whole train of events may be due to a sudden liberation of epinephrine. Taylor and Cameron found their experiments to be followed by a glycosuria but failed to find the expected hyperglycemia—a failure which they very reasonably believe may be due to the timing of their blood samples. On the other hand, with such a mechanism one might expect a delay of a quarter minute for the epinephrine to pass from the adrenal body to the peripheral resistance and the coronary system. One might also expect a measurable after-effect. In one at least of our subjects the onset and cessation of the acceleration were very considerable within 10 seconds of the command (Fig. 5).

Moreover, Euler and Liljestrand[5] found that subcutaneous doses of epinephrine great enough almost to double the cardiac output produced no increase in diastolic pressure; in fact, in every case they show a very small decrease. Similar findings are reported by Elliot and Nuzum.[4] Pickering and Kissin,[12] Gordon and Levitt,[8] and Fatheree and Hines[6] find similar effects with small doses intravenously though larger doses commonly produced a rise. Grollman explains the findings of Euler and Liljestrand by suggesting that in the dosage in question the vasodilator effects of epinephrine are greater than its vasoconstrictor effects to an extent sufficient to balance the increased cardiac output. It might be argued that we are dealing with a much larger liberation of epinephrine and consequently have a predominantly constrictor effect. Should this be so, it becomes necessary to consider whether the direct action of epinephrine on the heart can increase the cardiac output in the presence of a generalized vasoconstriction which must tend, by diminishing the *vis a tergo*, to lower the venous pressure and thus the filling of the heart.

Blood Mobilization. It would seem that the most likely way in which such a filling insufficiency could be compensated would be by the liberation of blood from the blood stores. Although such a liberation would in no way help to explain the rapidity of the phenomena, it would provide a possible explanation for the raised diastolic pressure. Let us then consider the possibility that such an increase could have been brought about in part by the rapid mobilization of some of the blood ordinarily stored in the cutaneous, splenic, mesenteric, and pulmonary vascular beds. With the exception of the last, these bloods are all venous, and in the case of the splenic, the most mobile store, of a very high oxygen capacity. The extra oxygen intake in the first minute necessary to saturate this extravenous blood would appear in the oxygen analyses as excess oxygen removed from the first 9-liter period. That this circulatory acceleration may have had a greater effect in removing oxygen from the alveoli than the hyperventilation had in providing it, is suggested by the fact that the oxygen content of the expired air was always lower during the beginning of the period of voluntary acceleration than later (see Table 1), whereas with uncomplicated hyperventilation it is usually increased immediately (Soley and Shock[14]).

In order to assess the quantity of oxygen which might be taken up in this way it would be necessary to have a quantitative measurement of the amount of excess ventilation, but a rough attempt may nevertheless be interesting. Such an attempt, based on the data available, would involve certain assumptions. Fortunately it seems possible to approach the problem in two different ways, the necessary assumptions for which are entirely different. In the case of "A.L." (4/11/38) we see (Table 1) that an immediate effect of the command to accelerate is a fall in the oxygen content of the expired air in spite of the hyperventilation already discussed. Since the highest oxygen content of expired air is usually at the beginning of hyperventilation, we may be justified in assuming that such might have been the case here. On the basis of the two subsequent figures, let us assume that the maximum oxygen content attained by the subject can be attributed entirely to hyperventilation. The difference between this maximum value (17.63) and the value observed (16.48) during the first period might well have been due to the passage of oxygen into the newly released venous blood during this time. This difference (1.15) multiplied by the volume of air expired during the period (8.3 liters*) gives a value of 95 cc. for the amount of oxygen which might have been expected in the expired air but was not found. To regard this as stored in the newly released blood involves the further assumption that the metabolic rate does not vary greatly during the period of cardiac acceleration—an assumption which seems reasonable in view of the way in which the levels of increased pulmonary and circulatory activity are maintained.

* This figure, 8.3, is the volume of the "9-liter" spirometer (9.19 liters) corrected for pressure, temperature, and water vapor.

TABLE 3.—CALCULATION OF "OXYGEN STORAGE."

	Experiment.			
	I.	II.	III.	IV.
Subject	A. L.	A. L.	A. B.	A. B.
Date of experiment	4/11/38	5/2/38	4/4/38	4/18/38
Method I:				
Observed % O_2 in expired air attributed to hyperventilation	17.63	17.21	17.63	17.49
Observed % O_2 in expired air of Period I	16.48	16.50	16.17	16.87
Difference	1.15	0.71	1.46	0.62
Cc. O_2 "stored"	95	59	121	52
Method II:				
Volume of air in lungs (assumed)	3000	3000	3000	3000
Change in % O_2 in expired air (basal—Period I)	0.43	0.41	−0.60	−0.68
O_2 disappearing from lungs due to change in gas composition (cc.)	13	12	−18	−20
Measured excess O_2 uptake during Period I (cc./min.)	142	139	143	131
Total O_2 taken up in Period I (cc.)	155	151	125	111
Total excess metabolism (cc.)	240	320	330	95
Excess metabolism for Period I	79	104	99	29
O_2 "stored" (cc.)	76	47	126	82

The other method of attacking this problem involves the assumption that the amount of air in the lungs is not greater than 3 liters and that its mean oxygen content does not change any more than that of the expired air. These limiting assumptions, as will be seen, give us a figure which is of the same order of magnitude as that already obtained. Thus in the same experiment, the quantity of oxygen absent from the lungs during the first 9-liter period of acceleration is not greater than $3000 \times .0043 = 13$ cc. oxygen. The increased oxygen intake in this period as compared with the resting period amounts to 142 cc. Adding this to the 13 cc. which have left the lungs, as shown by the changed oxygen content of the expired air, we have 155 cc. of oxygen to account for the excess metabolism during this period as well as the saturation of the newly liberated blood. From Table 2 we find that the total excess metabolism of the acceleration period was 240 cc. If we assume that the metabolic increase, like the circulatory displacements, was evenly distributed over the experimental period, we may divide it between the three acceleration collections in proportion to their duration. It then appears that 79 cc. excess oxygen were used during the period which is under consideration. After these 79 cc. have been subtracted from the 155 cc. above, there remain 76 cc. for the saturation of the stored blood. Considering the two calculations just made we may say that an amount of about 75 to 100 cc. of oxygen has left the lungs in excess of that which appears to have been required for metabolic needs. Table 3 shows the results of similar calculation on all 4 of the experiments.

It is not beyond the bounds of possibility that in "A.B." the increase of systolic without much change in diastolic pressure might have been achieved by a moderate vascular relaxation (of the epinephrine type, perhaps, since this subject reacted more slowly) which in turn would have increased the cardiac output and frequency without calling on the blood stores to any great extent. This suggestion is further supported by the somewhat less sudden appearance and disappearance of the phenomena in his case.

If the excess oxygen found in these calculations is taken up, as has been suggested by blood newly liberated from blood stores, it would be sufficient to saturate between $\frac{1}{2}$ and 2 liters of pulmonary artery blood. The former of these figures is based upon one of the higher figures for pulmonary arteriovenous oxygen difference (130 cc. per liter) observed by Grollman; the latter and larger figure upon an arteriovenous oxygen difference of 50 cc. per liter. The blood released from the stores, however, would almost certainly differ from pulmonary artery blood in being more venous and having a higher oxygen capacity since at least some of it would come from the spleen. This makes it possible that the oxygen in question might be absorbed by considerably less than a liter of mobilized blood. Barcroft[1] regards his estimate of a 1-liter blood mobilization for man as conservative.

The above considerations with respect to oxygen uptake are borne out by the indications of the carbon dioxide analyses (Table 1) which indicate excess carbon dioxide elimination, but it does not seem possible to deal with this gas on a similar basis because of the vast stores of it throughout the body and of the ease with which it may be liberated in response to the production of fixed acids.

It might be suggested, on the other hand, that the sudden uptake of oxygen and evolution of carbon dioxide might be more simply explained by a sudden burst of metabolism. The gas analysis figures offer no absolute refutation of this contention, but it is difficult to visualize a sudden violent metabolic activity of about 1 minute's duration without any corresponding disturbance appearing in the cardiovascular and pulmonary system.

Discussion. The incidence of the ability voluntarily to accelerate the heart has been much discussed in the 9 reports cited, but there does not appear to have been any attempt to investigate its frequency in adequately large groups. The extreme view is that of Van de Velde,[18] based on a series of 5 cases, of which 4 showed the phenomenon, which holds that the power is almost universal and that acceleration can be induced by anyone after a little practice.

The pharmacology of the phenomenon has been studied, particularly with reference to atropine, in attempts to throw light on the efferent nervous or chemical mechanisms involved, but the uncertainty as to the adequacy of any given dosage has made for confusion in the findings. Tarchanoff observed that the power was in abeyance under the influence of subanesthetic doses of nitrous oxide. Oddly enough, it was while submitting to this drug that one of our subjects (A. B.) first noticed his ability to alter his pulse rate.

Dilatation of the pupil, increased blood pressure,[16] perspiration, increased oxygen consumption,[3] and hyperventilation[16] have all been noted as phenomena concomitant with voluntary acceleration of the pulse, but it is not recorded that acceleration has ever been produced without the accompaniment of them.

Summary. Two subjects are discussed; both of them are able voluntarily to accelerate their pulse rates at the word of command. This acceleration is accompanied by increased systolic and diastolic blood pressures, increased rate of respiration, ventilation volume, oxygen uptake and carbon dioxide output. The onset and disappearance are abrupt.

The quantitative consideration of the magnitude and time relations of these functions indicates that there is a true hyperventilation with excess elimination of carbon dioxide, a true increase in metabolism with excess oxygen utilization, and a hypercirculation, or circulation in excess of the metabolic requirement. Epinephrine might be responsible for the cardiac acceleration and the increased blood pressure and, by discharge of blood reservoirs, for the increased circulation. But it is believed that the phenomenon appears and disappears so rapidly that epinephrine cannot be solely held to account. In the 9 reports cited, covering some 15 other cases of voluntary acceleration of the heart, there is no instance in which the acceleration was not accompanied by one or more of the other phenomena described.

Grateful acknowledgment is hereby made to Mrs. Kathryn Long and Mr. Theodore Chernikoff for technical assistance, to the W.P.A. (O.P. No. 665-08-3-30, Unit A-8) for clerical and statistical work, and to the Board of Research of the University of California for financial support.

REFERENCES.

(1.) **Barcroft, J.**: Features in the Architecture of Physiological Function, London, Cambridge University Press, p. 169, 1934. (2.) **Boothby, W. W.**, and **Sandiford, I.**: Laboratory Manual of the Technique of Basal Metabolic Rate Determination, Philadelphia, W. B. Saunders Company, 1920. (3.) **Carpenter, T. M., Hoskins, R. G.**, and **Hitchcock, F. A.**: Am. J. Physiol., **110**, 320, 1934. (4.) **Elliot, A. H.**, and **Nuzum, F. R.**: Am. J. Med. Sci., **189**, 215, 1935. (5.) **Euler, U. v.**, and **Liljestrand, G.**: Skandin. Arch. f. Physiol., **52**, 243, 1927. (6.) **Fatherree, T. J.**, and **Hines, E. A.**: Am. Heart J., **16**, 66, 1938. (7.) **Favill, J.**, and **White, P. D.**: Heart, 6, 175, 1915–17. (8.) **Gordon, W.**, and **Levitt, G.**: J. Clin. Invest., **14**, 367, 1935. (9.) **Koehler, M.**:

Pflüger's Arch. f. d. ges. Physiol., **158**, 579, 1914. **(10.) Liljestrand, G.**, and **Zander, E.**: Ztschr. f. d. ges. exp. Med., **59**, 105, 1928. **(11.) Pease, E. A.**: Boston Med. and Surg. J., **120**, 525, 1889. **(12.) Pickering, G. W.**, and **Kissin, M.**: Clin. Sci., **2**, 201, 1936. **(13.) Read, J. M.**, and **Barnett, C. W.**: Arch. Int. Med., **57**, 521, 1936. **(14.) Soley, M.**, and **Shock, N. W.**: Am. J. Med. Sci., **196**, 840, 1938. **(15.) Tarchanoff, J. R.**: Pflüger's Arch. f. d. ges. Physiol., **35**, 109, 1885. **(16.) Taylor, N. B.**, and **Cameron, H. G.**: Am. J. Physiol., **61**, 385, 1922. **(17.) Tuke, D. H.**: Illustrations of the Influence of the Mind upon the Body in Health and Disease Designed to Elucidate the Action of the Imagination, 2d ed., Philadelphia, Henry C. Lea's Son & Co., p. 372, 1884. **(18.) Van de Velde, T. H.**: Pflüger's Arch. f. d. ges. Physiol., **66**, 232, 1897. **(19.) West, H. F.**, and **Savage, W. E.**: Arch. Int. Med., **22**, 290, 1918.

Cardiac Arrest through Volition

C. M. McClure

SINCE DISCOVERY of the heartbeat, yogis and fakirs have claimed to be able to control it at will, but there are no documented cases in the medical literature of cardiac arrest through volition, without physical manipulations of any kind. The following case is presented because it is unusual, perhaps unique.

REPORT OF A CASE

A 44-year-old aircraft mechanic of Danish descent was admitted to Lindsay Municipal Hospital, April 24, 1958, because of a cold with cough of two weeks' duration. He said that in the previous 20 years he had had six episodes of upper respiratory tract infection, and that during these periods he had found that by sitting quietly, relaxing completely and "allowing everything to stop," he could induce progressive slowing of the pulse until cessation of heart action would occur, then a feeling of impending loss of consciousness. After a few seconds of this sensation, he would take a deep breath and normal heart action would resume. These occurrences resulted in the patient's developing a fear of sleeping, lest his heart stop and not start again. In 1953 and several times afterward the author verified this story by auscultating the heart and palpating the radial pulse while the patient induced several seconds of cardiac arrest. At these times his color would become the ashen grey of sudden circulatory failure, and partial loss of consciousness would ensue. However, no cardiac irregularities were ever observed during either normal sleep or general anesthesia. Cardiac arrest occurred only when the patient deliberately induced it.

Submitted October 30, 1958.

The patient stated that at the age of seven years he had had rheumatic fever, then was bedfast for a long time and took digitalis for five years thereafter. Since that time he had had a cardiac murmur but no arthralgia, dyspnea, orthopnea, fever or chills. He underwent tonsillectomy at age 12 and cholecystectomy in January 1958, both under general anesthesia, without incident.

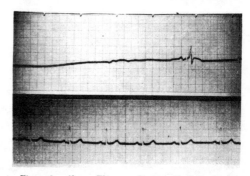

Figure 1.—*Above:* Electrocardiogram (lead I) showing cardiac arrest through volition. *Lower:* Normal tracing (lead I) for same patient.

On physical examination, the oral temperature was 99°, the pulse rate 60, respirations 18 per minute and blood pressure 134/74 mm. of mercury. The patient was muscular, intelligent, somewhat tense and anxious. There was no icterus, cyanosis or edema, but there was clubbing of all fingers. The only other abnormality noted in a complete examination was a grade I systolic murmur heard over the aortic area, becoming grade II along the left sternal border, the sound transmitting poorly to the apex. The rhythm was regular. Upon x-ray examination it was observed that the size and contour of the heart were within normal limits.

Electrocardiographic leads were connected and the patient was asked to induce slowing of the heart. This actually occurred with no physical manipulations except lying very quietly and allowing respiration to become quite shallow. The electrocardiogram showed slowing of the sinus rate progressively to the point of sinus arrest for a period of five seconds, followed by several atrioventricular nodal beats, and then resumption of sinus bradycardia at a rate of about 55. At several points in the record, occasional atrioventricular nodal beats were observed. An electrocardiogram an hour later with the patient at rest was normal.

A consultant who examined the patient and read the electrocardiograms suggested the use of sympathomimetic drugs, but subsequent use of atropine, ephedrine, amphetamine and aminophylline at different times did not change the arrest mechanism. Even after recovery from the respiratory tract infection, the patient found that he could still induce bradycardia and brief periods of cardiac arrest almost at will.

DISCUSSION

It was felt that the clubbing of the fingers noted in this patient was familial, since it was present also in his son, who was healthy. The underlying cardiac change is believed to be well compensated rheumatic heart disease with aortic valvulitis. The bradycardia and cardiac arrest are probably manifestations of exaggerated vagotonia, induced through some mechanism which, although under voluntary control, is not known to the patient himself. Careful observation did not reveal any breath-holding or Valsalva maneuver in connection with the cessation of heartbeat. Apparently the patient simply abolished all sympathetic tone by complete mental and physical relaxation.

SUMMARY

A case is presented of a patient with old rheumatic heart disease, who is able to produce cessation of heartbeat, apparently by volition alone.

II

OVERVIEW

The Psychophysiology of Private Events

Ralph F. Hefferline and Louis J. J. Bruno

I shall not begin, as Socrates would require, by defining my terms. Instead, as a kind of discursive preparation for the more tightly organized material to come later, I shall recount briefly the stages by which my students and I reached our current perspective on the psychology of private events. First, though, let me say something in advance about what this perspective is. We believe that the scientific approach to private events is best made by means of electrophysiological techniques. Furthermore, we believe that feedback to the subject of what the instruments "see" may turn out to be a way to implement the ancient injunction, "know thyself," and to provide both a rationale and a method for effective training and retraining across what potentially may be the full spectrum of ontogenetic behavior.

My investigation of private events began when, as a graduate student, I tackled the problem of why a person may cling to behavior which no longer seems to serve a purpose. By this I do not mean why it is hard to break a "bad habit." The answer there, of course, is that in the short run it is so enjoyable! What I am referring to is why a person may avoid, with utmost urgency, a situation which, to the disinterested observer, poses no threat. One answer is that it is not the situation per se which disturbs, but what it tempts the person to do, namely, what he actually did or attempted on past occasions *with dire punishment as the consequence*! Now, of course, he withdraws automatically without even remembering why.

Obviously, this answer is nothing more than a crude version of psychoanalytic trauma-theory minus the usual appeal to repression to account for the failure to remember. However, what I was looking for was not a complete explanation of avoidance; instead, I wanted some basis upon which the problem could be approached in the laboratory. Since I was at Columbia University, the inevitable choice was to start out and go as far as possible with rats.

[1] Preparation of this paper was supported in part by Grant MH-13890 from the National Institute of Mental Health to Ralph F. Hefferline, Principal Investigator.

Reprinted with permission of Academic Press, Inc. from *The Psychology of Private Events* edited by Alfred Jacobs and Lewis B. Sachs. Copyright © 1971 by Academic Press, Inc.

Estes, at Indiana, had not long before completed an elaborate study of punishment using rats as subjects (Estes, 1944). He concluded:

> It is clear ... that a disturbing or traumatic stimulus arouses a changed state of the organism of the sort commonly called "emotional" and that any stimulus present simultaneously with the disturbing stimulus becomes a conditioned stimulus capable of itself arousing the state on subsequent occasions. ... When punishment is correlated with ... response, emotional reaction can ... become conditioned to the incipient movements of making the response [p. 36].

On the strength of this concordance between clinical theory and laboratory findings, I—and, later, two of my students—designed and carried through experiments on avoidance behavior in the rat (Eldridge, 1954; Hefferline, 1950; Winnick, 1956). I shall mention here only enough of this early work to show how it laid the basis for further pursuit of the problem at the human level.

An Animal Prototype of the Human Problem of Letting Go

A person with a broken arm in a sling learns quickly not to move the injured member. Even after the arm mends and the sling is removed he tends at first to hold the arm flat against his chest with elbow bent. This behavior, after it has largely disappeared, may be temporarily restored in full force if the sling is put on again and the person asked to move through a crowd. This happens despite the fact that the person "knows" that it will not hurt to move his arm or have it jostled.

In what I came to call the "holding experiment," I conditioned somewhat similar behavior in the albino rat (Hefferline, 1950). Lacking eye pigment, this animal is light sensitive, and given the chance, will turn off a bright light. For this purpose I placed a small lever switch in the cage: When the rat pressed it down, the light went out; as soon as he released it, the light returned. Under these conditions, the rat learned to hold the lever down without once releasing it for as long as 45 min. Of course, after such lengthy stretches of lever holding, it is likely that muscular strain eventually became so punishing as to override other variables. In fact, the observed behavior was that the animal would release the lever, race about the cage for a few seconds, and then fling himself once more on the lever.

Later, after the light circuit was disconnected, the rat could roam freely without penalty. Occasionally, if he happened to touch the lever, he would press it down; then, it was as if he became trapped there, unable to let go. Pressing the lever actually did, so far as he was concerned, put him back in precisely the situation in which he had been when release of the lever reinstated the light.

I assumed that the rat did not sit there reminding himself, as a person might, that if he let go the light might return. Instead, I inferred that the animal made many incipient movements of letting go; that is, it would begin to let go, but since this would, in Estes' terms, produce an "emotional" state conditioned to the proprioceptive and tactual stimuli generated by releasing, it would resume pressing to remove the state.

Since my apparatus worked on an on–off basis and could not reveal incipient lever-releasing movements, a follow-up study was done in which the animal, instead of pressing a lever, pushed against a small vertical panel hinged at the top (Winnick, 1956). The panel could be pushed beyond the point necessary to turn off the light and a pen writing on a moving paper tape indicated extent of movement. Early in conditioning, the animal wavered back and forth, frequently allowing the light to return momentarily. Later, he kept it off for long periods. The panel, however, did not remain stationary. The recording pen indicated an irregular drifting movement toward the position where the light would resume, but, before this occurred, the animal would, repeatedly, make a corrective thrust back to a "safe" position.

While Winnick's study demonstrated that the holding response is one of continuing activity, with incipient movements occurring and being corrected for, it was still merely hypothetical that the proprioceptive stimuli involved in "letting go" set up an emotional state. To test for this objectively, Eldridge (1954) designed an experimental variation which permitted respiration rate to be recorded, since breathing irregularity has long been thought to be a sensitive index of autonomic upset (Lindsley, 1951).

The rats were trained to stand on a perch several feet above the floor. Directly over the animal's head was a wooden block, counter-weighted to rise slightly if the animal pushed upward with a force of about 5 gm. Upward movement turned off the light. Lowering the head allowed it to come back. Breathing rate was measured with due observance of a number of appropriate controls. If the rate before incipient letting go was about 80 per minute, then the first two breaths after movement were likely to be more shallow, although not necessarily, and the rate invariably accelerated to anywhere from 100 per minute to—although this was rare—400 per minute.

These three experiments comprise a unit. I conditioned the holding response and gave an inferential analysis of the controlling variables. Winnick followed with a demonstration that such holding was not immobilization, but vacillation between release movement and compensatory renewals of forward thrust. Eldridge, finally, produced evidence for the presence of an emotional state of "anxiety" generated in the animal by his own proprioceptive and tactual stimuli.

While this work was in progress, I became involved also in a nonlaboratory approach to the same and related problems at the human level. This will be described in the following section.

A Phenomenological Excursion into "Somatic Psychology"

In 1951 I collaborated with Fritz Perls and Paul Goodman in a book called *Gestalt Therapy: Excitement and Growth in the Human Personality.*[2] The portion of *Gestalt Therapy* which I wrote was a kind of do-it-yourself manual, subtitled "Mobilizing the Self." It attempted to provide a means of objectively approaching, among other things, one's hand-me-downs of chronic or recurrent muscle tensions, aches, pains, and assorted subjective discomfitures. These were viewed as somatic residuals, pointless perhaps in the "here and now," but persisting from earlier life periods when they had done service with some effectiveness in holding back strong, but punishable behavior from overt expression. The theoretical and practical background was a mixed one and only partly developed, as will appear shortly, but the basic therapeutic formula for self-application was simply this: Since conflict consists of some degree of clenching of antagonistic but *voluntary* muscles, the thing to do is to cultivate awareness of just *how* one produces the physical hang-up, experiment with it, even try to make it momentarily more severe, while remaining alert to possible anxiety and bits of fearful fantasy of what might happen if one let go.

A partial statement of procedure is quoted here from a previous summary account (Hefferline, 1958):

As part of his assignment the subject was encouraged to make what amounted to a systematic proprioceptive survey of his own body, conducted in private with minimum external distractions. A person neither dead nor in flaccid paralysis presumably should be able to discriminate the tonic condition of all or of any part of his skeletal musculature. The first report of a subject is likely to be that he can do this. He can feel, he says, every part of his body. When further inquiry is made, it often turns out that what he took to be proprioceptive discrimination of a particular body part was actually a visualization of the part or a verbal statement of its location. Or else, to discriminate the part, he may have had to intensify proprioception by making actual movements.

With further work, if he can be persuaded to continue, the subject may report certain parts of his body to be proprioceptively missing. Suppose it is his neck. He may discriminate a mass that is his head and a mass that is his trunk with what feels like simply some empty space between. At this stage the subject is apt to remember more important things to do and his participation in the silly business ends.

Some subjects, however, apparently made curious by blank spots and hopeful of recovering some lost degrees of freedom in their system of voluntary control, do whatever is involved in paying closer attention to and acquiring interest in this peculiar private situation. A blank spot, they say, may gradually fill in. Or it may suddenly become the locus of sharp pain, paresthesias of one sort or another, "electric" sensations, or the unmistakable ache of muscular cramp.

Then what formerly was a blank may become as demanding of attention as an aching tooth. Further and more detailed discriminations may be made. It soon becomes imperative to relax the cramp, but the subject says that he does not know how to do so; he is concerned with so-called voluntary muscles, but these are reportedly not under voluntary

---

[2] The term *gestalt* went into the title and was kept there at the insistance of Fritz Perls, who convinced himself and the publisher that *Gestalt Theorie* still flourished in the United States. As an adjective descriptive of the therapy, *Gestalt* was not exactly a misnomer. Figure-ground terminology did permeate the book, and in a recent chapter on "the visionary sociology of Paul Goodman," Roszak mentions "the mystical 'wholism' which the therapy inherits from Gestalt theories of perception [1968, p. 187]." However, when prepublication copies of *Gestalt Therapy* were presented to Wolfgang Köhler and Molly Harrower, they disavowed its being in any legitimate line of descent from Gestalt doctrine.

control. The subject is somewhat in the position of the elementary psychology student who is assured that he has the voluntary muscles needed to wiggle his ears. The difference, of course, is that in the case of ear-wiggling he has never acquired control, whereas, in the case of the cramped muscles he has somehow lost control.

As soon as the question becomes one of how to acquire or regain control, problems for investigation sprout in all directions. A host of variables, direct or indirect in their effect, become relevant. One involves merely the instruction to continue to pay sustained attention to the blocking, regardless of discomfort, and to be on the alert for subtle changes of any sort. This seems to give rise to what previously was called ideomotor action, and may in itself bring a loosening of the reported muscular clinch.

Another instruction is to increase the clinch deliberately, if possible, and then, while relaxing from this added intensity, to learn something about relaxing still further. Also relevant are procedures that make use of proprioceptive facilitation—for instance, those used in training polio victims, partial spastics, and others to make better use of whatever healthy muscle they still possess. Kabat (1950) has developed these methods systematically.

When a muscular block is definitely resolved, it is frequently claimed by the subject that there occurs vivid, spontaneous recall of typical situations, perhaps dating back to childhood, where he learned to tense in this particular manner [pp. 747-748].

The monthly reports submitted to me by over a thousand undergraduates who undertook the "experiments in self-awareness" for point credit, gave substantial basis for concluding that the experiments were deemed beneficial by those who had been able to make contact with critical aspects of their own malfunctioning and thus had discovered the manner in which they blocked or interfered with their own supposedly highly motivated efforts. However, these tended to be persons who were already in pretty good shape. Those more severely conflicted might simply drop the whole thing—as they were privileged to do at any time—or else continue it while attacking the procedures as utterly fatuous, or immoral, or else attacking me as a "concentration camp doctor" who dangerously experimented with helpless people while not assuming responsibility for what would probably be disastrous outcomes.

The criticism eventually most bothersome was one actually anticipated in the material itself (Perls, Hefferline, & Goodman, 1951):

As you work along with us, you will be inclined from time to time to question statements that we make, and you will demand, "Where is your proof?" Our standard answer will be that we present nothing that you cannot *verify for yourself in terms of your own behavior*, but ... this will not satisfy you and you will clamor for "objective evidence" of a verbal sort, *prior to* trying out a single non-verbal step of the procedure [p. 7].

To meet this criticism, I shifted back to the laboratory, this time to study the "king-sized rat"—man—by *electropsychological* means. That is, I intended to apply *electro*physiological techniques to the analysis of *psychological* problems—a pursuit later called "psychophysiology": "The general goal of psychophysiology is to describe the mechanisms which *translate* between psychological and physiological systems of the organism [Ax, 1964, p. 8]." The idea was to make public, through instrumentation, the private, physiological events which were the focus of self-awareness and self-control in *Gestalt Therapy*. Made public, these events could not only be scrutinized experimentally, but as we shall see later, they could, when fed back to the subject in visual or auditory analog, furnish a powerful method for somatic surveys and retraining.

The Electropsychological Analysis of Private Events

In shifting back to the laboratory, the presumption was, as Skinner (1953) put it, that "a private event may be distinguished by its limited accessibility but not, so far as we know, by any special structure or nature [p. 257]." What makes a private event private is that it occurs in the region beneath the skin; what can make it public are instruments which detect the same events as do the organism's internal receptors, that is, the instruments of electrophysiology. Of course, even with electrophysiological detection, "we are still faced with events which occur at the private level and which are important to the organism without instrumental amplification [Skinner, 1953, p. 282]," but we are now in a position to deal with these events objectively. Thus, the private events which were the object of phenomenological report in *Gestalt Therapy* became, in electropsychological analysis, the variables—dependent and independent (cf. Hefferline & Bruno, 1971)—in conditioning and psychophysical experiments.

Response Control

Our initial efforts in electropsychology, reflecting our special interest in tension patterns, dealt with covert muscular contractions. Using both negative reinforcers (Hefferline & Keenan, 1961; Hefferline, Keenan, & Harford, 1959) and positive secondary reinforcers (Hefferline & Keenan, 1963; Sasmor, 1966), we showed that a response so small as to require electromyographic techniques for its detection could, nevertheless, be controlled through instrumental conditioning procedures. Since these experiments, which we have reviewed in detail elsewhere (Hefferline, Bruno, & Davidowitz, 1971), were conducted without the *S*'s observation of the response, that is, with the variables controlling his verbal behavior rendered temporarily inoperative, they gave support both to the general position "that there is no qualitative difference between the behavior of man and that of subhuman organisms, the seeming difference being produced by the fact that man alone possesses verbal behavior [Hefferline, 1962, p. 126]," and to the particular position that private events are not distinguished "by any special structure or nature." In other words, our experiments gave reason to believe that the learning principles which apply to the public behavior of human and subhuman species would also find application in the study of private events.

Of course, there had been many earlier indications that learning principles apply to private events; these, however, had nearly all come from studies carried out under classical procedures. In a sense, this was as it should be. Since private events are by definition inaccessible to the general community, we might, on quasi-evolutionary grounds, expect to find them amenable to classical procedures, which require only that two exteroceptive stimuli be paired consistently, but intractable under instrumental procedures, which specify that the reinforcing stimulus be presented in consequence of an emitted response. At point here is the once-certain dichotomy between classical and instrumental conditioning (cf. Kimble, 1961): Classical procedures apply to elicited, involuntary, autonomic responses (most of which qualify as private events); instrumental procedures apply to emitted, voluntary, skeletal responses (most of which are clearly public). Considered in this context, our conditioning studies were somewhat enigmatic. Our procedures were surely instrumental, but the responses conditioned were hardly "voluntary": In one study, the group which had been "informed that the effective response was a tiny twitch of the left thumb ... kept so busy producing voluntary thumb-twitches that the small reinforceable type of response had little opportunity to occur [Hefferline et al., 1959, pp. 1338-1339]."

The enigma, as it turned out, was as false as the dichotomy between classical and instrumental "responses." The procedures, of course, are unquestionably different—classical reinforcements are contingent upon the prior occurrence of a stimulus; instrumental reinforcements, in contrast, are response-contingent—but there are reasonably clear indications that they do not apply to different types of responses (cf. Miller, 1969). For example, there is now evidence obtained in man to show that heart rate (Engel & Chism, 1967; Engel & Hansen, 1966), systolic blood pressure (Shapiro, Tursky, Gershon, & Stern, 1969), vasoconstriction (Snyder & Noble, 1968), salivation (Brown & Katz, 1967), and electrodermal activity (Shapiro & Crider, 1967) are all susceptible to instrumental control, although each qualifies as "elicited, involuntary, and autonomic," and each has been the object of uncounted studies in classical conditioning. On the other side of the fence, there is, of course, longstanding evidence that classical procedures can be applied to the control of skeletal responses (Bekhterev, 1913), although, to our knowledge, there is only one study (Van Liere, 1953) which dealt with responses so small as to make it unlikely that their occurrence could, in instrumental fashion, modify the effects of the unconditional stimulus (cf. Rescorla & Solomon, 1967). In short, the weight of the evidence, even after taking the artifact and mediation arguments into account (cf. Katkin & Murray, 1968; Smith, 1954), leans to the position that classical and instrumental procedures apply in common to most responses, a position which suggests that there may be only one not two, varieties of learning (cf. Hefferline et al., 1971; Herrnstein, 1969; Miller, 1969; Rescorla & Solomon, 1967; Schoenfeld, 1966).

The enigma of our conditioning studies has thus been handsomely resolved: Instrumental procedures do, in fact, apply to "involuntary" responses, and, it would seem, to most of the responses which might be categorized as private events, cortical activity not excepted (Fetz, 1969; Rosenfeld, Rudell, & Fox, 1969). The resolution deals us two high-valued cards—the first logical, the second methodological. Logically, it is now possible to extend to psychosomatic symptoms, that is, to symptoms of the autonomic system, the type of analysis formerly applied only to symptoms thought to be mediated by the central nervous system (cf. Dollard & Miller, 1950; Miller, 1969). Methodologically, the fact that instrumental procedures may be used to control private events means that they can be manipulated experimentally as independent variables (Hefferline & Bruno, 1971), and that they can be manipulated clinically in systematic programs of self-education and re-education.

While the logical card has only just been turned up, the methodological is now clearly in play. Hand in hand with the experiments demonstrating that instrumental control applies to one and then another private event has come another series of experiments, inaugurated in our laboratory (Hefferline, 1958), which have developed effective and public means for controlling private events in a practical way. These experiments, which have adapted the "engineering" procedures of augmented or supplementary feedback, provide the S with an auditory or visual analog, typically continuous, of the private event which is the object of study. The experiments, whose efficacy seems to derive from their status as a variety of discriminative instrumental conditioning (Hefferline *et al.*, 1971), have now been carried out successfully with covert muscular responses (Basmajian, 1963; Hefferline, 1958), cardiac variables (cf. Brener, Kleinman, & Goesling, 1969), salivation (Delse & Feather, 1968), electrodermal responses (Stern & Kaplan, 1967), and with the alpha frequencies of the electroencephalogram (Kamiya, 1969). An application and some of the implications of these experiments are given in later sections of this paper (see "Response Bias" and "Bio-Feedback: Controlling the 'Uncontrollable' ").

Response Discrimination

The experiments in response control made in our laboratory and in others serve, we think, to provide "objective evidence" that the self-control of private events discussed in *Gestalt Therapy* is in fact attainable. But what of self-awareness—can this too be given an experimental foundation? To answer this question, we turned again to muscular tension, first with a design modified from animal studies of response discrimination (Hefferline & Perera, 1963), and then with procedures used in the psychophysical laboratory to study sensory magnitude (Bruno, Hefferline, & Suslowitz, 1971).

In the initial study (Hefferline & Perera, 1963), "our specific objective was to train the subject, although he might remain otherwise unresponsive to an occasional minute twitch in his left thumb ... nevertheless to 'report' its occurrence ... by pressing a key with his right index finger [p. 834]". To do this, we presented a tone as quickly as possible after each thumb-twitch and asked the *S*, who was not told the specific objective, to press the key each time he heard the tone. Then, after several sessions with this procedure, we gradually reduced the intensity of the tone until it was in fact no longer presented.[3] The result was that the *S* complained that the tone was "getting hard to hear," but he continued to press the key! More exactly, 72% of the thumb-twitches were followed within 2 sec by a key-press, and 80% of the key-presses correctly reported or discriminated the prior occurrence of a thumb-twitch. Apparently, in fading out the tone, we had transferred discriminative control of the key-press from the tone to the thumb-twitch in much the same way that Terrace, in the animal laboratory, transfers control of key-pecking in pigeons from a colored stimulus to one of line orientation (1963).

Our results clearly established that a private event—the covert thumb-twitch—can serve as a discriminative "stimulus" in the same sense that response-produced cues are said to provide stimuli in studies of animal learning (cf. Berryman, Wagman & Keller, 1960; Notterman & Mintz, 1965; Rilling & McDiarmid, 1965). We say "in the same sense" because we had, in effect, succeeded in teaching the *S* a discrimination at what might be called the "animal level." It was clearly not a "conscious" discrimination: Although we would say that a *S* had learned to discriminate thumb-twitches, when questioned after the experiment, the *S*—who had never been told anything other than to respond to tones—explained that he continued to press the key because he still heard the tone.

> The apparent discrepancy between our account and the subject's is not, however, difficult to resolve when we recall that our instructions to the subject had explicitly equated, at least within the experimental context, the status of two responses—the overt key-press and the verbal report "I heard it." We had, in effect, told the subject to report with a key-press the same events which ordinarily he would report by saying "I heard the tone." If we assume, with Schoenfeld and Cumming (1963), that what is reported is the occurrence of a mediating event, and not the occurrence of the stimulus *per se,* then it makes sense to presume that the effect of our procedure was to enable the thumb-twitch to enter significantly into the control of those events originally evoked primarily by the tone, with the result that the twitch, via the mediating events, acquired discriminative control of *both* the key-press and the verbal report [Hefferline *et al.,* 1971].

In other words, we had, not unwittingly, demonstrated the discriminative properties of the covert thumb-twitch by conditioning an "hallucination". Of course, the drama is not necessary to the demonstration, provided one is willing to work at the "conscious" rather than at the "animal" level. Antrobus and Antrobus (1967), for example, have shown that *S*s can be taught to discriminate Stage 1 REM from Stage 2 sleep, while Kamiya (1969) demonstrated that the presence and absence of alpha rhythms in the electroencephalogram are

---

[3] The procedure for this study was actually somewhat more complicated. A full description and the rationale are given in detail elsewhere (Hefferline *et al.,* 1971).

perfectly discriminable. The wonder, in fact, is not that these private events are discriminable, but that so few others have ever been studied (cf. Hefferline, 1962).

Although basic work is still needed on a host of obvious variables—heart rate, stomach contractions, and respiration among them—our particular interest in tension patterns led us, in our second experiment on response discrimination, to investigate sustained muscular contractions. While contractions of this sort have been interpreted perceptually as a manifestation of set (Allport, 1955) and have been implicated clinically as a factor in functional disorders (Whatmore & Kohli, 1968), little is known of man's ability to report on them verbally. What is known, however, suggests that perceived tension grows as a power function of the magnitude of muscular contractions (cf. Hefferline *et al.,* 1971; J. C. Stevens & Mack, 1959). Accordingly, we chose a psychophysical method—magnitude estimation (cf. Stevens, 1956)—suitable for perceptual continua governed by the "power law" (cf. Stevens, 1961).

The method proved simple and direct. The S sat in a reclining chair with his right hand resting palm down on a board which held two upright bars. The tip of his little finger was placed between the bars, as shown in Fig. 1, and the S was asked to press his finger against the appropriate bar to produce adductive or

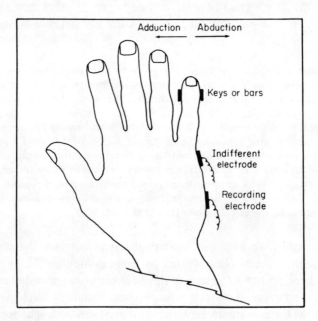

Fig. 1. Electrode placement on the hypothenar eminence and the position of the keys or bars used to define adductive and abductive tensions.

abductive tensions. An intermediate, "comfortable" tension was selected as the standard and it was assigned the value "10." On each trial the S was first required to reproduce the standard tension, and then, afer a 5-sec period of relaxation, was asked to produce, with verbal guidance, a preselected comparison level of tension. At this point, the S was told to relax and then to report a number which stood in proportion to "10" as the comparison tension stood in proportion to the standard tension.

The S's numerical estimates of the comparison tensions were recorded and compared to electromyographic measurements of the same tensions, quantified in decibels relative to the energy produced by a 200-Hz sine wave at 10 μV peak-to-peak amplitude. The latter measurements were made using surface electrodes, placed as in Fig. 1, to record the electrical activity from the muscles of the hand which adduct and abduct the little finger (cf. Bruno, Davidowitz, & Hefferline, 1970). Figure 2 shows, for an experienced S, the relationship

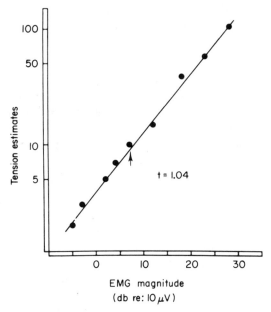

Fig. 2. Magnitude estimation of abductive tension. The function is for an individual observer giving one determination per point. The arrow marks the standard stimulus.

between perceived tension and the magnitude of abductive muscular contractions. The straight line drawn through the data points is the graph of a power function whose exponent, t, is approximately unity. Similar data were obtained from another S scaling abductive tensions and from two others who scaled adductive tensions, although naive Ss gave somewhat less orderly data spanning a narrower range of tensions.

Several aspects of the data deserve comment. Chief among these is the fact that these are individual data taken from Ss giving only one estimate for each tension level. These are not long-term, averaged data collected from a group of Ss, but rather, to put it loosely, these are the data of "the here and now." They are, nevertheless, reliable—both in the sense that the same S gave comparable results on different measurement occasions and in the sense that different Ss gave comparable data for the same muscle—and valid as determined by cross-modality matches (cf. Stevens, 1959) made between tension and tonal loudness (Bruno *et al.,* 1971).

The data are of particular interest in the present context because they demonstrate that tensions are perceived according to the same rule whether they be large or small, public or private. In fact, we have collected many tension functions from different observers, for different muscles, and under different procedures, but we have never found one that demonstrated a discontinuity between public and private tensions. For tension, at least, the discontinuity is not in the mind of the perceiver.

Before we turn to other matters, another word is in order about the lack of studies dealing with response discrimination. Surely the problem is not methodological: Even monkeys, who neither speak our language nor understand psychophysical procedures, can be taught to report on the visual motion (spiral) aftereffect, that is, on an "illusion" (Scott & Powell, 1963). Perhaps the lack of studies reflects a lack of interest. Here we will be the first to concede that a parade of studies demonstrating response discrimination of one private event and then another, *ad nauseam,* is not very interesting, and in any case, to paraphrase the retort of George Bernard Shaw's black girl to Pavlov after being treated to a demonstration of conditional reflexology, "we already know *that*" (Shaw, 1933). What we have in mind, however, are studies directed to problems somewhat meatier than the question of whether such-and-such a private event is discriminable. At the very least, we might ask how the perception of a private event relates to its frequency or magnitude. Or we might attempt to discover the extent to which state variables or payoff or accessory exteroceptive stimuli can influence the detection of a private event, that is, influence self-perception. Or, if the psychophysical aspects seem uninteresting, perhaps it would be worthwhile to explore conditioning approaches for new techniques to heighten self-awareness. Certainly, the possibilities are many and rich; clearly the problem is deserving of attention.

Response Bias

In the last two sections ("Response Control" and "Response Discrimination"), we dealt electropsychologically with the principles underlying the topics of self-control and self-awareness which had been examined phenomenologically in *Gestalt Therapy.* In this section, we intend to deal with a topic, response bias, whose phenomenological counterpart has been variously given as attitude, predisposition, postural organization, etc. (cf. Hefferline, 1962, p. 116ff.). The topic, which in part covers the material traditionally collated under the rubric of set, has both a long history—one which in fact is nearly as long as the history of experimental psychology itself (cf. Woodworth & Schlosberg, 1954)—and a sizeable literature, some of which was considered in an earlier paper (Hefferline, 1962). However, our present concern with the topic is not historical, but methodological and analytic; specifically, our concern is to describe first a new variety of research (Hefferline & Bruno, 1971) and then an extension of signal detection theory (Bruno & Hefferline, in preparation), both of which have important application to the study of response bias.

To see what's "new" about the new variety of research, let us look first at an example—a good example—of the old variety. Davis (1952), in one of several similar experiments, presented on each trial two tones of equal loudness, the second 5 sec after the first. He asked his S to close a telegraph key with the forefinger of one hand if the second tone seemed "stronger" than the first, but to close the key placed under the forefinger of the other hand if the second tone seemed "weaker". Throughout the experiment he recorded the electromyogram from electrodes placed over the extensor digitalis of each arm. He found that the proportion of "weaker" judgments, which varied in this and similar experiments from 30 to 70%, was a direct "function of the action potential difference in the two arms just prior to the second stimulus [p. 387]". He described these results as follows:

> It is proposed that the action potentials from the two arms represent the activity state of two competing response systems; that both are excited by the first stimulus but have differing response curves and initial levels; that, other things being equal, the system which has the advantage at the time of the second stimulus will further build up its activity until enough force is generated to close a key [p. 390].

The design was indeed most ingenious: Since the two stimuli on each trial were subjectively identical, the particular outcome of a trial could only reflect the influence of "setting or biasing" factors. While ingenious, the design was nevertheless of the old or traditional variety in the sense that the private, physiological event (prestimulus tension) was, like the probability of "weaker" judgments, a dependent variable. To be sure, the design exemplifies precisely the traditional experiment in psychophysiology described by Sternbach:

> Mental or emotional behavioral activities are made to occur while physiological events are being observed; correlations between these activities and the observed physiological events are noted, and then some intervening internal event is postulated [1966, p. 2].

Of course, there is nothing intrinsically "wrong" with the traditional variety of research. On the contrary, it has brought to light many interesting correlations between private events and public behavior—witness the many correlations between reaction time and events such as heart rate, tension, alpha rhythm, respiration—and it has generated many interesting hypotheses concerning the nature of set or response bias (cf. Allport, 1955). But what of these hypotheses? How do we set about determining their validity? We might, as Davis (1952) did, replicate the originating experiment directly, by repeating it, or systematically (Sidman, 1960), by varying "nonessential" details such as the modality of the stimulus or the nature of the response. However, replication of this sort, while it may serve to support the original hypothesis, or to generate new ones if the replication "fails," cannot provide the grounds necessary to accept or reject the hypothesis. What is needed is another form of systematic replication which can provide a convergent operation with respect to the hypotheses elaborated by traditional means (cf. Garner, Hake, & Eriksen, 1956; Stoyva & Kamiya, 1968). This, in fact, is what the new variety of research does.

The new variety is distinguished from the old in that it, unlike the old, manipulates the private, physiological event as an independent variable. For example, if we were to repeat Davis' experiment using the new variety of research, we would manipulate or vary prestimulus tension rather than simply measuring it. Of course, we would compute the same set of correlations between prestimulus tension and response probability, and we would compare these to Davis' results, but now our position would be considerably improved. Since the traditional experiment and the new variety form converging operations—they are neither identical nor orthogonal experiments, but together they converge to permit us to accept or reject a particular experimental hypothesis—if Davis' results and ours differ we can reject his notion, if they are the same we can accept the notion that "the system which has advantage at the time of the second stimulus will further build up its activity until enough force is generated to close a key."

The power for this new variety of research, which of course is not really "new" (cf. Courts, 1942; Meyer, 1953), derives from our new-found ability to control private events other than skeleto-muscular via exteroceptive feedback loops (see "Response Control" and "Bio-Feedback: Controlling the 'Uncontrollable' "). While in the past it was possible to control, and therefore manipulate as a variable, only large-scale, public events (force of handgrip, for example), it is now possible, with feedback, to manipulate events as private and subtle as cortical activity (cf. Rosenfeld *et al.*, 1969; Spilker, Kamiya, Callaway, & Yeager, 1969).

To illustrate the new variety of research, we have performed a series of studies in which covert prestimulus tension was manipulated during reaction-time, absolute-threshold, and stimulus-generalization paradigms (Hefferline & Bruno, 1971). Since all of the studies used comparable procedures, only the generalization study will be described here. This study was carried out in three phases: feedback training, discrimination training, and generalization testing.

The object of the first phase was to teach the S to control, with feedback, covert activity in the muscles of the hand which adduct and abduct the little finger. To do this, we placed the finger between the response bars, as illustrated in Fig. 1, and recorded the electromyogram from surface electrodes placed on the hypothenar eminence. Since the electromyogram taken from this placement contains, during all degrees of contraction, short waveforms during adductor activity and long waveforms during abductor activity (Bruno et al., 1970), it was possible, by displaying the waveforms to the S on a triggered oscilloscope, to train him to produce adduction and abduction at covert levels. The training procedure, which was similar to that used by Basmajian (1963; 1967, p. 103ff.), has been described in detail elsewhere (Bruno et al., 1971; Hefferline & Bruno, 1971). The result of the training was that the S learned to produce covert adduction (50 μV peak-to-peak short waveforms) when a red pilot lamp was lit, and covert abduction (200 μV long waveforms) when a green lamp was lit.

In the second phase, discrimination training, each trial began with the presentation of the red or green light requiring covert adduction or abduction, respectively. When the tension appropriate to the light had been sustained for about .5 sec, one of two tones was presented via earphones. The S was instructed to close the adductive key (which replaced the bar in that position—see Fig. 1) whenever he heard the loud (104-dB) tone, and to close the abductive key when he heard the soft (86-dB) tone. This procedure, which was carried out for 200 trials, was adapted from one previously used by Cross and Lane (1962), as was the procedure of the next phase—generalization testing.

Generalization testing, which was carried out for 704 trials, was identical to discrimination training except that the tone intensity on each trial was chosen at random from a set of 11 intensities ranging from 80 dB to 110 dB in 3-dB steps. The S was asked to close the adductive key when the test tone resembled the loud training tone, and to close the abductive key when it resembled the soft tone.

Figure 3 shows, for an individual S, how the probability that a test tone was categorized as "soft"—i.e., the probability of an abductive key closure—varied with tone intensity and with the pretone covert tension. (The probability that a tone was called "loud" is the complement of the probability graphed.) In general, the probability of "soft" judgments was high in the region of the training stimulus for that response, decreased systematically over intermediate intensities, and was low in the region of the training stimulus for the opposite response. However, "soft" judgments were consistently more probable when the test tone was presented after appropriate rather than inappropriate covert tensions; that is, the S was more likely to call a tone "soft" if he was already "saying soft" rather than "loud" covertly.

The experiment, in effect, constitutes an eleven-fold systematic replication of Davis' 1952 study. For each of the 11 test stimuli, we have purposely given advantage on half the trials to the "soft" response system and given advantage to the "loud" system on the remainder of the trials. Like Davis, we found "that, other things being equal, the system which has the advantage ... will further build up its activity ... to close a key." Except that "other things," as usual, are not "equal," so that the magnitude of the effect appears to vary with the intensity of the test tone—although the prestimulus tensions were fixed in

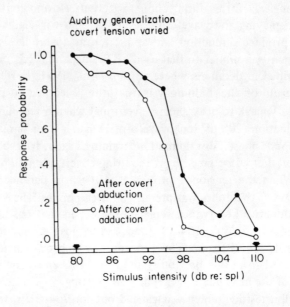
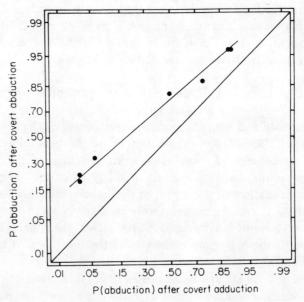

Fig. 3. Upper panel: probability of an abductive key closure to generalization stimuli presented during covert adduction or abduction, after training to respond adductively to 104-dB and abductively to 86-dB tones. Lower panel: iso-bias contour for the same data obtained by plotting, in double probability coordinates, the upper psychometric function against the lower.

magnitude and did not vary from tone to tone. Actually, the effect of the tensions was "constant" across tone intensity, as we might intuitively expect, but to show this we must detour briefly through a signal detection analysis of the experiment.[4]

Figure 4 depicts, in much simplified and idealized form, the hypothetical mechanisms operating in the stimulus-generalization experiment. The strong assumption is made that the physical dimension of stimulus (tone) intensity has a monotonically related correlate in psychological space. This is shown as the abscissa in Fig. 4. Along the abscissa are plotted 11 Gaussian density functions of equal variance. Since these represent the perceptual effects of the 11 test stimuli, the means of the distributions, which, quite unrealistically, are shown as

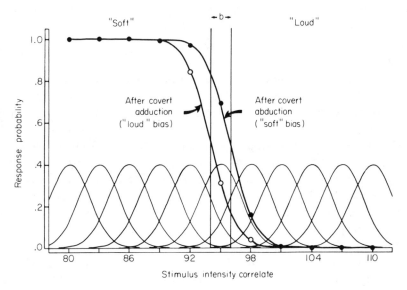

Fig. 4. Diagram of the hypothetical mechanisms correspondent to an 11-stimulus generalization experiment conducted under the method of constant stimuli. The superimposed ogives represent idealized psychometric functions plotted in psychological space.

equi-distant in psychological space, are labeled with the intensity (in decibels) of the corresponding physical stimulus. Probability density functions rather than single values along the abscissa have been used to represent the perceptual effects of the stimuli because, for organismic (cf. Corso, 1967, p. 410ff.) or for physical reasons (cf. Hecht, Schlaer, & Pirenne, 1942), the effect of a stimulus in fact varies from presentation to presentation. Consequently, different physical stimuli can give rise to the same perceptual effect, a circumstance portrayed by showing the density functions as overlapping distributions.

[4] A lucid account of signal detection concepts may be found in Green and Swets (1966).

Because different stimuli may generate the same effect, the observer asked to categorize the test tones as either "loud" or "soft" must adopt a decision rule which will allow him to operate despite the fact that he cannot "recognize" the tones by their effects. We assume that he establishes a decision point or criterion—two such points are indicated by the vertical lines in Fig. 4—and presses the abductive key to say "soft" when he observes a perceptual effect to the left of the criterion, but "says loud" to effects to the right of the criterion. Under this rule, the probability of a "soft" judgment is given for each test stimulus by the area under its corresponding density function which falls to the left of the criterion. As shown by the ogival curves in Fig. 4, when the probability of a "soft" judgment is plotted for each stimulus above the mean of its corresponding distribution, the result, for each location of the criterion, is the usual psychometric function spread out along an axis of psychological rather than physical distance (cf. Boneau & Cole, 1967).

In Fig. 4, the idealized psychometric function labeled "after covert abduction" is the one which would be generated if the S adopted the right-hand criterion; the function labeled "after covert adduction" would come from adopting the left-hand criterion. Notice that these idealized functions are not unlike the actual functions displayed in the upper panel of Fig. 3. In fact, we assume that a S required to sustain a covert abductive tension, that is to covertly say "soft," would adopt the liberal right-hand criterion which, because more area lies to the left of it, makes "soft" judgments (abductive key closures) consistently more probable than they would be if the stricter left-hand criterion were adopted. Similarly, we assume that requiring covert adduction—saying "loud" covertly—leads to the adoption of the left-hand criterion, making "soft" judgments less probable. In short, we assume that manipulating covert prestimulus tension in the muscles about to make an overt perceptual response is equivalent, in our model, to manipulating the S's decision point or criterion.

In a sense, we have come full circle: The idealized functions in Fig. 4, like the actual functions in Fig. 3, seem to show a biasing effect whose magnitude varies with stimulus intensity. But our model, confirming our intuitions, suggests that the effect is in fact "constant" when its magnitude is judged not from the separation of the psychometric functions, but rather from the distance, "b," which separates the two criteria. What is needed now is a way to estimate this distance from actual data.

Our iso-bias contour provides the way (Bruno & Hefferline, in preparation). The lower panel in Fig. 3 shows the iso-bias contour which results when the psychometric function taken "after covert abduction" is plotted in double-probability coordinates (which "straighten out" Gaussian ogives) against the function taken "after covert adduction." Had there been no difference between the functions, the data points would lie along the major diagonal. The functions, however, were different, and what distinguished them was the psychological distance between the two decision points or criteria which were adopted to generate them. A measure of this distance, in standardized units, is reflected by calculating b', which is the difference between the two criteria divided by the standard deviation common to each of the distributions of perceptual effects. The calculation, which in this case gives a value of about 1.0 standard deviation units, is described in detail elsewhere (Bruno & Hefferline, in preparation). It is sufficient here to say that it is comparable to the d' calculation for the common iso-sensitivity contour (cf. Green & Swets, 1966, p. 58ff.), and that it, like d', depends on whether or not the equal variance assumption holds.[5] It is sufficient because the value of b' is reflected graphically by the degree to which the iso-bias contour is displaced away from the major diagonal toward the upper left-hand corner of the figure: The greater the displacement the greater is b'.

As its use here suggests, the iso-bias contour, so-called because it links points of equal bias or criterion difference, applies whenever data are taken under the method of constant stimuli (Woodworth & Schlosberg, 1954, p. 201) using operations which cause the S to adopt two or more distinct criteria (n contours result for $n+1$ criteria). For example, we have applied the contour successfully to absolute threshold data taken with visual stimuli under the rating method (Barlow, 1956), to visual increment threshold data taken when payoff was varied (Swets, Tanner, & Birdsall, 1961), to pain and thermal thresholds taken under the rating method to examine the placebo effect (Clark, 1969), and to wavelength generalization data collected from pigeons on test trials given before and after reinforcement (Boneau, Holland, & Baker, 1965). In each case, when the psychometric function presumably taken under a liberal criterion was plotted, in double-probability coordinates, against the function presumably taken under a strict criterion, the result was a straight line—that is, an iso-bias contour—which told us immediately that the two psychometric functions were distinguished by a difference in response bias (criterion) and not a difference in sensory capacity. Here, in distinguishing bias from capacity, we see the power of the iso-bias contour (and of signal detection theory itself): It renders explicit variables which in classical psychophysics were implicit or ignored. Of course it, not incidentally, also provides the means for quantifying one of those variables, viz., response bias.

The implications of the iso-bias contour for the study of private events are, perhaps, obvious. Since it provides a way to identify changes in response bias, it allows us to hunt out those private events which influence the S's readiness or unreadiness to respond. Then, having found a biasing event, if we manipulate it—as in the new variety of research described previously—the contour provides the means to quantify the effect of our "private" independent variable.

In this part of our presentation, dealing with the electropsychological analysis of private events, we have tried to sketch in some of the methods and some of

---

[5] Where the assumption holds, the iso-bias contour has a slope of unity. Where it does not, but variance increases with intensity, as in the present data, the slope is less than unity.

the data which provide the experimental underpinnings for an all-out study of the "universe ... enclosed within the organism's own skin [Skinner, 1953, p. 257]." In the next section, we take up in detail one of the experimental methods which seems particularly promising for the future of this private part of the universe.

Bio-Feedback: Controlling the "Uncontrollable"

In some of the studies described in previous sections, we shared with the S our "sight" or "sound" of his muscle-action potentials as electronically transduced into an oscilloscopic display or an auditory output from a speaker system. We discovered, as have many others more recently, that such an opportunity fascinates the S and that it also constitutes a powerful tool for self-training and self-control. Until recently it has been called "augmented feedback," since it supplies an external feedback loop which augments the internal loop via proprio- or interoceptors; but this term, which is often inappropriate, as when applied to signals of cortical origin, will now presumably give place to "bio-feedback," since the many workers presently joining the field are organizing under the name of the Bio-Feedback Research Society.

What has spurred this sudden rallying around a single experimental stratagem is, of course, the fact that it has effectively exploded the doctrine, widely held for so long, that autonomic activities are inaccessible to instrumental training procedures (see "Response Control"). At the present time, however, not only have they proved to be trainable in instrumental fashion, but, from the humanistic viewpoint, a host of activities previously deemed to be necessarily involuntary have come within the jurisdiction of voluntary control. A broad new vista spreads before us, with both foreseeable and unforeseeable opportunities and responsibilities. For better or worse, the internal environment has been thrown open to "civilizing" influences.

For example, "externalization" by means of bio-feedback of otherwise private events makes it possible for the first time to label and reinforce such events precisely by bringing the E and the S into common possession of the key properties of the behavior. Gardner Murphy wrote, afer a visit to our laboratory during which he learned to control three different muscles via feedback of their surface-recorded waveforms (cf. Bruno *et al.*, 1970):

> [Hefferline] and his collaborator attached electrodes to my left little finger and left ear lobe and asked me to watch an oscilloscope as I carried out small directed movements with my left little finger. Whenever I extended my finger to the left, this produced, in the midst of a shower of "noise," a well-defined peak on the oscilloscope which, as I practiced, became more and more clear—the Matterhorn, as I called it. A countermovement, flexing rather than extending the finger, produced another pattern, which I called the Jungfrau. A third task—and a very difficult one—was to push the little finger straight out as if it slipped in a socket, producing a flat table top which I called a mesa. Soon I was producing at will the Matterhorn, Jungfrau, and mesa patterns [Murphy, 1964, p. 105].

With a little more time to spend "shaping up" his own behavior by means of the visual feedback, Murphy would have learned, as do our regular Ss, to get along without the external "crutch." He would have become able upon his own command or that of the E to produce any of the labeled responses with the same clarity and sharpness of definition that he had achieved through practice with the scope.

Our finding that bio-feedback, once it has served its purpose of permitting a selected response or sequence of responses to be sharply discriminated, and perhaps "named," can usually be "faded out" (cf. Hefferline et al., 1971), has been confirmed by Basmajian (1963, 1967), who has worked intensively on training Ss to control individual neuromuscular motor units. Basmajian states that his Ss "were provided with two modalities of 'proprioception' that they normally lack; namely, they heard their motor unit potentials and saw them on monitors [1963, p. 440]." Within 15-30 min of practice most had learned to relax the whole muscle upon command and then to recruit the activity of a single motor unit, keeping it active for as many minutes as desired. Activity at first consisted of a steady train of pulses; later, the S learned to "turn off" the unit or even to produce single firings upon request. Similar control of other units might be acquired, some Ss being able to recruit up to a fourth or fifth isolated unit. After an hour or more, a S could become so skilled, Basmajian reports, that he could produce "various gallop rhythms, drum-beat rhythms, doublets, and roll effects [p. 441]," and then might even become able to perform such tricks in the absence of the feedback necessary to establish control in the first place!

While the trained S giving "gallops" and "doublets" makes it all look easy, the voluntary singling out of an individual motor unit from the sizeable array of those that are ordinarily active in a muscle "at rest," is, presumably, not so easy. Phillips, at Oxford, comments as follows:

... The refined degree of minute localization that is evidently possible in voluntary control ... must require a marvelously subtle routing of activity in the outer cortical layers to pick up, in significant functional groupings, the required corticofugal neurons, which are scattered and intermingled with unwanted ones which may be suppressed [1966, p. 402].

It is paradoxical that what might seem to be the simplest possible response, namely, the firing of a single motor unit, should require for its production the highly selective facilitatory and inhibitory effects postulated by Phillips. Sperry in 1955 supposed that something of this sort was involved as "the neural basis of the conditioned response," and here he was speaking of something that might be as simple (?) as a particular leg movement. He argued for elaborate patterns of central nervous facilitation and/or inhibition, which he said were the neural counterparts of and derivatives from psychological expectancies and anticipatory sets. In 1958 he related the "functional settings of the brain" more intimately to the experimental psychologist's preoccupation with skeletal behavior in the following passage:

... Perhaps it would be more accurate and fruitful to picture the differentiation and interrelations of facilitory sets, not in terms of tree limbs and branches, but directly in terms of the potential postures and movements of the vertebrate body. The "postural sets" then have direct implication for the implicit operations of perception and thought processes, depending on the closeness or remoteness of these latter to motor adjustment [Sperry, 1958, pp. 416-417].

Today a certain expertise in bringing about subtle manipulations of the "mind-body relation" has arisen among coaches, physical therapists, psychotherapists with a somatic approach, relaxationists, and various others. We believe that much useful order could be brought to this area—and to the whole of psychology—through the systematic study of central-peripheral interactions now discernible through the use of sophisticated instrumentation for multichannel recording of patterns of bodily activity.

An ambitious but most circumspect step in this direction has been taken by George B. Whatmore, of Seattle (Whatmore & Kohli, 1968). His medical practice consists of the diagnosis and treatment of what he calls "dysponesis"—defined as misdirected effort. Daily life, he says, is made up of performing, bracing, representing and attending efforts. Performing covers the vast range of movements outwardly observable. Representing, "a form of self-signaling," is made up of the covert behaviors which produce imagery (not restricted to the visual modality). Attending is the "effort" which provides impulses from various sense organs with differential access to the nervous system, in large degree by gross or subtle eye efforts.

Bracing efforts we have met before as "holding responses," conflict, and unresolved tensions. Whatmore himself says that by bracing efforts "we hold the body, or a part of the body, rigid or 'on guard'." He states further that "often one effort is pitted against another, one muscle contracted against its antagonist", and that "these efforts can be made in any or all parts of the body such as the extremities, the breathing musculature, the neck, the back, the jaw, the tongue, and the throat [1968, p. 103]."

Whatmore quantifies dysponesis by electromyometry, allows his patient to hear his muscular effort by electromyophony and to see it by electromyoscopy. His published case studies report good results for a wide range of troublesome conditions. He avoids psychopathological terminology and describes his procedures as simply "a form of neurophysiologic engineering wherein basic principles of neurophysiology are used to carry on a retraining within the nervous system [1968, p. 102]."

In a personal communication, Whatmore (1969) reports that norms for his work were established "by measuring from persons 'relatively free of functional disturbances' and from patients and control subjects after effort training." Standard electrode positions are on the forehead, jaw, forearm, and leg. Special positions pertinent to the given patient may be on the neck, back, or other region.

It appears from Whatmore's papers that dysponesis is most frequently hyperponesis, that is, excessive muscular tension. The therapeutic goal, as in Jacobson's earlier method of progressive relaxation (1938), is to move in the direction of lowered tension, although Whatmore indicates that "some patients need more training in various forms of 'going on with the power' than they do in 'going off with the power' [1968, p. 117]." How to proceed with a given case is based, he states, as far as possible "upon laboratory and clinical evidence and not upon philosophic concepts [p. 116]." Psychoanalysis and other forms of psychotherapy "with treatment limited to interview techniques," he regards as having no correcting effect upon dysponesis. He does add, however:

> Psychotherapy can be used, when there are indications for it, along with definitive treatment of dysponesis, but it is a separate procedure that attacks [sic] a different aspect of the individual and does not alter dysponetic tendencies [p. 115].

Despite his sharp dichotomizing of the results achievable by his physiologic approach and those obtainable by "treatment limited to interview techniques", Whatmore does not take the militant stand against psychodynamics that characterizes such behavioral therapists as Eysenck (1952) or Wolpe (1962). The statement quoted above would seem to acknowledge not only a basis for co-existence, but even, on occasion, for co-function. He looks forward, of course, to the further development of his own method and asserts that "the task for the future is one of developing still better training methods of neurophysiologic engineering wherein instrumentation and instrumented learning will undoubtedly play an increasing role [1968, p. 119]."

We, too, share wholeheartedly this interest in improved "neurophysiologic engineering." One interesting possibility in this direction involves what Walter (1964) called the "contingent negative variation"—a phenomenon better described as the "dc shift," since it consists of a shifting potential difference between, typically, an electrode at the vertex and a second one at an inactive site such as the ear lobe. The dc shift has been shown to be indicative of cerebral processes and to be controlled by a wide variety of both psychological and state variables. A parametric study is under design in our laboratory in which hopefully the dc shift will prove to be another of the private events trainable by bio-feedback and possibly useful as an index of central-peripheral interactions otherwise accessible only in a cumbersome way by multichannel recording at the periphery (Camp, 1970).

Unlike Whatmore, we see it as only a temporary state of affairs that he can talk as plausibly as he does of the independent use of "psychotherapy" and "neurophysiologic engineering". There is, to be sure, the large contingent of behavioral therapists relying heavily on the process of "desensitizing" the patient, and effecting "symptom removal" by what essentially consists of arranging a situation in which "undesirable" conditioned emotional reactions may be extinguished, or suppressed, perhaps, by conditioning stronger incompatible behavior. It can hardly be questioned that, at this level, the method is "successful" (Bandura, 1969).

On the other hand, Freudian psychoanalysis and the "depth therapies" derivative from it all take the position that the symptom is exactly that—a trait or condition symptomatic of the more comprehensive systemic malfunction—with the consequence that personality alterations of any profundity are premised upon conceptually elaborate strategies for resolving conflict. As Whatmore has said, the treatment may indeed be limited to interview techniques. This does not, however, mean per se that the effects are merely verbal ones. On the contrary, strong emotion is generated in the therapeutic situation, with consequences, inevitably, for the neurophysiologic systems with which Whatmore deals. The trouble is that one has to guess at what is going on from moment to moment in the analytic session and be continually uncertain about how and when to intervene for optimal results.

Some psychoanalysts, however, concern themselves in various degree with the physiological version of the psychological problem (Braatøy, 1954; Christiansen, 1963; Fagan & Shepherd, 1970; Lowen, 1958; Shatan, 1963). This may involve no more than encouraging the patient to experience vividly the functioning of his body; or it may include direct physical manipulations by the doctor. This combined approach has its share of partisan enthusiasts, but its superiority, if it is superior, has not been clearly established. Nevertheless, if it avails itself of bio-feedback techniques, it seems likely that its therapeutic power will increase by several orders of magnitude.[6]

Finally, in modern group therapy there is implicit a kind of crude psychophysiology of private events. Leaders of sensitivity and encounter groups set high value on various kinds of physical contact as "sensory awakeners" and "releasers." The literature to date is semipopular, but instructive (Gustaitis, 1969; Howard, 1970; Shepard & Lee, 1970; Schutz, 1967). Bio-feedback applied here could supply a new dimension. Perhaps we may close with a quotation from Mulholland which, although it comes from a most sober paper, has elements of blithe fantasy (Mulholland, 1968):

> Studies of interpersonal and group processes might achieve a new dimension using feedback methods. If the physiological response for one person were information to another, and his physiological response were information to the first, the behavior of both might be modified. For instance, if heart beat were indicated by a flashing light worn, say, on the lapel, and a man and woman were introduced, the man's light might flash more quickly. The woman, perceiving this, might be slightly embarrassed, and her light would begin flashing more quickly. The man would perceive this, and his light would flash more, etc., etc. Soon both lights may be flashing at a fast rate. I would hope that both would laugh when this occurred, restoring equilibrium [p. 436]!

---

[6] A dramatic type of sensory feedback, made available by television technology, is now employed in many psychiatric hospitals to provide videotape self-confrontation. A patient may see and hear himself in action when he has played back for him a tape made earlier. By means of an additional closed circuit his present appearance may be shown at the same time as an insert in a corner of the monitor; in effect, he may watch himself watching himself (Berger, 1970).

References

Allport, F. H. *Theories of perception and the concept of structure.* New York: Wiley, 1955.
Antrobus, J. S., & Antrobus, J. S. Discrimination of two sleep stages by human subjects. *Psychophysiology,* 1967, **4,** 48-55.
Ax, A. F. Goals and methods of psychophysiology. *Psychophysiology,* 1964, **1,** 8-25.
Bandura, A. *Principles of behavior modification.* New York: Holt, 1969.
Barlow, H. B. Retinal noise and absolute threshold. *Journal of the Optical Society of America,* 1956, **46,** 634-639.
Basmajian, J. V. Control and training of individual motor units. *Science,* 1963, **141,** 440-441.
Basmajian, J. V. *Muscles alive: Their functions revealed by electromyography.* Baltimore: Williams & Wilkins, 1967.
Bekhterev, V. M. *Objective psychologie.* Leipzig and Berlin: Teubner, 1913.
Berger, M. M. (Ed.). *Videotape techniques in psychiatric training and treatment.* New York: Brunner/Mazel, 1970.
Berryman, R., Wagman, W., & Keller, F. S. Chlorpromazine and the discrimination of response-produced cues. In L. Uhr & J. G. Miller (Eds), *Drugs and behavior.* New York: Wiley, 1960. Pp. 243-249.
Boneau, C. A., & Cole, J. L. Decision theory, the pigeon, and the psychophysical function. *Psychological Review,* 1967, **74,** 123-135.
Boneau, C. A., Holland, M. K., & Baker, W. M. Color-discrimination performance of pigeons: Effect of reward. *Science,* 1965, **149,** 1113-1114.
Braatøy. T. *Fundamentals of psychoanalytic technique.* New York: Wiley, 1954.
Brener, J., Kleinman, R. A., & Goesling, W. J. The effects of different exposures to augmented sensory feedback on the control of heart rate. *Psychophysiology,* 1969, **5,** 510-516.
Brown, C. C., & Katz, R. A. Operant salivary conditioning in man. *Psychophysiology,* 1967, **4,** 156-160.
Bruno, L. J. J., Davidowitz, J., & Hefferline, R. F. EMG waveform duration: A validation method for the surface electromyogram. *Behavior Research Methods and Instrumentation,* 1970, **2,** 211-219.
Bruno, L. J. J., & Hefferline, R. F. An iso-bias contour for the method of constant stimuli. In preparation.
Bruno, L. J. J., Hefferline, R. F., & Suslowitz, P. D. Cross-modality matching of muscular tension to loudness. *Perception and Psychophysics.* 1971. In press.
Camp, J. A. Some factors controlling cortical dc shifts in man: A review. Electropsychology Report MH-13890-10, Columbia Univ., 1970.
Christiansen, B. Thus speaks the body. Oslo: Institute for Social Research, 1963.
Clark, W. C. Sensory-decision theory analysis of the placebo effect on the criterion for pain and thermal sensitivity (d'). *Journal of Abnormal Psychology,* 1969, **74,** 363-371.
Corso, J. F. *The experimental psychology of sensory behavior.* New York: Wiley, 1967.
Courts, F. A. Relations between muscular tension and performance. *Psychological Bulletin,* 1942, **39,** 347-367.
Cross, D. V., & Lane, H. L. On the discriminative control of concurrent responses: The relations among response frequency, latency, and topography in auditory generalization. *Journal of the Experimental Analysis of Behavior,* 1962, **5,** 487-496.
Davis, R. C. The stimulus trace in effectors and its relation to judgment responses. *Journal of Experimental Psychology,* 1952, **44,** 377-390.
Delse, F. C., & Feather, B. W. The effect of augmented sensory feedback on the control of salivation. *Psychophysiology,* 1968, **5,** 15-22.

Dollard, J., & Miller, N. E. *Personality and psychotherapy.* New York: McGraw-Hill, 1950.
Eldridge, L. Respiration rate change and its relation to avoidance behavior. Unpublished doctoral dissertation, Columbia Univ., 1954.
Engel, B. T., & Chism, R. A. Operant conditioning of heart rate speeding. *Psychophysiology,* 1967, **3,** 418-426.
Engel, B. T., & Hansen, S. P. Operant conditioning of heart rate slowing. *Psychophysiology,* 1966, **3,** 176-187.
Estes, W. K. An experimental study of punishment. *Psychological Monographs,* 1944, **57,** (3, Whole No. 263).
Eysenck, H. J. The effects of psychotherapy: An evaluation. *Journal of Consulting Psychology,* 1952, **16,** 319-324.
Fagan, J., & Shepherd, I. L. (Eds.). *Gestalt therapy now. Theory, techniques, applications.* Palo Alto: Science and Behavior Books, 1970.
Fetz, E. E. Operant conditioning of cortical unit activity. *Science,* 1969, **163,** 955-958.
Garner, W. R., Hake, H. W., & Eriksen, C. W. Operationism and the concept of perception. *Psychological Review,* 1956, **63,** 149-159.
Green, D. M., & Swets, J. A. *Signal detection theory and psychophysics.* New York: Wiley, 1966.
Gustaitis, R. *Turning on.* Toronto: Macmillan, 1969.
Hecht, S., Schlaer, S., & Pirenne, M. H. Energy, quanta, and vision. *Journal of General Physiology,* 1942, **25,** 819-840.
Hefferline, R. F. An experimental study of avoidance. *Genetic Psychology Monographs,* 1950, **42,** 231-334.
Hefferline, R. F. The role of proprioception in the control of behavior. *Transactions of the New York Academy of Sciences,* 1958, **20,** 739-764.
Hefferline, R. F. Learning theory and clinical psychology—An eventual symbiosis. In A. J. Bachrach (Ed.), *Experimental foundations of clinical psychology.* New York: Basic Books, 1962. Pp. 97-138.
Hefferline, R. F., & Bruno, L. J. J. The physiological event as an independent variable. *Psychophysiology,* 1971. In press.
Hefferline, R. F., Bruno, L. J. J., & Davidowitz, J. Feedback control of covert behavior. In K. J. Connolly (Ed.), *Mechanisms of motor skill development.* New York: Academic Press, 1971. In press.
Hefferline, R. F., & Keenan, B. Amplitude-induction gradient of a small human operant in an escape-avoidance situation. *Journal of the Experimental Analysis of Behavior,* 1961, **4,** 41-43.
Hefferline, R. F., & Keenan, B. Amplitude-induction gradient of a small-scale (covert) operant. *Journal of the Experimental Analysis of Behavior,* 1963, **6,** 307-315.
Hefferline, R. F., Keenan, B., & Harford, R. A. Escape and avoidance conditioning in human subjects without their observation of the response. *Science,* 1959, **130,** 1338-1339.
Hefferline, R. F., & Perera, T. B. Proprioceptive discrimination of a covert operant without its observation by the subject. *Science,* 1963, **139,** 834-835.
Herrnstein, R. J. Method and theory in the study of avoidance. *Psychological Review,* 1969, **76,** 49-69.
Howard, J. *Please touch: A guided tour of the human potential movement.* New York: McGraw-Hill, 1970.
Jacobson, E. *Progressive relaxation.* Chicago: Univ. of Chicago Press, 1938.
Kabat, H. Central mechanisms for recovery of neuromuscular function. *Science,* 1950, **112,** 23-24.
Kamiya, J. Operant control of the EEG alpha rhythm and some of its reported effects on consciousness. In C. T. Tart (Ed.), *Altered states of consciousness.* New York: Wiley, 1969. Pp. 507-517.

Katkin, E. S., & Murray, E. N. Instrumental conditioning of autonomically mediated behavior: Theoretical and methodological issues. *Psychological Bulletin,* 1968, **70,** 52-68.
Kimble, G. A. *Hilgard and Marquis' conditioning and learning.* 2nd ed. New York: Appleton, 1961.
Lindsley, D. B. Emotion. In S. S. Stevens (Ed.), *Handbook of experimental psychology.* New York: Wiley, 1951. Pp. 473-516.
Lowen, A. *Physical dynamics of character structure: Bodily form and movement in analytic therapy.* New York: Grune & Stratton, 1958.
Meyer, D. R. On the interaction of simultaneous responses. *Psychological Bulletin,* 1953, **50,** 204-220.
Miller, N. E. Learning of visceral and glandular responses. *Science,* 1969, **163,** 434-445.
Mulholland, T. Feedback electroencephalography. *Activitas Nervosa Superior,* 1968, **10,** 410-438.
Murphy, G. Communication and mental health. *Psychiatry,* 1964, **27,** 100-106.
Notterman, J. M., & Mintz, D. E. *Dynamics of response.* New York: Wiley, 1965.
Perls, F. S., Hefferline, R. F., & Goodman, P. *Gestalt therapy: Excitement and growth in the human personality.* New York: Julian Press, 1951.
Phillips, C. G. Precentral motor area. In J. C. Eccles (Ed.), *Brain and conscious experience.* New York: Springer-Verlag, 1966. Pp. 389-421.
Rescorla, R. A., & Solomon, R. L. Two process learning theory: Relationships between pavlovian conditioning and instrumental learning. *Psychological Review,* 1967, **74,** 151-182.
Rilling, M., & McDiarmid, C. Signal detection in fixed-ratio schedules. *Science,* 1965, **148,** 526-527.
Rosenfeld, J. P., Rudell, A. P., & Fox, S. S. Operant control of neural events in humans. *Science,* 1969, **165,** 821-823.
Roszak, T. *The making of a counter culture.* New York: Doubleday, 1968.
Sasmor, R. M. Operant conditioning of a small-scale muscle response. *Journal of the Experimental Analysis of Behavior,* 1966, **9,** 69-85.
Schoenfeld, W. N. Some old work for modern conditioning theory. *Conditional Reflex,* 1966, **1,** 219-223.
Schoenfeld, W. N., & Cumming, W. W. Behavior and perception. In S. Koch (Ed.), *Psychology: A study of a science.* Vol. 5. New York: McGraw-Hill, 1963. Pp. 213-252.
Schutz, W. C. *Joy: Expanding human awareness.* New York: Grove Press, 1967.
Scott, T. R., & Powell, D. A. Measurement of a visual motion aftereffect in the Rhesus monkey. *Science,* 1963, **140,** 57-59.
Shapiro, D., & Crider, A. Operant electrodermal conditioning under multiple schedules of reinforcement. *Psychophysiology,* 1967, **4,** 168-175.
Shapiro, D., Tursky, B., Gershon, E., & Stern, M. The effects of feedback and reinforcement on the control of human systolic blood pressure. *Science,* 1969, **163,** 588-590.
Shaw, G. B. *The adventures of the black girl in her search for God.* New York: Dodd, Mead, 1933.
Spilker, B., Kamiya, J., Callaway, E., & Yeager, C. L. Visual evoked responses in subjects trained to control alpha rhythms. *Psychophysiology,* 1969, **5,** 683-695.
Shatan, C. Unconscious motor behavior, kinesthetic awareness and psychotherapy. *American Journal of Psychotherapy,* 1963, **17,** 17-30.
Shepard, M., & Lee, M. *Marathon 16.* New York: Putnam, 1970.
Sidman, M. *Tactics of scientific research.* New York: Basic Books, 1960.
Skinner, B. F. *Science and human behavior.* New York: Macmillan, 1953.
Smith, K. Conditioning as an artifact. *Psychological Review,* 1954, **61,** 217-225.

Snyder, C., & Noble, M. Operant conditioning of vasoconstriction. *Journal of Experimental Psychology*, 1968, **77**, 263-268.

Sperry, R. W. On the neural basis of the conditioned response. *British Journal of Animal Behavior*, 1955, **3**, 41-44.

Sperry, R. W. Physiological plasticity and brain circuit theory. In H. F. Harlow & C. N. Woolsey (Eds), *Biological and biochemical bases of behavior*. Madison: Univ. of Wisconsin Press, 1958.

Stern, R. M., & Kaplan, B. E. Galvanic skin response: Voluntary control and externalization. *Journal of Psychosomatic Research*, 1967, **10**, 349-353.

Sternbach, R. A. *Principles of psychophysiology*. New York: Academic Press, 1966.

Stevens, J. C., & Mack, J. D. Scales of apparent force. *Journal of Experimental Psychology*, 1959, **58**, 405-413.

Stevens, S. S. The direct estimation of sensory magnitudes—loudness. *American Journal of Psychology*, 1956, **69**, 1-25.

Stevens, S. S. Cross-modality validation of subjective scales for loudness, vibration, and electric shock. *Journal of Experimental Psychology*, 1959, **57**, 201-209.

Stevens, S. S. The psychophysics of sensory function. In W. A. Rosenblith (Ed.), *Sensory communication*. New York: Wiley, 1961. Pp. 1-33.

Stoyva, J., & Kamiya, J. Electrophysiological studies of dreaming as the prototype of a new strategy in the study of consciousness. *Psychological Review*, 1968, **75**, 192-206.

Swets, J. A., Tanner, W. P., & Birdsall, T. G. Decision processes in perception. *Psychological Review*, 1961, **68**, 301-340.

Terrace, H. S. Errorless transfer of a discrimination across two continua. *Journal of the Experimental Analysis of Behavior*, 1963, **6**, 223-232.

Van Liere, D. W. Characteristics of the muscle tension response to paired tones. *Journal of Experimental Psychology*, 1953, **46**, 319-324.

Walter, W. G. The convergence and interaction of visual, auditory, and tactile responses in human nonspecific cortex. *Annals of the New York Academy of Sciences*, 1964, **112**, 320-361.

Whatmore, G. B. Personal communication, 1969.

Whatmore, G. B., & Kohli, D. R. A neurophysiologic factor in functional disorders. *Behavioral Science*, 1968, **13**, 102-124.

Winnick, W. A. Anxiety indicators in an avoidance response during conflict and non. *Journal of Comparative and Physiological Psychology*, 1956, **49**, 52-59.

Wolpe, J. The experimental foundations of some new psychotherapeutic methods. In A. J. Bachrach (Ed.), *Experimental foundations of clinical psychology*. New York: Basic Books, 1962. Pp. 554-575.

Woodworth, R. S. & Schlosberg, H. *Experimental psychology*. (rev. ed.). New York: Holt, 1954.

Learning Mechanisms 7

Leo V. DiCara

CONDITIONING AND LEARNED RESPONSES IN VISCERAL CONTROL

Classical conditioning is important to the study of psychophysiologic disorders because it is assumed to provide the basic model for emotional learning. In the Russian laboratories, however, classical conditioning has had much broader significance, partly because of Pavlov's particular theories of the neurophysiological events behind conditioning, but also because of the diverse experimental observations that the Russian investigators have made. In general, the Russians have been more deeply aware of the importance of higher cortical integrative activity in visceral and somatic activity than have their American counterparts. The Russians have concentrated on three aspects of conditioning: interoceptive conditioning, semantic conditioning, and theories of neurosis and personality that arise from Pavlov's work. In experiments on interoceptive conditioning, the Russian investigators have shown that all internal organs are under neural control and can have their functions altered by conditioning. More recently, American investigators have demonstrated that visceral responses also can be directly modified by trial-and-error learning, a form of learning much more flexible than classical conditioning (see below).

AUTONOMIC NERVOUS SYSTEM

Peripheral Aspects

The peripheral organization of the autonomic nervous system ensures the autonomy and automatism of the various visceral functions. Central connections ensure that visceral control is adaptive to external conditions and responsive to the highest levels of neural integration. Peripheral organization is found to some extent in the effector organ itself, as in smooth muscle or glands, as well as autonomic ganglia and plexuses.

The peripheral autonomic nervous system consists of a sympathetic, or thoracolumbar, division and a parasympathetic, or craniosacral, division. The division is based on anatomical, pharmacological and physiological grounds, the most important of which include the facts that the anatomical distributions of the nerve fibers in the two divisions are distinct from each other, and that synaptic transmission in the two divisions is accomplished, for the most part, by the action of different neurotransmitters. The effects of the two divisions on the organs they innervate are often antagonistic to each other.

Ganglia of the sympathetic division lie in close proximity to the spinal cord (Fig. 30-1). The sympathetic fibers leave the cord

Reprinted with permission from *Pathophysiology: Altered Regulatory Mechanisms in Disease* edited by Edward D. Frohlich. Copyright © 1972, by J. B. Lippincott Company.

through the anterior roots of the spinal nerve and pass into the sympathetic chain, where they synapse. From here, fibers travel in two directions; some pass into visceral sympathetic nerves that innervate the internal organs, and the others return through the gray ramus back into the spinal nerve. The latter fibers then travel throughout the body along the spinal nerves. Since the synaptic connections in the sympathetic division are made near the spinal cord, the preganglionic fibers are short and the postganglionic fibers are long.

Although parasympathetic fibers may be found in several cranial nerves, the greatest number travel in the tenth cranial (vagus) nerve (Fig. 30-2). The cell bodies of the preganglionic neurons are in the brain stem or sacral cord, and their fibers usually run all the way to the organ that they innervate, where they synapse with the postganglionic neurons. This is quite different from the sympathetic system, whose postganglionic cells are located in the sympathetic ganglia at relatively great distances from their respective target organs.

The parasympathetic division contains two anatomically distinct outflow sections, the cranial and the sacral. The cranial outflow comprises visceral motor fibers from several brain stem nuclei (e.g., vagus and glossopharyngeal), whose projections reach postganglionic neurons in the walls of the viscera. The sacral outflow contains visceral efferents from lower cord nuclei which project directly to these postganglionic neurons in pelvic organs.

Several neurotransmitters are involved in visceral regulation. Acetlycholine is apparently the transmitter between preganglionic and postganglionic sympathetic neu-

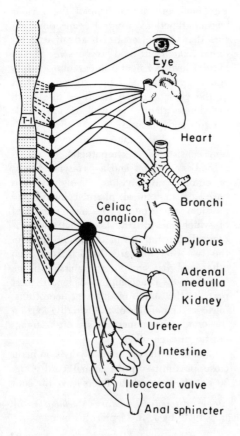

FIG. 31-1. Oversimplified anatomy of peripheral sympathetic nervous system.

rons. Several excitatory and inhibitory autonomic neurotransmitters have been identified, including acetylcholine, norepinephrine, serotonin and histamine.

The neurotransmitter is synthesized within the neuron and is stored in small "packages" or quanta at the nerve terminal, to be released upon excitation of the neuron. The sympathetic and parasympathetic divisions work together to achieve a great variety of reflex and generally adaptive patterns of response. The neurotransmitter substances act on the target tissue in a localized and precise fashion by way of "receptors" discretely sensitive to one or another transmitter. A neuron that elaborates norepinephrine at its terminal may, for example, either constrict or dilate an arteriole, depending on whether its connection at the vessel wall is an alpha or a beta receptor.

Sympathetic and parasympathetic divisions affect various organ functions quite differently. Thus, sympathetic activation prepares the body for action, raises arterial pressure, increases heart rate, dilates bronchi, and releases glucose from the liver. Parasympathetic activation, for the most part, results in conservation of energy although, in the gastrointestinal tract, stimulation is associated with increased acid secretion and motility. However, the sympathetic and parasympathetic systems do not operate independently of one another. Normally, impulses are transmitted continuously through the fibers of both divisions, thereby allowing for maintenance of tonic and dynamic activity. This allows each system to exert both facilitating and inhibiting effects, thus modulating the activity of the organ concerned. Psychosomatic disorders have been viewed as reflecting a disturbed balance of autonomic function or as representing inappropriate patterns of autonomic response.

Central Aspects of the Autonomic Nervous System

Central regulation of autonomic function is discussed in Chapter 27. It is important to emphasize that interactions among frontal, limbic, hypothalamic, thalamic and cerebellar structures regulate responses integrated at brain stem, spinal and peripheral levels. The nature of the effector function depends, therefore, on myriad potential

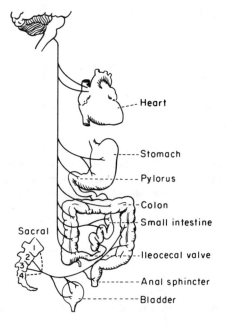

Fig. 31-2. Oversimplified anatomy of peripheral parasympathetic nervous system.

influences from circuitry concerned with individual past experience, learning, attitudes, values, goals and aspirations. An appreciation of the complexity of the central integrative process helps to explain the question, "Why does one person react to "stress" in a fashion widely different from another person?"

Certain areas of the neocortex have a direct cardiovascular function, as demonstrated by techniques of electrical stimulation and ablation. Thus, stimulation of the motor cortex can effect renal vasoconstriction severe enough to produce nephron necrosis. These changes depend on intact renal nerves, suggesting that cortical autonomic control may be capable of activating the renin-angiotensin system involved in hypertension.

Connections between higher integrative levels and visceral effector pathways in the brain stem are concentrated in the hypothalamus. In addition, the hypothalamus provides access to hormonal regulators by way of "releasing" hormones transported through portal veins from the hypothalamus to the anterior hypophysis, which is attached to the hypothalamus by the infundibulum. In this way, the hypothalamus controls the secretion of most anterior pituitary hormones and, thereby, many bodily metabolic functions.

Interest in the connections between the hypothalamus and superior limbic structures, often called the visceral brain, has been intense since Papez' classical studies and his hypothesis that activity in this system provides the autonomic basis for emotional experience. Thus, electrical stimulation in certain parts of the limbic system can produce fully integrated patterns of behavior in the awake animal that seem identical with the behavior normally occurring in situations in which the animal is required to fight or to defend itself. These results of brain stimulation in animals have been confirmed clinically in man by observing patients who are conscious during certain brain operations. In such patients, for example, electrical stimulation in or near the amygdala is reported to result in feelings of fear. Other lines of research converge and support the notion that the limbic system is involved in emotional and visceral behavior.

LEARNING AND CONDITIONING AND VISCERAL PATHOPHYSIOLOGY

Learning is the outcome of experience, and it is the adaptive use of experience in the life of the organism. Conditioning is the description of very elementary types of learning. The extent to which any creature learns and the kinds of learning it executes are the result both of the capacity for learning in that animal and the opportunities for learning presented by the environment. In recent years, there have been a number of attempts to derive rational methods of symptomatic treatment of certain disorders such as tics, and phobias, based on the assumption that these symptoms are learned patterns of behavior.

Pathological Model Systems

A major feature of the pathophysiology of most, if not all, psychosomatic conditions is profound disturbance in vegetative function. These dysfunctions include excessive lability and range as their primary characteristics, and exaggerations or disturbances of integrative patterns of autonomic mobilization responses and homeostatic mechanisms are also importantly involved.

It has been postulated that chronic diseases such as peptic ulcer, hypertension, hyperthyroidism, ulcerative colitis, and rheumatoid arthritis result when disturbances in the activity of the autonomic nervous system are prolonged.

In order to understand better the natural history of such diseases, recent experiments have attempted to produce pathological model systems in animals by special conditioning procedures. Except for work by Gantt and Masserman in the United States, the major part of previous work on this problem has been reported by Soviet scientists. In some of these experiments the investigators have been able to produce symptoms of cardiovascular pathology by conditioning. For example, sustained hypertension accompanied by histologically verified subendocardial necrotic foci, frequently resulting in fatal congestive failure, has been observed in dogs who had developed an experimental neurosis as a result of conditioning procedures in which mutually competitive responses were trained to the same stimulus. Furthermore, it has been reported that not only do conditioned disturbances in higher nervous function give rise to disorders in cardiovascular regulation, but also that the reverse holds true; namely, that experimentally induced and naturally occurring cardiovascular disorders have been found to give rise to concomitant parallel changes in those central nervous system structures participating in cardiovascular regulation. On the basis of such results and of experience with natural and other experimental vasoneuroses as well as the evidence of psychogenic influence on arterial pressure, a corticovisceral theory of hypertension has been formulated.

Experimental Neurosis and Individual Differences

The origins of experiments to produce experimental neurosis and pathophysiological effects date back to Pavlov, who noticed in his studies on conditioned reflexes that certain dogs behaved unpredictably and were difficult to handle. Systematic studies disclosed that experimental neuroses and altered physiology could be produced by conflict between emotions during conditioning or by requiring the animal to establish too fine a sensory discrimination as well as by extremes in any of the aspects of the training procedure.

Recent experiments have confirmed and extended Pavlov's original demonstration. In summary, these experiments indicate that several procedures can lead to physiological disturbances. Exposure to severe and unexpected pain, emotional excitement in a life-threatening situation, and subjection to a conflict between strong innate drives and learned reactions can lead to behavioral and physiological disturbances.

Pavlov observed that the basic temperament and autonomic reactivity of his dogs were important in determining the degree of susceptibility to, as well as the type and severity of, "neurotic" disturbance. When animal subjects with different temperaments are exposed to identical stresses, they tend to develop different types of disorders. In a somewhat analogous situation, Lacey and Lacey have clearly demonstrated that human beings who were experimentally stressed showed consistent response tendencies which differed among individuals. This consistency of autonomic reactivity or autonomic constitution could be the product of a number of different mechanisms which might predispose the individual, as clinical observation suggests, to specific types of behavior. Thus, the same objective conditions may impose different amounts of stress on different individuals, and the responses of different organ systems may vary with its intensity. In addition, genetic differences may predispose to different responses in different individuals. Moreover, experiences in infancy may affect the physical growth and the physiological activity of specific systems, providing a basis for subsequent differences in susceptibility. Finally, differences in the response of visceral systems may be attributable to the effects of classical conditioning by early learning experiences.

Instrumental Learning of Visceral and Glandular Responses

Previous work in the Soviet Union has shown the surprising degree to which a number of glandular and visceral responses under the control of the autonomic nervous system can be modified by classical conditioning. The traditional view has been that these responses can be modified only by an inferior form of learning called Pavlovian, or classical, conditioning and not by the superior form called trial-and-error, or instrumental, learning, employed in the training of skeletal responses. The difference between the two forms of learning is illustrated in Figure 31-3.

In classical conditioning, the reinforcement must be by an innate or unconditioned stimulus that already elicits the specific response to be learned, for example, the sight and smell of food, which elicits a salivatory response. In trial-and-error or instrumental learning, the reinforcement, called a reward, has the property of strengthening any immediately preceding response. Therefore, the possibility for reinforcement is much greater than is the case in classical conditioning, since a given reward may reinforce any one of a number of different responses and a given response may be reinforced by any one of a number of rewards. The old belief that trial-and-error learning is possible only for voluntary responses mediated by skeletal muscles and, conversely, that the involuntary responses mediated by the autonomic nervous system can be modified only by classical conditioning has been used as the main argument for the notion that instrumental learning and classical conditioning are two basically different phenomena with different neurophysiological substrates. Recent experiments, however, have disproved this belief by the demonstration that visceral responses can be modified by trial-and-error learning. These experiments have deep implications for theories of learning, for psychosomatic medicine and, possibly, for an increased understanding of homeostasis.

A major problem encountered in research on the instrumental modification of visceral responses is that the majority of such responses are altered by voluntary responses such as tensing of specific muscles or changing the rate or pattern of breathing. One way to circumvent this problem, at least in animals, is to abolish skeletal muscle activity by curarelike drugs, such as d-tubocurarine; these interfere pharmacologically with the transmission of the nerve impulse to the skeletal muscle but do not affect autonomically mediated responses.

Curarized subjects cannot breathe and must be maintained on artificial respiration, and, because they cannot eat or drink, the

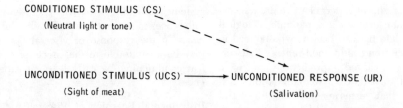

PAVLOVIAN OR CLASSICAL CONDITIONING

After several pairings the initially neutral CS substitutes for the UCS, producing the UCR

CONDITIONED STIMULUS (CS)
(Neutral light or tone)

UNCONDITIONED STIMULUS (UCS) ⟶ UNCONDITIONED RESPONSE (UR)
(Sight of meat)   (Salivation)

INSTRUMENTAL OR TRIAL AND ERROR LEARNING

After several reinforcements of the CR, the CS serves as a signal to perform the learned response (CR) in order to obtain reinforcement

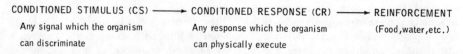

CONDITIONED STIMULUS (CS) ⟶ CONDITIONED RESPONSE (CR) ⟶ REINFORCEMENT
Any signal which the organism   Any response which the organism   (Food, water, etc.)
can discriminate   can physically execute

FIG. 31-3. Difference between classical conditioning and instrumental or trial-and-error learning.

possibilities for rewarding them in a training situation are somewhat limited. It is well known, however, that training of instrumental skeletal muscle responses can be accomplished either by using direct electrical stimulation of rewarding areas of the hypothalamus or by allowing escape from and/or avoidance of mildly noxious electric shock. Recent experiments in which these techniques have been used with curarized animals have shown that, by instrumental procedures, either increases or decreases can be produced in visceral responses such as heart rate, intestinal motility, arterial pressure, vasomotor responses, urine formation and contractions of the uterus. Several investigators have reported similar instrumental learning of heart rate and arterial pressure in man and have started to employ these techniques in therapy of certain clinical cardiovascular disorders.

Other experiments have shown that visceral learning can be quite specific, so that learned changes in heart rate do not involve systematic changes in intestinal contractions and learned changes in intestinal contractions do not involve systematic change in heart rate. Similarly, learned changes in renal blood flow and rate of urine formation have been found to be independent of changes in arterial pressure, heart rate or peripheral vasomotor responses. In another experiment, a marked dissociation of learned arterial pressure responses and heart rate was obtained in experimental animals but not in yoked controls. Taken as a group, these results indicate that visceral learning in rats is not mediated by a generalized reaction such as a shift in overall level of parasympathetic or sympathetic arousal. The striking degree of specificity to which a visceral response can be trained is shown by the results of an experiment in which opposite vasomotor responses were learned in the two ears of the curarized rat.

The demonstration that visceral and glandular responses are subject to instrumental learning and that this learning can have significant behavioral and physiological consequences means that it is theoretically possible for learning to produce the tendency, experimentally demonstrated by Lacey and Lacey, for individuals to respond to stressful situations with consistently ordered hierarchies of visceral responses that vary among individuals. It is also theoretically possible that such learning can be carried far enough to create an actual psychosomatic symptom. Presumably, genetic and constitutional differences among individuals would also explain the differences in response of the different visceral systems. Moreover, it seems probable that, under circumstances that can reward different psychosomatic symptoms, the subject will be most likely to learn the one to which he is innately most predisposed.

The effectiveness of the technique used with animals—objectively recording glandular and visceral responses and then modifying them by immediately rewarding, first, slight and, then, progressively greater changes—suggests that this kind of procedure might be used to produce therapeutic learning in patients with psychophysiological disorders.

A New View of Homeostasis

The reactions through which the constancy of the internal environment is maintained are more complex and indirect than are the ordinary chemical and physical adjustments. They are essentially physiological reactions regulated through the visceral nerves. The traditional view has been that the processes that tend to maintain homeostasis are carried out automatically. They involve both relatively simple reflex mechanisms and integrating mechanisms of higher orders, the most important of which are located in the hypothalamus. According to classical views, the automatic regulation of the routine necessities for maintenance of homeostasis permits the functions of the brain that subserve manual skills, intelligence, imagination, and insight to find their highest expression. On the other hand, we now know that visceral responses are not always involuntary but can be learned in the same way as skeletal muscle responses. This raises the fundamental question as to whether or not the capacity for instrumental learning of visceral responses is in itself important to the maintenance of the homeostasis of the internal environment.

Since skeletal muscles operate mainly upon the external environment, the ability to learn responses that bring rewards such as food, water, or escape from pain has survival value. To what extent does regulation of the internal visceral responses by the autonomic nervous system have a similar adaptive value? Will an animal with experimentally altered internal homeostasis perform the instrumental tasks necessary to correct the imbalance? In a two-part experiment designed to test this question, Miller, DiCara and Wolf injected antidiuretic hormone (ADH) into rats if they selected one arm of a T maze and isotonic saline if they selected the other. One group of normal animals, loaded in advance with water by stomach tube, learned to select the saline arm; choice of the ADH arm would have prevented the secretion of excess water required for restoration of water balance. By contrast, a group of rats with diabetes insipidus, and loaded in advance with hypertonic NaCl by stomach tube, regularly chose the ADH arm. For this group the homeostatic consequences of the injections were reversed; ADH caused concentration of urine, thus promoting excretion of the excess NaCl, whereas isotonic saline would have merely produced further imbalance. The control rats received neither water nor NaCl and exhibited no preference in the maze.

These instrumental responses aimed at correcting internal imbalance appear to be no less adaptive than those learned skeletal responses that are directed toward altering the animal's relationship to its external environment.

Taken together with the other evidence that glandular and visceral responses can be instrumentally learned, these data raise the question of the degree to which animals normally learn the various autonomic responses required to maintain proper homeostasis. Whether or not such learning could contribute to homeostasis in ordinary day-to-day life experience would depend on the range of sensitivity or inertia of the control mechanism. Thus, a very small drift in the quantity of a physiological variable may be quickly compensated for under ordinary circumstances. But where such a control system may display greater inertia and require a larger drift before corrective change is triggered, visceral learning, rewarded by a return to homeostasis, might be available as a back-up mechanism.

ANNOTATED REFERENCES

Bykov, K. M.: The Cerebral Cortex and the Internal Organs (W. H. Gantt, translator and editor). New York, Chemical Publishing Co., 1957. (Presents evidence that glandular and visceral responses, controlled by the autonomic nervous system, can be modified by classical conditioning)

DiCara, L. V.: Learning in the autonomic nervous system. Sci. Am., 222:30, 1970. (Review of evidence that visceral responses can be modified by trial-and-error learning)

Figar, S.: Conditioned circulatory responses in men and animals. In: Hamilton, W. F. (ed.): Handbook of Physiology. Section 2, Circulation, Vol. 3, p. 1991. Washington, D. C., American Physiological Society, 1965.

Gantt, W. H.: Cardiovascular component of the conditioned reflex to pain, food and other stimuli. Physiol. Rev. (Supp.), 40(4): 266, 1960.

Gellhorn, E.: Autonomic Imbalance and the Hypothalamus. St. Paul, University of Minneosta Press, 1957.

and

Gellhorn, E., and Loofbourrow, G. N.: Emotions and Emotional Disorders. New York, Harper & Row, 1963. (Two of the most comprehensive and critical textbooks available on the subject of autonomic imbalance and psychosomatic disorders)

Haymaker, W., Anderson, E., and Nauta, W. H.: The Hypothalamus. Springfield, Ill., Charles C Thomas, 1969. (A comprehensive treatise on the hypothalamus, with emphasis on anatomy and *function* and the basis of *clinical syndromes*)

Hoff, E. C., Kell, J. F., and Carrol, M. N.: Effects of cortical stimulation and lesions on cardiovascular function. Physiol. Rev., 43:68, 1963.

Ingram, W. R.: Central autonomic mechanisms. *In:* Field, J. (ed.): Handbook of Physiology. Section 1, Neurophysiology, Vol. 2, p. 951. Washington, D. C., American Physiological Society, 1960. (Review of literature indicating that autonomic function is influenced by structures at all levels in the central nervous system)

Lacey, J. I., and Lacey, B. C.: Verification and extension of the principle of autonomic response stereotypy. Am. J. Physiol., *71:* 50, 1958.

MacLean, P. D.: Psychosomatics. *In:* Field, J. (ed.): Handbook of Physiology. Section 1, Neurophysiology, Vol. 3, p. 1723. Washington, D. C., American Physiological Society, 1960. (Comprehensive review concerning mechanisms through which emotion and stress alter function leading to psychosomatic illness)

Miminoshvili, D. I., Magakian, G. O., and Kokaia, G. I.: Attempts to obtain a model of hypertension and coronary insufficiency in monkeys. *In:* Utkin, I. A. (ed.): Problems of Medicine and Biology in Experiments on Monkeys. p. 103. New York, Pergamon Press, 1960. (The above four references present evidence in the experimental animal, as well as in man, that symptoms of cardiovascular pathology can be produced by conditioning.)

Miller, N. E., DiCara, L. V., and Wolf, G.: Body fluid homeostasis and learning in rats: T-maze learning reinforced by manipulating anti-diuretic hormone. Am. J. Physiol., *215:* 684, 1968.

Papez, J. W.: A proposed mechanism of emotion. Arch. Neurol. Psychiat., *38:*725, 1937.

Suggested Review Articles on Learning Mechanisms

Astrup, C.: Pavlovian Psychiatry. Springfield, Ill., Charles C Thomas, 1965.

Holland, B. C., and Ward, R. S.: Homeostasis and psychosomatic medicine. *In:* Arieti, S. (ed.): American Handbook of Psychiatry. Vol. 3. New York, Basic Books, 1966. (The above two references are suggested as excellent review articles, in addition to others already cited, concerned with learning and autonomic mechanisms in relation to pathophysiology.)

Learning of Glandular and Visceral Responses

Neal E. Miller

Postscript[1]

In the short space available, it will be impossible to do justice to the excellent work published by other laboratories since the appearance of the foregoing article; to get the complete story, you will have to read the studies cited.

Two other laboratories have confirmed and extended our experiments showing that changes in heart rate can be learned by rats paralyzed by curare and rewarded by electrical stimulation of the brain (Hothersall & Brener, 1969; Slaughter, Hahn, & Rinaldi, 1970). The first of these also demonstrated extinction during nonrewarded performance and progressive improvement during three sessions; the second demonstrated that discriminative pretraining in the bar-press situation is not essential. Black (1971) has demonstrated heart-rate learning by curarized dogs. Plumlee (1969) has demonstrated the avoidance learning of up to 60 mm Hg increases in the diastolic blood pressure of noncurarized monkeys over a fairly extended training period. Black, Young, and Batenchuk (1970) have extended Carmona's work on learned changes in the brain waves in curarized animals and on the relationship between such learning and changes of behavior in the noncurarized state. Kamiya (1969), and other workers inspired by his pioneering studies, have greatly expanded their investigation of the human learning of changes in brain waves.

A number of additional studies on rats have been completed in our laboratory. Pappas, DiCara, and Miller (1970) demonstrated that noncurarized rats could be rewarded to learn either small increases or small decreases in blood pressure. DiCara and Stone (1970) have shown that rats trained under curare to increase their heart rates have immediately thereafter an increased level of catecholamines in both the heart and the brainstem while those trained to decrease the heart rate have a decreased level. Fields (1970) used a computer to administer brief electric shocks to curarized rats whenever a specific interval of the ECG fell beyond the specified range. He was able either to increase or decrease, respectively, either the PR or PP interval of the ECG without appreciably changing the other interval. He was able to study a number of specific details of the course of learning. In experiments that are being prepared for publication, Pappas, DiCara, and Miller showed that curarized rats could learn either to increase or decrease, respectively, the contractions of the uterus in order to escape and/or avoid punishment. They also found that 33-day-old rats could learn either to increase or to decrease the heart rate in order to get rewarding stimulation of the brain but that this learning seemed to be poorer, and certainly was not better, than that of older rats. Furthermore, in contrast to older rats, there was no retention when these infants were retested at the age of 90 days. We hope to test the infant rat's ability

---

[1] Studies mentioned from the author's laboratory were supported by USPHS grants MH 13189, MH 19172, and MH 19183 from the National Institute of Mental Health.

Reprinted from *Current Status of Physiological Psychology:* Readings edited by D. Singh and C. T. Morgan. Copyright 1972 by Wadsworth Publishing Company, Inc. Reprinted by permission of the publisher, Brooks/Cole Publishing Company, Monterey, California.

to learn temperature changes, a type of control that may be more vital to them at that age, and to see whether or not there are any long-lasting effects of such learning on the subsequent pattern of either visceral or emotional responses.

Finally, DiCara, Braun, and Pappas (1970) succeeded in classically conditioning curarized neodecorticate rats to change heart rate in one experiment and intestinal contraction in another, but failed to change either of these responses by instrumental shock-avoidance training. This experiment raises the question of a possible fundamental difference between the mechanisms of classical conditioning and instrumental learning. Our demonstration that, with the intact brain, visceral responses can be instrumentally learned as well as classically conditioned removed one of the strongest arguments for a fundamental difference between these two types of learning, but, of course, it did not prove that they are identical. While the decortication study opens up the possibility for a difference in brain mechanisms, it does not prove such a difference. For example, under the particular conditions used in these experiments, the technique of classical conditioning may be a more effective training method than that of instrumental learning. One of the reasons for the effectiveness of classical conditioning could be the fact that the unconditioned stimulus forces a large correct response to occur on every trial (Miller & Dollard, 1941). But this would not necessarily mean that two fundamentally different types of learning of different levels of the brain are involved any more than does the fact that some instrumental training techniques are easier than others—for example, it is easier to train a pigeon to peck a key than to press a bar. On the other hand, it is possible that further work will show that a fundamental difference is involved.

We hope to use the highly specific control over visceral functions that can be achieved by training as a means of testing the effects of drugs and of analyzing further the neuroanatomical and physiological mechanisms involved in visceral functions.

In the especially important and difficult area of human instrumental learning of visceral responses, since Kimmel and Hill (1960) and Shearn (1961) revived the issue, the results have continued to become more impressive. There have been a number of ingenious studies showing changes in a rewarded visceral response but not in an unrewarded one.[2] While it is just barely conceivable that the results of these particular studies might be explained by assuming that, under the particular conditions used, the unrewarded response was not as sensitive to change as the rewarded one, this kind of explanation cannot be applied to a recent experiment of the Harvard group. Using a design analogous to that of Miller and Banuazizi (1968), they measured both blood pressure and heart rate, showing in one experiment that subjects rewarded for changes in blood pressure learned such changes but did not learn changes in heart rate (Shapiro et al., 1969); in a second experiment, under similar conditions, subjects rewarded for changes in heart rate learned such changes but not changes in blood pressure (Shapiro et al., 1970). With the accumulation of more beautiful demonstrations of this type of specificity, it becomes increasingly harder to argue that the human visceral changes are mediated by skeletal movements or even by emotional thoughts.

Some of the strongest evidence for the human ability to learn a visceral response instrumentally has been available but neglected for some time. The control of urination, especially in the female, is mediated by the autonomic nervous system. Under the good learning conditions provided by strong motivation and immediate knowledge of results, virtually everyone learns this form of visceral control. That this learned response can indeed be executed by autonomically innervated smooth muscle is convincingly demonstrated by a heroic experiment in which Lapides and his associates (1957) paralyzed by curare or succinylcholine the skeletal muscles of 15 human subjects maintained on artificial respiration. Without additional practice, these subjects were able to initiate and stop urination on command, although the control was slower, especially for stopping, than in the normal state.

[2] Shapiro, et al., 1964; Rice, 1966; Gavalas, 1967; Snyder and Noble, 1968.

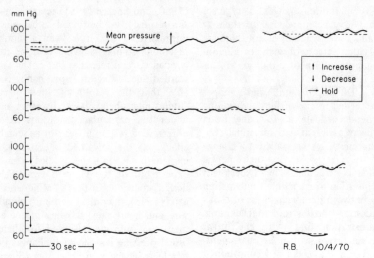

Figure 1. Performance of learned voluntary control over blood pressure. Broken lines represent the integrated diastolic pressure for the first and last minutes during each training trial. The arrows represent the direction in which changes were requested and rewarded.

Engel, who has been a pioneer in the therapeutic applications of visceral learning, has taught patients impressive control over premature ventricular contractions (Weiss & Engel, 1971).[3]

As one of the conditions on which to try visceral training therapy, we have selected the early variable states of hypertension. Clinical studies have shown that using drugs to reduce even borderline hypertension increases life expectancy. If training works, we would expect it to have less side effects than drugs; there is no reason why it should be more likely to produce symptom substitution.

Some of the complications and difficulties as well as the encouraging preliminary results of our current work on therapeutic training are illustrated by a case of hypertension. This intelligent 33-year-old girl had had a stroke producing brainstem damage (a complication) approximately four months before she entered the New York University Medical Center for rehabilitation of disabling motor deficits.[4] There, a graduate student, Barry Dworkin, an internist, Saran Jonas, and I gave her training in voluntary control over her high blood pressure. A special device indirectly recorded diastolic blood pressure automatically on a beat-to-beat basis and sounded a tone that served as a reward by signalling to her that she had achieved the correct change in her blood pressure (Miller et al., 1970). First, we rewarded her for a small decrease and then changed to reward for a small increase so that she could again succeed in producing another decrease. We also hoped that the contrast of successive reversals would help to teach her voluntary control.

At first she was able to produce changes of only 5 to 6 mm Hg. As she learned, we required larger changes. Then, in order to be sure that the change was not produced by any transient maneuver of skeletal muscles we required her to hold it. After 10 weeks of 3/4-hour training five days a week, she had achieved the degree of control shown in Figure 1. The average integrated diastolic blood pressure for the first baseline minute was 76 mm Hg. When instructed to increase,

---

[3] Somewhat similarly, Budzynski, Stoyva, and Adler (1970) have used recording of the EMG from the frontalis muscle to train patients to relax that skeletal muscle and relieve the pain of tension headaches.

[4] We thank Dr. Clark T. Randt, Professor and Chairman of the Department of Neurology, New York University Medical Center, for making available facilities for this study.

she did and held the diastolic pressure for a minute at an average of 94 mm Hg. During the rest period before the next trial, the pressure dropped to 74 mm Hg; during the last minute of the third trial of being rewarded for decreases, it averaged 65 mm Hg, a range of control totalling 29 mm Hg. It seems clear that she has gradually learned an unusual degree of voluntary control over her blood pressure. She does not use any obvious maneuver of skeletal muscles or breathing; changes in her EEG are not consistently related to her changes in blood pressure, and the EMG from her deltoid muscle tends to be somewhat higher during increases than during decreases, but this is by no means invariable.

Figure 2 shows the changes in the baseline diastolic pressures of this patient during the course of training. In spite of daily medication with 750 mg of the antihypertensive drug Aldomet (alpha-methyldopa) plus 100 mg of the antidiuretic drug Hydrodiuril, her diastolic pressure averaged 97 mm Hg during the 30 days in the hospital before training started. During this period, the pressure was variable but there was no appreciable trend. When training started her pressure came down, but another doctor at the hospital put her on l-dopa (a complication) temporarily to see whether it would control a tremor. We did not feel that we could withhold other therapies just to make this experiment neater. After her blood pressure had come down to normal, she was taken off medication. During the three days while the effects of the Aldomet were wearing off, her pressure rose but then decreased to an average level of 76 mm Hg during the last 30 days before Figure 2 had to go to press. You should note that the blood pressure remained down despite discharge from the hospital and return to an unusually stressful environment. Furthermore, during training, a number of observers reported that she became much more relaxed; she reported becoming able to prevent the frustrations in her life from getting to her as they used to.

The foregoing changes during training could have been placebo effects (for example, a transference cure) caused by the impact of the apparatus plus the personal relationship with the training therapist. That this was not entirely the case is suggested by the unusual degree of voluntary control she acquired over her blood pressure and by her

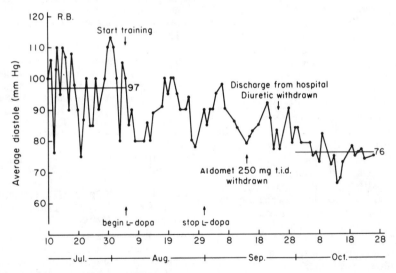

Figure 2. General reduction in diastolic blood pressure concurrent with learning of voluntary control. The horizontal line at the upper left represents the average blood pressure while the patient was in the hospital prior to training and on the antihypertensive drugs indicated. The horizontal line at the lower right represents the average blood pressure during 30 days without drugs and in a more stressful living situation outside of the hospital.

explicit intent to use this control to reduce her hypertension. Nevertheless, it remains to be seen whether similar results can be secured with other patients and how long they will last.

It also remains to be seen whether we can develop methods to achieve with noncurarized human patients the same rapid visceral learning that we have achieved with curarized rats. Perhaps we can approximate the favorable conditions of curarization by using a computer to train the subject first to relax and breathe regularly and then to reward only those visceral changes that are not accompanied by changes in breathing or other skeletal responses.

Another means of facilitating visceral learning (or achieving other therapeutic results) may be to train subjects directly to recognize visceral changes that are now not noticed or are wrongly identified. A child who calls a cow a dog has this mistake corrected. He receives no such training in visceral identification because there has been no way to check on what is happening inside of him. Therefore, when it comes to accurate visceral perception, we are all novices. But, with the help of modern instrumentation, perhaps we can be trained to become experts. That a great deal of perceptual learning can occur is demonstrated by the difference between the novice and the expert in many fields—for example, the histologist looking at sections under the microscope, the electroencephalographer identifying wave patterns in the EEG, or a pilot detecting drift while landing a plane. That fine visceral discriminations can be learned is shown by work demonstrating that a dog can react differently to the inflation of tiny balloons in the intestine that are approximately an inch apart (Ádám, 1967).

If a patient can learn to discriminate small enough changes in a visceral response, such as blood pressure, he can dispense with further instrumentation and be rewarded for producing changes in the proper direction in the same way that he is rewarded by success when he observes that his golf ball has dropped into the cup.

Neal E. Miller

References

Ádám, G. *Interoception and behaviour*. Budapest: Akadémiai Kiadó, 1967.

Black, A. H. Autonomic aversive conditioning in infrahuman subjects. In F. R. Brush (Ed.) *Aversive conditioning and learning*. New York: Academic Press, 1971.

Black, A. H., Young, G. A., & Batenchuk, C. The avoidance of hippocampal theta waves in Flaxedilized dogs and its relation to skeletal movement. *Journal of Comparative and Physiological Psychology*, 1970, 70, 15-24.

Budzynski, T., Stoyva, J., & Adler, C. Feedback-induced muscle relaxation: Application to tension headache. *Journal of Behavior Therapy and Experimental Psychiatry*, 1970, 1, 205-211.

DiCara, L. V., Braun, J. J., & Pappas, B. A. Classical conditioning and instrumental learning of cardiac and gastrointestinal responses following removal of neocortex in the rat. *Journal of Comparative and Physiological Psychology*, 1970, 73, 208-216.

DiCara, L. V., & Stone, E. A. Effect of instrumental heart-rate training on rat cardiac and brain catecholamines. *Psychosomatic Medicine*, 1970, 32, 359-368.

Fields, C. Instrumental conditioning of the rat cardiac control systems. *Proceedings of the National Academy of Science*, USA, 1970, 65, 293-299.

Gavalas, R. J. Operant reinforcement of an autonomic response: Two studies. *Journal of the Experimental Analysis of Behavior*, 1967, 10, 119-130.

Hothersall, D., & Brener, J. Operant conditioning of changes in heart rate in curarized rats. *Journal of Comparative and Physiological Psychology*, 1969, 68, 338-342.

Kamiya, J. Operant control of the EEG alpha rhythm and some of its reported effects on consciousness. In C. T. Hart (Ed.), *Altered states of consciousness*. New York: Wiley, 1969.

Kimmel, H. D., & Hill, F. A. Operant conditioning of the GSR. *Psychological Reports*, 1960, 7, 555-562.

Lapides, J., Sweet, R. B., & Lewis, L. W. Role of striated muscle in urination. *Journal of Urology*, 1957, 77, 247-250.

Miller, N. E., & Banuazizi, A. Instrumental learning by curarized rats of a specific visceral response. intestinal or cardiac. *Journal of Comparative and Physiological Psychology*, 1968, 65, 1-7.

Miller, N. E., DiCara, L. V., Solomon, H., Weiss, J. M., & Dworkin, B. Learned modifications of autonomic functions: A review and some new data. *Supplement 1 to Circulation Research*, 1970, **26,27**, I-3 to I-11.

Miller, N. E., & Dollard, J. *Social learning and imitation*. New Haven: Yale University Press, 1941. Also issued as Yale Paperbound, 1962.

Pappas, B. A., DiCara, L. V., & Miller, N. E. Learning of blood pressure responses in the noncurarized rat: Transfer to the curarized state. *Physiology and Behavior*, 1970, **5**, 1029-1032.

Plumlee, L. A. Operant conditioning of increases in blood pressure. *Psychophysiology*, 1969, **6**, 283-290.

Rice, D. G. Operant conditioning and associated electromyogram responses. *Journal of Experimental Psychology*, 1966, **71**, 908-912.

Shapiro, D., Crider, A., & Tursky, B. Differentiation of an autonomic response through operant reinforcement. *Psychonomic Science*, 1964, **1**, 147-148.

Shapiro, D., Tursky, B., Gershon, W., & Stern, M. Effects of feedback and reinforcement on the control of human systolic blood pressure. *Science*, 1969, **163**, 588.

Shapiro, D., Tursky, B., & Schwartz, G. E. Differentiation of heart rate and systolic pressure in man by operant conditioning. *Psychosomatic Medicine*, 1970, **32**, 417-423.

Shearn, D. W. Does the heart learn? *Psychological Bulletin*, 1961, **58**, 452-458.

Slaughter, J., Hahn, W., & Rinaldi, P. Instrumental conditioning of heart rate in the curarized rat with varied amounts of pretraining. *Journal of Comparative and Physiological Psychology*, 1970, **72**, 356-360.

Snyder, C., & Noble, M. Operant conditioning of vasoconstriction. *Journal of Experimental Psychology*, 1968, **77**, 263-268.

Weiss, T., & Engel, B. T. Operant conditioning of heart rate in patients with premature ventricular contractions. *Psychosomatic Medicine*, 1971, **33**, 301-321.

The Operant Conditioning of Central Nervous System Electrical Activity

A. H. Black

I. Introduction

The subject matter of this paper is the operant conditioning of central nervous system (CNS) electrical activity. This is a relatively new area of research, all of the published papers on this topic having appeared within the last ten years (see Table I for references). Because the field is of such recent origin, I shall focus on questions that new research areas such as this seem to provoke.

One set of questions concerns the operant conditioning process. First, are the observed changes in the electrical activity of the brain really produced by operant conditioning? If this question is answered positively, a second issue is often raised. Were the changes produced directly by the operant conditioning of CNS electrical activity, or indirectly by the operant conditioning of some other response which mediated the observed changes in CNS electrical activity? Given satisfactory answers to both of these questions, one is naturally led to consider features of the procedure that lead to success or failure in conditioning, and about optimal conditioning procedures for the electrical activity of the brain as compared to other more familiar responses.

A second set of questions concerns the significance of the research. Does it provide useful information? If so, does it add to the information provided by more familiar procedures for analyzing and controlling the electrical activity of the brain?

I shall discuss the operant conditioning process in Section II of this chapter, and the significance of the research in Sections III and IV.

[1] The preparation of this paper and the research from my laboratory described in it were supported by Research Grant 258 from the Ontario Mental Health Foundation, Research Grant 70-476 from the Foundations' Fund for Research in Psychiatry, and by Research Grant APA-0042 from the National Research Council of Canada. I would like to thank A. Dalton, G. Young, L. Grupp, and F. Brandemark who collaborated in the research. I would also like to thank my colleagues for their helpful comments on the manuscript.

Reprinted with permission from *The Psychology of Learning and Motivation: Advances in Research and Theory* edited by Gordon H. Bower. Copyright © 1972, by Academic Press, Inc.

II. The Operant Conditioning Process

A. Factors Responsible for Changes in CNS Electrical Activity

The first question that we must consider is this. When some pattern of CNS electrical activity is reinforced, are the changes in the probability of that pattern produced by operant conditioning, or by some other procedure? I shall illustrate the discussion of this question with research from my own laboratory.

The essential feature of the operant conditioning procedure is usually assumed to be the contingency between response and reinforcer. We can ask, therefore, whether the probability of the reinforced response increased because of this contingency, or because of some other variable such as the noncontingent presentations of the reinforcer, the contingency between a discriminative stimulus (S^D) and reinforcer, and so on. Only when the first of these alternatives is correct can we conclude that we have operantly conditioned neural activity.

Although a variety of methods have been devised to distinguish between these alternatives (for a discussion of these procedures, see Black, 1967, 1971a), the one most commonly employed is the *bidirectional procedure*. In the bidirectional procedure, we operantly condition two groups in which all the relevant procedural variables (noncontingent presentations of the reinforcer, etc.) ideally have the same value in both groups, except for the response that is reinforced. The response that is reinforced in one group is usually mutually exclusive of the one reinforced in the other group. For example, one group of subjects could be operantly reinforced for increasing the rate of a given response, while another group could be reinforced for decreasing the rate of that response. If a difference was found between the two groups in the rate of the reinforced response, it could be attributed to the effects of the contingency between response and reinforcer, provided, of course, that other relevant variables were constant in the two groups.

One example of research employing this bidirectional procedure is provided by an experiment from our laboratory on the operant conditioning of hippocampal electrical activity in dogs. In this experiment, electrodes were chronically implanted in the dorsal hippocampus. After recovery from surgery, each dog was reinforced for a given pattern of hippocampal electroencephalographic (EEG) activity. The reinforcer was food for two dogs, and brain stimulation for two dogs. On each trial, the S^D (an auditory stimulus) was presented. When a series of EEG waves of the appropriate frequency was

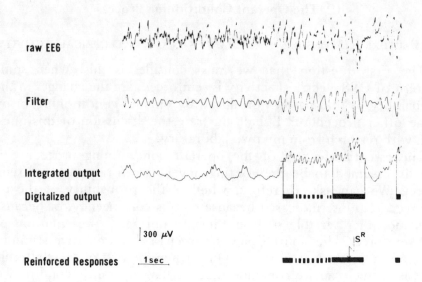

Fig. 1. System employed for reinforcing RSA. Top channel, raw EEG (lower and upper limits of band-pass filter at 1 and 35 Hz usually). Second channel, filtered EEG. Third channel, integrated EEG. Fourth channel, periods during which integrated EEG voltage is above criterion level. Fifth channel, one period during which the integrated output was above the criterion long enough to obtain reinforcement (S^R).

detected, the dog was reinforced. A brief time out followed the reinforcement; during the time out no reinforcements could be obtained. Then the S^D was turned on, and the process repeated. Every tenth trial was a test trial. On test trials, the S^D was presented for a fixed 5-second period during which no reinforcements were administered.

The method for determining when the reinforcer was to be presented is illustrated in Fig. 1. The EEG was passed through a band-pass filter, and the output of the filter integrated. When the voltage of the integrated output reached a predetermined level, a Schmitt trigger was fired. If the voltage was maintained in this state for a fixed period of time, a reinforcement was administered. This procedure is based on that employed by Wyrwicka and Sterman (1968).

The system was programmed to reinforce hippocampal RSA (rhythmic slow activity). This is a relatively high amplitude almost sinusoidal waveform between 4 and 7 Hz in the dog. The lower and upper limits of the band-pass filter were usually set at 4.5 and 5.5 Hz, respectively. The dogs were trained for a series of daily sessions until a session occurred in which the median latency of the reinforced response was less than 10 seconds. Then reinforcement was

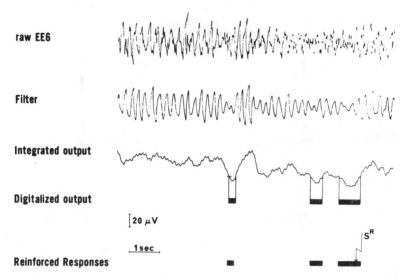

Fig. 2. System employed for reinforcing non-RSA. Top channel, raw EEG (lower and upper limits of band-pass filter at 1 and 35 Hz usually). Second channel, filtered EEG. Third channel, integrated EEG. Fourth channel, periods during which integrated EEG voltage is below criterion level. Fifth channel, one period during which the integrated output was below the criterion long enough to obtain reinforcement (S^R).

made contingent on non-RSA which could include a variety of different patterns such as the SIA (small amplitude irregular activity) described by Stumpf (1965). The method for determining when the reinforcer was to be presented is shown in Fig. 2. The procedure was identical to that employed in reinforcing RSA, but in this case, the frequency limits of the band-pass filter were set at 2 and 12 Hz for two dogs, and at 3 and 7 Hz for the other two dogs. Also, the voltage had to drop *below* a criterion level for a fixed period of time for reinforcement to occur. After the training session in which the median latency of response was less than 10 seconds, the dog was shifted back to reinforcement for RSA. This procedure was repeated several times for each dog.

Examples of EEG responses and power spectra on test trials are shown for each of the four dogs in Fig. 3. Data are shown for the final session of one RSA and one non-RSA training phase of the experiment. When RSA was reinforced, the probability of RSA was higher than non-RSA, and vice versa. Furthermore, on the terminal day of each stage, the latencies of response and the number of reinforcements were much the same. Therefore, its seems reasonable to conclude that these patterns of hippocampal electrical activity were modified by the contingency between response and reinforcer.

Similar data on other patterns of CNS electrical activity have been

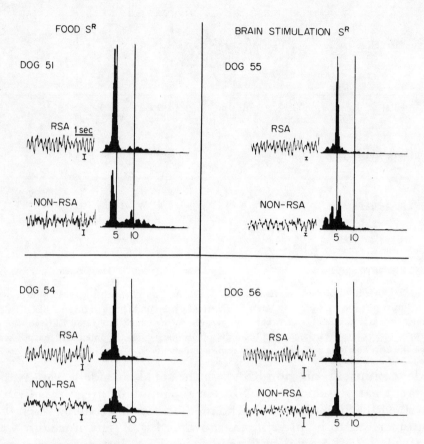

Fig. 3. Sample EEG records and power spectra for two dogs reinforced with food, and two dogs reinforced with brain stimulation. A sample of the EEG recorded for a 5-second test trial when RSA was being reinforced, and for a 5-second test trial when non-RSA was being reinforced, is shown for each dog. Power spectra taken over the last three to five test trials on the final training session for RSA and non-RSA are also shown.

obtained. In Table I, the experiments on CNS operant conditioning are classified according to type of CNS electrical activity, location of the recording electrodes, and type of S. In addition, experiments which have employed bidirectional controls or some equivalent procedure are indicated. This Table reveals that for each type of electrical activity, at least one experiment employed bidirectional or equivalent controls.[2] Therefore, all the types of CNS electrical activity that have been studied so far can be modified by response-reinforcer contingencies.

[2] Two further points that are revealed by this Table should be mentioned in passing. First, the major concentration of research effort has been on spontaneous EEG patterns; second, fewer types of CNS electrical activity in fewer brain locations have been operantly conditioned in human Ss than in infrahuman Ss. This is understandable, since it is difficult to study single-cell activity and to record from subcortical structures in human Ss.

B. MEDIATION

The second procedural question focuses attention on the possibility that operantly conditioned neural events are mediated. Instead of having operantly conditioned changes in CNS electrical activity directly, perhaps we operantly conditioned some other response inadvertently, with feedback from the occurrence of this other response reflexively eliciting the observed changes in CNS activity.

There are certain criteria that a response must meet if it is to be a mediator:

1. The mediating response must be correlated with the conditioned pattern of CNS electrical activity, and it should precede it. (Unfortunately, we can't always apply the temporal criterion because of our inability to measure temporal sequences accurately.)

2. The correlation, by itself, is usually considered to be insufficient evidence for mediation; we must demonstrate also that the mediating response is necessary for the occurrence of the mediated pattern of CNS electrical activity in the operant conditioning situation. If, for example, one blocks the mediating response, the mediated response should not occur. (One must, of course, take into account the possibility that a given CNS event might be mediated by several interchangeable responses, in which case blocking would not cause a disappearance of the mediated response.)

3. Finally, a third criterion requires considerable discussion, not because it is more important than the others, but because it is more obscure. There seems to be some agreement that all of the events that precede the reinforced pattern of CNS electrical activity and are necessary for its occurrence should not be labeled mediators. For example, the firing of certain motoneurons may precede and be necessary for the contraction of skeletal muscles, but one would not want to call the former a mediator of the latter. Unfortunately, the criteria that are employed in making this distinction between mediators and nonmediators are not clear. It is obvious that only those responses that can be operantly conditioned can act as mediators. But this does not limit the category of mediators very much. Another possibility is the following: responses which are considered to be components of the same neural control systems are not considered to be mediators of each other; responses which are components of different neural control systems are possible mediators of each other. According to this criterion, activity of spinal motoneurons would not be considered a mediator of activity of skeletal muscles, but a change in respiratory activity could be considered a mediator of changes in the activity of the cardiac muscula-

TABLE I
CLASSIFICATION OF EXPERIMENTS ON OPERANT NEURAL CONDITIONING

Subject	Electrode location	Type of conditioned neural activity		
		Single cell	"Spontaneous" EEG	Evoked potentials
Infrahuman	Cortex	Motor cortex Fetz (1969)[a] Fetz and Finocchio (1971)[a, b]	High- and low-voltage EEG Carmona (1967)[a, b] Sensorimotor rhythm, postreinforcement synchronization, desynchronization Chase and Harper (1971) Sterman et al. (1969c)[a] Sterman et al. (1969b) Sterman et al. (1969a) Wyrwicka and Sterman (1968)[a]	Elicited by visual stimuli Fox and Rudell (1968)[a] Fox and Rudell (1970)[a] Elicited by movement Rosenfeld (1970)[a]
	Subcortical structures	Limbic system Olds and Olds (1961)[a] Olds (1965, 1967, 1969)	Hippocampal RSA and non-RSA Black (1971b)[a, b] Black et al. (1970)[a, b] Dalton (1969)	

		Amygdala spindling	
		Delgado et al. (1970)[a]	
Human	Scalp	Alpha, beta, theta waves	Elicited by auditory stimuli
		Beatty (1971)[a]	Rosenfeld et al. (1969)[a]
		Brown (1970, 1971)[a]	
		Dewan (1967)[a]	
		Green et al. (1970a, 1970b)	
		Kamiya (1968, 1969)[a]	
		Lynch and Paskewitz (1971)[c]	
		Mulholland (1968, 1969, 1971)[c]	
		Nowlis and Kamiya (1970)[a]	
		Paskewitz and Orne (1971)	
		Paskewitz et al. (1970)[a]	
		Peper (1970)[a]	
		Peper and Mulholland (1970)[a]	
		Spilker et al. (1969)	

[a] These experiments employed bidirectional control procedures or their equivalent.
[b] These experiments employed curare-like drugs or dissociative conditioning procedures to rule out peripheral mediation.
[c] These are theoretical or review articles.

ture. Unfortunately, it is difficult to make this distinction because we know so little about the neural circuits of which reinforced events are components.

The purpose of most research on mediation has been to demonstrate that we can operantly condition neural events directly by ruling out the possibility that we have inadvertently conditioned some other response. The fewer the potential mediators of the reinforced pattern of CNS electrical activity, the surer we can be that we have operantly conditioned that pattern directly. In this sense, mediators are "a bad thing"; the fewer the better.

One might ask, at this point, why we are so concerned with showing that we can operantly condition neural events directly by demonstrating an absence of mediators. The reason, it seems to me, is the desire to show that the results of operant neural conditioning are novel. If we can reject the mediation hypothesis, then we can also reject the following familiar argument: "Some observable response, whose operant conditioning has been demonstrated repeatedly in the past, has been conditioned inadvertently in experiments on operant neural conditioning, and therefore, operant neural conditioning is not really a new phenomenon."

This attempt to rule out mediation in order to demonstrate that operant neural conditioning is novel is, in some ways, an odd enterprise. As was made clear in the description of the criteria that are employed in identifying mediators, one must gather data on the neural and behavioral changes that are correlated with the changes in the reinforced pattern of CNS electrical activity in order to deal with the problem of mediation. But such data could also be employed to help us understand the structure and function of neural and behavioral systems of which the reinforced pattern might be a component. They could provide information about the neural and behavioral systems that are activated when a given reinforced pattern occurs, and, therefore, about the mechanisms that are involved in performing a given operantly conditioned neural response. They could provide information on the specificity of conditioning; that is, on whether a response is conditioned as part of a global behavioral pattern, or as an individual unit over which we have precise and specific control. These issues are, I think, more important than the demonstration that one has conditioned a novel response.

Furthermore, if one is really interested in determining whether the results of reinforcing patterns of CNS electrical activity are different from the results of reinforcing observable responses, one ought to compare the changes in the neural and behavioral systems related to each of the reinforced events. Such a comparison will reveal more subtle and interesting differences than will the attempt to rule out mediation.

In summary, the attempt to find out if operantly conditioned neural responses are mediated provides a relatively gross approach to a question of relatively limited importance. I shall, therefore, discuss research designed to answer the mediation question only insofar as it deals with obvious potential mediations such as observable skeletal responses. In Section IV, the same data will be discussed in relation to what I consider a more fundamental problem: the attempt to understand the neural and behavioral systems that are uniquely related to reinforced patterns of CNS electrical activity, and to compare the systems that are related to different CNS patterns.

One procedure for dealing with potential mediators, as defined by the first two criteria that were listed above, is to block or prevent changes in the potential mediator while operantly conditioning a pattern of CNS electrical activity. If the pattern cannot be operantly conditioned while the potential mediator is blocked or held in a constant state, the mediation hypothesis would be supported. On the other hand, if the pattern can be operantly conditioned while the potential mediator is blocked or held in a constant state, the mediation hypothesis would not be supported.

An obvious method for dealing with peripheral skeletal mediators is the curarization procedure. Curare-like drugs, such as d-tubocurarine chloride or Gallamine, block transmission at the neuromuscular junction. Sensory input and central processing of that input can go on, as well as transmission of information from the brain to the spinal motoneuron. The information flow, however, is blocked at the junction between nerve and muscle so that no skeletal movement and no movement-produced feedback can occur. Therefore, if the operant conditioning of neural responses is peripherally mediated by skeletal responses, operant conditioning should not occur under curare-like drugs.

We have studied the operant conditioning of both RSA and non-RSA patterns of hippocampal electrical activity in completely curarized dogs (Black, Young & Batenchuk, 1970). Data for seven dogs reinforced for RSA and six dogs reinforced for non-RSA patterns are shown in Fig. 4. The appropriate response was learned in both groups. Therefore, the hypothesis that these hippocampal patterns are mediated by overt skeletal movement can be rejected.

A question naturally arises, at this point, as to the possibility of the mediation of hippocampal EEG patterns in curarized dogs by events other than peripheral skeletal responses. Mediation by autonomic responses, which is one such possibility, is an unlikely one.

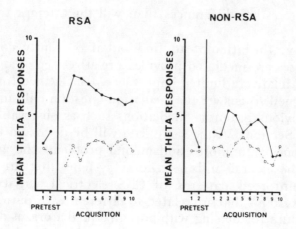

Fig. 4. Mean number of RSA responses for a group trained to make RSA, and for a group trained to make non-RSA under Gallamine paralysis. Data are averaged for blocks of two trials. During pretest, data are presented for S^D ●—● and blank ○---○ trials. (On blank trials, no stimuli were presented.) During acquisition, data are presented for the first 5 seconds of the S^D on test trials and for the 5-second blank trials. (From Black et al., 1970.) Copyright 1970 by the American Psychological Association and reproduced by permission.

Our observations indicate that curare-like drugs can produce fairly extreme changes in patterns of autonomic responding without correlated changes in operantly conditioned patterns of hippocampal electrical activity. Another possibility is that perhaps some other CNS process was operantly conditioned under curare, and this mediated the hippocampal EEG patterns. Fortunately, we do not have to worry about this possibility. The original purpose of this research on mediation was to support the view that the results of operant neural conditioning were novel, in the sense that we could operantly condition changes in central neural events independently of changes in some peripheral response whose operant conditioning has been repeatedly demonstrated in the past. For this purpose, it does not matter whether central mediation occurs. When it does not occur, the reinforced central event is operantly conditioned; when it does occur, some other central event is operantly conditioned. In both cases, we have demonstrated the operant conditioning of a central event without the involvement of peripheral responses.

I do not want to give the impression that research on the central events that precede and are necessary for the occurrence of the reinforced neural process is unimportant. As I made clear earlier in this section, such research provides information on the neural systems of which the reinforced pattern of CNS electrical activity is a component. This is very useful information. My only point is that, for the methodological question that we are considering, it is not necessary to concern oneself about these central events as mediators.[3]

Table I lists experiments which have ruled out peripheral skeletal mediation by blocking procedures (either by curare, or by dissociative conditioning), in which the S is reinforced only if the neural pattern occurs while the potential mediator is unchanged. (These procedures are discussed in more detail in Section IV.) The number of experiments is surprisingly few. There are enough however, to indicate that at least some neural events can be operantly conditioned without skeletal mediation.

C. EFFICIENCY OF CONDITIONING: FEEDBACK AND CONSTRAINTS ON CONDITIONING

Having established that at least some of the changes in reinforced patterns of CNS electrical activity in operant conditioning situations can be attributed to the direct effects of the response-reinforcer contingency, we can now consider the procedural manipulations that might facilitate or hinder the operant conditioning of neural events. This discussion will, of necessity, be brief because there are no experiments in which parametric analyses of the CNS conditioning procedure have been carried out. The research on human subjects provides what could best be described as a few hints. Nowlis and Kamiya (1970) have shown that more Ss could be conditioned to increase and decrease the number of alpha waves per unit time when they were trained with eyes open than with their eyes closed. Peper (1970) indicated that the same differences were found when his results were compared to those of Waitzkin (personal communication, cited in Peper, 1970). Also, once Ss have learned to increase and decrease the density of alpha waves, they can continue to do so for a brief period of time after feedback and reinforcement have been omitted (Peper & Mulholland, 1970).

It is somewhat surprising that so little formal research has been carried out on variables related to the efficiency of CNS conditioning. There are two classes of variables that are noteworthy in this respect. The first, which concerns the effects of feedback about response state, is important because there seems to be some question

---

[3] This is different from the situation that is found in research on operant autonomic conditioning in which we seem unable to refrain from worrying the mediation issue (Black, 1971a; Crider, Schwartz & Shnidman, 1969; Katkin & Murray, 1968; Katkin, Murray, & Lachman, 1969; Miller 1969).

about its role in the operant neural conditioning of human Ss.[4] The second, which concerns variables which provide limitations on the operant conditioning of neural events, is important for practical applications. If the situations in which an S can be conditioned or can perform a previously conditioned response are very limited, practical applications of the technique will be correspondingly few. I shall discuss each of these issues briefly.

1. Feedback about Response State and Reinforcement

The main question concerning the addition of feedback after a response is whether it is a sufficient condition for reinforcement. Consider a standard experiment in which a rat must press a lever in the presence of an auditory S^D in order to obtain a food reinforcement. The S^D is presented, and the rat presses the lever. This is followed by added feedback (e.g., the noise of a switch that is operated when the lever is depressed). The operation of the switch is followed by a second noise that is produced by the food magazine. The rat then leaves the vicinity of the lever and approaches the food cup. Finally, it picks up the food and eats.

The stimuli in this sequence can have a variety of functions. First, they can act as reinforcers, increasing the probability of responses which they follow. Second, they can act as discriminative stimuli; their presentation can lead to the occurrence of a response. Third, they can provide information about the state of the response. Two types of feedback about response state can be distinguished—naturally occurring feedback (e.g., proprioceptive feedback from movement), and feedback added by the experimenter (e.g., the switch noise that follows a lever press).

While it is difficult to specify the conditions that must be met for a stimulus to be a reinforcer, we can do so roughly in at least some cases. Stimuli which provide some consequence that is important for the internal economy of the organism are reinforcers (e.g., food in the above example). Also, stimuli which originally did not have reinforcing powers can acquire them by being paired with other stimuli that are reinforcers. They can become "conditioned reinforcers" (e.g., the noise of the switch and of the magazine in the above example). Arranging to have a stimulus provide feedback about response state, however, does not seem to be a sufficient

---

[4] A great deal, of course, has been done on feedback produced by cortical EEG activity in nonoperant conditioning procedures (Mulholland, 1968, 1969, 1972; Peper, 1970). Also, considerable research has been carried out on the role of feedback produced by electromyographic (EMG) and autonomic responses in operant conditioning situations (e.g., Basmajian & Simard, 1967; Brener, Kleinman, & Goesling, 1969; Green, Green, & Walters, 1970a).

condition for making it a reinforcer. If, in the example described above, one had simply added the noise of the switch after each lever press without arranging to have food follow it, there would have been no increase in the probability of the lever press. It might be, of course, that added feedback will only act as a reinforcer for internal responses which have no naturally occurring feedback. But this does not seem to work either. We have added clicks and tones after the occurrence of hippocampal theta waves in rats, and this, by itself, has not produced any apparent learning.

One might argue that human Ss are different from infrahuman subjects, and that for human subjects, added feedback is a sufficient condition for reinforcement. This seems to be the implicit assumption in much of the human research where the training procedure is described as "feedback control" or "biofeedback" as often as "operant conditioning." Clear-cut evidence on this point is not available. Nevertheless, it seems to me that the most likely explanation of the operant conditioning of neural activities in human Ss when the only apparent reinforcer is feedback about response state, is that other subtle reinforcers are established by instructions and by the previous history of the S. Human Ss are usually motivated to cooperate with the requirements of the E (Orne, 1962), and therefore, feedback could be a reinforcer, not so much because it provides information about response state, but because it indicates successful performance. Obviously, it would be extremely useful to have more data on this point. As Peper and Mulholland (1970) have pointed out,

The status of the feedback stimulus is ambiguous. Is it analogous to the (a) proprioceptive, visual, tactual and acoustic feedback stimuli which inform the monkey that it has pressed a key? Or is it (b) reinforcement, or (c) both? This is a fundamental point that can be examined experimentally [Peper & Mulholland, 1970, p. 12].

That feedback about response state may not be sufficient in itself to produce learning does not, of course, rule out the possibility that it plays an important ancillary role in operant conditioning. The unwritten lore of operant conditioning laboratories passes the dictum on from generation to generation that added feedback about the response facilitates operant conditioning in certain situations. Such feedback is also considered very important in the training of complex motor skills (Bilodeau, 1969). While it is usually assumed that the ancillary role of such added feedback is to facilitate operant conditioning, this need not always be the case. Rosenfeld, Rudell, and Fox (1969) employed money to reinforce increases and decreases in the voltage of late components of an auditory evoked potential in human Ss. One group of Ss received further information about their responses by watching the evoked potential on an oscilloscope, while a second group did not receive such additional information. They found no obvious differences in performance between the two groups. In fact, some of the Ss complained that feedback from the oscilloscope distracted them from the task.

In research on operant neural conditioning in human Ss, there is a further role assigned to feedback about response state—that is, to make the S aware of his internal responses. The purpose of much of the research on human Ss is to train them to obtain voluntary or self-control over their own internal processes. (In this discussion, I shall treat voluntary control as a more complex phenomenon than simple operant conditioning. See Section IV, D.) Although there are many ways of defining voluntary control, the view seems to be accepted in this research that voluntary control in human Ss implies a conscious and deliberate decision to perform a response and an awareness of the response as it occurs. Given this view, one essential step for achieving voluntary control is to make the S aware of the response when normally he is not aware of it.

There are at least two distinct conditioning methods that might be employed to make an S "aware" of his internal responses. One is to employ the internal response as an S^D in an operant conditioning situation. An experiment on human Ss that was described by Kamiya (1969) illustrates this approach. In this experiment, the presence and absence of alpha waves served as discriminative stimuli. When alpha waves were occurring, the Ss were reinforced only if they made the appropriate identifying response—A. When alpha waves were not occurring, the Ss were reinforced only if they made the correct identifying response—B. After Ss had learned this discrimination, they were asked to produce or refrain from producing alpha, and they could do so. It would seem, then, that employing an internal pattern of electrical activity as an S^D led to voluntary control over that pattern. It would be interesting to see whether classical discriminative conditioning procedures, such as those in which internal autonomic states were employed as conditioned stimuli (Adam, 1967; Bykov, 1959; Razran, 1961), would lead to the same sort of voluntary control.

The second method for making an S aware of his own internal responses is to add feedback after the response. We have discussed this procedure at length and have questioned whether it is sufficient to produce operant conditioning. The same questions can be raised about its role in the establishment of voluntary control.

It would seem that many assumptions concerning the role of feedback, especially in human research on operant neural conditioning, are on shaky grounds. We require more data in order to determine whether added feedback about response state is a sufficient condition for reinforcement in human Ss. We require more data on the ancillary function of such feedback in the training of complex responses. When does it facilitate, and when does it hinder conditioning? We need similar data about the role of added feedback in establishing voluntary control.

2. Constraints on the Operant Conditioning of Neural Events

The second area in which more parametric research is needed concerns the constraints or limitations on the operant conditioning of neural events. The assumption has often been made that, if we can condition a given response in one situation, we can condition any other response in that situation. As has been emphasized in a number of recent papers (Black & Young, 1972; Bolles, 1970; Seligman, 1970; Shettleworth, 1972), this assumption concerning the interchangeability of elements of the conditioning situation is not correct for certain observable responses. A given set of procedural conditions can interfere with the acquisition or performance of certain types of responses but not of others. For example, rats that were deprived of water were trained to press a lever to avoid shock in the presence of one S^D and to drink water to avoid shock in the presence of a second S^D (Black & Young, 1972). The rats were trained until they were avoiding shock regularly, both by lever pressing and drinking. Then they were satiated. The rats continued to press the lever to avoid shock, but the performance of the drinking response deteriorated. In this case, the drinking response was more constrained than the lever-pressing response with respect to conditions under which successful performance could occur.

Similar constraints on the operant conditioning and performance of CNS electrical activity have been observed. Paskewitz and Orne (1971), for example, first operantly conditioned Ss to produce cortical alpha waves. They then required the Ss to continue performing the conditioned alpha wave response while counting backward by sevens. This produced a deterioration in alpha-wave performance. If other operantly conditioned neural patterns do not show the same deterioration under the same conditions, one might conclude that the performance of operantly conditioned alpha waves is constrained by the concurrent behavior of the Ss. Similar constraints have been shown for hippocampal EEG (Black, 1971b). Rats could be operantly conditioned to make high frequency RSA waves while moving, but not while holding still; certain other EEG patterns do not seem to be limited in this manner.

Without further knowledge of such constraints, we will not be able to accurately assess the extent to which the operant conditioning of particular neural responses can be used in practical situations. If an operantly conditioned neural response, for example, could be performed only under laboratory conditions, it would be of little use in practice.

D. SUMMARY

The first question discussed in this section of the chapter asked whether the changes that were observed in reinforced patterns of CNS electrical activity were produced by the response-reinforcer contingency, or by some other variables. The answer to this question is clear. The evidence indicates that the changes in the patterns of CNS electrical activity that have been studied were produced by the contingency between response and reinforcer.

The second question concerned the attempt to find out whether changes in patterns of neural activity were operantly conditioned directly, or were mediated. The possibility of peripheral mediation by skeletal responses was ruled out in at least a few cases. The data that were employed in the attempt to rule out mediation can also be employed in the attempt to understand the neural and behavioral systems that are related to operantly conditioned patterns of CNS electrical activity. The latter issue is, I think, more important than the former.

The third question concerned the variables that control the process of conditioning. There is a surprising lack of parametric research on this question, especially with respect to the role of feedback about response state, and constraints on operant neural conditioning. Perhaps a concern with the response-reinforcer contingency and with mediation, while understandable during the early stages of development of this research area, has captured our attention for too long, and has prevented us from progressing as rapidly as we should have on the third question. At this stage, we should know more about the learning process and about efficient training procedures.

III. Significance of the Research: Goals

In this section, I shall deal with the significance of the research on operant neural conditioning by describing its goals, and the questions that must be answered if we are to assess the extent to which these goals have been achieved. In Section IV, I shall discuss the experimental results which provide answers to these questions.

A. The Control of Internal Neural and Psychological Processes

The goal that probably is most familiar is to obtain control over the internal neural and psychological processes of others, or to train Ss to obtain voluntary or self-control over their own internal processes. This goal, which is most frequently expressed in research on human Ss, produces optimistic and pessimistic reactions whose only common feature is the intensity with which they are expressed. On the one hand, we have the hope that the method (especially when it increases self-control) will lead to a better future. As Green, Green, and Walters (1970b) put it:

The importance to our culture of this now-developing methodology for enhancing voluntary control of internal states can hardly be overstated. . . . Without stretching the imagination, the long-range implications and the effects for society of a population of self-regulating individuals could be of incalculable significance [Green *et al.*, 1970b, pp. 1-2].

On the other hand, there is fear that the method will lead to a worse future in which it will provide a powerful technique for thought control and the like. Krutch (1953), for example, has labeled future societies in which we have "the scientific ability to control men's thoughts with precision" as "ignoble utopias."

Those who are interested in this goal assume that obtaining control over the electrical activity of the brain gives us control over important internal neural and psychological processes. Suppose, for example, that one operantly conditions two EEG patterns recorded from the same location over the cortex, employing the same reinforcer and schedule of reinforcement. During the performance of the first EEG pattern, the subject is angry, tense, and agitated; during the performance of the second pattern, he is serene, calm, and relaxed. The conclusion usually drawn from such data is that control over these specific EEG patterns also gives control over the related mood states.

So, the first question that we must answer in order to evaluate operant neural conditioning as a method of control is the following:

Question 1. Over what neural and psychological processes, if any, do we obtain control when we operantly condition a particular pattern of CNS electrical activity?

The attempt to answer this question leads to a further question. How do we know which internal neural and psychological processes are controlled, when we control a particular pattern of CNS electrical activity? One can refer to the literature on the neural and behavioral processes that are related to such patterns of electrical activity in order to obtain the required information, or, if it is not available there, try to provide the information by employing traditional methods for studying brain-behavior relationships—correlating neural activity with behavior in natural situations, analyzing the effects of brain stimulation and lesions, etc. We cannot be sure, however, that processes which are related to a pattern of CNS electrical activity in natural situations will also be related to the pattern in operant conditioning situations. It would seem necessary, therefore, to examine the relationship between the reinforced pattern of CNS electrical activity and other measures of neural activity and behavior in the operant conditioning situation. I shall discuss this approach in the next section, since it plays a role in a second major goal of research on operant neural conditioning.

B. THE ANALYSIS OF THE FUNCTIONAL SIGNIFICANCE OF PATTERNS OF CNS ELECTRICAL ACTIVITY

The second goal of research on operant neural conditioning, which is, perhaps, less familiar than the first, is to employ operant neural conditioning procedures to obtain information about the neural, behavioral, and psychological processes that are represented by particular patterns of CNS electrical activity.

Successful operant conditioning of a pattern of CNS electrical activity, by itself, will not provide the required information. It does not tell us what these processes are; it tells us only that the reinforced pattern is part of some neural system that can be operantly conditioned. Several modifications of the operant conditioning procedure have been employed in attempts to obtain the required information. These will be described in Section IV. The main feature which these procedures share is that they examine the relationship between the reinforced pattern of CNS electrical activity and other measures of neural activity and behavior.

Data on such relationships are employed in several ways. The simplest, which might be labeled *the relational approach,* is to attempt to infer the nature of the internal processes to which the reinforced pattern of CNS electrical activity is related by looking for measures which are highly correlated with the reinforced pattern. In the hypothetical example described in Section III, A that relates cortical EEG patterns to mood states, one might have employed correlations between introspective reports of internal feelings and patterns of electrical activity in order to conclude that the patterns were related to mood states.

Another approach, which can be called the *systems approach,* employs the same types of data to make what is, perhaps, a more ambitious analysis. A given EEG pattern, for example, can be studied in at least two ways—first, as a summated index of the electrical activity of individual neurons near the recording site, and second, as representing processes which are components of functionally important neural circuits or systems. In the systems approach, one is concerned with the latter of these two possibilities, and attempts to make inferences about the structure and function of the system of which a reinforced pattern of CNS activity is a component (i.e., whether it is a system concerned with sensory, attentional, motivational, motor, or other functions). Also, one might attempt to identify the function of processes represented by a given pattern of electrical activity within the systems of which they are components. For example, after identifying a particular pattern of spinal neuron activity as a component of a motor system, we might go on to show that its function within the system is to relay information from muscle receptors to cortical motor areas.

For the sake of brevity, the attempt to make such relational and systems analyses will be referred to as the attempt to understand functional significance of patterns of CNS electrical activity. The question that we must answer, therefore, in order to evaluate operant neural conditioning as an analytic tool can be stated as follows:

Question 2. What information, if any, does the operant conditioning of patterns of CNS electrical activity provide about their functional significance?

There is one further point that should be made about this question. As Fox and Rudell (1970) and Mulholland (1969) have pointed out recently in papers on operant neural conditioning, the analysis of the functional significance of patterns of CNS electrical activity is a complex and difficult task. For example, we can take a number of different measures of the electrical activity in a particular location of the brain, and it is not always clear which best represents significant neural processes in that location. Also, a given pattern of neural electrical activity could function in different neural systems, and different patterns could play the same role in a given neural system The question, therefore, is more difficult to answer than one might expect at first glance.

C. Comparisons of Operant Neural Conditioning with Other Procedures

One might infer from the preceding discussion that the control of CNS electrical activity by operant conditioning is a new achievement.

But this inference is not necessarily true; we have always been able to bring neural processes under operant control. For example, whenever we operantly condition the ubiquitous bar-pressing response, we bring under operant control processes in neural systems which control the reinforced movements as well as processes in related CNS systems. Therefore, one is curious as to whether there is any difference in the control over internal neural and psychological processes that we achieve when we operantly condition the electrical activity of the brain as compared to observable responses.

This comparison brings to mind another. A variety of techniques are already available for controlling internal neural and psychological processes—drugs, electrical stimulation of the brain, and so on. Questions naturally arise, therefore, about the differences between operant neural conditioning and other methods for controlling these internal processes. We can ask if there are differences in the *type of process* that can be controlled by each method; we can also ask if there are differences in *type of control* that can be exerted by each method. If the operant conditioning of neural activity is not different from, and in some sense "better" than other methods of control, there would be little point to its use. Therefore, we must also answer the following question:

Question 3. Does the control provided by the operant conditioning of CNS electrical activity differ from that obtained by other methods for controlling internal neural and psychological processes, and if so, how?

As was noted above, many methods have been devised to study functional significance—brain-behavior correlations in natural situations, analysis of the effects of lesions and of electrical stimulations, and so on. If the operant conditioning of neural activity does not complement these more familiar methods and, hopefully, provide some new insights, there would be little point in employing it. Therefore, we must answer the following question:

Question 4. Does the operant conditioning of CNS electrical activity provide information on the functional significance of the reinforced patterns of electrical activity that is different from the information obtained by more traditional methods, and if so, how?

D. THE ANALYSIS OF THE OPERANT CONDITIONING PROCESS

A third goal is to employ operant neural conditioning to inform us about the process of operant conditioning itself (Olds, 1965). Perhaps operant neural conditioning can provide information on the locus of the neural changes that are necessary for operant conditioning, on the neural circuits by means of which the reinforcement produces these changes, and on the properties shared by responses that are amenable to operant conditioning.

One intriguing question concerns the operant conditioning of a neural event that is involved in the process of operant conditioning. Suppose, for example, that a given pattern of neural electrical activity represents some process that is necessary for operant conditioning. Could we operantly condition this pattern of neural activity which seems to be part of a neural system that produces operant conditioning? One pattern of operantly conditioned neural electrical activity that might be related to neural circuits that are involved in the conditioning process is the postreinforcement synchronization recorded from the posterior cortex of cats (Sterman, Wyrwicka, & Roth, 1969c). This EEG pattern is a fairly regular waveform between 4 and 12 Hz that occurs during and just after the consumption of a liquid reward. Because of its occurrence in conjunction with reinforcement, the pattern could be related to neural circuits by means of which reinforcement changes the probability of the response. It could, however, just as easily be related to motivational level rather than to circuits involved in learning. There are, unfortunately, no other examples of operantly conditioned patterns of CNS electrical activity that might be related to the conditioning process. Because of this paucity of relevant data, the attempt to understand more about the operant conditioning procedure will not be discussed further.

E. Summary

The above discussion suggests that we should attempt to answer four questions in order to assess the extent to which the goals of research on operant neural conditioning have been achieved. The first is concerned with the use of operant neural conditioning to obtain control over internal processes. The second is concerned with the use of operant neural conditioning to obtain information on the functional significance of patterns of CNS electrical activity. A comparison of the first two questions reveals that they are very closely related. When we obtain information about the functional significance of the reinforced CNS pattern, we are also obtaining information about the neural and behavioral processes over which we gain control when we operantly condition that pattern. The third and fourth questions concern the comparison of operant neural conditioning with other methods for obtaining control and for studying functional significance. Each of these questions will be discussed in Section IV.

IV. Significance of the Research: Success in Achieving Goals

The purpose of this section is to determine the extent to which research on operant neural conditioning has achieved its goals by answering the four questions that were posed in Section III. The section is divided into three parts. The first deals with the information on functional significance that is provided by research on operant neural conditioning. The second deals with the processes over which we obtain control by means of operant neural conditioning. The third deals with the comparison between operant neural conditioning and other procedures for analyzing functional significance and for controlling internal processes.

A. THE FUNCTIONAL SIGNIFICANCE OF OPERANTLY CONDITIONED CNS ELECTRICAL ACTIVITY

What information, if any, does the operant conditioning of patterns of CNS electrical activity provide about their functional significance?

As was pointed out in the previous section, we require information beyond the fact that we have operantly conditioned a pattern of CNS electrical activity in order to deal with its functional significance. Two modifications of the basic operant conditioning procedure have been employed to provide the required information—the concomitant measures procedure, and the transfer procedure. I shall discuss data obtained by each method separately.

1. Concomitant Measures

In the *concomitant measures* procedure, which is the most commonly employed, we analyze the relationships between the reinforced pattern of electrical activity and other concomitantly measured patterns of neural activity and behavior. The steps in this approach are: first, to identify the concomitantly measured patterns of neural activity and behavior that are highly correlated with the reinforced patterns of CNS electrical activity; second, to study the nature of the relationship among these events; and third, to make inferences from these and other available data, about the functional significance of the reinforced pattern of electrical activity.

a. Hippocampal EEG in Dogs and Rats. One example that illustrates this method is our research on hippocampal EEG patterns (Black, 1971b; Black *et al.*, 1970; Dalton, 1969). In order to identify some of the concomitant measures that might be related to reinforced patterns of hippocampal electrical activity, we videotaped overt skeletal behavior and recorded heart rate and cortical EEG.

During the occurrence of operantly conditioned hippocampal RSA (see Section II, A, 1), cortical desynchronization was observed, heart rate was high, and the dogs moved about a great deal (turning their heads, lifting their legs, etc.). One might be tempted to conclude that all of these changes were related to the RSA response. But this conclusion could be wrong as the following discussion will make clear.

Figure 5 presents schematic diagrams which illustrate in an extremely oversimplified manner the types of changes that might occur after different responses had been reinforced in the presence of an S^D. In Fig. 5A, the reinforced response is some skeletal movement. The S^D is presented; circuits involved in the reception of the S^D are activated; central integrating circuits process the input; then the motor system controlling the response is activated. In Fig. 5B, the reinforced response is some pattern of CNS electrical activity. The same circuits are activated after the presentation of the S^D except for the final one which is a circuit of which the reinforced pattern is a component. (The situation in Fig. 5C, where the reinforced event is a component of the integrative system involved in the learning process, provides some interesting complexities, but, fortunately, the data have not yet compelled us to deal with this alternative.) Certain neural processes which occur during the performance of the operant-

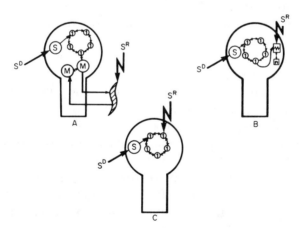

Fig. 5. Schematic diagrams illustrating the possible connections between the discriminative stimulus (S^D) and the response that was followed by reinforcement (S^R) after operant conditioning has taken place. The keyslot form represents the central nervous system; the small squares and circles within it represent structures of the central nervous system that might be involved in the stimulus-response connection. Figure 5A illustrates the situation that might exist if some observable skeletal response had been reinforced. Figures 5B and 5C illustrate the situation that might exist if some patterns of central nervous activity had been reinforced. S-circuits are involved in the input of information. I-circuits are involved in central processing. M-circuits are involved in the control of movement.

ly conditioned response are the same when both observable responses and neural events are reinforced; others are different because they are associated with a particular response that is reinforced. It is the latter in which we are interested. If we were interested in the former, there would be little point to the use of operant neural conditioning; we could study the former just as well by reinforcing more familiar observable responses.

The problem, of course, is to distinguish between those changes to the S^D in concurrently measured events that are uniquely related to the reinforced response, and those that are not. One way of making this distinction is to reinforce two different patterns of CNS electrical activity, and to compare the concomitant measures in the presence of the two patterns of electrical activity. Presumably, only those concomitant measures which are related to the type of response would be different.

We have carried out several experiments in which two different hippocampal EEG patterns were reinforced, RSA and non-RSA (Black, 1971b; Black & de Toledo, 1972; Black et al., 1970). Overt behavior differed in the presence of the two responses. As is illustrated in Fig. 6 where movements during RSA and non-RSA for a single dog are shown, RSA was accompanied by skeletal movements such as head turning, struggling, and so on, and non-RSA was associated with either holding still or licking. Heart rate increased during the S^D for the RSA response and remained the same or decreased during the S^D for the non-RSA response when non-RSA was accompanied by holding still. But when non-RSA was accompanied by consumatory and instinctive responses, such as drinking or body licking, heart rate tended to be high. There were no apparent differences in cortical electrical activity; desynchronization was observed during both RSA and non-RSA. It would seem, then, that the cortical EEG and heart-rate changes were not related to the reinforced CNS response, while overt behavior was.

Our next step was to attempt to analyze the relationships between these concomitant measures of overt behavior and the reinforced patterns of CNS electrical activity. One question which we considered was whether the relationship was symmetrical with respect to the administration of reinforcement. The reinforcement of hippocampal EEG patterns led to correlated changes in observable behavior. Would the reinforcement of overt behavior of the appropriate type lead to correlated changes in hippocampal electrical activity? The answer is yes. Reinforcement of pedal pressing and lever pressing was accompanied by high-frequency RSA, and reinforcement of holding still or drinking was not (Black & Young, 1972a, b).

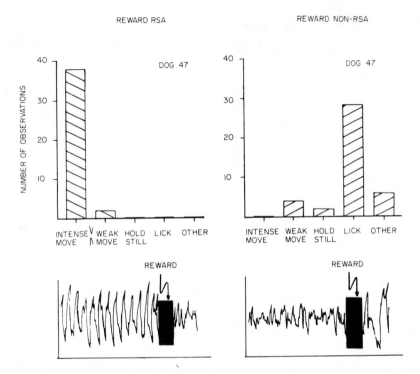

Fig. 6. The frequency of various types of activity while one dog was being rewarded for specific patterns of hippocampal electrical activity. Data when the dog was being reinforced for RSA are presented on the left; data for the same dog when it was being reinforced for non-RSA are presented on the right. Examples of RSA and non-RSA patterns of electrical activity are given under the appropriate graphs. Forty observations were made for each graph. The categories of movement are as follows: *Intense Move*—Clear-cut movement of the head, body, and limbs; *Weak Move*—Slight movements of the head, twitches, and eye blinks; *Hold Still and Lick*—These labels are self-explanatory. Most of the responses in the *Other* category were yawning. (From Black, 1971b.)

The symmetry of these results suggests that the patterns of hippocampal electrical activity are closely related to overt behavior; it does not tell us, however, whether one is necessary for the occurrence of the other. One method for determining whether they are necessarily related in the operant conditioning situation is to carry out what may be called *dissociative* conditioning—that is, to block or hold constant one of the related responses while operantly conditioning the other.[5] The research that we described in Section II, B on

---

[5] Schwartz (1971), in research on operant autonomic conditioning, has discussed similar procedures. He refers to the operant conditioning of increases in one response and no change or decreases in another, as *differentiation*, and to the simultaneous operant conditioning of changes in the same direction in the two responses as *integration*.

When these methods are employed, a problem of interpretation arises when one has demonstrated the successful operant conditioning of a given response while holding constant

mediation provided one example of this method (Black *et al.*, 1970). RSA was conditioned, while overt skeletal responding was blocked by Gallamine; therefore, we can conclude that the occurrence of overt behavior is not necessary for the operant conditioning of the hippocampal RSA pattern. The possibility still remains that the central components of the neural circuits controlling overt skeletal movements are necessarily related to the RSA patterns, because curare-like drugs do not block activity in these central circuits. Another dissociative conditioning procedure which deals with this possibility is based on the assumption that holding still keeps at least some central components of the system controlling skeletal movement in a steady state. If an S can learn to make hippocampal RSA responses while simultaneously holding still, then hippocampal RSA and central movement control systems would not necessarily be related. We have attempted to operantly condition rats to hold still and make hippocampal RSA. The reinforcement was water. The results indicate that we can operantly condition rats to make low-frequency RSA while holding still, but not high-frequency RSA. Fig. 7 presents examples of EEG records for one rat. Examples are shown of RSA responses which occurred during the final stage of training when the rat was reinforced for making RSA responses while holding still, and when it was reinforced for making RSA responses while moving. In each condition a shaping procedure was employed to increase the frequency of the RSA response. Four reinforced RSA responses during each phase of training are shown. The form of the RSA wave is less regular and the frequency of RSA lower when the rat is required to hold still while making RSA than when it is required to move while making RSA. This result, then, suggests that certain types of skeletal movement are necessarily related to high-frequency RSA, at least in the very limited situations that we have explored so far.

The next step in this approach is to attempt to make some inferences about the functional significance of the reinforced pattern of electrical activity. We should, of course, discuss data in addition to those obtained in the operant conditioning experiments described above. Space limitations do not permit us to deal with this extensive literature which has been reviewed recently by Bennett (1971), Gray

---

a second response. If the two responses were correlated before conditioning, we do not know whether the dissociative procedure simply *revealed* a lack of relationship that had not been demonstrated before, or actually *modified* the nature of the relationship between the two responses. In the latter case, the dissociative procedure would not provide information on the nature of the relationship between the two responses before conditioning. This issue is discussed further in Section IV, C, 2, a.

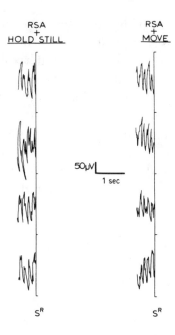

Fig. 7. Examples of reinforced RSA responses. The four examples on the left are from a session in which a rat was required to simultaneously hold still and produce three consecutive RSA waves in order to obtain water reinforcement (S^R); the four examples on the right are from a session in which the rat was required to simultaneously move and produce three consecutive RSA waves in order to obtain water reinforcement. The rats were shaped to make higher frequency RSA under each condition by increasing a criterion level above which the frequency of RSA had to be for reinforcements to be administered. The criterion level was increased during each session until fewer than six reinforcements per 100 seconds occurred. The first four responses are shown for the period of each session during which the criterion level was at its highest. The criterion was 6.0 Hz in the hold still condition, and 7.6 Hz in the move condition.

(1970), and Vanderwolf (1969, 1971). It is worth mentioning, however, some examples of apparent exceptions to the relationship between high-frequency RSA and skeletal responses such as lever pressing. Gray reported hippocampal RSA in the 7.5–8.5 Hz range in rats that were holding still. Bennett also reported that trains of relatively high-frequency hippocampal RSA waves occur during periods in which animals were holding still. Still other exceptions could be described. Whether these are related to species differences, differences in the function of different parts of the hippocampus, problems in recording behavior, or whether they are actual exceptions to the hypothesis, has yet to be determined.

The relationship between higher frequency RSA and skeletal responses such as bar pressing, if it is confirmed by further research, is consistent with two types of hypotheses. The first is the hypothesis that the neural processes represented by both RSA and skeletal behavior belong to the same system. Vanderwolf (1969, 1971), for example, has suggested that RSA represents processes in a motor system that controls voluntary behavior, and Klemm (1970) has suggested that RSA represents processes in the system that produces EMG changes in skeletal muscles. The second hypothesis is that high-frequency RSA belongs to a neural system that is different from the system that controls skeletal responses such as bar pressing, and that the two systems are necessarily related. For example, hippocampal RSA may represent some event in a neural system controlling attention or orientation to external stimuli, and this system is activated only during certain types of movement. Another possibility is that RSA may be involved in producing corollary discharge for certain skeletal movements. The present results, then, are consistent with hypotheses that suggest that high-frequency hippocampal RSA represents processes in neural systems controlling certain types of skeletal movement, or in other neural systems that are congruent with the systems that control these types of skeletal movement.

b. Motor Cortex Single-Cell Activity in Monkeys. One particularly elegant application of the concomitant measures method is provided by the research of Fetz and Finocchio (1971). They operantly conditioned single-cell activity of neurons in the precentral cortex of monkeys using food reinforcement. The monkey was kept in a restraining chair with a special cast holding one of its arms so that the arm could be moved at the elbow, or held rigidly in place. In addition to single-cell activity and arm movement, electromyographic activity (EMG) was measured from four arm muscles—the two major flexors and extensors of the wrist, and the two major flexors and extensors of the elbow.

When the arm was moved at the elbow for food reinforcement, EMG activity in the flexor and extensor muscles and activity of the single unit were related in the manner that is illustrated in Fig. 8A. Unit activity was most closely related to flexor activity of the biceps muscle. The cast was then locked firmly in place so that the arm could not be moved. Unit activity was reinforced, and correlated changes in isometric EMG activity were observed. As is shown in Fig. 8B, firing of the unit was correlated with a change in biceps EMG activity. Next, an attempt was made to dissociate unit activity and muscle activity. One of the patterns for which the monkey was reinforced was to fire the unit while refraining from EMG activity. As can be seen from Fig. 8C, the monkey learned to fire the unit while keeping EMG activity at practically zero level. Then the reverse dissociation of cell and biceps EMG activity was attempted. The monkey was reinforced for isometric biceps EMG activity while refraining from firing the unit. As is shown in Fig. 8D, the monkey failed to learn to suppress unit activity completely. There was a change in the reinforced direction, but the total suppression of unit activity during biceps EMG activity was not achieved. Because of the length of the training session, however, it is not clear whether this failure to produce dissociation was due to some physiological breakdown (Fetz, 1971 personal communication). The complex relationship between unit activity and EMG activity of the biceps that was

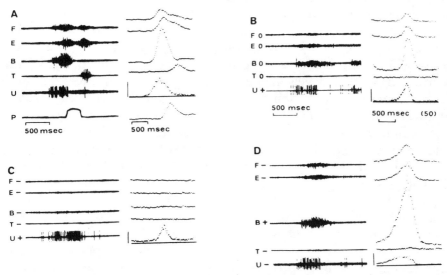

Fig. 8. Responses of precentral cell and arm muscles during active elbow movements. Successive lines from top to bottom show activity of flexor carpi radialis (F), extensor carpi radialis (E), biceps (B), triceps (T), cortical unit (U), and the position of the elbow (P). A single trial is shown at left, and the averages over 60 successive trials at right. All EMG averages were computed at identical gains. Time histogram of cell activity is shown with a zero baseline and a vertical calibration bar for 50 impulses per second. Upward deflection of the position monitor represents flexion. B, C, D: Operant reinforcement of patterns of neural and muscular activity under isometric conditions. Muscles and unit are labeled + if their activity produced reinforcement, − if their activity prevented reinforcement, and 0 if their activity was not included in the reinforcement contingency. In B and C, the monkey was reinforced for bursts of cortical cell activity, first with no contingency on the muscle (B), then requiring simultaneous suppression of all muscle activity (C). In D, the monkey was reinforced for isometric contractions of the biceps, requiring simultaneous suppression of all cortical activity. Averages were computed for 50 successive responses, with identical vertical scale on EMG averages. Vertical bars on time histograms of unit activity represent 50 impulses per second. (From Fetz and Finocchio, 1971.) Copyright 1971 by the American Association for the Advancement of Science.

revealed by this experiment provides a very convincing demonstration of the power of the dissociative conditioning method in exploring the relationship between reinforced neural events and concomitant measures of behavior.

c. Evoked Potentials to Light in Cats. Both examples of the concomitant measures method that I have just described carried out a series of steps, ending with explicit dissociative conditioning of two responses that were observed to be related during the early stages of the procedure. Very useful information on functional significance can be obtained, however, even when only the first stages of the procedure are carried out. This is especially true of those cases in which one fails to find a relationship between the reinforced pattern of neural activity and concurrently measured responses that is expected on the basis of previous data. Research by Fox and Rudell (1968, 1970) illustrates this point. Fox and Rudell first determined the form of the average evoked potentials to light in cats. They then chose a brief interval at a fixed period after the onset of light, and reinforced voltage levels higher and lower than the average level observed before conditioning began. They examined the relationship of the reinforced component of the evoked potential to earlier and later components. They found that in some cases they could reinforce an increase or decrease in the voltage of the evoked potential at a given time after the onset of the light without correlated changes in the voltages before or after the reinforced component. In other cases, there seemed to be some relationship between the reinforced change and other components of the evoked potential. Fox and Rudell suggested that the independent reinforcement of components of the average evoked potential indicates that the evoked response does not represent some unitary process, nor does it represent some series of processes that are dependent on each other. This is an unexpected result if one assumes that the various components of the average evoked potential represent a sequential series of highly dependent events such as those that occur during the transmission of impulses from the periphery to the cortex.

These results obviously provide information on the neural processes represented by components of the average evoked potential in sensory systems. They also illustrate that a failure to find a relationship between the reinforced event and concomitantly measured processes can be especially informative when the relationship is expected on the basis of previous data.

d. Alpha Wave Conditioning in Humans. The patterns of neural activity that have received the most attention are the synchronous

EEG rhythms of the human cortex. In this research, the concomitant measures were usually introspective reports, although others, such as eye movements, have also been employed.

A typical experiment (Kamiya, 1969) showed that human Ss could learn to change the density of alpha waves (i.e., the number of alpha waves that occurred during a fixed period of time). The data are presented in Fig. 9.

Because different mechanisms might be involved in decreasing and increasing alpha density, I shall discuss these two responses separately. One proposal suggests that the blocking of alpha waves is produced by processes which lead to cortical desynchronization, and that these processes involve the activation of visual motor systems (Dewan, 1967; Mulholland, 1968, 1969, 1971, 1972). The S may focus his eyes, track some object, or even imagine some visual object, and this results in alpha blocking. A more general hypothesis of the same type has been proposed by Lynch and Paskewitz (1971). They suggest that decreases in alpha wave density are produced by paying attention to any stimulus. Another possible mechanism is that one can produce decreases in the density of alpha waves by becoming drowsy and producing slow-wave activity (Mulholland, 1971). Much of the data is consistent with the visual motor hypothesis. For example, alpha control is better in Ss whose eyes are open when it would be easier, presumably, to control the visual system (Nowlis & Kamiya, 1970). Also, Ss who are paying attention to stimuli but are not employing the visual motor apparatus do not show alpha blocking (Mulholland, 1972). This visual motor hypothesis has, however, received some criticism. Chapman, Cavonius, and Ernest (1971), for example, suggested that the occurrence of alpha and alpha blocking

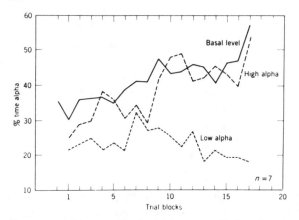

Fig. 9. Data for seven subjects each of which was studied under three conditions: (1) reinforced for increasing alpha density (high alpha); (2) reinforced for decreasing alpha density (low alpha); (3) resting. (From Kamiya, 1969.)

in eyeless Ss is inconsistent with the hypothesis.

While there is considerable agreement about the neural and psychological processes that are related to alpha blocking, there is less agreement about the processes that occur when the density of cortical alpha waves is increased. As is indicated in Fig. 9, the Ss displayed a gradual increase in the relative frequency of alpha waves during the rest period that were very similar to the changes that occurred during the period in which the Ss were trained to increase alpha-wave density. This similarity in performance during training and rest periods has raised questions about what was learned during training to increase alpha-wave density in this experiment.

A variety of hypotheses have been proposed to account for this result. According to Mulholland (1972), refraining from visual motor activity is a necessary but not sufficient condition for the occurrence of alpha activity, since alpha activity does not necessarily begin for some time after an S begins to refrain from visual motor activity. It may be, nevertheless, that the Ss learn to refrain from employing the visual motor system when they are being trained to increase the density of alpha waves, and this permits alpha-wave density to gradually increase during the experimental period. If the Ss were also gradually decreasing visual motor activity during the rest period, we would see a similar increase in alpha density. Lynch and Paskewitz (1971), and Paskewitz, Lynch, Orne, and Costello (1970) have presented a similar hypothesis in which they suggest that Ss learn to refrain from paying attention to features of the experimental situation, and this permits the alpha density to increase during both training and rest periods. A third proposal is that cortical alpha activity is related to a psychological state very similar to that which occurs during Zen or Yoga meditation, and that Ss learn to produce this state (Brown, 1970, 1971; Kamiya, 1968, 1969). Kamiya (1969) accounted for the gradual increase in alpha-wave density during the rest period in the following manner: The psychological state related to alpha activity is pleasant for some Ss. Once such an S learns to produce this state efficiently, he enters it during both the rest period and the training period because it is preferred to the non-alpha state. A fourth explanation of the production of alpha waves is that the alpha state is produced by self-induced anoxia (Watanabe, Shapiro, & Schwartz, 1971). The Ss might decrease the level of oxygen content by controlling respiration, and this could produce both the pleasant state associated with alpha and the increase in alpha density.

It would seem that the functional significance of human cortical alpha waves that occur in operant conditioning situations is still not well understood. Perhaps more research employing the concomitant measures procedure might help to resolve some of the problems that have been encountered.

2. Transfer Studies

The second operant conditioning method for studying the functional significance of reinforced patterns of CNS electrical activity is the *transfer procedure*. In the concomitant measures approach, one studies patterns of neural activity and behavior that are measured concomitantly with the reinforced pattern of CNS electrical activity; in this approach, one studies the effects of the reinforced pattern of electrical activity on some response that does not occur when the pattern is being conditioned. The procedure is as follows: Operant conditioning of a particular pattern of CNS electrical activity is carried out in the presence of an S^D. Then the S^D is presented during the occurrence of some other apparently unrelated response, and the effects of its presentation (and of the reinforced pattern which the S^D evokes) are assessed. An attempt is made to infer the types of neural and psychological processes that are related to the reinforced pattern of electrical activity from its effects on the other response.

The same precautions that were taken in studies which employed the concomitant measures method must be taken in studies which employ the transfer design. One must be sure that the effects of the superimposed S^D are related to the reinforced response rather than to some other aspect of the conditioning process. Again, one can attempt to do this by operantly conditioning two responses, keeping the reinforcer and schedule of reinforcement the same. If the presentation of the S^D for one response has different effects from the presentation of the S^D for the other response, the effects of the S^D might be related to the reinforced pattern. If, however, the presentation of the two S^Ds has the same effect, it would be unlikely that this effect was related to the reinforced pattern (see the discussion of Fig. 5).

An example of the transfer design is provided by the work of Sterman on the sensorimotor rhythm in the cortex of cats. This rhythm is a regular sinusoidal EEG pattern between 12 and 20 Hz. The sensorimotor rhythm seems to be related to processes involved in the inhibition of movement (Chase & Harper, 1971; Roth, Sterman, & Clemente, 1967; Sterman *et al.*, 1969c). Sterman, Howe, and MacDonald (1969a) operantly conditioned the sensorimotor rhythm and then cortical desynchronization in one group of cats, and the opposite sequence in a second group. They then studied the effects of the presentation of the S^D for each type of conditioned EEG pattern on sleep. (One must assume that the experimental situation acted as the S^D in this case.) Sensorimotor rhythm training increased the percent of spindle-burst activity during sleep and the duration of quiet sleep epochs.

Sterman, Lopresti, and Fairchild (1969b) also studied the effects of reinforcing the sensorimotor rhythm on reactions to the injection of a convulsion-producing drug, monomethylhydrazine. In this experiment, a group trained to make the sensorimotor rhythm was compared to a nonconditioned control group. They found that the onset of convulsions after a 9 mg per kg injection of the drug was delayed in the group reinforced for the sensorimotor rhythm as compared to the control group.

These results are consistent with the hypothesis that the sensorimotor rhythm is related to processes involved in the inhibition of ongoing motor behavior. It is this inhibition which presumably facilitated certain aspects of sleep behavior and also protected the animals from some of the effects of the convulsion-producing drug.

Although this transfer design has been employed extensively in research on the operant conditioning of skeletal behavior, it has been employed rarely in research on operant neural conditioning (Black *et al.*, 1970; Paskewitz & Orné, 1971; Spilker, Kamiya, Callaway, & Yeager, 1969).

3. Conclusion

The examples that I have described make it clear that both the concomitant measures and transfer methods have provided information on the functional significance of operantly conditioned patterns of CNS electrical activity. In some cases, such as the sensorimotor rhythm, most of the data on functional significance was obtained from research employing operant neural conditioning procedures. In other cases, such as the hippocampal EEG, motor cortex single-cell activity, and sensory evoked potentials, a great deal of information has been provided by more traditional methods. But even in these cases, the operant conditioning procedures have provided additional information, especially by the use of dissociative conditioning methods. In most cases, of course, a great deal of information is still required for an adequate understanding of the functional significance of these patterns. Operant neural conditioning procedures have already made a significant contribution, and will, I think, continue to do so in this further research.

B. The Processes Controlled by Operant Neural Conditioning

Over what neural and psychological processes, if any, do we obtain control when we operantly condition a particular pattern of CNS electrical activity?

This question can be dealt with relatively quickly, since the analysis of functional significance in Section IV, A provided a major source of information about the processes that are controlled in operant neural conditioning.

The data on certain neural patterns are too limited to permit us to guess their functional significance with any accuracy. The electrical activity of the amygdala was studied in only one S by Delgado, Johnston, Wallace, and Bradley (1970). Olds (1965, 1967, 1969) and Carmona (1967) reported no clear correlations between the reinforced CNS patterns and concomitant measures of neural activity and behavior. As one begins to classify the remaining patterns, the outcome is rather surprising. Most of the reinforced neural events seem to be related to motor control systems. There are two neural patterns whose involvement in motor control systems would be expected: single-unit activity of the motor cortex (Fetz, 1969) and sensory-evoked potentials resulting from movement (Rosenfeld, 1970). Also, high-frequency hippocampal RSA (Black, 1971b) and the cortical sensorimotor rhythm (Chatrian et al., 1959; Chase & Harper, 1971) seem to be related to motor processes—RSA to the occurrence of certain types of movement, and the sensorimotor rhythm to the inhibition of movement. Finally, alpha blocking in human Ss seems to be produced by the activation of the visual motor system (Mulholland, 1971; Peper & Mulholland, 1970).[6] There is, of course, some disagreement about this interpretation of alpha blocking, and even more disagreement about alpha production where a variety of mechanisms have been proposed—from self-induced anoxia through refraining from visual motor activity and inattention, to entering a meditative state similar to that employed by Zen and Yoga practitioners. Similar problems of interpretation are found when we consider other human cortical EEG patterns (Brown, 1970, 1971).

There are two examples of operantly conditioned CNS patterns that do not seem to be related to motor processes; one is the postreinforcement synchronization which seems to be related to reinforcement and motivational processes (Sterman et al., 1969c), and the other is the visual evoked potential (Fox & Rudell, 1968).

One might be tempted to speculate further, as was suggested by Olds (1965), that success has been achieved with patterns of neural activity that are related to motor processes because they are easier to

[6] Both the alpha and the sensorimotor rhythms have similar frequency ranges and are correlated with the inactivation of specific motor control systems (occipital alpha with visual motor activity, and the more anteriorly recorded sensorimotor rhythm with general skeletal movement). This tempts one to speculate that all rhythms of this type have the same functional signficance. Each might occur during the inactivation of a specific motor system which in turn is related to a specific cortically localized sensory system.

condition than patterns that are related to non-motor processes. There is one bit of evidence that could provide some support for the hypothesis that motor-related processes are easier to condition. Fox and Rudell (1970) operantly conditioned changes in late components of the visual cortical-evoked potential in cats. Rosenfeld (1970) operantly conditioned changes in components of a sensory-evoked potential in the primary somatosensory cortex of cats that was elicited by feedback from movement. When cats were trained to make the visual evoked potential more positive, they required from 6 to 10 days to depart from baseline. When cats were trained to make the evoked potential elicited by feedback from movement more positive, they began to depart from baseline in the first session (Rosenfeld, 1971 personal communication). Therefore, it could be easier to operantly condition sensory-evoked potentials elicited by movement than sensory-evoked potentials elicited by visual stimulation. Such an interpretation must be treated with caution, however, because the experiments did differ procedurally. For example, the cats in the Rosenfeld experiment received extensive pretraining in which they were reinforced for movement, and the cats in the Fox and Rudell experiment did not.

In summary, it would seem that most of the CNS patterns of electrical activity over which we have obtained control by means of operant conditioning are related to motor processes. It may be that so many motor-related patterns have been conditioned because they are prewired for easy learning, or because transfer from previous motor learning makes them easy to learn. But the alternative is equally, if not more likely, that motor-related neural patterns have been conditioned frequently simply because the Es chose to study such patterns.

C. Comparisons of Operant Neural Conditioning with Other Procedures

In this section, the questions concerning the comparison between operant neural conditioning and other procedures for studying functional significance and for controlling internal processes will be discussed.

1. Operant Neural Conditioning as a Method for Studying Functional Significance

Does the operant conditioning of CNS electrical activity provide information on the functional significance of the reinforced patterns of electrical activity that is different from the information obtained by more traditional methods of analysis, and if so, how?

In dealing with this question, there is one obvious comparison that we must make. It is between operant neural conditioning methods, such as the concomitant measures and transfer procedures, and observations of the relationship between particular patterns of CNS activity and other neural and behavioral events in natural situations. There are, I think, two advantages to operant neural conditioning methods. First, they permit us to control the CNS pattern in which we are interested so that we can make it occur when we want to. We do not have such control if we simply observe the pattern in natural situations. This advantage is most apparent in the transfer procedure which permits us to obtain information on the interaction of the reinforced pattern with other apparently unrelated responses. Second, there are a great many neural and psychological processes that could be related to a given pattern of electrical activity. By keeping the reinforcer, deprivation level, schedule of reinforcement, etc., constant when we condition different patterns of neural electrical activity, we can reduce the number of processes that we have to consider, as was pointed out in the discussion of Fig. 5. If, for example, we could operantly condition Ss to increase and decrease the probability of a certain EEG pattern while holding deprivation level constant, we would be unlikely to conclude that there was a simple one-to-one relationship between this EEG pattern and deprivation level.

If we compare operant neural conditioning to other methods such as stimulation, extirpation of neural tissue, etc., the most powerful operant conditioning method would seem to be dissociative conditioning—the identification of two correlated responses followed by the attempt to condition the simultaneous performance of one response and the nonperformance of the second response (Black, 1971b; Fetz & Finocchio, 1971). This method provides information on the nature of the relationship between responses in a given experimental situation; such information is, I think, more difficult to obtain by other methods. Even though the dissociative conditioning method seems to be the most powerful one, the other conditioning procedures that I described also provide useful information. The research by Fox and Rudell (1968, 1970) on the relationship among the components of the average evoked potential illustrates that the simple measurement of concomitant responses during the operant conditioning of a neural response can provide extremely useful data when the relationship between the reinforced response and the concomitant response is different from that obtained by other methods. Finally, the transfer procedures will, I think, provide especially interesting data once they have been properly exploited.

2. Operant Neural Conditioning as a Method of Control

Does the control provided by the operant conditioning of CNS electrical activity differ from that obtained by other methods for controlling internal neural and psychological processes, and if so, how?

In order to answer this question, we must deal with differences in the types of neural processes that can be controlled by each method, as well as differences in the type of control exerted by each method. I shall discuss each of these issues separately.

a. Differences in the Types of Processes That Are Controlled. There are a variety of ways in which one could approach this topic. The simplest, I think, is to ask whether we have been able to produce patterns of CNS electrical activity by operant conditioning that we had not been able to produce by other methods. The data seem to indicate that we have not reinforced novel patterns of electrical activity with the possible exception of certain components of the evoked potential (Fox & Rudell, 1970). This outcome, however, may be a function of the behavior of the Es rather than a property of reinforceable patterns of CNS electrical activity. The patterns that have been studied so far obviously were not chosen at random; they were probably chosen because the E had information that led him to believe that they were related to important psychological or neural processes. It may very well be that one might be able to reinforce patterns of CNS electrical activity that have never been observed before by shaping, but, to my knowledge, this has not been done with the exception noted above (Fox & Rudell, 1970).

A number of examples can be cited in which the operant conditioning procedure, rather than producing novel patterns of electrical activity, seems to have resulted in apparently novel relationships among neural and behavioral events. Examples are provided by the research on the relationship of components of the average evoked potential (Fox & Rudell, 1970), on the relationship of a single-unit activity in the motor cortex to EMG activity (Fetz & Finocchio, 1971), and on the relationship of the cortical sensorimotor rhythm to the effects of convulsion-producing drugs (Sterman *et al.*, 1969b). Effects of this sort very often play a prominent role in discussions of the specificity of operant neural conditioning. Before conditioning, the reinforced event was observed to be correlated with a number of other responses. After conditioning, especially dissociative conditioning, the correlation is no longer observed. There is specific control over the reinforced response without accompanying changes in other responses.

One must interpret results such as these with caution. One could infer that the operant conditioning procedure *revealed* the presence of certain relationships that had not been seen before. One might say, for example, that the interconnections of the elements of a given neural system were not changed; rather, the operant conditioning procedure simply produced a normal mode of functioning in this system that no one had bothered to elicit before. Alternatively, one could infer that the operant conditioning procedure *produced* new relationships that had not been seen before. For example, one might say the operant conditioning procedure produced a fundamental change in the relationships among elements of a given neural system, so that components which were related before are no longer related, and vice versa.

The latter interpretation attributes greater power to the control achieved by operant conditioning than does the former. At the same time, we know so little about the circuits that relate many of the neural events and processes in which we are interested that we have little support for either interpretation. Perhaps it is wiser to avoid taking any position on this issue until more evidence is available.

b. *Differences in the Type of Control.* There is a possibility that we might obtain control over neural and psychological processes by the operant conditioning of patterns of CNS electrical activity that is more efficient, more powerful, or has fewer side effects than we have been able to obtain by other methods of control. In order to deal with this possibility, one has to compare neural operant conditioning with these other methods of control. The most obvious comparison is with the operant conditioning of observable behavior. It might be, for example, that some patterns of electrical activity that are related to overt skeletal responding might be controlled better by reinforcing skeletal responding than by reinforcing the neural patterns. Another comparison is with control achieved by drugs, electrical stimulation, etc.

This question is unsettled at the moment because explicit comparisons between the reinforcement of neural events and other methods of control have not been made. Consider, for example, the research on the reinforcement of the sensorimotor rhythm and its effect on sleep (Sterman *et al.*, 1969b). The sensorimotor rhythm is postulated to be related to the inhibition of movement. Would reinforcement for holding still have had the same effect on sleep? Similarly, in research on the mood states associated with alpha rhythm, could the same effect have been obtained by training *S*s to lower their respiratory rate, to refrain from visual motor activity, or, perhaps, to relax?

One especially important claim about the advantage of the control achieved by the operant conditioning of CNS electrical activity over other methods for controlling internal processes is that it leads to voluntary or self-control of internal processes. The term "voluntary control" is difficult to define.

One definition which equates operant with voluntary control can be questioned, however. While the reinforcement of a response may be necessary for establishing voluntary control, it does not seem to be enough. In fact, certain types of operant conditioning seem to decrease voluntary control when, for example, they lead to compulsive behavior, as is illustrated in the work of Solomon, Kamin, and Wynne (1953) on the difficulty of extinguishing certain types of operantly conditioned avoidance responses. It seems reasonable, therefore, to think of voluntary control as a more complex phenomenon than operant control.

A set of minimum requirements for recognizing voluntary control might be the following: To achieve effective voluntary control over some behavior, we must demonstrate not only that we can operantly condition Ss to make the response to stimuli which normally do not elicit it, but also that we can condition them to refrain from making the response to stimuli which normally elicit it. (We must, of course, employ the same reinforcers and schedules of reinforcement in both cases.) We would be loath to say that we have voluntary control if the S could not refrain from performing a response as well as perform it, or could not switch easily back and forth from one to the other. Furthermore, this stimulus control must be conditional on the behavior of the S. That is, if required to, he should be able to perform a response when his own behavior produces the S^D for the response, and refrain from performing the response when the S^D is presented by some external agent. Also, we usually assess the precision of voluntary control, which could range from the awkward initiation or inhibition of a response at one extreme, to skilled control over the amplitude and direction of the response at the other. Finally, the fewer the constraints on the conditioning and performance of the response, the more voluntary control we would judge an S to have.

All of the neural patterns that have been operantly conditioned in experiments employing a bidirectional procedure meet the first criterion. It is only in the operant conditioning of cortical EEG patterns in human Ss that some of the other criteria have been met, and the data is still sketchy. Dewan's (1967) attempt to train Ss to send Morse Code messages by varying the duration of bursts of alpha activity indicates that considerable voluntary control can be achieved, but more work needs to be done on precision of control and also on constraints. Furthermore, as was pointed out in Section II, C, 1, it is not clear that the operant conditioning of patterns of CNS electrical activity is a better procedure for establishing voluntary control than making the CNS pattern an S^D for some other response.

c. *Summary.* In comparing operant neural conditioning with other methods of control, one can look for differences in the types of processes that can be controlled by each method, and for differences in the type of control that can be exerted by each method. With respect to the types of processes that are controlled, there is little evidence that operant neural conditioning produces control over novel patterns of CNS electrical activity. Operant conditioning procedures do seem to result in new relationships between the reinforced pattern of electrical activity and other responses. It is not clear, however, whether the procedure produces such relationships or simply reveals their presence. With respect to the type of control achieved, operant neural conditioning seems to differ from other control procedures in that it seems to establish voluntary or self-control in human *S*s. The data on this question, however, are inadequate, and we still cannot assess accurately the extent to which control by operant conditioning is different from control achieved by other methods.

V. Conclusion

The first set of questions which were considered in Section II focused attention on the operant conditioning process. The first question was concerned with the factors responsible for changes in the reinforced patterns of CNS electrical activity. The data made it clear that these were produced by the response-reinforcer contingency. The second question was concerned with mediation. Peripheral mediation of operantly conditioned CNS electrical activity by skeletal and some autonomic responses was ruled out in a few cases. It would seem, then, that one can operantly condition CNS electrical activity directly. The effects of variables which influence the rate of conditioning were discussed—in particular, the role of feedback about response state and constraints on conditioning. There was a surprising lack of data on this topic. Perhaps an overconcern with the demonstration of the importance of the response-reinforcer contingency and the lack of mediation hindered the acquisition of information about effective training procedures. One can question the concern with the mediation issue especially.

The second set of questions which were considered in Sections III and IV dealt with the significance of the research. One question was concerned with the use of operant neural conditioning procedures to study the functional significance of patterns of CNS electrical activity, and another with the comparison between operant neural conditioning and more familiar methods for the analysis of functional significance. There is, I think, no doubt that operant neural conditioning procedures provided information on the functional significance of the reinforced patterns. Furthermore, the methods employed in this research, such as the concomitant measures and transfer procedures, complement more familiar methods for studying functional significance. Dissociative operant conditioning seemed to be especially useful in providing information about relationships between the reinforced pattern of CNS electrical activity and concomitantly measured events. One might also mention that these same procedures provided information on a variety of other issues that were mentioned—the mediation of the reinforced response, the specificity of the operantly conditioned response, and the analysis of constraints on the operant conditioning of the response.

Two further questions concerned the use of operant conditioning procedures to obtain control over internal neural and psychological processes, and the comparison between operant neural conditioning and other methods of control. The data on the functional significance of operantly conditioned neural events, along with data from other sources, indicated that one could employ the operant conditioning procedure to obtain control over psychological processes related to the reinforced pattern of CNS electrical activity. It was somewhat surprising, however, to find that most of the reinforced patterns were related to motor processes. Furthermore, it is not well established that the control achieved by reinforcing patterns of CNS electrical activity is significantly different from that which would be achieved by other methods of control, such as the reinforcement of observable behavior, or drugs, direct stimulation, etc. Some results indicate that the relationships between the reinforced neural event and other events are different after operant conditioning. Also, there is evidence that operant neural conditioning will provide control that we could not achieve by other means, especially with respect to voluntary or self-control, but not enough formal experimentation on this question has been carried out.

Further research on the use of the operant conditioning of CNS electrical activity to obtain control over neural and psychological processes is especially important for practical applications. We need more information on the operant conditioning of neural patterns that are related to non-motor processes. We need more information on the range of patterns of neural activity that can be operantly conditioned: it is incorrect to assume that we can condition all patterns because we can condition some of them. We need more information on the variables that influence the rate of conditioning; no parametric studies on this topic have been carried out. We need more information on the constraints on the operant conditioning of particular patterns of neural electrical activity, and about the conditions under which the operantly conditioned brain-wave patterns will generalize beyond the training laboratory.

Perhaps we should wait until more of this information is available before we permit ourselves to be carried away by passionately expressed hopes and fears for the future that were mentioned in Section III. In this respect, it is somewhat ironic that many discussions of operant neural conditioning have focused on its role in the control of neural and psychological processes in real-life situations when the literature reveals that its main contribution so far has been as an analytic tool for studying brain-behavior relationships.

References

Adam, G. *Interoception and behaviour: An experimental study.* Budapest: Publishing House of the Hungarian Academy of Sciences, 1967.

Basmajian, J. V., & Simard, T. G. Effects of distracting movements on the control of trained motor units. *American Journal of Physical Medicine,* 1967, 46, 1427.

Beatty, J. Effects of initial alpha wave abundance and operant training procedures on occipital alpha and beta wave activity. *Psychonomic Science,* 1971, 23, 197-199.

Bennett, T. L. Hippocampal theta activity and behaviour—a review. *Communications in Behavioral Biology,* 1971, Part A, 6, 1-12.

Bilodeau, E. A. (Ed.) *Principles of skill acquisition.* New York: Acadmic Press, 1969.

Black, A. H. A comment on yoked control designs. Technical Report No. 11, September 1967, McMaster University, Department of Psychology.

Black, A. H. Autonomic conditioning in infrahuman subjects. In F. R. Brush (Ed.), *The aversive control of behavior.* New York: Academic Press 1971. Pp. 3-104. (a)

Black, A. H. The direct control of neural processes by reward and punishment. *American Scientist,* 1971, 59, 236-245. (b)

Black, A. H., & de Toledo, L. The relationship among classically conditioned responses. In A. H. Black and W. F. Prokasy (Eds.), *Classical conditioning II: Current theory and research.* New York: Appleton, 1972. Pp. 290-311.

Black, A. H., & Young, G. A. Constraints on the operant conditioning of drinking. In R. M. Gilbert & J. R. Millenson (Eds.), *Reinforcement: Behavior analyses.* New York: Academic Press, 1972. Pp. 35-50. (a)

Black, A. H., & Young, G. A. Electrical activity of the hippocampus and cortex in dogs operantly trained to move and to hold still. *Journal of Comparative and Physiological Psychology,* 1972, 79, 128-141. (b)

Bolles, R. C. Species-specific defense reactions and avoidance learning. *Psychological Review,* 1970, 77, 32-48.

Brener, J., Kleinman, R. A., & Goesling, W. J. The effect of different exposures to augmented sensory feedback on the control of heart rate. *Psychophysiology,* 1969, 5, 510-516.

Brown, B. Recognition of aspects of consciousness through association with EEG alpha activity represented by a light signal. *Psychophysiology,* 1970, 6, 442-452.

Brown, B. Awareness of EEG-subjective activity relationships detected within a closed feedback system. *Psychophysiology,* 1971, 7, 451-464.

Bykov, K. *The cerebral cortex and the internal organs.* Moscow: Foreign Languages Publishing House, 1959.

Carmona, A. Trial and error learning of the voltage of the cortical EEG activity. *Dissertation Abstracts,* 1967, 28, 1157B-1158B.

Chapman, R. M., Cavonius, C. R., & Ernest, J. T. Alpha and kappa electroencephalogram activity in eyeless subjects. *Science,* 1971, 171, 1159-1161.

Chase, M. H., & Harper, R. M. Somatomotor and visceromotor correlates of operantly conditioned 12-14 c/sec. sensorimotor cortical activity. *Electroencephalography and Clinical Neurophysiology,* 1971, 31, 85-92.

Chatrian, G. E., Magnus, M. D., Petersen, C., & Lazarte, J. A. The blocking of the rolandic wicket rhythm and some central changes related to movement. *Electroencephalography and Clinical Neurophysiology,* 1959, 11, 497-510.

Crider A., Schwartz, G. E., & Shnidman, S. On the criteria for instrumental autonomic conditioning. *Psychological Bulletin,* 1969, 71,, 455-461.

Dalton, A. J. Discriminative conditioning of hippocampal electrical activity in curarized dogs. *Communications in Behavioral Biology,* 1969, 3, 283-287.

Delgado, J. M. R., Johnston, V. S., Wallace, J. D., & Bradley, R. J. Operant conditioning of amygdala spindling in the free chimpanzee. *Brain Research,* 1970, 22, 347-362.

Dewan, E. M. Occipital alpha rhythm, eye position and lens accommodation. *Nature* (London), 1967, 214, 975-977.

Fetz, E. E. Operant conditioning of cortical unit activity. *Science,* 1969, 163, 955-957.

Fetz, E. E., & Finocchio, D. V. Operant conditioning of specific patterns of neural and muscular activity. *Science,* 1971, 174, No. 4007, 431-435.

Fox, S. S., & Rudell, A. P. Operant controlled neural event: Formal and systematic approach to electrical coding of behavior in brain. *Science,* 1968, 162, 1299-1302.

Fox, S. S., & Rudell, A. P. Operant controlled neural event: Functional independence in behavioral coding by early and late components of visual cortical evoked response in cats. *Journal of Neurophysiology,* 1970, 33, 548-561.

Gray, J. A. Sodium amobarbital, the hippocampal theta rhythm, and the partial reinforcement extinction effect. *Psychological Review,* 1970, 77, 465-480.

Green, E. E., Green, A. M., & Walters, E. D. Self-regulation of internal states. In J. Rose (Ed.), *Progress of cybernetics: Proceedings of the International Congress of Cybernetics, London, 1969.* London: Gordon & Breach, 1970. (a)

Green, E. E., Green, A. M., & Walters, E. D. Voluntary control of internal states: Psychological and physiological. *Journal of Transpersonal Psychology,* 1970, 2, 1-26. (b)

Kamiya, J. Conscious control of brain waves. *Psychology Today,* 1968, 1, 57-60.

Kamiya, J. Operant control of the EEG alpha rhythm and some of its reported effects on consciousness. In C. Tart (Ed.), *Altered states of consciousness: A book of readings.* Wiley: New York, 1969.

Katkin, E. S., & Murray, E. N. Instrumental conditioning of autonomically mediated behavior: Theoretical and methodological issues. *Psychological Bulletin,* 1968, 70, 52-68.

Katkin, E. S., Murray, E. N., & Lachman, R. Concerning instrumental autonomic conditioning: A rejoinder. *Psychological Bulletin,* 1969, 71, 462-466.

Klemm, W. R. Correlation of hippocampal theta rhythm, muscle activity, and brain stem reticular formation activity. *Communications in Behavioral Biology,* 1970, Part A, 3, 147-151.

Krutch, J. W. *The measure of man.* New York: Grosset, 1953.

Lynch, J. J., & Paskewitz, D. A. On the mechanisms of the feedback control of human brain wave activity. *Journal of Nervous and Mental Diseases,* 1971, 153, 205-217.

Miller N. E. Learning of visceral and glandular responses. *Science,* 1969, 163, 434-445.

Mulholland, T. B. Feedback electroencephalography. *Activitas Nervosa Superior* (Prague), 1968, 4 410-438.

Mulholland, T. B. Problems and prospects for feedback electroencephalography. Paper presented at the meetings of the Feedback Society, Los Angeles, October, 1969.

Mulholland, T. B. Can you really turn on with alpha? Paper presented at the meeting of the Massachusetts Psychological Association, Boston College, May 1971.

Mulholland, T. B. Occipital alpha revisited. *Psychological Bulletin,* 1972, in press.

Nowlis, D. P., & Kamiya, J. The control of electroencephalographic alpha rhythms through auditory feedback and the associated mental activity. *Psychophysiology,* 1970, 6, 476-484.

Olds, J. Operant conditioning of single unit responses. *Excerpta Medica Foundation International Congress Series,* 1965, 87, 372-380.

Olds, J. The limbic system and behavioral reinforcement. In W. R. Adey & T. Tokizane (Eds.), *Progress in brain research. Vol. 27. Structure and function of the limbic system.* Amsterdam: Elsevier, 1967.

Olds, J. The central nervous system and the reinforcement of behavior. *American Psychologist,* 1969, 24, 114-132.

Olds, J., & Olds, M. E. Interference and learning in paleocortical systems. In J. F. Delafresnaye (Ed.), *Brain Mechanisms and Learning.* Oxford: Blackwell's, 1961.

Orne, M. T. On the social psychology of the psychological experiment: With particular reference to demand characteristics and their implications. *American Psychologist,* 1962, 17, 776-783.

Paskewitz, D. A., Lynch, J. J., Orne, M. T., & Costello, J. The feedback control of alpha activity: Conditioning or disinhibition? *Psychophysiology,* 1970, 6, 637-638.

Paskewitz, D. A., & Orne, M. T. Cognitive effects during alpha feedback training. Paper presented at the annual meeting of the Eastern Psychological Association, New York, April 1971.

Peper, E. Feedback regulation of the alpha electroencephalogram activity through control of the internal and external parameter. *Kybernetik,* 1970, 7, 107-112.

Peper, E., & Mulholland, T. Methodological and theoretical problems in the voluntary control of electroencephalographic occipital alpha by the subject. *Kybernetik,* 1970, 7, 10-13.

Razran, G. The observable unconscious and the inferable conscious in current Soviet psychophysiology: Interoceptive conditioning, semantic conditioning and the orienting reflex. *Psychological Review,* 1961, 68, 81-147.

Rosenfeld, J. P. Operant control of a neural event evoked by a stereotyped behavior. Unpublished doctoral dissertation, University of Iowa, 1970.

Rosenfeld, J. P., Rudell, P. A., & Fox, S. S. Operant control of neural events in humans. *Science,* 1969, 165, 821-823.

Roth, S., Sterman, M. B., & Clemente, C. D. EEG correlates of reinforcement, internal inhibition, and sleep. *Electroencephalography and Clinical Neurophysiology,* 1967, 23, 509-520.

Schwartz, G. E. Operant conditioning of human cardiovascular integration and differentiation. Unpublished doctoral dissertation, Harvard University, 1971.

Seligman, M. E. P. On the generality of the laws of learning. *Psychological Review*, 1970, 77, 406-418.

Shettleworth, S. J. Constraints on learning. In D. S. Lehrman, R. A. Hinde, & E. Shaw (Eds.), *Advances in the study of behavior*. Vol. 4. New York: Academic Press, 1972, in press.

Solomon, R. L., Kamin, L. J., & Wynne, L. C. Traumatic avoidance learning: The outcomes of several extinction procedures with dogs. *Journal of Abnormal and Social Psychology*, 1953, 48, 291-302.

Spilker, B., Kamiya, J., Callaway, E., & Yeager, C. R. Visual evoked responses in subjects trained to control alpha rhythms. *Psychophysiology*, 1969, 5, 683-695.

Sterman, M. B., Howe, R. C., & MacDonald, L. R. Facilitation of spindleburst sleep by conditioning of electroencephalographic activity while awake. *Science*, 1969, 167, 1146-1148.(a)

Sterman, M. B., Lopresti, R. W., & Fairchild, M. D. Electroencephalographic and behavioral studies of monomethylhydrazine toxicity in the cat. Report No. TR-69-3, 1969, Aerospace Medical Research Laboratory, Wright-Patterson Air Force Base, Ohio.(b)

Sterman, M. B., Wyrwicka, W., & Roth, S. Electrophysiological correlates and neural substrates of alimentary behavior in the cat. *Annals of the New York Academy of Sciences*, 1969, 157, 723-739.(c)

Stumpf, Ch. Drug action on the electrical activity of the hippocampus. *International Review of Neurobiology*, 1965, 8, 77-138.

Vanderwolf, C. H. Hippocampal electrical activity and voluntary movement in the rat. *Electroencephalography and Clinical Neurophysiology*, 1969, 26, 407-418.

Vanderwolf, C. H. Limbic-diencephalic mechanisms of voluntary movement. *Psychological Review*, 1971, 78, 83-113.

Watanabe, T., Shapiro, D., & Schwartz, G. E. Meditation as an anoxic state; a critical review and theory. Paper presented at the meeting of the Psychophysiology Society, St. Louis, October, 1971.

Wyrwicka, W., & Sterman, M. B. Instrumental conditioning of sensorimotor cortex EEG spindles in the waking cat. *Physiology and Behavior*, 1968, 3, 703-707.

CONSCIOUSNESS AND CREATIVITY

10
Some Applications of Biofeedback Produced Twilight States

Thomas H. Budzynski

Biofeedback research has been under way only a few short years and yet its techniques and preliminary results, embryonic though they may be, have captured the imagination of both laymen and researchers. Even the word "biofeedback" has somehow acquired a charismatic quality of its own. The interest and excitement is due primarily to the suggestion that biofeedback may enable man to exercise voluntary control over many of the physiological functions that had been considered to be almost totally beyond such control. It would appear that this brainchild of psychologists, physiologists, engineers, and physicists is a true product of the times. Emerging from a *Zeitgeist* generated by state-of-the-art electronics, psychophysiology, operant conditioning procedures, and a desire to explore "inner space," biofeedback may yet realize its greatest potential in the applied areas of psychotherapy, behavior therapy, psychosomatic disorders, education, and attitude and value change.

Research in our laboratory, although initially focused on studies which attempted to demonstrate learning effects (Budzynski, 1969; Budzynski and Stoyva, 1969), has been concentrated on the application of EMG and EEG biofeedback procedures to anxiety problems and psychosomatic disorders (Budzynski, Stoyva, and Adler, 1970, 1971; Stoyva and Kamiya, 1968; Stoyva, 1970). Since our work generally involves the production of low arousal levels in individuals trained with biofeedback techniques, we have become familiar with the experiential as well as the physiological aspects of this state. I would like, therefore, to compare some of the characteristics of this biofeedback-produced low arousal state with the "twilight" states of drowsy presleep and sleep itself. Where possible, I also wish to emphasize the applied aspects of these two states. Following this I hope to draw your attention to some interesting potential applications of a twilight state produced and sustained by biofeedback techniques.

Drowsy, Presleep States

Consider first that period in our daily lives which always occurs before sleep but also may manifest itself at other times as well. This drowsy period, sometimes called the sleep onset, hypnagogic, or reverie state has been divided by Foulkes (1966) into two periods defined by the EEG and EOG (eye movement signal). These periods are the alpha rhythm (usually slowed) with SEM (slow eye movements) and descending stage 1 sleep with low voltage theta waves. During these periods Foulkes noted that subjects often reported heightened awareness of bodily sensations, bodily positions, and states of muscular fatigue (muscular relaxation?), and so on. Some of the typical bodily sensations are shrinkage or swelling of limbs, or feelings of constriction about the waist, and floating sensations. Generally, both of these drowsy periods also are characterized by hallucinatory, dreamlike experiences which are more discontinuous and brief than those dreams associated with REM (rapid eye movement) sleep. Foulkes and Vogel (1965) questioned subjects upon awakening them from sleep-onset periods. Their replies to questions about control over mentation and loss of contact with the external world indicated that loss of volitional control over mentation tended to occur first; then loss of awareness of surroundings; and finally loss of reality testing

---

Paper presented at the 1971 American Psychological Association Convention in Washington, D.C.

Thomas H. Budzynski is Assistant Professor of Psychology, University of Colorado Medical Center, Denver, Colorado.

occurred. Vogel, Foulkes and Trosman (1966) also scored subject reports for two ego functions: the degree of maintenance of nonregressive content and the maintenance of contact with the external world. Report content was rated as nonregressive if the mentation was plausible, coherent, realistic and undistorted. Examples of regressive content would be: single isolated images, a meaningless pattern, an incomplete scene or bits and pieces of a scene, bizarre images, dissociation of thought and images, and magical thinking. The results showed that there was a statistically significant tendency for each EEG state (alpha, stage 1, and stage 2) to be associated with a different combination of ego functioning. In the first combination (ego state), usually found during alpha EEG, the ego maintains both functions, or, at most, showed an impairment of only one function. A second ego state in which both functions were impaired was associated with descending stage 1. The third ego state usually occurred during stage 2 and was characterized by a return to less regressive content; however, contact with reality was completely lost.*

One of the more interesting findings in these studies was the relationship of personality variables to the amount of dreamlike fantasy reported by subjects. Those individuals who showed lesser amounts of such fantasy material during sleep onset expressed a rigid, moralistic, and repressive outlook on life. They seemed less able to "let go" and express inner feelings and thoughts than did those subjects reporting a great deal of fantasy. Those high on fantasy tended to be more tolerant of shortcomings in themselves and others, and less dogmatic in their beliefs. Scores from TAT revealed that subjects with the ability to exercise their imaginations freely in waking life tended to report more vivid fantasies during sleep onset.

It is tempting at this point to speculate about the possible change in personality in an individual who could be trained to produce more sleep onset fantasy material; however, let us save that for later and instead summarize some of the more relevant findings of the work of Foulkes, Vogel and Trosman:

1. As individuals become drowsy and pass into sleep, their brain rhythms change from predominantly alpha, to fragmented alpha, to low-amplitude theta.
2. Paralleled (though not perfectly) with these EEG patterns are three ego states showing an increasing impairment of ego functioning (as defined above).
3. Individuals with rigid, repressive, dogmatic personality traits report less sleep onset fantasy material.

Whereas Foulkes et al. examined their subjects as they passed through the drowsy and sleep stage at their normal bedtimes, other investigators have employed more deliberate means to enhance the drowsy conditions. For example, Henri Gastaut (1967) used sleep deprivation to induce the drowsy state. His sleep-deprived subjects were then asked to read a book until they simply fell asleep. Gastaut found that beta activity predominated in the EEG until the moment the subject stopped reading. At this point steady alpha appeared only to be replaced by a fragmented alpha pattern soon thereafter. Finally, all the alpha waves disappeared and low voltage theta predominated. It is interesting to note that this time the EMG showed a considerable diminution of tonic muscular activity. (In our laboratory we have often seen the sudden decrease in forehead EMG with the appearance of theta waves in the EEG.)

Bertini, Lewis and Witkin (1969) have also studied the hypnagogic state but with the goal of developing an experimental technique which could facilitate drowsiness, reverie and free association. They reasoned that, "the transitional nature of the hypnagogic state makes it an especially fertile period for the production of primary process material. Loosened controls partly resulting from the drowsy state seem to make the primary process thinking more accessible to observation (p. 94)." Their induction technique involved the generation of a monotonous white noise sound and the use of "ganzfeld" glasses made up of halves of ping pong balls to produce a homogeneous visual field. Besides employing this mild sensory deprivation technique, Bertini et al. also stimulated their subjects with highly emotional material before they went into the hypnagogic state. Subjected to these stimulus conditions, individuals were reported to have recalled experiences, images and feelings from childhood, woven in with thoughts about current events in their lives. Their thoughts, images and feelings were often disjointed and incoherent, leading the investigators to state that the situation produced ". . . an associational flow relatively removed from ordinary conscious control (p. 108)."

It is important to note that even though these results were obtained at other times as well, the strongest effects of this technique were seen at bedtime.

The Bertini, Lewis and Witkin study describes a procedure which can be used to *induce* a hyp-

---
* It should be noted that Foulkes et al. did find occasional exceptions to the rule, e.g., some individuals would pass through the first two ego states while still in the alpha SEM period indicating that the psychological and physiological measures are not perfectly parallel.

nagogic state. In essence, it is a mild form of sensory deprivation which does enhance the production of hypnagogic imagery. However, it is almost a necessity that an experimenter be present since the subject might otherwise drift into sleep. (This point will become more relevant later on.) Bertini et al. suggest the application of this technique to enhance free association in the context of psychotherapy.

While on the subject of sensory deprivation it might be well to consider one study of the many that have dealt with this phenomenon. It is well known that severe sensory deprivation conditions can produce a whole host of bizarre experiential effects such as grotesque body image changes and vivid visual hallucinations along with slowed brain rhythms. However, another aspect of this condition would appear to be a facilitation of attitude change. Heron (1961) found that a group of sensory deprived subjects showed significantly greater changes in attitude toward psychic phenomena than did a control group which was not sensory deprived. Both groups heard a record which argued for believing in various types of psychical phenomena. Bogardus-type scales divided into five sections dealing with telepathy, clairvoyance, ghosts, poltergeists, and psychical research were used to assess changes in attitude. Three or four days after termination of the experiment several experimental subjects reported that they had borrowed library books on psychic research. Some said they were now afraid of ghosts!

Now let us turn to some research which involves biofeedback. Green, Green and Walters (1970) have also dealt with the problem of producing the hypnagogic state. In fact, a major focus of these researchers is the documentation of the link between alpha and theta rhythms in the brain wave and reverie-and-hypnagogic imagery, as well as the relationship between reverie-and-hypnagogic imagery and creativity. Green examined the EEGs of three demonstrably creative individuals (a professor of physics, a psychiatrist, and a psychologist) as they maintained a state of deep reverie. A high percentage of 6-8½ Hertz waves was seen in their records indicating a slowing alpha and an increase in theta production over that of a relaxed, alert EEG pattern. In order to assist subjects in the production of hypnagogic imagery, Green has developed an alpha/theta feedback system that provides the subject with information concerning the percentage of alpha, frequency of alpha and percentage of theta.

Since the material experienced during the hypnagogic state is generally forgotten rapidly, Green has devised a mercury switch finger ring which closes a switch sounding a chime whenever the subject's forearm begins to deviate from a vertical position as balanced on the elbow.* If the subject's consciousness diminishes below a certain level, the forearm will begin to tilt. This closes the mercury switch and sounds a chime that brings the subject back to a level of consciousness sufficient for the verbalization of the imagery which preceded the loss of balance. This retrieval procedure combined with the biofeedback-assisted production of the alpha/theta twilight state should prove to be a powerful technique for the study of creativity enhancement in particular, and the hypnagogic state in general. Like Bertini et al., Green and his co-workers also have suggested the possibility of applying experimentally produced hypnagogic states to psychotherapy.

Although the studies described above represent a sampling of the studies dealing with the drowsy, hypnagogic period, they do help to establish the important relationship between brain rhythm frequency and the associated experiential state. Two of the studies are unique in that they deal with techniques that can be used to produce and sustain the hypnagogic state.

If this twilight state can be recognized by the generation of certain brain rhythms, low muscle tone, bizarre disjointed imagery, and a loosening of the reality-oriented frame of reference, it probably can also be characterized by an increase in suggestibility, since the drowsy individual is less able to marshal the usual defenses and critical faculties of a more alert state. Some evidence for this assumption was provided by T.X. Barber (1957) in a study which demonstrated that subjects were just as suggestible when in a light sleep or in a drowsy state as when hypnotized. One subject who had followed suggestions when drowsy said, "I was just sleepy enough to believe what you were saying is true. I couldn't oppose what you wanted with anything else (p. 59)." In his concluding remarks Barber mentioned that at the therapeutic level it is possible that suggestions could be presented to people while they sleep for purposes of helping obese persons reduce, getting heavy smokers to cut down, and helping timid persons gain confidence.

Sleep Learning

Even though Barber achieved some startling results with sleeping and drowsy subjects, the area of sleep-learning has been fraught with controversy. Initial reports of positive results in

* The subject is lying flat on his back while maintaining his arm in a vertical position balanced on the elbow so that it stays up with a minimum of effort.

the 1930's and '40's were followed by a series of "null" results in the '50's (see Simon and Emmons, 1955, 1956). Interestingly, as the insignificant results brought research in sleep learning to a near halt in this country, Soviet sleep-learning research began to flourish. However, there have been some important differences between Soviet and American studies. Whereas the Russian research involves training during drowsy, presleep periods and light stages of sleep, with special emphasis on preparatory suggestions and training periods of weeks or months, almost all non-Russian work has emphasized single night training during deeper levels of sleep, with little in the way of preparatory suggestions to learn (Rubin, 1968).

In spite of the lack of positive results during deeper states of sleep, several American investigators saw the drowsy state as useful for learning. Simon and Emmons (1956) noted that, "It may be that in the drowsy state preceding sleep, the individual is more susceptible to suggestion; perhaps one's attitudes or habits can be modified during this presleep period when criticalness is minimized (p. 96)." Stampfl (1953) and Leuba and Bateman (1952) suggested that the intermediate point between waking and deep sleep might be optimal for sleep learning.

Recent experiments by Levy, Staab and Coolidge (1971) at the University of Florida have indicated that subjects can "learn to learn" while in deep sleep (REM and Stage 4). All subjects were given a suggestion prior to sleep *and* before each sleep presentation, that they would learn and remember the material. (This use of a suggestion period prior to sleep is an important feature of the Russian technique.) Subjects improved on recognition tasks on the same material presented on two successive nights indicating that repetition of materials across nights was facilitative. They also improved their scores on succeeding nights when different material was presented each night, suggesting general learning-to-learn effects.

The positive results of Levy, et al., together with the reports of the successful implementation of the Russian techniques, will probably influence a renewed interest in sleep-learning research in this country. However, it might be more correct to label this learning method as "twilight learning" since the most effective period for such learning would seem to be sleep onset. However, let us take a further look at the two states.

Sleep is similar to the hypnagogic presleep state in that it is a state of low arousal (with the exception of REM periods), with slow EEG frequencies, and characterized by a high degree of suggestibility due to the disintegration of the reality-oriented frame of reference and a lowering of certain defenses. Orne (1969), for example, has noted that an individual in stage 1 sleep is capable of carrying out purposive behavior in response to suggestions administered while he is asleep without any evidence of physiological arousal. Learning, as we have seen, can take place during sleep, but like that during the drowsy state, it is of a qualitatively different nature from learning occurring during alert waking conditions. Russian researcher Svyadoshch (1968) has said that, "Speech assimilated during sleep, in contrast to that assimilated during waking state, is not subjected during assimilation to the critical processing, and is experienced on awakening as a thought of which the source remained outside consciousness, to some extent, therefore, it is as if it belonged to an alien personality (p. 112)."

Now, it is generally recognized that most perceptual defense studies showed that emotionally toned words have higher thresholds than neutral words; however, this conclusion was drawn from experiments performed during the alert waking state. Rubin (1968) has noted that one of the leading Russian sleep-learning experts, L. A. Bliznitchenko, believes that just the opposite happens during sleep. It would be interesting to test the hypothesis that emotionally toned words have higher thresholds than neutral words during the alert waking state, equal thresholds in the drowsy state, and lower thresholds during sleep. Perhaps one could use the same paradigm to test for uncritical acceptance of information leading to significant attitude and opinion change under these conditions.

With regard to the implementation of such a paradigm it is probable that biofeedback procedures may be of great value. The waking and sleeping conditions are not too difficult to establish and maintain; however, the drowsy state can be difficult to manage. In the natural course of events, people tend to pass through this state rather quickly when falling asleep and when awakening. Bertini's technique might be useful in producing the state but does not guarantee that the subject will not pass into sleep. The use of alpha and theta feedback, with an arousing system (the mercury finger switch and chime) such as Green, et al. have devised, may be useful for maintaining the drowsy state and ensuring that the subject does not fall asleep. However, this procedure assumes that the slow-alpha-theta pattern can be successfully shaped without undue difficulty (frustration) and/or undue duration of training. If the production of the drowsy state is to take place at, or near, the subject's usual bedtime, there may be sufficient slow-alpha/theta activity for successful shaping and maintenance. On the other hand, it may be much more difficult (without considerable training) to produce the desired result during the alert waking state and in a laboratory situation. Under these latter conditions the subject is quite likely to show a minimum of theta or slowed-alpha EEG. Perhaps the shaping process could be accelerated by combining elements of Bertini's procedure with biofeedback. In this case the white noise and ganzfeld glasses might increase the amount of slowed-alpha/theta (the base operant in learning terms) so that the biofeedback learning could proceed at a faster rate. Once the desired EEG pattern was produced, the prolongation of the state could be assured by the continuous feedback plus the mildly arousing chime (or some other signal) that would guard against passage into sleep.

Biofeedback and the Drowsy State

Several studies have already demonstrated that biofeedback procedures can be used to increase the amount of alpha activity in the EEG (Kamiya, 1962, 1968; Kamiya and Nowlis, 1970; Hart, 1968; Mulholland, 1967; Brown, 1970a); however, there has been little in the way of published results on theta enhancement through biofeedback. Brown (1970a) showed that through biofeedback, subjects could produce and sustain various EEG rhythms long enough to report subjective impressions associated with beta, alpha, theta, and combinations of these (beta/alpha and alpha/theta). The alpha condition was described as calm, peaceful, pleasant, at ease, and neutral, among others. This is consistent with reports from other labs on the subjective experience during alpha. The alpha/theta condition was experienced as conjuring up, wish fulfillment, passive, and sleepy. Theta, however, was described as vacillating or problem solving. This may have been due to the fact that the task of producing a predominantly theta EEG during normally alert waking conditions may have been somewhat frustrating. In our own laboratory we have found that the appearance of theta rhythm in the EEG and reports of drowsiness are associated with the deep relaxation produced by EMG feedback from the forehead musculature. (Green, et al. (1970) have obtained similar reports from subjects who were receiving EMG feedback from the forearm extensor.) This observation led us to conclude that a possible procedure for training people with sleep onset insomnia would consist of two or three phases depending upon the amount of resting alpha. Patients would receive EMG feedback training initially until they had learned to thoroughly relax their musculature. Following this they would receive alpha feedback if they still showed a predominantly beta rhythm. A final phase would be theta feedback. Hopefully, the patient would be able to progress through each phase by retaining and using the learning acquired in each earlier phase. Although pilot work is just under way, thus far, two of three patients have completed training. In these two cases there has been a complete elimination of the sleep-onset insomnia even though one of the two retained an early-awakening problem. The third patient terminated because results were not forthcoming. Noteworthy in this pilot study was the fact that the patients reported that they were able to control the theta to some extent; i.e., they could produce an increase in the theta feedback signal although their attention at times wavered between the feedback and inner thought processes.

Our research with EMG feedback has led us to believe that low EMG levels are more characterized by theta and slowed-alpha activity than by normal alpha rhythms. Subjects report that their attention at these low levels does shift back and forth between "inward-looking" and the external EMG feedback signal. Reports of body image changes are quite common, especially if a high level of muscle tension is present at the start of the feedback session.* In a study of tension headache, patients trained in frontalis relaxation with EMG feedback reported more body image changes than did a control group which was subjected to the same experimental conditions with the exception of a yoked feedback signal. Due perhaps to their lower EMG levels, the experimental patients also reported more hallucinatory imagery (Budzynski, Stoyva, and Adler, 1970b).

Reports from subjects and patients who have been thoroughly trained in our laboratory to lower their forehead EMG levels sound very much like experiences associated with a drowsy, presleep state. Perhaps then, the production of such a state might be facilitated by EMG feedback from the forehead area.

A Biofeedback Approach to Twilight State Generation

Having considered several approaches to the induction and prolongation of the drowsy state, it is now possible to specify the characteristics of what might be the ideal biofeedback approach to this problem.

To begin with, the subject would be placed in a mild sensory deprivation environment such as

---

* This contrast effect seems to be an important factor in whether the subject experiences a mild "high." It is known that the same phenomenon tends to occur when a normally low alpha subject, through training, produces a sudden and sustained increase in alpha activity.

described by Bertini. This situation should serve to enhance the development of the hypnagogic phenomena. Along with the constant white noise and the ganzfeld goggles, the subject would hear a feedback signal that would change as a function of EEG frequency. Thus, he would be made aware of the slowing of alpha and the appearance of theta activity. The circuitry could be designed so that as theta frequency decreased toward delta, a very noticeable qualitative or quantitative change in the feedback signal would produce a slight increase in cortical arousal thus "moving the trainee" into the alpha/theta borderline again.

The feedback could also be made to bracket the high frequency or alpha end of the drowsy zone, e.g., a noticeable change in feedback would warn of high alpha output.

In the event the subject still could not approach a drowsy state, a shift to forehead EMG feedback might allow him to succeed.

Possible Applications of a Biofeedback Twilight State

Given that this "ideal" system could be used to generate a prolonged drowsy state, how might it be used?

There are two general situations in which a biofeedback twilight system could be of value. The first would be the enhancement of hypnagogic imagery and its retrieval, for the purpose of studying the "inner man." This situation, therefore, would consist of a focusing on *internally* generated stimuli.

The second situation would deal with the presentation of *external* stimuli for purposes of assimilation during this peculiarly receptive state.

In reference to the first case, several applications have already been suggested by Bertini et al. and Green et al. An easily produced and controlled reverie state would delight therapists of analytic persuasion. The increased production of primary process material and free-associations should allow an acceleration of the therapy process.

The same biofeedback-produced hypnagogic state also might be used to enhance creativity. One can easily conceive of the captains of industry taking their daily "creativity breaks." The technique would certainly be useful to those in the arts as well as the sciences.

Regular practice at hypnagogic imagery might even produce positive personality change as suggested earlier in the discussion of the research of Foulkes et al. These trainees hopefully would become more tolerant of others, less dogmatic, and better able to express feelings.

Finally, consider those seekers of new inner experiences who might be persuaded to give up "heavier" agents such as LSD or other drugs, if they could achieve interesting highs through a biofeedback twilight procedure which could be used at home. The exploration of biofeedback-produced experiential states will probably become as commonplace as Yoga is today.

Turning now to the second situation referred to earlier, that of presenting external material to the drowsy subject, it is easy to imagine some of the educational possibilities. Soviet scientists have, through the years, refined sleep learning to the point where it is now used in many of the Russian schools. They have concluded that the material is best assimilated during drowsy and light sleep stages (Rubin, 1970). A biofeedback twilight state could be established at any time of the day. Thus, the user of such a system could take advantage of this special, hypersuggestible learning state whenever he desired (without the problem of losing sleep). Computer-assisted teaching programs augmented by a biofeedback produced twilight state might constitute an educational breakthrough.

Therapists should be able to produce gradual and controlled positive personality changes through the introduction of carefully worded suggestions as the patient maintains himself in this hypersuggestible state. The technique could prove to be particularly effective with cases of depression. In fact, suggestions to increase self-esteem, confidence, and possibly assertiveness might be efficiently imparted to a wide variety of patient types.

Cassette tape systems could be used to present suggestions to patients while they trained at home with biofeedback units. Perhaps one day therapists will have at their disposal libraries of such tapes which can be used by the patients in conjunction with home biofeedback systems. At certain phases in therapy these general-topic tapes could prove to be quite helpful. Therapists would be able to handle larger numbers of patients at a reduced cost to the patient. This could mean a partial solution to the problem of providing low-cost, short-term therapy for the masses.

References

1. Barber, T. X., "Experiments in Hypnosis," *Scientific American*, 1957, 196, 54-61.
2. Bertini, M., Lewis, H. B., & Witkin, H. A., "Some Preliminary Observations with an Experimental Procedure for the Study of Hypnagogic and Related Phenomena." In C. T. Tart (Ed.), *Altered States of Consciousness*, 1969 New York: John Wiley & Sons, pp. 93-115.

3. Brown, B. B., "Recognition of Aspects of Consciousness through Association with EEG Alpha Activity Represented by a Light Signal," *Psychophysiology,* 1970, 6, 442-452.
4. Budzynski, T. H., "Feedback-induced Muscle Relaxation and Activation Level." Unpublished doctoral dissertation, University of Colorado, 1969.
5. Budzynski, T. H., & Stoyva, J. M., "An Instrument for Producing Deep Muscle Relaxation by Means of Analog Information Feedback," *Journal of Applied Behavior Analysis,* 1969, 2, 231-237.
6. Budzynski, T. H., Stoyva, J. M., & Adler, C. S., "Feedback-induced Muscle Relaxation: Application to Tension Headache," *Behavior Therapy and Experimental Psychiatry,* 1970(a), 1, 205-211.
7. Budzynski, T. H., Stoyva, J. M., & Adler, C. S., "The Use of Feedback-induced Muscle Relaxation in Tension Headache: A Controlled-outcome Study." Annual meeting of the American Psychological Association, Miami Beach, Florida, September 3-8, 1970(b).
8. Budzynski, T. H., & Stoyva, J. M., "Biofeedback Techniques in Behavior Therapy and Autogenic Training." Submitted for publication, 1971.
9. Foulkes, D., *The Psychology of Sleep.* New York: Scribners, 1966.
10. Foulkes, D., & Vogel, G., "Mental Activity at Sleep Onset," *Journal of Abnormal Psychology,* 1965, 70, 231-243.
11. Gastaut, H., "Hypnosis and Pre-sleep Patterns." In L. Chertok (Ed.), *Psychophysiological Mechanisms of Hypnosis,* 1969, New York: Springer-Verlog, pp. 40-44.
12. Green, E., Green, A., & Walters, D., "Voluntary Control of Internal States: Psychological and Physiological," *Journal of Transpersonal Psychology,* 1970, 1, 1-36.
13. Hart, J. T., "Autocontrol of EEG Alpha," *Psychophysiology,* 1968, 4, 506. (abstract).
14. Heron, W., "Cognitive and Physiological Effects of Perceptual Isolation." In P. Solomon, P. Kubzansky, P. Leiderman, J. Mendelson, R. Trumbull, and D. Wexler (Eds.), *Sensory Deprivation,* Cambridge, Mass.: Harvard University Press, 1961, pp. 6-33.
15. Kamiya, J., "Conditional Discrimination of the EEG Alpha Rhythm in Humans." Paper presented at the Meeting of the Western Psychological Association, San Francisco, April, 1962.
16. Kamiya, J., "Concious Control of Brain Waves," *Psychology Today,* 1968, 1, 57-60.
17. Kamiya, J., & Nowlis, D., "The Control of Electroencephalographic Alpha Rhythms through Auditory Feedback and the Associated Mental Activity," *Psychophysiology,* 1970, 6, 476-484.
18. Leuba, C., & Bateman, D., "Learning During Sleep," *American Journal of Psychology,* 1952, 65, 301-302.
19. Levy, M., Staab, L., & Coolidge, F., "Associative Learning During EEG-Defined Sleep." Submitted for publication, 1971.
20. Mulholland, T., "The Concept of Attention and the Electroencephalographic Alpha Rhythm." Paper presented at the National Physical Laboratory on "The Concept of Attention in Neurophysiology," Teddington, England, October, 1967.
21. Orne, M., "On the Nature of Posthypnotic Suggestion." In L. Chertok (Ed.), *Psychophysiological Mechanisms of Hypnosis,* New York: Springer-Verlog, 1969, pp. 173-192.
22. Rubin, F. (Ed.), *Current Research in Hypnopaedia,* London: Macdonald, 1968.
23. Rubin, F., "Learning and Sleep," *Nature,* 1970, 226, 477.
24. Simon, C., & Emmons, W., "Learning During Sleep?" *Psychological Bulletin,* 1955, 52, 328-342.
25. Simon, C., & Emmons, W., "Responses to Material Presented During Various Levels of Sleep," *Journal of Experimental Psychology,* 1956, 51, 89-97.
26. Stampfl, T., "The Effect of Frequency of Repetition on the Retention of Auditory Material Presented During Sleep." Unpublished master's thesis Loyola University, Chicago, 1953.
27. Stoyva, J., "The Public (Scientific) Study of Private Events." In E. Hartmann (Ed.), *Sleep and Dreaming,* Boston: Little, Brown, 1970, pp. 353-368.
28. Stoyva, J., & Kamiya, J., "Electrophysiological Studies of Dreaming as the Prototype of a New Strategy in the Study of Consciousness," *Psychological Review,* 1968, '75, 192-205.
29. Svyadoshch, A., "The Assimilation and Memorisation of Speech During Natural Sleep." In F. Rubin (Ed.), *Current Research in Hypnopaedia.* London: Macdonald, 1968, pp. 91-117.
30. Vogel, G., Foulkes, D., & Trosman, H., "Ego Functions and Dreaming During Sleep Onset, *Archives of General Psychiatry,* 1966, 14, 238-248.

Biofeedback for Mind-Body Self-Regulation: Healing and Creativity

Elmer Green

As people in this audience know, the revolution in consciousness we are now talking about was foretold a long time ago. It was thought of as the time when the sleeping giant, humanity, would awaken, come to consciousness, and begin to exert its power. British medical people began to get an inkling of the power of consciousness as long as 250 years ago when they began to study certain Indians who could do some very unusual and interesting things. These people, called yogis, apparently had phenomenal powers of self-regulation, both of mind and body. Of course, medical doctors as a whole did not believe it, but as the decades passed and reports became more numerous, some British and European physicians began the study of mind-body relationships. By the end of the nineteenth century the physiological phenomena of hypnotism, spiritualism, and various yogic disciplines had attracted some serious medical and philosophical attention, and by 1910 of this century, a mind-body training system, eventually called Autogenic Training (self-generated or self-motivated training), had begun to be developed by Dr. Johannes Schultz in Germany. This was at approximately the time that Freud gave up the use of hypnosis as a medical tool because it was unpredictable. It occurred to Schultz that perhaps hypnosis was an erratic tool because the patient often unconsciously resisted the doctor. If the patient were able to direct for himself the procedure being used, with the doctor acting as his teacher, then the control technique would come into the realm of self-regulation and perhaps be more effective.

It is an interesting fact that the first English translation of Schultz and Luthe's handbook, Autogenic Training, (1959), contains in its 604 references only 10 that are in English. In addition to telling us that there was much interest on the continent in healing by self-regulation, it also tells us something about the British and the Americans. Freud's ideas became very important in the United States, whereas the self-regulation techniques of Autogenic Training were largely confined to Europe. It is also interesting to note that because of his interest in

---

Reprinted from *The Varieties of Healing Experience: Exploring Psychic Phenomena in Healing, A Transcript from the Interdisciplinary Symposium of October 30, 1971.* © Copyright 1972 by the Academy of Parapsychology and Medicine.

self awareness, Schultz included in his training system some psychological disciplines which he called "meditative" exercises. These exercises gradually lead into a kind of self-awareness in which the person develops both physiological and psychological self-knowledge.

It was because of our interest in both consciousness and volition that my wife, Alyce, and I decided in 1965 to test Autogenic Training and find out whether or not people who used the exercises actually could develop some of the physiological controls Schultz talked about. Our research program began with 33 housewives. They practiced autogenic exercises for only two weeks and at the beginning and end of their training experience we measured most of the physiological variables that were easy to get, such as brain waves, heart rate, breathing rate, skin potential, skin resistance, blood flow in the fingers, and the temperatures of both hands, front and back. Some of the ladies failed to achieve much temperature control but a couple of them succeeded so well that we decided to continue studying Autogenic Training. It was clearly a training technique in which volition "entered" into the psychosomatic domain. It was also clear that if psychosomatic disease really existed, then it was logically necessary to hypothesize the existence of psychosomatic health. Both the literature of Autogenic Training and our research indicate that psycho-physiological processes definitely could be self-regulated and to allow for the existence of psychosomatic disease without postulating its opposite, psychosomatic health, would be an absurdity. If we can make ourselves sick, then we must also be able to make ourselves well.

Since physicians are saying these days that about 80% of human ailments are psychosomatic in origin, or at least have a psychosomatic component, it seems reasonable to assume that about 80% of our disabilities can be cured, or at least ameliorated, by the use of special training programs for psychosomatic health.

Gardner Murphy was very much interested in these research implications and one day suggested that we include biofeedback in our research methodology. He had been interested since 1952 in the possibility of using electrophysiological instrumentation for measuring and presenting to a person some of his own normally unconscious physiological processes, that is, processes of which a person is normally unaware. He knew, for instance, that muscle tension problems were often extremely difficult to handle, and when the doctor says, "Your problem is that you are too tense," he does not give much new information nor does it help much. Generally, the medical pronouncement is followed by a prescription, but what Gardner was suggesting was that if a person could "see" his tension, could look at a meter and observe its fluctuations, then perhaps he could learn to manipulate the underlying psychophysiological problem. He could practice "making the meter go down" and its behavior would tell him immediately if he was succeeding. In essence this is an application of the engineering principle of feedback, the servo-mechanistic principle by means of which automatic machines are controlled. A furnace and its thermostat, for instance, form a closed (self contained) feedback system. In this analogy, the human who adjusts the thermostat represents volition entering into the system from an energy source outside the closed loop.

Murphy's idea seemed useful, so we combined it with Autogenic Training and developed in 1967 a system of psychosomatic self-regulation that we called Autogenic Feedback Training. By coupling biofeedback with Autogenic Training, progress in controlling physiological variables is often highly accelerated. The self-suggestion formula of Autogenic Training (such as "I feel quite quiet") tells the unconscious section of mind, or brain, the goal toward which the person wishes to move, and the physiological feedback device immediately tells him the extent to which he is succeeding. The objective fact of a feedback meter, as a "truth" detector, has a powerful though not entirely understood effect on a person's ability to control normally involuntary physiological processes. For example, if a person's heart is malfunctioning from psychosomatic causes it is certainly not "all in his head," but knowing that the cause is psychosomatic does not tell him what to do about it. When his heart rate is displayed on a meter, however, he can easily and objectively experiment with the psychological states that influence the rate. While using any kind of biofeedback device as a tool for learning something about yourself, it is interesting and instructive to experimentally induce in yourself a feeling of anxiety and nervousness, then calmness and tranquility. You can play with anger and with peacefulness. Then you can experiment with muscle tension, relaxation, slow deep breathing, etc. Learning to manipulate physiological processes while seeing the meter (or listening to it if it has an auditory output) is quite similar to learning to play a pinball machine with your eyes open. If you had to learn blindfolded it would be difficult, but if you use your eyes (employing visual feedback), it is easy.

Since time is limited, I will quickly summarize the work we have been doing in the Voluntary Controls Project at The Menninger Foundation and later mention our research with Swami Rama, an Indian Yogi who demonstrated in the laboratory some of the results of his own psychosomatic training program. If you wish to receive more information on any of this work, a note or postcard to us will be sufficient.

As already mentioned, our first project was to train a group of housewives to increase the temperature of their hands, using only Autogenic Training methods. After that we began an ambitious program in which an attempt was made to train eighteen college men to control three physiological variables simultaneously, using Autogenic Feedback Training. Feedback meters showed muscle tension in the right forearm, temperature of a finger on the right hand, and percentage of alpha rhythm in the brainwave pattern (over the preceding ten-second interval of time).

Muscle tension was picked up from an electrode attached with salt paste to the skin. Temperature was obtained from a thermister taped lightly to the middle finger of the right hand, and brainwave (EEG) signals were obtained from an electrode pasted to the left occiput (the back of the head). The instruments were adjusted so that if the subject could relax his forearm completely the meter would "rise to the top." Complete relaxation means that there is a complete absence of muscle fiber "firing" and the electrical signal resulting from muscle tension "goes to zero." In this situation, the meter is wired to go to the top, showing complete success in relaxation. If the subject's finger showed an increase in <u>temperature</u> of ten degrees Fahrenheit, the temperature meter would <u>rise to the top</u>. It is interesting that if a subject tried to force the temperature to rise, by active volition, it invariably went down. But if he relaxed, "told the body" what to do, and then detached himself from the response, the temperature would rise. This is a passive volition. A ten degree decrease would cause the meter to go to the bottom. In other words, we set the meter in the center of its scale regardless of the absolute temperature of the hand, and studied only the temperature variations associated with the training program. The third meter showed <u>percentage of alpha</u> rhythm in the visual (occipital) area of the brain. After a period of training in relaxation and temperature control we would say to the subject, "Now while you keep the relaxation and warmth meters up try to make the third meter, the alpha meter, go up without closing your eyes." It is common knowledge that about 90% of the population produce alpha waves when the eyes are closed, so if subjects closed their eyes they could be expected to produce alpha waves, but we wanted them to generate, or bring about, an increase in the percentage of alpha while their eyes were open. One of our research objectives was to test the hypothesis that a person's success in remembering would be a function of, or correlated with, the percentage of alpha waves present while he was trying to remember. The eyes-open condition was useful because it served the purposes of keeping the subject from getting drowsy, permitted the use of a simple visual feedback device (though we could have used auditory feedback if desired), and it also enhanced the subject's "coupling" to the outside world. We wanted the subject to be able to look at the outside world and to answer questions without destroying his alpha rhythm, so it seemed useful to train him in awareness of both internal and external "worlds" at the same time by using a visual feedback system for brainwaves. Results showed that in the students with whom we worked, the ability to remember was indeed positively correlated with the percentage of alpha ($r=.54$, $p=0.1$), and it indicated that an alpha training program might be of great value in assisting students to overcome "mental blocks" during examinations.

Before continuing, however, it is useful to examine the accompanying diagram (see next page). It shows the major brainwave frequency bands and their relation to conscious and unconscious processes, in the general population.

When people focus attention on the outside world they usually produce only beta frequencies. If they close their eyes and think of nothing in particular, they generally produce a mixture of alpha and beta. If they slip toward sleep, become drowsy, theta frequencies often appear and there is less of alpha and beta. Delta waves are not normally present except in deep sleep. For example, at this moment I am predominantly in the beta state, and so are you--at least I think so.

Now alpha waves can be made to appear when the eyes are open if attention is turned inward, away from the outside world, but the way to make the alpha meter go up while looking at it is to learn to "observe without looking." Perhaps this sounds paradoxical, but it is not as difficult as answering the Zen koan, "What is the sound of one hand clapping?" In any event, we wanted students to be able to produce alpha waves with their eyes open and also to be able to answer questions while in this state.

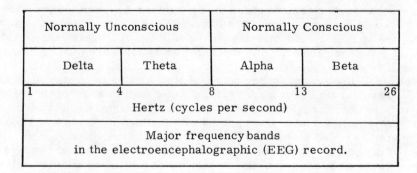

Major frequency bands in the electroencephalographic (EEG) record.

The attempt to develop simultaneous self-regulation of three physiological variables in five weeks proved to be overly ambitious but students usually learned to control one or another of the original three variables quite well, and sometimes could control two at once, but it was not easy for anyone to control all three at the same time.

As we worked with the college group, we also ran pilot subjects in a theta training program and soon we noticed that the psychological state associated with theta contained, in a number of subjects, very clear hypnagogic-like imagery. Pictures or ideas would spring full blown into consciousness without the person being aware of their creation. The theta "reverie," as we began to call it, was definitely different from a daydreaming state and much to our surprise we found that it seemed to correspond with descriptions given by geniuses of the past of the state of consciousness they experienced while being their most creative. From these observations and from our experiences in training college students we developed our present research project called "alpha-theta brainwave feedback, reverie, and imagery."

Without going into details about the machinery and procedures of theta training, in order to signal the presence of both alpha and theta frequencies in the occipital (visual) brain rhythm, we use auditory feedback (musical tones) rather than visual feedback. As already mentioned, alpha production can easily be learned with the eyes open, but theta production is generally possible only with the eyes closed. The imagery associated with theta is often so tenuous that open eyes drive it away. In our presently used EEG feedback devices a low frequency tone signifies theta and a higher frequency signifies alpha. (These machines are manufactured under licensing agreement from the Menninger Foundation, by a small electronics firm in Lawrence, Kansas.) For training away from the lab we have developed portable alpha-theta "home trainers," but in the lab we use an auditory feedback system that detects the presence of various brainwaves and multiplies their frequencies by two hundred (up to the audible range). By this procedure, beta waves are made to produce a "piccolo" type of music. Alpha sounds like a flute, theta like an oboe, and delta like a bassoon. This combination results in an interesting and not unpleasing quartet. We have constructed two identical sets of amplifiers, filters, multipliers, and associated hardware, so that two brainwave channels can be studied simultaneously, and when we attach electrodes to the two sides of the head, it is possible to feed back to the subject musical information concerning the simultaneous electrical activity of the two cerebral hemispheres.

The right ear "listens" to the right side of the brain and the left ear listens to the left side. We have not used this elaborate laboratory feedback system except in pilot research, but we sometimes claim that we are going to use the machine to train a subject to play "The Star Spangled Banner," with the hope that it will encourage the federal government to release additional funds. That, at least, would demonstrate a high level of control.

In working with research subjects we obviously do not say that the physiological functions they are going to control are involuntary -- because if they really believed that, the training would not work. In actuality, it is the "belief" of the subconscious (or unconscious) that is the controlling factor in learning to manipulate a so-called involuntary process. Feedback meters are remarkably powerful in training the autonomic nervous system because seeing is believing, even for the unconscious. There is little room for skepticism or disbelief concerning the practicality of temperature control when the temperature (blood flow) meter is used. Unconscious skepticism, as a factor in the psychophysiological matrix, seems to cause the hand temperature to drop, but this response indicates to the subject that the meter really does tell something, even if at first he does not know what it is. After a bit of practice with a feedback device, the situation is rather like learning to drive a car. The student driver does not question whether the car will move if he steps on the accelerator. And if he thinks that turning the steering wheel to the left will make the car turn to the right, he soon discovers his error.

Without getting involved in details, it seems that a hierarchy of attitudes (or sets) is involved in learning to control normally involuntary processes--first, hypothesis--then, belief--and finally, knowing. A person may start with the hypothesis that the temperature of his hand will rise, but because of his previously conditioned response to introspection, his temperature may rapidly drop when he attempts to raise it. This response tells him, however, that something significant is going on in his nervous system in response to his efforts. Then, when he relaxes and begins to think of something else, becomes detached, the temperature begins to rise. So he begins to believe that the training system will work. Eventually he knows he can control the process and he knows what is happening in himself (in regard to raising or lowering hand temperature) whenever he turns his attention on the matter.

Quite often with beginners the hand temperature will rise at first, but then an insidious thought will creep in, such as, "it may work with other people, but it probably won't work for me." This precipitates vasoconstriction in the hands due to the activation of the sympathetic nervous system and blood flow in the hands is appropriately reduced. Within a few seconds temperature of the hands begins to drop.

From our various research experiences with Autogenic Training, biofeedback, and with highly trained persons such as Swami Rama, and lately with Jack Schwarz, we are beginning to believe, or at least hypothesize, that "any physiological process that can be detected and displayed in an objective fashion to the subject, can be self-regulated in some degree." Blood pressure, blood flow, heart rate, lymph flow, muscle tension, brainwaves, all these have already been self-regulated through training in one laboratory or another. Where is the limit to this capacity for psychosomatic self-regulation? Nobody knows, but research indicates that the limits lie much farther out than was at first suspected by most of those interested in biofeedback.

It is useful here to draw attention to the major areas of the brain and discuss their relation to the process we are calling voluntary control of internal states. The item of particular significance in the accompanying diagram is that it is divided vertically into conscious and unconscious domains, on the right and left respectively. The conscious side contains both the cerebral cortex, which someone in the American Medical Association called the "screen of consciousness," and the craniospinal nervous system, roughly, the voluntary muscular system. The unconscious side includes both the subcortical brain, the "old" lower brain structures that man shares with most of the animals, and the autonomic nervous system, the involuntary nervous system, which lies outside of the brain and brainstem, and which controls, among other things, the skin, the internal organs and glands, and the vascular system of the body. The paleo-cortex, the old brain, includes a section called the limbic system which has been given a name of particularly great significance for understanding psychosomatic self-regulation, namely, "visceral brain." It is quite clear from recent research that electrical stimulation of the visceral brain and related neural structures through implanted electrodes causes emotional changes in humans. Conversely, it is well known that perceptual and emotional changes are followed by neural changes, or responses in the limbic system of the brain, though most of this work was performed with animals. Putting together some of the pieces of the mind-body system as observed both physiologically in the nervous system and behaviorally (through Autogenic Feedback Training) we have found it convenient to postulate a psychophysiological principle which goes as follows, "Every change in the physiological state is accompanied by an appropriate change in the mental-emotional state, conscious or unconscious, and conversely, every change in the mental-emotional state, conscious or unconscious, is accompanied by an appropriate change in the physiological state." This closed-loop statement obviously does not allow for volition or free will in humans any more than the furnace-and-thermostat system of your house is "allowed" to have a will of its own, but even as the thermostat is manipulated by a force from outside, namely your hand (which would have to be categorized as a meta-force in the furnace-and-thermostat system), so also the psychophysiological principle, or its expression in the psychosomatic unity of mind and body, is manipulated by volition, which at present is of indeterminate origin. Whatever volition is, it at least exhibits some of the characteristics of a meta-force. These ideas, incidentally, are quite clearly put forth in the Vedas, sacred scriptures of India, and lie behind the system of Raja Yoga. These basic concepts are well considered in Aurobindo's Integral Yoga, especially in his book called The Synthesis of Yoga.

But to return to the diagram, the dashed line represents at a specific moment the actual, rather than theoretical, division between conscious

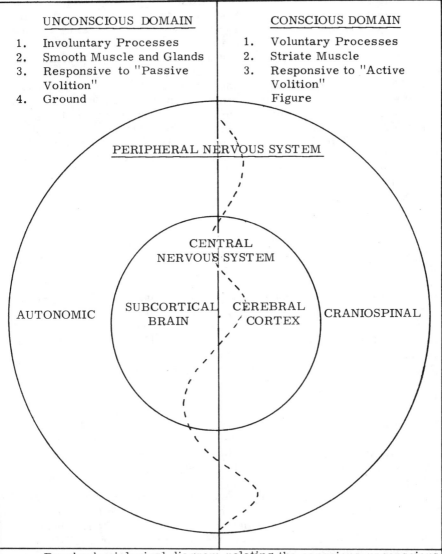

Psychophysiological diagram relating the conscious-unconscious psychological domain to the various sections of the voluntary-involuntary physiological domain. The solid vertical line separates the central and peripheral nervous systems into functional subregions. The dashed line (conceptually visualized to be in continuous undulatory movement) separates the conscious and unconscious areas.

and unconscious functions in the nervous system. We must visualize the dashed line as continually shifting and undulating between various brain structures as attention shifts from one thing to another. It is clear that when we learn to drive a car, every movement must at first be worked over in the conscious domain. But after we have learned to drive well enough it is possible, when we are thinking of something else, to drive well all the way through town without being aware of stopping at stop signs. In other words, what was once conscious became unconscious for a time. In other words, the dashed line moved from left to right and extended for a period of time over neurological structures in the neocortex and in the craniospinal system. On the other hand, when one learns through biofeedback training to control the flow of blood in his hands, he is obviously extending conscious control for a period of time over an area which lies to the left of center in the diagram. That is, some of the subcortical brain structures and some of the neural circuits in the autonomic nervous system have come under conscious control. It is interesting that when patients using Autogenic Training reported to Dr. Schultz that they could not achieve control over some physiological process, he would say that they had not made "mental contact" with that part of the body. In other words, again, they had not extended the dashed line of the diagram far enough to the left in a specific neural pathway.

Now to discuss our research with Swami Rama. Dr. Daniel Ferguson, at the Veterans Administration Hospital in St. Paul, Minnesota, became aware of the "voluntary controls" work at the Foundation and one day near the end of 1969 he wrote to me saying that if we were interested in making some tests of an Indian yogi who could control a number of normally uncontrollable physiological processes, he would arrange a meeting in Topeka. The upshot of this was that Swami Rama (of Rishikesh and the Himalayas) and Dr. Ferguson visited us for three days in March, 1970. The Foundation for the Study of Consciousness, Philadelphia, Pennsylvania, provided research funds for this visit. Later we obtained funding from both The Millicent Foundation, Vancouver, Washington, and from the Joseph and Sadie Danciger Fund, Kansas City, Missouri, to work with the Swami over an extended period of time.

During the first visit, Swami Rama demonstrated that he had exquisite differential control over arteries in his right hand. We had "wired" him for brain waves, respiration, skin potential, skin resistance, heart behavior (EKG), blood flow in his hands, and temperature. While thus encumbered he caused two areas a couple of inches apart on the palm of his right hand to gradually change temperature in opposite directions (at a maximum rate of about $4°F$ per minute) until they showed a temperature difference of about $10°F$. The left side of his palm, after this performance (which was totally motionless), looked as if it had been slapped with a ruler a few times, it was rosy red. The right side of his hand had turned ashen gray. During the last session of this visit, he made the comment that according to theory a swami could not be upset or distracted, but that it was a good thing that none of his students were here to demonstrate their powers of physiological control, because he doubted if they would be able to succeed in such a strange scientific setting. Other demonstrations included the speeding and slowing of his heart rate, and finally we concluded the tests, we thought. But during dinner, on the evening before he and Dr. Ferguson were to return to Minneapolis, he suddenly said he was sorry he had not demonstrated "stopping" his heart and that he would do it in the morning.

Both Alyce and I objected to this because he had just finished telling us that in order to demonstrate some of the more physiologically serious controls, such as stopping the heart, it was necessary to fast for a couple of days, taking nothing but fluids. I could well believe this because according to the Swami he would stop his heart in this particular demonstration by control of the vagus nerve. Since the vagus nerve also has an important control function over the stomach and other visceral organs, it could logically cause a serious case of indigestion, to say the least.

The Swami's answer to our objection, however, was that anything he could do in three minutes his Guru could do in three seconds and that he wanted to perform this experiment in order to test himself. Finally, having satisfied himself that he had answered or demolished our arguments, and having said that if necessary he would sign papers to the effect that The Menninger Foundation was not in any way responsible for anything that might happen to him, he said that he could stop his heart in this way for three or four minutes, and how long would we need for an adequate test. I said that ten seconds would be quite impressive and he agreed to this limitation.

The next day we hurried to the lab at nine o'clock (we had previously scheduled a lecture for him at ten o'clock) and wired him for the demonstration. Before starting his "inner focussing" procedure, however, he said that when his heart stopped he wanted Alyce to call over the intercom from the control room and say "that's all." This would be the signal for timing the duration of his demonstration and would also remind him not to "go too far." He said that he did not want to interfere with the functioning of his "subtle heart," the one that lay behind the workings of his "physical heart." Having explained this, he made a few trial runs at speeding and slowing his heart, then said, "I am going to give a shock, do not be alarmed." To me this meant that he was going to give himself some kind of neural shock, but later I learned that he was going to shock the research personnel and doctors who were watching the paper records and polygraph pens in the control room, and they were being told not to be alarmed. After about twenty seconds of motionless silence I heard Alyce say, "That's all." At this, the Swami pulled in his stomach muscles for a few seconds, then he relaxed. From his look I could see that he felt the test had been a success, so I began asking questions about the "internal" process he used to accomplish such a thing. While he was answering, Alyce called over the intercom and said that the heart record was not what we had expected and suggested that I look at it before going any further.

To my surprise, the heart rate instead of dropping to zero had jumped in one beat from about 70 per minute to about 300 per minute. I returned to the experimental room and described the record to the Swami. He seemed somewhat surprised and bothered and said, "You know that when you stop the heart in this way, it still trembles in there," and

he illustrated with fluttering hands. I speculated then that we had
recorded some kind of fibrillation, but later was told by Dr. Marvin
Dunne, cardiologist and professor at the Kansas University Medical
Center (Kansas City, Kansas), after he had examined the records, that
it was a case of "atrial flutter," a state in which the heart fires at its
maximum rate without blood either filling the chambers properly or the
valves working properly. He showed me similar records obtained from
patients and asked what happened to the Swami, he should have passed
out, but I had to answer that we quickly "unwired" him so he could get
to his lecture on time.

The atrial flutter actually lasted for an interval between 17 and 25
seconds. The exact duration could not be determined from the record
because when the Swami drew in his stomach the resulting electrical
signal from muscle firing caused the EKG pen to go off the edge of the
paper, and after it returned the heart rate was normal again. I asked
him why he had moved his stomach and he said that he had established a
"solar plexus lock," by means of which the heart condition could be
maintained for quite a long time if desired. It is interesting to note
that when the heart began to flutter the people in the control room had
a hurried consultation amongst themselves to decide if this was what
the Swami meant by "stopping" his heart. After about eight seconds
they decided that whatever it was looked dangerous and decided to
give the "that's all" signal.

In summary, we may say that the Swami stopped his heart from
pumping blood for at least 17 seconds. This was his technique we
discovered for obliterating his pulse during examination by medical
doctors. The "other" kind of heart stopping, he said, involved a hiber-
nation-like state that he might be prepared to demonstrate on some
other occasion.

The importance of Swami Rama's demonstrations did not lie in the
performances themselves but in their implications. I do not intend to
practice stopping my heart or to try to teach anyone else according to
Swami's instructions, but the fact that it can be done is of major scientific importance. Aside from supporting the phsychophysiological
theory previously discussed, it more importantly gives us additional
reason to believe that training programs are feasible for the establishment and maintenance of psychosomatic health. If every young student
knew by the time he finished his first biology class, in grade school,
that the body responds to self generated psychological inputs, that blood
flow and heart behavior, as well as a host of other body processes, can
be influenced at will, it would change prevailing ideas about both physical and mental health. It would then be quite clear and understandable
that we are individually responsible to a large extent for our state of
health or disease.

Perhaps then people would begin to realize that it is not life that
kills us but rather it is our reaction to it, and this reaction can be to a
significant extent self-chosen.

Later in the year Swami Rama returned to the Foundation for another
series of experiments, especially for correlating internal psychological
states (phenomenological or existential states) with brain wave patterns.
At first the experiments appeared disastrous for our general theory of

"focus of attention and brain wave correlates." No matter what psychological state we asked the Swami to demonstrate, from a list he had supplied, the only definite brain rhythm was in the beta frequency band, which was presumably associated with (at least in our understanding of the matter) attention on outside-world sensory processes or attention on intense internal activation of a stressful nature. But one day, after about two weeks of sessions, the Swami said he had some news for us. All the records would have to be thrown away because he had not successfully entered any of the subjective states we had been attempting to study. Since I had not shown him the records, and had not discussed my misgivings with him, this unsolicitied announcement came like a ray of light through dark clouds. I asked what the problem seemed to be and he answered that it would have been much better if I had not told him that the polygraph paper cost $16 per box. All he could think of, he said, was the terrible expense involved and all the people watching the paper shoot out of the machines.

After that came out, we assured him that if necessary we could run paper for twenty-four hour periods without bankrupting the project, and it was agreed that henceforth he would take as long as necessary to move into a particular state of consciousness and that at the appropriate time he would come out of the state and tell us, essentially, that the last five minutes of the record contained what we were looking for. The result of this conversation was a considerable lessening of tension and Swami was subsequently able to enter various states (evidenced by remarkable changes in brain wave patterns) in no more than fifteen minutes, and usually in five minutes.

In five 15-minute brainwave feedback sessions he was able to tie together in his mind the relationship between the tones produced by activation in the various brain wave bands and the states of consciousness he had learned in a Himalayan cave. Then he produced 70% alpha waves over a five-minute period of time by thinking of an empty blue sky "with a small white cloud" sometimes coming by. After a number of alpha-producing sessions the Swami said, "I have news for you, alpha isn't anything. It is literally nothing." This did not surprise us, because we had already observed that the best way to produce alpha was to close the eyes and think of nothing in particular, but it would have provided a shock, I suppose, to the many mind-training researchers who are telling people all over the country that when in the alpha brainwave state you can get rid of your diseases, get the most wonderful ideas, and best of all be telepathic. This kind of talk is nonsense to those clinicians and technicians who work in EEG labs. Whether they accept the possibility of telepathy or not, they know that about 90% of the population of the United States produce alpha waves when the eyes are closed, and the majority of our people are certainly not telepathic nor can they rid themselves of disease merely by being in an alpha brainwave state.

In any event, the Swami next produced theta waves by "stilling the conscious mind and bringing forward the unconscious." In one five-minute period of the test he produced theta waves 75% of the time. I asked him what his experience was and he answered that it was an unpleasant state, "very noisy." The things he had wanted to do but did not do, the things he should have done but did not do, and associated images and memories of people who wanted him to do things, came up in a rush and began shouting at him. It was a state that he generally kept turned off, he said, but it was also instructive and important to look in once in a while to see what was there. From what he said I could well understand that his life in India, of rigorous discipline, sublimation, and strenuous practice, had also involved a measure of suppression and that his reverential attitude toward his Guru, as a being in whom conflicts were resolved, was partly based on his understanding of the difficulty of creating a true synthesis of forces in oneself. The perfected Guru according to Indian tradition is a liberated being who, among other things, has consciously examined all parts of his nature, conscious and unconscious, and has established tranquility and harmony there.

After producing theta waves, the Swami said he knew exactly how the inner states of awareness were arranged in respect to the brain wave frequency bands. Then he said, "Tomorrow I will consciously make delta waves for you." I replied that I doubted that he would succeed in that because he would have to be sound asleep in order to produce delta. He laughed at this and said that I would think that he was asleep but that he would be conscious of everything that occurred in the experimental room.

Before this test he asked how long I would like to have him remain in the delta state. I said that 25 minutes would be all right and he said he would bring himself out at that time. After about five minutes of meditation, lying down with his eyes shut, the Swami began producing delta waves, which we had never before seen in his record. In addition, he snored gently. Alyce, without having told Swami that she was going to say anything (she was in the experimental room observing him during this test) then made a statement in a low voice, "Today the sun is shining, but tomorrow it may rain." Every five minutes she made another statement and after 25 minutes had passed the Swami roused himself and said that some one with sharp heels had walked on the floor above and made a click, click, click noise during the test, and a door had been slammed twice somewhere in the building, and that Mrs. Green had said--and here he gave her statements verbatum, except for the last half of the fourth sentence, of which he had the gist correct though not the words. I was very much impressed because in listening from the control room, I had heard her sentences, but could not remember them all, and I was supposed to have been awake. Dale Walters, our colleague in this research for the last five years didn't remember much more than I, but we reminded each other that our attention was supposed to be on the physiological records, not on what Alyce was saying.

The Swami said that this "yogic sleep," as he called it, was extremely beneficial. He said fifteen minutes of it was as good as an hour's normal sleep. Most people, he continued, let their brains go to sleep while their minds were still busy worrying over various matters, with the result that they woke up tired. It is necessary for the mind and brain to sleep at the same time, he explained. In the delta state he had just produced, he said, he told his mind to be quiet, to not respond to anything but to record everything, to remain in a deep state of tranquility until he activated it. He also said that this kind of sleep was called "dog sleep," because a good dog can leap up from a sound sleep and chase after something without any apparent signs of having to reactivate, a very Zen-like condition it seemed to me.

We did not complete all the experiments planned with Swami Rama. He became involved in giving lectures and seminars around the country and eventually went to India to attend the "yoga and science" conference at New Delhi (December 1970). It was interesting that he carried with him two of our biofeedback machines and gave a lecture called "Yoga and Biofeedback Training." His attitude about biofeedback was that it would accelerate the training of young yogis, up to the point where machines could no longer follow. The machines would also eliminate fakers (not fakirs, please note) by the dozens. Hopefully we will again, one day, have a chance to do more psychophysiological work with Swami Rama.

In our research with the Swami we naturally focused a good deal of attention on physiological data because they are easy to put into graphical form, and it is easier to get research money for projects that come out with red ink on green graph paper. Some of Swami Rama's other accomplishments were of utmost interest, however. For instance, we observed that he could diagnose physical ailments very much in the manner of Edgar Cayce, except that he appeared to be totally conscious, though with indrawn attention for a few seconds while he was "picking up" information. His training program will hopefully be made available to medical doctors, he says, but not, I gathered, before 1973.

Before concluding, it should be mentioned that when the Swami produced alpha, he did not cease the production of beta. And when he produced theta, both alpha and beta were retained, each about 50% of the time. Likewise, when he produced delta he was also producing theta, alpha, and beta during a relatively high percentage of the time. Perhaps this tells us something important. Since alpha is a conscious state, it may be necessary to retain it when theta is produced if one wishes to be aware of the hypnogogic imagery which is often associated with theta. This idea was supported by some of our pilot research with a group of adults who were interested in being experimental subjects for theta research, using brainwave feedback and also some of Swami Rama's breathing exercises for tranquilizing the autonomic nervous system. The main exercise consisted of deep and slow rhythmic breathing, at a constant rate both in and out, with no pauses at the bottom or top of the respiration cycle. After four or five months had passed, with at least ten days a month including a "breathing and meditation" period (at home), the breathing rate could be comfortably slowed to once or twice per minute, for a period of ten minutes. These findings are probably consistent with Wallace's observation, reported in Science, that a significant drop in basal metabolism rate (BMR) accompanied the practice of "transcendental meditation" by a group of college students. We did not have metabolism-measuring equipment in our lab, but we might suspect that if the BMR did not drop significantly in our subjects that it would be very difficult for them to breathe at such an "inadequate" rate without experiencing involuntary diaphragmatic gasping. The Swami's instruction, to allow "no jerks," was eventually complied with, much to our surprise.

As a final word, it seems increasingly certain that healing and creativity are different pieces of a single picture. Both Swami Rama and Jack Schwarz, a Western Sufi whom we recently had a chance to work with, maintain that self healing can be performed in a state of deep reverie. Images for giving the body instructions are manipulated in a manner very similar to that used by Assagioli for personality and transpersonal integration, as in his Psychosynthesis. But this "manner" of manipulation of images is also the same as that in which we find ideas being handled creatively (by two pilot subjects) for the solution of intellectual problems. What an interesting finding! Creativity in terms of physiological processes means then physical healing, physical regeneration. Creativity in emotional terms consists then of establishing, or creating, attitude changes through the practice of healthful emotions, that is, emotions whose neural correlates are those that establish harmony in the visceral brain, or to put it another way, emotions that establish in the visceral brain those neurological patterns whose reflection in the viscera is one that physicians approve of as stress resistant. Creativity in the mental domain involves the emergence of new and valid ideas, or a new and valid synthesis of ideas, not by deduction, but springing by "intuition" from unconscious sources.

The entrance, or key, to all these inner processes we are beginning to believe, is a particular state of consciousness to which we have given the undifferentiated name "reverie." This reverie can be approached by means of theta brainwave training in which the gap between conscious and unconscious processes is voluntarily narrowed, and temporarily eliminated when useful. When that self-regulated reverie is established, the body can apparently be programmed at will and the instructions given will be carried out, emotional states can be dispassionately examined, accepted or rejected, or totally supplanted by others deemed more useful, and problems insoluble in the normal state of consciousness can be elegantly resolved.

Perhaps now, because of the resurgence of interest in self-exploration and in self-realization, it will be possible to develop a synthesis of old and new, East and West, prescience and science, using both yoga and biofeedback training as tools for the study of consciousness. It is also interesting to hypothesize that useful parapsychological talents can perhaps be developed by use of these reverie-generating processes of yoga and biofeedback. Much remains to be researched, and tried in application, but there is little doubt that in the lives of many people a penetration of consciousness into previously unconscious realms (of mind and brain) is making understandable and functional much that was obscure and inoperable.

Alpha and the Development of Human Potential

Robert M. Nideffer

In the past two years the popular press has carried stories about man being able to control the electrical activity of his brain. Particular patterns of brain wave activity have been said to be associated with specific levels of feeling and/or functioning. One of the electrical states of the brain, the alpha state, has been linked to the meditative states achieved by the practitioners of Zen Buddhism and Yoga (Kamiya, 1969). This purported similarity has led several companies to produce portable units designed to allow people to monitor, and learn to control, their brain wave functioning. Emphasis has been placed on being able to develop and maintain the alpha state. Individuals in private enterprise are offering courses designed to teach people to develop their alpha. Publicity advertising these courses makes a variety of promises concerning the benefits that accrue from such training, e.g., an instant Zen experience, more rapid learning, increased interpersonal happiness, development of extra sensory perception (ESP), and control over body physiology and disease processes.

As "far out" as these claims sound, reputable companies are sending their personnel through such training procedures and are reporting positive changes as a result. The following is a quote from the director of training and personnel at Stirling Homex. "We know from our experiments thus far that the speed of learning and degree of retention of material from courses which are given to students in an 'alpha' condition far exceed normal expectations." Among the courses being presented in alpha are "speed reading, electrical assembly, pipefitting, mudding and taping, and carpet buying." In addition to increased learning speed, claims are made that there is a "discernable improvement in the attitude and enthusiasm of students" as well as improvement in terms of "cooperativeness and leadership ability..." (Pappas, 1972).

If such claims are founded, the development of individual control over the production of alpha would be a most worthwhile pursuit. The purpose of the present chapter is to evaluate those claims on the basis of research which has been conducted on the alpha rhythm. Such an evaluation should also suggest directions for future research and training on the control of brain wave activity.

Reprinted with permission of the author. Copyright 1973 by Robert M. Nideffer.

The activity of the human brain has been divided into different types on the basis of frequency and amplitude of the EEG signals. This chapter focuses on alpha activity, first described by Berger in 1929 (Wells, 1963). Alpha is a regular oscillating potential having a frequency between 8 and 13 cycles per second (CPS) and an amplitude between 25 and 100 microvolts (CV) (Glaser, 1963). Initially, Berger believed that the alpha rhythm was the result of syncronous firing of the entire brain. However, in 1934 Adrian and Matthews found that these patterns of activity were generated in the visual association and occipital areas of the brain (Walter, 1963). Localization of alpha in the visual areas of the brain suggested that this neurological pattern must be associated with functioning of the visual apparatus.

Since the studies by Berger, and by Adrian and Matthews, many attempts have been made to examine the relationship between alpha and various physiological and psychological variables. Information from these studies is critical if an attempt is to be made to understand what the alpha state is and what learning how to control the state has to offer. In examining these data, however, it is important to bear in mind that most of these studies are correlational in nature rather than causal. Thus although relations may be found to exist between alpha and various physiological and psychological processes, this is not to say that alpha causes any of these, or that they cause alpha.

In 1929 Berger noted that the alpha wave desyncronized (blocked) when subjects saw visual stimuli. Since then, many investigators have argued that alpha depends on an absence of visual stimuli (at least patterned stimuli) for its existence (Brazer, 1958; Gardner, 1963; Proler, 1963). In addition to being interfered with by visual stimuli alpha was also seen as subject to disruption by attentional processes. Short and Walter (1954), for example, suggested that alpha was a "non-focal attentive state, that would be blocked by attention to either real or imagined visual stimuli." Other investigators have generalized this view to encompass any focal attentive state be it auditory, tactile, or visual (Salamon and Post, 1965). Research has shown, however, that while attention, particularly to visual stimuli, affects alpha, the relation is less clearly defined than originally suggested. Mundy and Castle (1957) noted that there is not a one-to-one correlation of alpha with either attention or visualization. In fact, considerable mental activity and attention would appear to be possible without blocking alpha (Dewan, 1967). Since both attention and visual focus are important in learning, and because one claim for the alpha state is increased learning speed, the relation between alpha and each of these variables needs to be examined.

In 1935 Durup and Fessard hypothesized that the alpha rhythm was related to ocular accommodation. Subsequently, Mulholland (1967) expanded this hypothesis by arguing that "occipital alpha is 'blocked' or activation occurs when the triad of ocular accommodation is not stabilized." Stabilization of the accommodation triad is achieved by a coordinated action of the ocular system. The triad involves minimization of the angle between the visual target and the fovea, minimization of target blur, and control of the pupils to optimize the level of stimulus input. According to Mulholland when these three things occur, alpha is blocked. Support for this hypothesis is found in the following studies.

It has been demonstrated that some $\underline{S}$s will produce alpha simply by having them tilt their eyes into an upward position (Mulholland, 1967; 1968). Dewan (1967) showed that some $\underline{S}$s could control the production of alpha so well by rolling their eyes upward, that they could send Morse Code messages to a computer. Movement of the eyes upward resulted in sending alpha wave dots or dashes to a computer which was programmed to interpre them, and to print out the appropriate letter. Fenwick and Walker (1969) also found Ss who generate alpha when they tilt their eyes upward. These authors hasten to point out, however, that this is not true for all Ss. Dewan suggested that when alpha occurs under these conditions, it is because the eyes' upward position results in a defocus and a relaxation of occular convergence. Such an interpretation obviously supports Mulholland's hypothesis.

Mulholland (1967) studied the amount of alpha in subject's EEG records under three conditions, where subjects were instructed to: 1) focus on, and track, a moving visual stimulus; 2) blur the stimulus and continue tracking, and 3) blur the image and refrain from tracking. Alpha production, as expected, was greatest in the third condition, and most attenuated in the first. In another study of visual tracking Peper (1970) reported that Ss were totally unable to either initiate tracking or continue to track without blocking alpha.

Available data thus suggests that eye movement, accommodation and convergence is related to the presence or absence of alpha, at least for some Ss. Focusing of the eyes via the subject's ocular motor apparatus disrupts alpha. This appears to be true in virtually every case. Eye position though correlated with alpha in some Ss is not perfectly correlated. For example, Ss can develop alpha with their eyes looking straight to the front. Kamiya (1969) reported that one of his Ss who had learned to discriminate alpha from non-alpha did so at first, at least in part, on the basis of tilting his eyes upward. After additional training, however, he was able to develop and recognize alpha without tilting his eyes.

The eye movements that have been discussed thus far are associated with a more generalized "attentional" response called the orienting reflex. Mulholland (1967) suggested that the apparent relation between alpha and attentional processes is a secondary effect of the eye movements that are associated with attention per se. If this is true, when it is possible to attend without stabilizing the accommodation triad, alpha should still be present. There is evidence to suggest that this is so.

Mulholland (1967, 1968) has shown that alpha will persist in Ss who have their eyes open and tilted up, even when a visual stimulus is introduced. In addition, Ss can examine and talk about a visual after image that they are seeing with their eyes in the upward position without blocking alpha. Dewan (1967) maintains that Ss in his study must have been attending in order to send Morse Code, in alpha, to the computer, since the task requires that Ss think about what they are doing if they are to send a coherent message. Peper (1970) cites a study by Pollen in which Ss were able to read text and still produce alpha. Pollen inactivated the accommodation system and used an artificial lens to maintain visual focus for the Ss. Ss were thus able to perceive patterned visual stimuli without focusing their eyes, and they could maintain alpha. Oswald (1959) reported that intense auditory alertness could be maintained along with the production of alpha provided there was a loss of ocular fixation and accommodation.

These data strongly suggest that under some conditions alpha may be present even with sustained attention. Caution should be maintained with respect to accepting these data at face value however. Control of the visual apparatus does appear to be important with respect to alpha. It can be pointed out that although eye movements and accommodation are often related to the focus of attention they are also related to other things. Anxiety, and affective arousal also cause changes in eye accommodation. It may be that these variables are more central to the development of alpha than eye accommodation. This possibility will be discussed in more detail after considering the matter of subjective feelings that develop when a S learns to control the alpha state.

The first indication that the alpha state could be altered by learning occurred in 1935. Quite by accident, Durup and Fessard (1935) noticed that the alpha rhythm could be classically conditioned to block. These investigators were using a camera to record changes in alpha activity as a function of the presentation of a bright light. Initially, the click of the camera shutter had no effect on the presence of alpha. After the click was paired with the light stimulus several times, however, the click appeared to acquire some of the value of the light stimulus. The light would block alpha and was seen as an unconditioned stimulus. Eventually, the click after being associated with the light served to block alpha by itself. Several other investigators have replicated Durup and Fessard's findings (Loomis, Harvey and Hobart, 1936; Jaspar and Shagass, 1941). Alpha has been shown to be classically conditionable not only because it could be blocked by a previously neutral stimulus, but by pairing a neutral stimulus with conditions that facilitate alpha (e.g., darkness), bursts of alpha could be conditioned to appear.

These demonstrations failed for several reasons to capture the interest of researchers. First, the conditioned responses were unstable (i.e., the learning extinguished rapidly). Second, Ss did not have voluntary control of the response. Thus, although a click briefly blocked or even developed a burst of alpha the subject had no control over this response.

It was long felt that involuntary, autonomically mediated body functions could not be consciously controlled. This idea was dramatically altered in the 1960's when Neal Miller presented evidence that through operant or instrumental conditioning stable control over "autonomic functions" could be learned (Miller, DiCara, Solomon, Weiss and Dworkin, 1971). These investigators showed that if rats were provided with feedback about various physiological processes, such as heart rate and blood pressure, and were rewarded for learning to alter those processes in a particular direction they would learn to control them. It was through the use of this type of learning paradigm that Ss were first taught to discriminate and control their different brain wave states.

Much of the recent alpha rhythm research has been stimulated by the studies of Joseph Kamiya. Initially, Kamiya was interested in seeing if his Ss could learn to recognize what brain wave state they were in at a given time. He attempted to teach Ss to differentiate the feelings associated with alpha from those they had when they were in a beta state (Kamiya, 1962).

On the surface, such a discrimination would seem to be relatively easy to learn. The alpha state had been generally accepted as a relaxed, non-focal attentive state, as indicated in the following description by Salamon and Post (1965). "...a state of resting wakefulness. When attention is focused on a stimulus, whether this be visual, auditory, tactile or cognitive the alpha waves disappear or are markedly reduced in amplitude." By contrast, the lower amplitude higher frequency, beta state occurred when the S's attention was focused and the S was aroused.

Kamiya first connected Ss to the EEG and told them he wanted to see if they could learn to recognize two different brain wave states. He instructed the Ss that he would watch the activity of their brains and then occasionally ring a bell. Sometimes the bell would be presented when they were in state A (alpha) and sometimes when they were in state B (beta). When they heard the bell ring, they were to call out state A or state B, whichever they felt they were in. Each time they called out Kamiya would tell them if they were correct or not. At first, Ss responded to this procedure with only chance level performance (i.e., they were correct in their guesses about 50 percent of the time). After about three hours of this procedure, though, most Ss improved to about 75 percent accuracy. In fact, some of the Ss were able to name the brain wave state they were in every single time (Kamiya, 1968).

Having shown that Ss could be trained to recognize their brain wave state, the next step was to discover whether this knowledge allowed them to alter and control brain waves. Kamiya did this by simply saying to the subject, "Obviously, you've learned how to make this discrimination quite well; let's see if you can produce those states that you have been calling 'A' and 'B' upon our command." He found that once Ss had successfully gone through the discrimination training they seemed to have gained control of their brain wave patterns. They were able to produce the alpha and beta states on command.

Next, an attempt was made to find out if Ss could learn to control brain wave states without having first learned to discriminate between alpha and beta. In these later studies (Kamiya, 1968; Nowlis and Kamiya, 1970) Ss were connected to the EEG and told that a tone would come on and remain on as long as they were in the alpha wave state. The subject's task was to learn to keep the tone on. Ss were seated in a moderately darkened room and a base level of alpha production, in an eyes-closed resting state, was taken. Ss with high resting-levels of alpha were asked to keep their eyes open during testing, whereas the Ss with a lower initial percentage of alpha were tested with their eyes closed. It is worthy of note that some individuals may have alpha 80 to 90 percent of the time in an eye closed resting state prior to any training. Presumably Ss with such high base levels of alpha have little room for improvement through training. This ceiling effect can be avoided by having Ss open their eyes. Visual stimulation greatly reduces alpha and thus allows more room for change.

Most of Kamiya's Ss increased alpha production as a function of the tonal feedback. Other investigators have replicated Kamiya's findings (Peper & Mulholland, 1970; Brown, 1970). In Brown's study, Ss were trained to produce more alpha using visual feedback instead of a tone. All of Brown's Ss were tested with their eyes open and alpha presence was indicated by a small light. It is interesting that Brown's Ss learned to control their alpha production more rapidly than Kamiya's. In agreement with this finding, Kamiya reported that among his Ss, those who were tested with their eyes open learned to control alpha better than those tested with their eyes closed. Both Brown (1970) and Peper (1970) have suggested that Ss who have their eyes open are more aware of the effects of eye position, focus, and accommodation on the generation of alpha, a suggestion that well fits Mulholland's views concerning the importance of accommodation in alpha production.

In addition to demonstrations that the alpha rhythm can be conditioned, Kamiya has reported being able to alter, via operant conditioning and feedback, the amplitude and frequency of the alpha rhythm. Apparently, Ss can be taught to slow the frequency of alpha, and to raise its amplitude (Kamiya, 1969). However, these findings have yet to be replicated.

Clearly, some Ss can learn to control their brain waves. However, some studies have suggested that Ss may increase the production of alpha even when they are receiving false feedback (tone on, when alpha may or may not be present). Such findings, to be discussed in detail later, suggest that we are currently not sure just how much learning is taking place. Nor is it clear how important particular feedback may be to the production of alpha. Prior to dealing with these issues it will be useful to review some of the research on the subjective states Ss report experiencing as a function of alpha training.

The fact that people could learn to control their brain waves was not the critical variable in stimulating the public's interest in the alpha state. What appeared to be most exciting were the subjective descriptions of feelings while in alpha. In the studies by Nowlis and Kamiya (1970), Brown (1970), and by Peper and Mulholland (1970) individuals who had learned to control their brain waves were asked to describe how they accomplished this, and how they felt in the various states (alpha and beta).

Reports of Ss who had learned to control alpha indicated that they blocked alpha and entered the faster beta wave state by developing intense visual imagery, and by looking at or examining their visual images in great detail. They were also successful at blocking alpha by focusing their eyes on the spots (phosphenes) that sometimes drift past the eyes when they are closed. Consistent with Mulholland's view, these reports indicate that even with their eyes closed Ss focus on and track images.

In contrast to the beta state, Ss increased alpha by developing a "noncritical" attitude about the environment per se and the experiment in particular. In Nowlis and Kamiya's study (1970), Ss said they calmed their minds, relaxed their mental apparatus, stopped thinking about the outside world and just let themselves be carried along by the tone. Brown's subjects reported similar feelings, i.e., they lost awareness of the light which was used for feedback and felt dissolved by the environment.

The Ss in these studies reported the alpha state to be very pleasant and indicated that they preferred it to beta. Thus, they began requesting the experimenter to allow them to turn on the tone more often (Kamiya, 1969). Kamiya noted that this stated preference for alpha by his Ss was accompanied by a spontaneous increase in their resting levels of alpha. That is, even when Ss were not being tested they began to spend more time in the alpha state. The descriptions of what Ss said they felt like in the alpha state resembled the feelings which are said to accompany meditative states. The relaxed "noncritical" attitudes of Kamiya's and Brown's Ss correspond quite closely to the ego loss and relaxed, noncritical mergence of mind and body described by Zen and Yogic meditators.

Since his Ss' reports seemed so similar to prior descriptions of the meditative state, Kamiya invited several Zen priests, practiced in meditation, to come into his laboratory to monitor their brain waves. He found that they learned to control alpha much more rapidly than most Ss and that they appeared to do so through meditation procedures (Kamiya, 1969). A number of studies on the physiological correlates of meditation support this observation by Kamiya (Bagachi and Wenger, 1957; Anand, Chhina, and Singh, 1961; Wallace, 1970; Kasamatsu and Hirai, 1966).

Anand, Chhina, and Singh (1961) have studied the physiological correlates of Yogic meditation. Investigations of these authors indicated that the meditative exercises practiced by the yogis they tested were accompanied by a high percentage of alpha rhythms. Generally, both alpha amplitude and percentage was quite high while the S was meditating. It was also noted that in contrast to normal Ss, introducing external stimuli failed to block the alpha production of these meditators.

Kasamatsu and Hirai (1966) studied brain wave changes that accompany Za Zen meditation. These investigators studied the EEGs of 48 priests and desciples of the Soto and Rinzai sects of Buddhism. Kasamatsu and Hirai's Ss ranged in age from 24 to 72 years and in meditation experience, from 1 to 20 years. EEG recordings were taken before, during and after meditation, with the eyes always open in all conditions. It was discovered that with respect to the brain wave functioning, Za Zen meditation could be divided into four stages. Stage one involved the development of prolonged bursts of alpha; stage two was an increase in the alpha amplitude; stage three was a decrease in the alpha frequency, and stage four was the development of trains of high amplitude theta waves. It was also found that the ability with Ss to develop these four stages depended on both the amount (number of years) of training and on the quality of the meditation (as rated by their instructors). Only the more highly trained meditators achieved the fourth stage.

An interesting feature of the study by Kasamatsu and Hirai has to do with the manner in which the Zen meditators responded to the introduction of a click stimulus. One priest was selected, and every 15 seconds during the stage of prolonged alpha activity a click was presented. This click blocked his alpha production for from two to three seconds each time it was introduced. With respect to alpha's responsiveness to external stimulation there is an interesting contrast between the normal, the Zen, and the Yogic meditative states.

When repetitive stimuli are presented to normal Ss while they are in the alpha state they respond by gradually reducing alpha blockage. Alpha production is blocked the first two or three times, after which the S habituates to the stimulus presentation and stops blocking alpha. Such habituation, as noted above, does not occur in the Zen meditator. The yogic meditator in contrast to both of these responses fails to block alpha at all. In this regard the Yogi appears unresponsive to external stimulation. The contrast between the Zen and Yogic meditators merit some comment. Peper (1970) has suggested that both speed and length of time of alpha blockage are cues to the Ss' interest in the intrusive stimulus. The philosophy of the Zen meditator is drastically different from that of the Yogi with respect to involvement with the here and now. Johnson (1970) suggested that the differences between Zen and Yogic meditators with respect to blocking of alpha can be attributed to their philosophical differences. Whereas the Zen meditator seeks to become totally involved with nature and responsive to his environment, the Yogic meditator seeks to transcend the external world in the hopes of achieving a higher state.

The development of alpha in the meditative state is not limited to Easterners. More recently Wallace and Benson (1971) studied the physiological effects of transcendental meditation in American Ss. Their Ss had been practicing meditation procedures for periods of between 1 month to 9 years, with the majority of the subjects having had from two to three years of "training" (i.e., two 15-20 minute sessions per day). In these exercises the S sits in a comfortable position with his eyes closed. The meditator then "perceives a suitable sound or thought. Without attempting to concentrate specifically on this cue, he allows his mind to experience it freely, and his thinking as the practitioners themselves report, rises to a 'finer and more creative level in an easy and natural manner.'" Wallace and Benson found these meditation procedures to be associated with (among other things) marked intensification of alpha waves in all Ss.

It is on the basis of the information that has been presented thus far that companies have been making claims for the benefits accrued through alpha conditioning. From this information it is not readily apparent how individuals can maintain that the alpha state is an instant

Zen experience. Nor is it clear how the other claims are made, such as alleged increased speed of learning, the development of extra-sensory perception, are all said to be learned as a function of alpha training.

Control over extra sensory phenomena and physiological processes has been attributed to some Yoga and Zen practitioners. A number of Zen meditators describe extra sensory awareness of their surroundings and the development (through meditation) of a "sixth" sense (Uyeshiba, 1968). Yogis have been reported to demonstrate extra sensory powers (Bagachi, 1969) as well as control over involuntary processes such as heart rate and respiration (Barber, 1970). It would appear as if these same abilities attributed to the yogi and zen masters have also been attributed to the "normal" individual who learns in a few hours to control his alpha production. Apparently since meditative states are associated with a high production of alpha and since normal subjects report some of the subjective feelings of the meditators it has been automatically assumed that control over alpha leads to these other developments.

Although these assumptions would appear to be tenuous at best, it may prove difficult to completely evaluate them. For example, ESP phenomena, though abundantly reported, have yet to be clearly established as a legitimate phenomena. It is difficult to maintain that alpha, or anything else, does or does not lead to a phenomenon that we cannot even prove exists. With respect to increases in learning speed while in the alpha wave state the type of material to be learned may be critical. In spite of these considerations some reasonable conclusions may be drawn.

Among other things the popular press has referred to alpha training as "instant yoga" and "electric Zen." The implicit assumption being made is that alpha training causes spiritual enlightenment or a "oneness with nature or the universe." As evidenced by $\underline{S}$s who have learned to control alpha, this is not what occurs. These individuals have not claimed that they become one with either the universe or nature. What does exist is a similarity between the description these $\underline{S}$s give of their environment, and that presented by Zen and yogic meditators. That is, these groups perceive the world as though they were relaxed and uncritical. They feel "fused" or immersed with the tone or light. However, the significance of this perception is far greater for the Yoga and Zen practitioner than it is for the normal. $\underline{S}$s may see the world in similar ways but their perceptions have tremendously different meanings. The Westerner feels alpha results in an alteration in his normal mode of thinking, a pleasant relaxed state. The Zen and Yoga practitioner attribute a great deal of religious and spiritual significance to the subjective state and see it as a union with nature or the universe. The impact that the subjective state has on the meditator is thus related to the amount and intensity of his spiritual and intellectual preparation. The alpha state does not create a oneness with the universe. It may allow a perception to develop which, given the appropriate religious training (years of study) can have a profound effect on an individual's life. Without that preparation, however, the $\underline{S}$ is only relaxed and for the moment noncritical. Similar arguments are presented by Sidney Cohen (1964) when he compares religious experiences induced by LSD-25 with ones which occur naturally after years of study and devotion.

Just as spiritual enlightenment depends upon spiritual preparation and training for its development, so does the control over bodily processes depend upon physical training. Many Yogis have combined various physical exercises with their meditative practices. These would include the practices of various postures and breathing procedures (Bagachi, 1969; Barber, 1970). As a result of rigorous self-discipline and practice some

Yogis are able to exhibit considerable control over the musculature of their bodies. It is this physical practice which must be relied on to explain their ability to slow heart rate, rather than an alpha state. A study by Wenger, Bagachi and Anand (1969) exemplifies the importance of physical training.

With respect to slowing or "stopping" the heart, Wenger, Bagachi, and Anand (1969) tested four Hatha Yogis to find out what was taking place. These investigators found that the Yogis were able to execute a Val-Salva maneuver, which involves holding the breath and increasing muscle tension. Through this maneuver the Yogis could increase their intrathoracic pressure for from 10 to 15 seconds. This movement results in a slowing of the return of venous blood to the heart. As the blood flow is slowed, the heart cannot be heard with a stethoscope and the pulse is not detectable. The electrocardiogram (EKG), however, indicates that during this time the heart does in fact continue to beat.

Though alpha in itself does not lead to control over the physiology of the body, it is possible that the alpha state may contribute to the execution of some existent skills by allowing the practitioner to remain calm in a stressful situation. This possibility will be discussed in more detail later.

It has been mentioned that it is difficult to evaluate the possibility that training in alpha leads to increases in ESP. It is easier to evaluate this claim by dividing ESP phenomena into two categories. On the one hand, reference can be made to "para-normal" powers such as walking through walls, teleportation, etc. On the other hand, some ESP phenomena might be described as simply an expansion of "normal" awareness (i.e., being aware of the movements of a person behind you in a fight, or knowing by minute cues when a person is going to pull the trigger on a gun that is aimed at you). Conclusions with regard to the contribution of the alpha state to these types of "ESP" phenomena vary according to the category and sensory mode being discussed. For the moment focus is on the development of "para-normal" powers.

If it is assumed that "para-normal" phenomena are possible (we see magicians do things like this all the time) it would appear to be ridiculous to attribute such phenomena to the production of alpha rhythms. A large number of individuals spontaneously produce a very high percentage of alpha without presenting any evidence at all that they can transform objects. Kamiya and Brown have yet to find a single subject who can walk through a wall or disappear and yet they have subjects who can control their alpha quite successfully. It would seem much more reasonable to attribute such phenomena, should they exist, to a gift of God, or a genetic accident, rather than to alpha.

The suggestion that learning occurs more rapidly in the alpha state is a complicated issue. To deal adequately with this question learning situations must be clearly defined. For the moment concern is with learning which would require visual fixation and accommodation on the part of Ss. According to some people (Pappas, 1972) individuals learn to read more rapidly by taking speed reading courses in an alpha state. On the basis of the influence of ocular fixation and accommodation on alpha this assumption cannot be accurate. Individuals must both focus and track visual stimuli in order to read. Such a process serves to block alpha (Dewan, 1967). In fact, alpha presence may be seen as interfering with this type of learning.

A study by Peper (1970) presents evidence which suggests that learning, via reading or visual observation, is more likely to occur in a non-alpha state. Peper noted that male Ss showed continued orienting (long periods of non-alpha) when a picture of a nude girl was presented. On the other hand, blocking of alpha was constrained when the picture (a flower) was not worth looking at. In view of these findings, Peper suggests that the initial attractiveness of a stimulus is reflected in the length of time it takes Ss to block alpha after the stimulus is presented and continued attractiveness of the stimulus is reflected in the length of the non-alpha interval. Mulholland (1968) maintains that for optimal learning to take place advantage should be taken of the S's interest. Thus, under optimal conditions visual stimuli would be presented to the S when he is in a non-alpha state and removed as he loses interest and drifts back into alpha.

In brief summary, research conducted would appear to indicate that there is little validity for a number of the claims made for the alpha state. Attainment of enlightenment, a "oneness with the universe" or "nature" requires dedication and study in addition to the presence of alpha. Likewise control over bodily processes such as those muscles required to slow the heart requires intense physical training. Such control is not automatically a gift of the alpha state. The development of "para-normal powers" has not been established either in or out of the alpha state. Finally, in many instances visual focus and convergence are necessary for learning to occur (i.e., reading). Since alpha is not present when visual focus and convergence occur it cannot increase the speed of this type of learning. In fact, the presence of alpha under these circumstances would disrupt the learning process. In spite of these negative findings alpha may have some very valuable uses. To examine this possibility, the relation of alpha production to affective arousal and "subjective" changes that occur as a function of alpha training are presented in more detail.

One of the most consistent descriptions of the alpha state is that while in it, Ss are "mentally calm." This "subjective" finding would indicate that, in addition to occular accommodation, level of arousal (particularly affective arousal) may be critical to the development and maintenance of the alpha state. Since changes in accommodation and convergence typically accompany changes in arousal it is difficult to estimate the relative importance of these two variables. Research needs to be conducted which will allow separation of the independent contributions of arousal and occular accommodation to alpha. Until such research is conducted, formulations such as the following must remain speculative.

It may be hypothesized that, within an individual, as affective arousal (emotional involvement) increases, alpha will decrease. That is, alpha should be produced so long as stimuli are not being reacted to in an ego-involved way. If attention is possible, as meditation procedures would suggest, without ego or personal involvement, and without changes in ocular accommodation, then alpha may be present. Such a suggestion does not preclude the possibility that independently, both of these variables (ocular accommodation, and arousal) will block alpha. In fact, it is explicitly maintained that for alpha to occur, the accommodation triad must not be stabilized and arousal level must be low. Research which follows is presented to illustrate some of the effects that arousal may have on alpha.

Support for the hypothesis that increased levels of arousal reduce alpha may be found in research conducted on schizophrenic patients. Examination of the cardiovascular and neuromuscular systems of schizophrenics points to a high level of arousal (Lang and Buss, 1965). If these patients have a higher base level of arousal than normals, and if arousal is related to alpha, then schizophrenics should evidence reduced alpha output relative to normals.

Goldstein, Murphree, Sugarman, Pfeiffer and Jenney (1963) designed a study to examine the "mean energy content" of the brain waves of schizoprenic and normal Ss. Schizophrenic patients employed in this study were all classified as poor premorbid. These investigators found two important differences between schizophrenic patients and normals. First, schizophrenics had a lower "mean energy content" than normals and second, they had less variability in the amount of energy present over time. "Mean energy content" would be increased by the presence of high amplitude low frequency waves such as alpha. Also, movement from the beta state to the alpha state would increase the variability of the mean energy content measure. Thus, the schizophrenic Ss may have had less alpha and other slow wave activity than normals. The lower "mean energy content" of schizophrenic Ss' EEGs was interpreted by the authors as indicating that they were more highly aroused than normals. In addition, the authors speculated about possible behavioral correlates of a reduced variability in the energy content of the EEG. It was suggested that variability in brain wave functioning (energy content) would be necessary for individuals to be able to adapt to changing environmental conditions. The reduced variability of the schizophrenics, then, would imply a reduced ability to adapt to changing situations (Goldstein, Sugarman, Stolberg, Murphree, and Pfeiffer, 1965).

Findings of reduction in alpha production by poor premorbid schizophrenic patients has been observed more directly by other investigators (Cromwell and Held, 1969; Nideffer, Deckner, Cromwell and Cash, 1971). In these studies schizophrenics were selected on the basis of the presence of alpha in an eyes-closed resting state. Assessment of the patients' premorbid status after they had been selected in this manner revealed that they were predominately good premorbid subjects. Such a subject distribution would not be expected on the basis of chance alone, since at least fifty percent of the patients screened for the studies were poor premorbid. The poor premorbid patients that were screened simply evidenced little or no alpha.

Some additional suggestions regarding the effects of tension of alpha come from observations regarding the type of individual who makes a good alpha training S. Kamiya (1969) described the good alpha S as a person who appears interested, relaxed, and comfortable. These individuals look you in the eye and feel comfortable in close interpersonal relationships. The description by Kamiya corresponds almost perfectly to Erickson's description of good hypnotic Ss (Erickson, 1967). In contrast to these descriptions, Saul and Davis noted that aggressive individuals were low alpha producers (Saul and Davis, 1949) and Glaser (1963) stated that "tension apprehension and anxiety lead to a decrease in alpha activity."

Further evidence that arousal affects alpha productivity may be found in some of the studies on hypnosis. Development of the hypnotic state is seen as dependent upon the S's trust in the hypnotist and upon his willingness to cooperate (Barber, 1970; Shapiro and Diamond, 1972). Good hypnotic Ss appear willing to relax their ego control, to allow themselves to respond to the hypnotist. Loss of ego control also appears to be a characteristic of the alpha state. It would be expected, then, that good hypnotic Ss will evidence greater alpha productivity and control than poor hypnotic Ss.

Earlier studies of hypnotized Ss found no significant difference between EEG records of Ss when they were in a waking state as compared to a hypnotic state (Weitzenhoffer, 1953). More recently, however, it has been suggested that a correlate of susceptibility to hypnosis is the amount of alpha Ss generate in a resting state. London, Hart and Leibovitz (1968) demonstrated that highly susceptible hypnotic Ss have significantly more alpha in both a resting eyes-closed state and when producing visual imagery than low susceptible Ss. The relation of alpha to hypnotic susceptibility was illustrated more fully in a sound study.

Engstrom, London, and Hart (1970) designed a study to see if training Ss to increase alpha would result in corresponding increases in their hypnotizability. Twenty Ss were assigned to an experimental group and ten to a control group. Both of the groups received feedback (tone) concerning alpha production. All of the Ss were instructed that the tone would be on when they were in the alpha state. In actuality, however, only the experimental group received real feedback concerning their brain wave state. The control Ss heard a tape recording of a person who had learned to gradually increase their alpha production over time. Results of the data analysis indicated the S's alpha base level (prior to feedback) was positively correlated with hypnotic susceptibility ($r = .79$). In addition, both groups increased their production of alpha and their hypnotic susceptibility as a function of the feedback training. The experimental group showed more change on both measures than the control group. The changes in alpha and hypnotic susceptibility for the control group (who received false feedback) are interesting and there are at least two possible explanations for it. First, Ss may increase alpha production because they become more comfortable and relaxed as the experiment progresses. Second, it may be that the tone in some way serves as an unconditioned stimulus for alpha. This hypothesis will be discussed in more detail later.

One of the ways that alpha could be valuable depends upon the extent to which its existence and development also reflects the existence and development of trust (i.e., the ability to give up some of our inter- and intrapersonal control) on the part of the S. A lack of trust results in increases in an individual's level of anxiety and arousal. The research by Engstrom et al. (1970), and London et al. (1968) suggests that alpha, trust, and hypnotic susceptability are all interrelated. Additional support for this notion comes from studies relating hypnotic susceptibility to therapy.

In many of the humanistic approaches to therapy a major focus is on the development of trust. Trust is seen as leading to richer interpersonal relations, greater tolerance and understanding for self and others, as well as increased self-confidence. Should trust, hypnotic susceptibility and alpha be related, as trust develops in therapy, changes in the hypnotic susceptability of Ss as well as changes in their base levels of alpha should occur. Information with respect to changes in alpha as a function of therapy is minimal and only correlational. Kamiya (1969) noted that Esalen Institute counselors (individuals who place great emphasis on the development of empathy and trust) are good alpha producers. There does appear to be more positive information concerning the relationship between therapy of the encounter group type (Esalen model) and hypnotic susceptibility.

Tart (1970) found there were significant increases in the hypnotic susceptibility of Ss who had participated in a nine-month Esalen training program. Shapiro and Diamond (1972) designed a study to assess the effects of encounter group experience on hypnotic susceptibility. They divided Ss into three groups and varied the amount of interpersonal communication in each group. Groups met once each week for eight, two-hour sessions; following the eight sessions each group participated in a ten-hour marathon encounter. Results showed that there were significant changes in the hypnotic susceptibility of the two groups which received the greatest amount of interpersonal communication. Shapiro and Diamond attributed these changes in susceptibility to increases in interpersonal trust.

Although direct evidence of the relation of alpha to level of arousal is lacking, there is enough correlational data available to justify some speculation. Subjective descriptions of meditative, hypnotic and alpha states are similar in that individuals appear relaxed and apparently relinquish some of the controls they normally exert over cognitive and perceptual processes. Good hypnotic $\underline{S}$s, good meditators, and good alpha producers all appear to have some common "personality" characteristics. These individuals are described as open, relaxed and trusting. Finally, changes in a $\underline{S}$'s level of responsiveness or skill in one of these areas is reflected in the others. Thus, as meditative skill increases or hypnotic susceptibility increases so does alpha production. Assuming the existence and development of alpha parallels the existence and development of a relaxed ego free state in $\underline{S}$s, a number of uses for alpha training can be postulated.

There would appear to be at least three ways in which $\underline{S}$s may benefit from alpha training: First, should the alpha state be directly related to physical relaxation it can be useful in the treatment of hypersensitive conditions and it may increase $\underline{S}$s' ability to withstand stress. Second, if the alpha state reflects lower levels of affective arousal and a relaxation of normal perceptual and cognitive control it may be useful in a variety of ways (i.e., to increase self-acceptance, tolerance, and understanding as well as empathy for others). Finally, the alpha state may be used to train individuals to recognize and experience a "noncritical ego-free" involvement. Then if this attitude can be generalized to other brain wave states (e.g., beta) improvement in physiological and psychological (cognitive) tasks which require focused visual attention may occur. It is possible to speculate about the changes that might occur in these three areas by examining some of the studies in the areas of hypnosis and meditation.

Schultz and Luthe (1959) have been using systematized relaxation training procedures (autogenic training) to treat a variety of physical and psychological problems. $\underline{S}$s are taught to passively concentrate on relaxing the muscles in various parts of their bodies. Procedures such as these and the ones used in hypnosis have been found to cause a variety of physiological changes, including dilation of peripheral blood vessels, changes in blood sugar levels, and a reduction in muscle tension (Luthe, 1969; Barber, 1970). Such changes may be very therapeutic particularly if the $\underline{S}$s exhibit some types of hypertensive pathology. As an example, relaxation procedures have been used to decrease the amount of insulin required by diabetic patients. It has been found recently that meditation procedures also result in some of the changes in physiology describe above.

Wallace and Benson (1972) studied the physiological correlates of meditative procedures on 36 American $\underline{S}$s. It was found that meditation was correlated with a reduction in blood pressure and heart rate, slower respiration, increased alpha production, and increased skin resistance. These authors suggested this indicated meditation procedures could be useful in combating diseases resulting from hypertension (e.g., heart problems, ulcers). In addition to physiological benefits, Maupin (1969) has suggested some psychological benefits associated with meditation, these included learning to be quiet and pay attention, learning to relax and store energy, thus being better prepared to cope with stressful situations.

Apart from physiological relaxation, the attitude that appears to be correlated with alpha presence may have some direct benefits for the trainee. There appears to be a breakdown both in alpha, and while meditating, with respect to ego boundaries and usual subject-object relations. This breakdown is referred to as deautomatization by Deikman (1969). According to Deikman the positive benefits which result from this process involve the way in which the world is seen after meditation (or to hypothesize, alpha). The individual, through his deautomatization experience, attains a fresh new perception of the environment, as though he were seeing it for the first time. This view allows him to form new conceptual relationships, and new ideas are stimulated. The world becomes more exciting and interesting and because of this the individual is happier and more involved with life.

The deautomatization which is hypothesized to occur both in meditation and alpha may be useful in treating neurotic disorders both as a means of increasing self-awareness and to improve interpersonal functioning. This can be illustrated most clearly by examining the neurotic from a Freudian framework. Freud noted that a great deal of the neurotic's psychic energy was expended in the service of the ego. The individual was seen as using defense mechanisms to avoid exposure to thoughts and feelings which were perceived as unacceptable. Thus, psychic energy had to be expended in an attempt to repress unacceptable thoughts and feelings. Energy used in this manner is not available for other purposes. In addition, the individual is isolating himself from a portion of life. Other defense mechanisms may also be employed, and often these result in distorted perceptions of and alienation from the environment. For example, unacceptable thoughts may be projected onto other people. The neurotic attributing his own unacceptable thoughts onto others begins to judge them and isolates himself. The alpha state might be useful in helping individuals lower their defense mechanisms. As defenses are lowered, $\underline{S}$s begin to see themselves and others more clearly and with assistance could use the information (insights) gained for constructive growth. Such a process appears to occur in meditation, provided the student has a teacher to assist him with his insights.

As the Zen or Yogi practioneer begins his meditative exercise, defense mechanisms are lowered and he is often confronted by unpleasant thoughts about himself, his attitudes, and his behaviors. Through the meditative training he gradually learns not to become emotionally upset by these thoughts. He learns to allow them to come into awareness, he accepts them and observes them, and then he lets them move on. Through his spiritual education and training, as well as his meditative practices, he learns to accept those things he cannot change in himself and to change those things than he can. He does this without castigating himself, he does it with a "mind like the water." He reflects and responds to a true image of both himself and the world (Uyeshiba, M., 1968). A study by Lesh illustrates how meditative (perhaps alpha) training might be used to increase self-awareness and understanding of others.

Lesh (1970) noted that therapists often appeared so affected by statements patients made, and by their own concerns about how they should respond to the patient, or where they should go next, that they often failed to hear what the patient was actually saying. They projected their own meaning onto the communication rather than listening to the patient. Their anxiety about performance distorted their perceptions of the world and interfered with their ability to respond appropriately. It was felt that this interference could be overcome by training the therapists in Za Zen meditation. Theoretically, these procedures would act to open $\underline{S}$s up to their own experiences. In becoming more aware of themselves they would remove some of the anxiety that was interfering with their performance. In addition, in knowing more about themselves they would be better able to empathize with the patient, because they would have experienced more of the things that the patient was talking about.

In his study Lesh had an experimental group which practiced Za Zen meditation 30 minutes each day, for four weeks. He also employed two control groups, one group had volunteered to meditate but were asked to wait. The second group did not believe in meditation but volunteered to act as control Ss. All groups were pretested to determine their ability to correctly recognize the emotions a video-taped patient displayed. This pretest gave an indication of the S's level of proficiency prior to training. At the end of four weeks all Ss were retested and again were asked to record the feelings they felt a video taped patient was expressing. Results showed that only the experimental group improved. In fact, the control group's performance was below what it had been prior to the initiation of the experimental procedures.

Lesh's study, though suggestive of possible uses for both alpha and meditative training, leaves a number of questions to be answered. First, would alpha training, as opposed to meditation, also have resulted in increased empathy? Next, how much of the increase in empathy was due to meditation procedures, and how much was a function of the set or expectancy of the Ss practicing the meditating? That is, Ss may have paid closer attention to feelings because that is what they saw as the purpose of the procedures, rather than having learned anymore about themselves. Were the Ss who meditated any better at recognizing feelings than a group would be who spent a similar amount of time studying pictures of people's faces and body positions as they express various feelings? Finally, would the training have generalized to the therapy situation itself? That is, if the subject had actually been the therapist and been in the line of fire, would he have still recognized the patient's feelings?

At the risk of becoming more speculative than the individuals claiming alpha increases ESP, I am going to suggest a way in which the development of alpha might be used to treat a highly aroused individual. The good premorbid schizophrenic patient[1] is described as hyper-aroused and over responsive to external stimuli (Shakow, 1967; Cromwell, 1968; Nideffer et al., 1971). Much of the pathological behavior of this type of patient can be seen (hypothetically) as resulting from stimulus overload, and from unsuccessful attempts to deal with excessive input. Theoretically, an inability to direct and focus attention or an inability to filter out irrelevant stimulation makes it difficult for the patient to respond appropriately to his environment. The schizophrenic becomes confused by all of the things around him, he adds to the excessive stimulation by becoming frightened, thus adding internal noise to the other interference. As he attempts to communicate he becomes incoherent and is responded to, by others, as though he were crazy. He may attempt to stop this confusion by withdrawing from everything, or he may lash out at the stimuli creating the confusion. To the extent that alpha is a relaxed state it may be possible to teach the patient another way of shutting out some of his excessive stimulus input. Also, alpha may alter his attitude so that he doesn't respond with fear and increased affective arousal as he begins to get over stimulated. Conceptually, the training process would be quite simple.

Initially the patient is connected to the EEG and in the dark stimulus-deprived atmosphere is taught to control alpha production.[2] As soon as the patient is capable of maintaining alpha under these conditions, he is requested to open his eyes. Each succeeding step is contingent upon the patients being able to maintain alpha under the preceding condition. Gradually the illumination in the testing room is increased. Following this neutral stimuli (lights and sounds) are introduced which the patient is taught to ignore. Slowly affective stimuli are introduced, the patient is called by name and comments are made about him. Now the procedure changes slightly. The patient is expected to block alpha to these stimuli; however, it is not necessary to continue to block as the stimuli cease. Thus, the patient is reinforced for his ability to relax immediately after the stimuli have been presented. Gradually he learns to decrease his level of arousal more quickly. Finally, the patient is taught to interact with others. He is engaged in role playing of traumatic situations (fights with his wife, etc.) and he is reinforced for being able to relax quickly. Such an idea is, for the moment, science fiction rather than science fact.

As a footnote to the above suggestion, I might mention that I attempted to condition alpha in a good premorbid schizophrenic patient. I was able to get him to increase his alpha in an eyes-closed resting state. I had told him (with considerable enthusiasm) that I was teaching him to relax, and that he should find himself better able to deal with all of the stimuli in his environment. He was extremely enthusiastic about the project and very cooperative. Unfortunately, the experiment was terminated when he eloped from the hospital one day because the staff had refused to give him a weekend pass. To illustrate his enthusiasm for the study, however, he called me on the phone and asked if there wasn't some way I could sneak him back into the hospital for his treatments. I told him that, unfortunately, there was no way we could continue.

Up to this point discusssion of the alpha wave state has centered on development of a relaxed trusting acceptance while in alpha. Perhaps the greatest benefits that might accrue from alpha training would occur if the attitudes that are associated with the alpha state could be generalized to other brain wave states. As suggested below, there is no reason to believe that the ego-free attitude must be limited to an alpha wave state.

Training in the martial arts (Aikido and Karate) has as a central focus the integration of mental (cognitive and spiritual) and physical functioning. According to masters of these forms of fighting, optimal performance is impossible unless an attitude, similar to, if not identical, with the attitude associated with alpha, is present. Supposedly, the Aikido or Karate expert is capable of maintaining a low level of affective arousal and an ego-free attitude even while being attacked by several opponents simultaneously (Uyeshiba, 1968). Although this attitude may be descriptively identical to the attitude experienced in alpha, research on the effects of visual accommodation would suggest that alpha would be absent during a fight.

It would appear to be impossible to respond to, and accurately perceive an opponent's moves if ocular convergence and accommodation did not occur. Images would be blurred, slight movements would not be seen, and coordinated responses, such as blocking a punch or kick, would be extremely difficult. All of this suggests that although the way in which the individual perceives (affectively responds to) the environment may be identical under these two conditions, the Aikido or Karate master must be in a different brain wave state. This might suggest $\underline{Ss}$ could be trained, via alpha, to recognize an ego-free way of responding to their environment, and then could be taught to maintain (with practice) this attitude while in other brain wave states.

The ability to accurately perceive what another person is feeling may depend upon the ability to maintain a reduced level of affective arousal (Lesh, 1970). Lesh pointed out that therapists as they became aroused were unable to accurately perceive patients' feelings. If the relaxation of the subject-object relationship, which occurs in alpha, can be maintained in a more focal attentive state, such as beta, this problem can be circumvented. The idea is to achieve a narrowed sharp attentional focus and yet keep ego involvement to a minimum. In this way, stimuli such as subject's facial expressions, posture, etc, are accurately perceived and not distorted by the therapist. Such an attitude is in no way limited to patient-therapist relationships but should hold equally well for any interpersonal interaction. For example, it would be an ability that could be very useful for a salesman. In accurately perceiving others he would be much more aware of how his sales pitch was being reacted to and could adjust his behavior accordingly.

As Tohei (1960) points out, the attitude being discussed is relatively easy to maintain in a meditative state. The difficulty comes as situations become potentially more threatening to the individual. As threat increases attention may be narrowed too much, and the focus may be on the individual's behavior or level of arousal rather than on relevant enviromental cues. Alpha training may provide a way of teaching individuals to avoid an excessive narrowing of their attentional process. It is the "alpha attitude" that expands Uyeshiba's (the founder of Aikido) awareness to the point that he is able to perceive the movements which indicate that an attacker is about to pull the trigger of a gun or attack with a sword. Uyeshiba then, is able to use the "alpha attitude" to maximize his fighting skills. Many of us have comparable skills in other areas involving physical performance (a particular sport or game) which could also be maximized by the development and maintenance of an "alpha attitude." In effect this attitude would allow us to direct all of our attention toward the performance of a task. Such a focus would eliminate the phenomena of "choking" in competitive situations.

Given our current level of knowledge the ideas that have just been presented are speculative and should be viewed with caution. Some of the earlier speculations concerning the alpha state (i.e., that it increases ESP, increases learning of visual material, leads to enlightenment) have been demonstrated to be incorrect. The ideas that alpha training may be beneficial to individuals who have emotional or physical problems that are complicated by high levels of arousal, or that the "alpha attitude" might be used to increase an individual's awareness and involvement with his surroundings are exciting. Prior to accepting such ideas as facts, however, several important factors need to be evaluated. A number of more basic questions remain to be answered and some of them are presented below.

It has been assumed that production of alpha can be brought under the conscious control of human S̲s̲. However, the finding in Engstrom et al.'s study that false feedback also resulted in a greater production of alpha might suggest that this is not completely accurate. It may be that increases in alpha production can be attributed to at least two other factors aside from learning. First, alpha production could increase as a function of becoming used to (reducing arousal) the experimental setting. Second, the tone used to provide feedback may assist in the development of alpha. It is possible that a tone may serve to inhibit intrusive thoughts which would interfer with alpha production. That is, the tone could serve to distract the S̲ and keep him from thinking about affectively ladden stimuli. It can be pointed out that meditation exercises are often practiced by focusing attention on a "pure tone" or "pure thought" in an attempt to avoid distracting thoughts.

Should the tone act to increase alpha production, secondary questions arise. For example, what is the net effect of an intermittent tone on learning? It might be expected that the more tone a subject receives initially the more alpha he will ultimately generate (a finding not uncommon in the literature). If this is true, it may be detrimental with respect to developing alpha to give the $\underline{S}$ actual feedback at first, particularly if he has a low base level of alpha to begin with. The tone in this instance rather than serving to relax the individual, would be likely to (with its sporadic beeps) arouse the individual and signal to him that he is failing. It may be more desirable to use a continuous tone, or some other stimulus at first to assist the subject in the development of alpha.

At the present time, we are still unaware of the long term effects of alpha training. Do permanent changes occur in individual's base-line levels of alpha? Is some follow-up to initial training necessary in order to maintain changes that may occur? Does physical relaxation result from the training as well as "mental calm"? If changes do occur over time, are they any greater than changes which might occur with other training methods? Are there certain types of individuals who cannot learn to control alpha? What is the optimal type of feedback for $\underline{S}s$?

The relationship between arousal level and alpha production needs to be investigated free from the confounds of ocular fixation and accommodation. Should the hypothesized relationship between alpha and level of affective arousal exist, how can the alpha state be used? For example, can alpha be used to desensitize an individual to anxiety-inducing situations? Does the presence of alpha have a prophylactic effect in that by having rested (in alpha), the person becomes more capable of tolerating stressful situations? What types of procedures can be developed to teach a person to generalize the "alpha attitude"?

Is the calm, relaxed, ego-free state ("alpha attitude") a direct accompaniment of alpha, or is it a function of the $\underline{S}$'s expectancy concerning the alpha wave state? Would a naive $\underline{S}$ who knew nothing about alpha and its theoretical relationship to the Zen experience describe his experience as a merging with the environment and as an uncritical state of awareness? Does this attitude result in an improved attentional focus? Can the attitude be used to improve interpersonal communication, or to maximize physiological and psychological functioning? We have evidence to suggest that many of these questions may ultimately be answered in a positive way. Before they can be accepted as truth, however, they must be tested in a critical and controlled environment.

Footnotes

1. Good premorbid refers to those schizophrenics who have evidenced normal social and sexual adjustment prior to hospitalization (Phillips, 1953, Ullmann & Giovannoni, 1964; Held & Cromwell, 1965).

2. To condition alpha a certain amount must be present in an eyes closed resting state. Good premorbid schizophrenics were selected for this example because their base level of alpha in an eyes-closed, resting state is similar to normals.

References

Adrian, E. E., & Matthews, B. The Berger rhythm: Potential changes from the occipital lobes in man. Brain, 57, 355, 1934.

Anand, B. K., Chhina, G. S., Singh, B. Some aspects of electroencephalographic studies in yogis. In Barber, T. X., DiCara, L. V., Kamiya, J., Miller, N. E., Shapiro, D., & Stoyva, J. (Eds.). Biofeedback & Self Control. Chicago: Aldine-Atherton, 1971.

Bagchi, G. Mysticism and Mist in India. In Barber, T. X., DiCara, L. V., Kamiya, J., Miller, N. E., Shapiro, D., & Stoyva, J. (Eds.). Biofeedback and Self Control. Chicago: Aldine-Atherton, 1971.

Bagchi, B., Wenger, M. Electro-physiological correlates of some yogi exercises. EEG and Clinical Neurophysiology, 1957.

Barber, T. X. LSD, Marihuana, Yoga, and Hypnosis. Chicago: Aldine Publishing Company, 1970.

Brazer, M. The electrical activity of the nervous system. New York: The MacMillan Company, 1958.

Brown, B. B. Recognition of aspects of consciousness through association with EEG alpha activity. Psychophysiology, 6, 442-452, 1970.

Cohen, S. The Beyond Within. New York: Atheneum, 1964.

Cromwell, R. L. Stimulus redundancy and schizophrenia. Journal of Nervous and Mental Disease, 146, 5, 350-375, 1968.

Deikman, A. J. Experimental Meditation. In Tart (Ed.). Altered States of Consciousness. New York: John Wiley & Sons, 1969.

Dewan, E. Occipital alpha rhythm eye position and lens accommodation. Nature, 214, 975-977, 1967.

Durup, G., & Fessard, A. "L'Electroencephalogramme de l'Homme. Observations Psycho-Physiologique relatives a l'Action des Stimuli Visuels et Auditifs," Ann. Psychol. Paris, 36: 1-36, 1935.

Engstrom, D. R., London, P., & Hart, J. T. Hypnotic susceptibility increased by EEG alpha training. Nature, 227, 1261-1262, 1970.

Erickson, M. H. A transcript of a trance induction with commentary. In Haley (Ed.), Hypnosis and Therapy. New York: Grune & Stratton, 1967.

Fenwick, P. B., & Walker, S. The effect of eye position on the alpha rhythm. In Evans & Mulholland (Eds.), Attention in Neurophysiology. New York: Appleton-Century-Crofts, 1969.

Gardner, E. Fundamentals of Neurology. Philadelphia: W. B. Saunders Company, 1963.

Glaser, G. H. EEG and Behavior. New York: Basic Books, 1963.

Goldstein, L., Murphree, H. B., Sugarman, A. A., Pfeiffer, C. C., & Jenney, E. H. Quantitative electroencephalographic analysis of naturally occurring (schizophrenic) and drug induced psychotic states in human males. Clinical Pharmacology & Therapeutics, 4, 10-21, 1963.

Goldstein, L., Sugarman, A. A., Stolberg, H., Murphree, G. B., & Pfeiffer, C. C. Electro-cerebral activity in schizophrenics and nonpsychotic subjects: Quantitative EEG amplitude analysis. Electroencephalography and Clinical Neurophysiology, 19, 350-361, 1965.

Held, J. M., & Cromwell, R. L. Premorbid adjustment in schizophrenia: The evaluation of a method and some comments. Journal of Nervous and Mental Disease, 146, 3, 264-272, 1968.

Jaspar, H., & Shagass, C. Conditioning of the occipital alpha rhythm in man. Journal of Experimental Psychology, 28, 373-287, 1941.

Johnson, L. C. A psychophysiology for all states. In Barber, T. X., DiCara, L. V., Kamiya, J., Miller, N. E., Shapiro, D., & Stoyva (Eds.), Biofeedback & Self Control. Chicago: Aldine, 1970.

Kamiya, J. Conditional discrimination of the EEG alpha rhythm in humans. Paper presented at the meeting of the Western Psychological Association, San Francisco, April, 1962.

Kamiya, J. Conscious control of brain waves. Psychology Today, 1, 57-60, 1968.

Kamiya, J. Operant control of the EEG alpha rhythm and some of its reported effects of consciousness. In Tart (Ed.), Altered States of Consciousness, 480-501, 1969.

Kasamatsu, A., & Hirai, T. An electroencephalographic study of the zen meditation (Za Zen). In Barber, T. X., DiCara, L. V., Kamiya, J., Miller, N. E., Shapiro, D., & Stoyva, J. (Eds.), Biofeedback & Self Control. Chicago: Aldine-Atherton, 1971.

Lang, P., & Buss, A. Psychological deficit in schizophrenia: II interference and activation. Journal of Abnormal Psychology, 70, 77-106, 1965.

Lesh, T. V. Zen meditation and the development of empathy in counselors. Journal of Humanistic Psychology, 10, 39-74, 1970.

London, P., Hart, J. T., & Leibovitz. EEG alpha rhythms and susceptibility to hypnosis. Nature, 219, 71-72, 1968.

Loomis, A., Harvey, E., & Hobart, G. Electrical potentials of the human brain. Journal of Experimental Psychology, 19, 249-279, 1936.

Luthe, Autogenic training: Method, research and application in medicine. In Tart (Ed.), Altered States of Consciousness. New York: John Wiley & Sons, 1969.

Maupin, E. On Meditation. In Tart (Ed.), Altered States of Consciousness. New York: John Wiley & Sons, 1969.

Miller, N. E., DiCara, L. V., Solomon, H., Weiss, J. M., & Dworkin, B.

Learned modification of autonomic functions: A review and some new data. In Barber, DiCara, Kamiya, Miller, Shapiro, & Stoyva (Eds.), Biofeedback & Self Control. Chicago: Aldine, 1970.

Mulholland, T. Feedback electroencephalography. In Barber, DiCara, Kamiya, Miller, Shapiro, & Stoyva (Eds.), Biofeedback & Self Control. Chicago: Aldine-Atherton, 1971.

Mulholland, T. The concept of attention and the electroencephalographic alpha rhythm. In Evans & Mulholland (Eds.), Attention in Neurophysiology. New York: Appleton-Century-Crofts, 1969.

Mundy, & Castle, A. EEG Clinical Neurophysiology, 9, 643, 1957.

Nideffer, R. M., Deckner, W., Cromwell, R. L., & Cash, T. The relationship of alpha activity to attentional sets in schizophrenia. Journal of Nervous & Mental Disease, 152, 5, 346-352, 1971.

Nowlis, D. P., & Kamiya, J. The control of electroencephalographic alpha rhythms through auditory feedback and the associated mental activity. Psychophysiology, 6, 476-484, 1970.

Oswald, I. The human alpha rhythm and visual altertness. Electroencephalographic Clinical Neurophsiology, 11, 601, 1959.

Pappas, V. Personal Communication, 1972.

Peper, E. Feedback regulation of the alpha electroencephalogram activity through control of internal and external parameters. In Barber, DiCara, Kamiya, Miller, Shapiro, & Stoyva (Eds.), Biofeedback and Self Control. Chicago: Aldine-Atherton, 1970.

Peper, E., & Mulholland, T. Methodological and theoretical problems in the voluntary control of electroencephalographic occipital alpha by the subject. In Barber, DiCara, Kamiya, Miller, Shapiro, & Stoyva (Eds.), Biofeedback & Self Control. Chicago: Aldine-Atherton, 1970.

Phillips, L. Case history data and prognosis in schizophrenia. Journal of Nervous & Mental Disease, 117, 515-535, 1953.

Proler, M. L. The alpha rhythm Part I. American Journal of EEG Technology, 3, 65-70, 1963.

Salamon, I., & Post, J. Alpha blocking and schizophrenia. Archives of General Psychiatry, 13, 367-374, 1965.

Saul, L., & Davis, H., & Davis, P. A. Psychologic correlations with the electroencephalogram. Psychosomatic Medicine, 11, 361, 1949.

Schultz, J., & Luthe, W. Autogenic Training. New York: Grune & Stratton, 1959.

Shakow, D. Understanding normal psychological function: Contributions from schizophrenia. Archives of General Psychiatry, 17, 1967.

Shapiro, J., & Diamond, M. J. Increases in hypotizability as a function of encounter training some confirming evidence. Journal of Abnormal Psychology, 79, 112-115, 1972.

Short, P. L., & Walter, W. G. The relationship between physiological variables and sterognosis. *Electroencephalographic Clinical Neurophysiology*, 6, 29, 1954.

Tart, C. *Altered States of Consciousness*. New York: John Wiley & Sons, 1969.

Tart, C. Increases in hypnotizability resulting from a prolonged program for enhancing personal growth. *Journal of Abnormal Psychology*, 75, 260-266, 1970.

Tohei, K. *Aikido*. Rikugei Publishing House, Tokyo, Japan, 1960.

Ullman, V., & Giovannoni, J. The development of a self-report measure of the process-reactive continuum. *Journal of Nervous & Mental Disease*, 138, 38-42, 1964.

Uyeshiba, M. *Aikido*. Tokyo, Hozanshi Publishing Company, 1968.

Wallace, R. K. Physiological effects of transendental meditation. *Science*, 167, 1751-1754, 1970.

Wallace, R. K., & Benson, H. The physiology of meditation. *Scientific American*, 1972.

Walter, W. G. *The Living Brain*. New York: W. W. Norton Co., 1963.

Weitzenhoffer, A. M. *An Objective Study of Hypnotism*. New York: Science Editions, John Wiley & Sons, 1963.

Wells, C. E. Alpha wave responsiveness to light in man. In Glasser (Ed.), *EEG and Behavior*. New York: Basic Books, 1963.

IV

CARDIOVASCULAR CONTROL

Operant Conditioning of Cardiac Function: A Status Report

13

Bernard T. Engel

ABSTRACT

The mechanisms of operant conditioning of cardiac function can be analyzed in terms of cardiodynamics, hemodynamics, neural regulations, and psychological regulations. These separate analyses help to suggest a number of experimental hypotheses. Analyses of data from early and late phases of conditioning show that performance at these two stages of training is different: Analyses of acquisition mechanisms must come from the early phase of training, however, analyses of control mechanisms should probably be derived from chronic studies.

DESCRIPTORS: Individual response specificity, Operant conditioning, Cardiovascular system, Antonomic nervous system, Heart rate, Blood pressure, Electroencephalogram. (B. T. Engel)

During the 1950's while I was still a graduate student, the *au courant* issue in psychophysiology was response specificity. It was natural, therefore, that I became interested in that problem, and in the course of the next several years I carried out a number of experiments on the general problems of specificity (see Engel & Moos, 1967). The focus of my research on specificity in the early 1960's was essential hypertension (Engel & Bickford, 1961; Moos & Engel, 1962).

The results of those studies pointed to the conclusion that one of the characteristics of patients with high blood pressure of unknown origin was that they were pressor hyper-reactors, i.e., they emitted a rather stereotyped pattern of autonomic responses which included a strong tendency for the maximal response to occur in blood pressure.

At this point I decided that there were two directions that my subsequent research might take. Either I could try to carry out longitudinal studies which might help explain how the pressor hyperactivity of the patients developed; or I could ignore the question of etiology temporarily and consider instead how I might go about trying to modify the pressor response. I chose the second course.

Factors such as the unavailability of practical devices for measuring arterial pressure on a beat-to-beat basis in man and such as the fact that hypertension is complicated because it is associated with structural as well as functional abnormalities, led me to search for an alternative model system which I might try to modify. I decided to work with cardiac function.

As is often the case with digressions, this research rapidly developed a life of

---

I want to express my appreciation to the many colleagues who contributed to the research I have described here. I also want to acknowledge my appreciation to Drs. Eugene R. Bleecker and George W. Ainslie, Jr. for their helpful criticisms of an early draft of this lecture. I want to thank Mrs. Estelle Carter for her forbearance during the typing of this manuscript.

Address requests for reprints to: Dr. Bernard T. Engel, Gerontology Research Center, Baltimore City Hospitals, Baltimore, Maryland 21224.

Reprinted with permission of the publisher and author from *Psychophysiology*, Vol. 9, 161-177. Copyright 1972, The Society for Psychophysiological Research.

its own, and for the last ten years the major thrust of my program has been directed at the study of cardiac function. What I am going to try to do in this hour is to describe some of the studies my colleagues and I have carried out in our efforts to modify cardiac function. Although much of our research is motivated by clinical interests, I am going to ignore the questions of clinical applications. Instead, I am going to concentrate on questions of mechanism.

I have two goals in mind. My first goal is to satisfy myself that it is possible to carry out an analysis of psychophysiological research which will fit within the conceptual frameworks of both psychology and physiology. And my second, and more important goal is to indicate some of the scientific questions about operant cardiac conditioning which are as yet unanswered.

In keeping with my first goal which is to make both physiological and psychological sense of the data, I intend to try to analyze the cardiac operant conditioning research at four distinct but interacting levels. Specifically, I plan to look at the research from the point of view of cardiodynamics, from the point of view of hemodynamics, from the point of view of neural regulations, and from the point of view of psychological regulations. Before I begin my analyses, I want to emphasize that these levels interact extensively. The cardiovascular system does not lend itself to cause-effect models.

Cardiodynamics

The function of the heart is clearly to pump blood. From a cardiodynamic point of view then, the first question to consider is: "What are the ways in which the pumping action of the heart can be expressed?" If we consider this question only in terms of intra-cardiac control mechanisms, then there are two ways in which cardiac output can be modified. The first way is by changing the rate at which the heart pumps, and the second way is by changing the nature or strength of each contraction. The first of these factors is what physiologists call chronotropic action and it is measured, of course, in terms of ventricular rate. The second function of heart action is what physiologists call inotropic action of the heart, and it is measured both in terms of the force of ventricular contraction and in terms of the synchrony of ventricular contraction. The question which I intend to consider here is: "How can these two functions be modulated through operant conditioning techniques, and what are the nervous mechanisms at the level of the heart through which this modulation takes place?"

Let us first consider ventricular rate. As you know, nervous impulses reach the heart at the sino-atrial (SA) node where they act to modulate the rate of depolarization of that pacemaker. From the SA node impulses are non-neurally transmitted by way of the atrial musculature to the atrio-ventricular (AV) node. Here too there are extrinsic autonomic nerves which act to modify pacemaker rhythmicity. Impulses which leave the AV node are conducted by way of the bundle branches to the ventricles where again there is considerable anatomic and physiologic evidence of autonomic innervation.

Thus, we see three regions where extrinsic autonomic nerves may affect ventricular rate: the SA node, the AV node, and the ventricle. Which of these intra-cardiac areas can be operantly conditioned, and what are the neural mechanisms by which this conditioning is mediated?

It is pointless to document the many experiments which have reported successful operant conditioning in the intact heart. As far as I have been able to determine, however, despite this plethora of research in normal subjects, there have been no studies which elucidate the autonomic mechanisms underlying the acquisition or performance of rate control in the intact heart.

There have been a few studies of autonomic control of the heart during classical conditioning, however. Dykman and Gannt (1959) have shown that the vagus is an important determinant of conditioned tachycardia in the dog, and Flynn (1960) has reported that the vagus is an important mediator of conditioned bradycardia in the cat. Obrist, Wood, and Perez-Reyes (1965) studied the acceleratory and deceleratory limbs of the conditioned cardiac response in man, and they concluded that both responses were primarily vagally mediated. They inferred that there was a significant outflow from the sympathetic nervous system during deceleration which was overridden by vagal stimulation. However, none of these three studies directly investigated the role of the beta-adrenergic system, and none of them tells us very much about control mechanisms in operant conditioning.

Cohen and Pitts (1968) did study both the vagal and sympathetic components of classically conditioned tachycardia in the pigeon. Two of their findings are especially interesting and relevant: 1. They showed that bilateral vagotomy and beta-adrenergic blockade abolished the acquisition of conditioned tachycardia whereas neither vagotomy nor propranolol alone was effective in blocking learning. Thus, the autonomic nervous system appears to be a necessary component in cardiac conditioning, and either branch is a sufficient factor in mediating learned cardio-acceleration. 2. They also showed that the most important determinant of response magnitude during early trials is increased sympathetic outflow. However, as training progresses vagal inhibition assumes greater importance. This is a conclusion to which I shall return: The physiological mechanisms of heart rate control are different early and late in training. Most of the published studies of operant conditioning of heart rate are based on single sessions experiments, and the results reported in these studies are probably going to give misleading if not wrong impressions about the physiological mechanisms of heart rate control.

Dr. Eugene Bleecker and I have recently completed a study which was designed to answer two questions: 1. Can a heart in which ventricular rate is not controlled by the SA nodal pacemaker but is controlled by the AV nodal pacemaker be taught rate control? In simple terms, can the AV node learn? and 2. If so, what are the mechanisms by which this learning is mediated?

In order to answer these questions we selected a group of 6 patients, each of whom was suffering from a cardiac arrhythmia called atrial fibrillation. In this arrhythmia the atrial muscles contract in an irregular, discoordinated fashion presenting anywhere from 300–600 impulses/min to the AV node. Ventricular rate in this arrhythmia, when it is under proper medical control with digitalis, ranges among patients from about 65 bpm to about 85 bpm. Within a given patient, variability in ventricular rate is rather high, usually ranging from about 12 bpm to about 20 bpm at rest. In contrast to this degree of variability, the standard deviation of heart rate in normal men at rest is usually less than 8 bpm.

Each of the 6 patients was trained to speed his ventricular rate and to slow his ventricular rate. After each patient had been trained in ventricular rate speeding and ventricular rate slowing, he was tested in what we called an alternating or differential training paradigm. During these sessions the subject was required to alternately speed and slow his ventricular rate relative to his baseline ventricular rate. The alternations in rate occurred every 256 sec, and each session lasted 1024 sec.

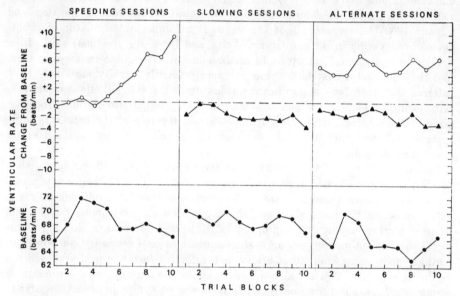

Fig. 1. Mean baseline ventricular rates (bottom graphs), and mean changes from baseline ventricular rates (top graphs) for all patients. Sessions were reduced to 10-trial blocks by averaging temporally contiguous session means. Symbols are: baseline rates, ●; changes during slowing sessions, ▲; changes during speeding sessions, ○.

Fig. 1 summarizes the results during each phase of the study. The patients were able to speed their ventricles, to slow their ventricles, and to alternately speed and slow their ventricles.

Furthermore, as can be seen on the bottom tracing, baseline ventricular rate remained stable throughout the study, which took about 2–3 weeks/patient.

The next question we tried to answer was, "what is the mechanism of this control?" And Fig. 2 answers this question. Isoproterenol, which is a beta-adrenergic stimulant, did not block the patients' abilities to speed or to slow their ventricles. Likewise, propranolol which is a beta blocker or edrophonium which is a vagomimetic drug did not affect the patients' abilities to differentially speed or slow their ventricles. Only atropine which blocks the vagus also blocks the behavior.

The results of this study seem reasonably clear: Ventricular rate control can occur at the level of the AV node, and the effect is mediated through the efferent pathway of the vagus nerve.

Having established that the intact heart and the AV node can be operantly conditioned, the next logical question in this sequence is obviously, "can the

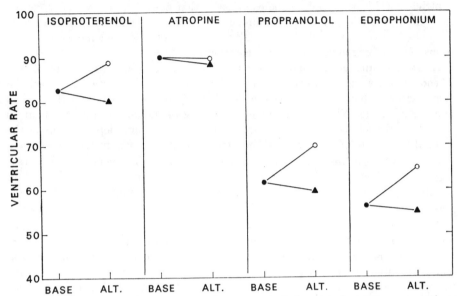

Fig. 2. Effect of autonomic drugs on voluntary control of ventricular rates among patients with atrial fibrillation. The values shown for alternating sessions (alt.) are the means for all patients who participated in that drug study. Symbols are: baseline rates, ●; rates during slowing phases of alternate sessions, ▲; rates during speeding phases of alternate sessions, ○.

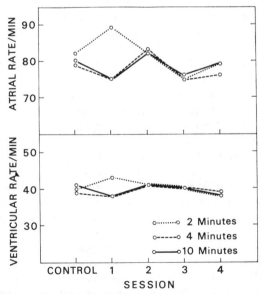

Fig. 3. Atrial and ventricular rates of a patient with 3rd degree heart block who was being trained to speed his ventricular rate. The graphs show performance during the second, fourth, and tenth minutes of each training session. Note the large increase in atrial rate during the second minute of the first session.

ventricle itself learn?" Before one can answer that question, however, it is necessary to be more explicit. As I pointed out in my introduction, the ventricle responds in two ways: Chronotropically, in terms of rate, and inotropically, in terms of something physiologists call contractility.

In order to answer the first question, can the ventricle learn rate control, Dr. Theodore Weiss and I studied a group of 3 patients with complete heart block. In complete heart block the ventricle is functionally disconnected from the atrium and the AV node. Whereas the atrium is beating at the normal rate of about 70 bpm, the ventricle beats anywhere from 30 bpm to 40 bpm depending upon the level of the lesion, and ventricular and atrial rhythms are completely dissociated. We were unable to teach any of the patients to speed their ventricles. Interestingly, one patient, whose data are presented in Fig. 3, speeded his atrium during the first session; however, this response, which was unreinforced, rapidly extinguished.

The results of this study then permit me to summarize what we know about intra-cardiac mechanisms of operant heart rate conditioning: The intact heart and the innervated AV node ventricle preparation can be taught rate control. The innervated ventricle cannot learn rate control. The autonomic mechanisms underlying operant conditioning of heart rate in the intact heart are not known; however, it is likely that either the vagus or the sympathetic nervous system can mediate learning. At the level of the AV node, however, the vagus is the mediator.

Now, how about inotropic control of the ventricle: can this be learned? The evidence at this time is only suggestive; however, I think the answer will be yes, inotropic activity of the ventricle can be operantly conditioned.

Most of the data on this point come from a study Weiss and I did with patients with premature ventricular contractions (PVCs). Since that report has now been published (Weiss & Engel, 1971), I will not review the results in detail. However, let us look carefully at the data from one subject who was able to differentially control his heart rate. Associated with this rate control, he also differentially controlled the prevalence of his ectopic beats. As you can see from Fig. 4, there was a strong and reproducible association between heart rate and PVCs such that slowing was associated with an increase in PVCs. However, we also looked at the relationship of PVCs to absolute heart rate during baseline and speeding and slowing. Specifically we selected those periods when absolute heart rates were similar across these three conditions. As you can see in Fig. 5, PVCs tended to be most frequent during slowing, intermediate during baseline, and least frequent during speeding. Thus, it appears possible that it is not heart rate *per se*, but some process which is correlated with rate control which mediates ventricular function.

Weiss and I also tried to identify the autonomic mechanisms associated with PVCs in the patients we trained. Among the patients we studied, there was one woman in whom PVCs were induced by isoproterenol. The effects of atropine and isoproterenol together were no different than isoproterenol alone, and atropine alone did not induce PVCs. Thus, in her case, it seems likely that her PVCs were associated with increased sympathetic outflow. In another patient, however, isoproterenol either had no effect on PVCs or reduced their prevalence whereas

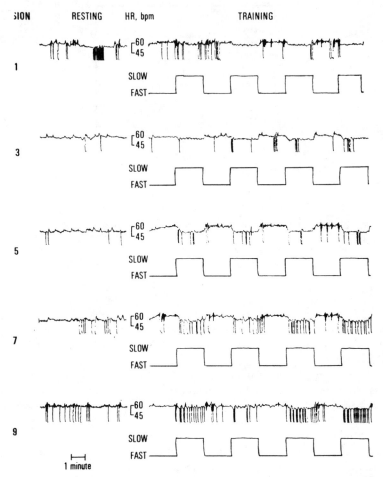

Fig. 4. Cardiotachometer tracings taken from experiments with a patient with premature ventricular contractions. The patient is consistently able to alternately speed and slow his heart rate. As training progresses, premature beats become progressively more concentrated in the slowing phases of the sessions. (reprinted from Weiss & Engel, 1971)

atropine greatly increased the prevalence of her ectopic beats. Phenylephrine, which is an alpha-sympathetic stimulant, and which produces a reflex vagal discharge by elevating blood pressure, abolished her PVCs. On the basis of these data, it seems likely that both the sympathetic nervous system and the vagus can mediate operant conditioning of ventricular action.

Hemodynamics

I would like to move now, from a consideration of intra-cardiac mechanisms, to a consideration of hemodynamics. If ever there was virgin territory, this is it. As far as I know, there are almost no studies of operant conditioning of heart rate which also include other, simultaneous cardiovascular measurements. I find this vacuum especially curious because so much has been written about respira-

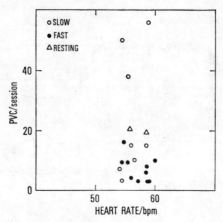

Fig. 5. Absolute heart rates during baseline periods, heart rate slowing periods, and heart rate speeding periods of alternating sessions. The data show a tendency for premature beats to be most frequent during periods of heart rate slowing, least frequent during periods of heart rate speeding, and intermediate during baseline periods even though absolute heart rates are similar among these periods. These data are from the same patient whose cardiotachometer tracings are presented in Fig. 4.

tory and somatic mediation of cardiac learning: it must be the case that hemodynamic adjustments are one major mechanism by which these effects will influence cardiac function.

All of the operant conditioning studies of heart rate suffer from the same weakness. None of us have included measures of cardiac output, and only a few of us have measured mean blood pressure. Thus none of us can estimate total peripheral resistance or stroke volume. The only data I can cite, therefore, are those which relate mean blood pressure to heart rate.

Despite these shortcomings, I think it is still possible to infer whether conditioned changes in cardiac rate are necessarily mediated by hemodynamic changes.

As you know, there have been a few reports of successful attempts to dissociate systolic blood pressure and heart rate by means of operant conditioning. As early as 1968 it was reported that rats could learn to modify their systolic pressures, and that heart rates did not differ significantly at the beginning or end of these sessions (DiCara & Miller, 1968). They also reported an insignificant, positive correlation between systolic pressure and heart rate in their experimental group.

The Boston group has also reported that it is possible to teach normal men to dissociate heart rate and systolic pressure. Last year, at these meetings, Gary Schwartz, from that group, reported that it was possible to teach subjects what he called integration and differentiation of heart rate and systolic pressure (Schwartz, 1971).

All of these studies are interesting in that they show that some degree of specificity is possible within the cardiovascular system. The presence of this degree of specificity seems to me to be a strong suggestive evidence that cardiac and vascular responses can be learned directly, and that whatever hemodynamic effects one may see are likely to be compensatory to, or correlative with the response being learned.

In 1970 Gottlieb and I reported that we were able to operantly condition monkeys to speed and to slow their heart rates (Engle & Gottlieb, 1970). That is, the same monkey could learn to emit both behaviors. We have recently extended those findings, and I want to share some of those data with you here.

Before reviewing our findings, however, let me briefly describe our procedures. Each monkey had a catheter permanently implanted in its external iliac artery which we were able to keep patent for periods ranging from 6 months to one year. During the period of study the monkey was chronically maintained in a primate chair. Because of the chronicity of our experiments the animals' performances were quite stable and their responses were very reliable.

In our experiments the monkey was required to slow its heart rate or speed its heart rate from a resting baseline level. Each training session was about 35 min. During the session the monkey had to maintain a relatively slow or fast heart rate or it would receive a 10 ma, .5 sec shock to its tail. The shocks were delivered every 8 sec until the monkey once again began to perform correctly.

There were 4 animals in the present study. Each animal received 10 training sessions (two/day) under one contingency, then 10 sessions under the second contingency; then 20 sessions under the first contingency; and finally, 20 sessions under the second contingency. Two animals were trained to slow first, and 2 animals were trained to speed first. I want to consider two stages of training: the first 10 sessions of each contingency, which I will call early training, and the last 10 sessions of each contingency, which I will call late training.

Table 1 presents some of the salient findings. I want to direct your attention to several data. Note that mean blood pressures consistently increased from baseline during speeding sessions but were inconsistent during slowing sessions. Note that heart rate changes and mean blood pressure changes are consistently, positively correlated during both slowing and speeding. The fact that these correlations are positive is quite interesting since it means that the changes in heart rate are not mediated by pressor reflexes. If this were true, the correlations would be negative. The correlations further suggest that total peripheral resistance does not change; however, confirmation of this interpretation will have to await direct measurement of cardiac output.

The second set of findings to which I direct your attention has to do with stage of training. Note that late in training the responses and the correlations between responses are different from what they are early in training. Heart rate responses are attenuated in both speeding and slowing sessions, and the correlations between heart rate changes and mean blood pressure changes increase during slowing from early training to late training whereas they remain unchanged during speeding from early training to late training. These data suggest two interesting possibilities: The first is that the physiological mechanisms underlying operant conditioning of heart rate slowing are different from the mechanisms underlying heart rate speeding—I want to underscore that point, the mechanisms are not mirror images of one another, they are different from one another—slowing is not merely unspeeding, and speeding is not merely unslowing. The second conclusion suggested by these data is that the hemodynamic mechanisms associated with

TABLE 1

Heart rate and mean blood pressure responses

Variable	Animal	Early Training		Late Training	
		Slow	Speed	Slow	Speed
Means of Responses[a]					
Heart Rate (bpm)	1	−5.7**	16.8**	−3.1**	6.5**
	2	−5.0**	11.6**	−2.7**	6.4**
	3	−10.9**	20.4**	−2.6*	15.3**
	4	−17.7**	1.6	—	6.4**
Mean Blood Pressure[b] (mm Hg)	1	−4.0**	3.6**	−0.5	2.3**
	2	0.3	5.3**	0.4	2.7**
	3	0.4	5.0**	1.2**	4.2**
	4	−0.6*	4.3**	—	5.2**
Correlations					
	1	.30**	.48**	.66**	.48**
	2	.46**	.34**	.69**	.48**
	3	.37**	.44**	.81**	.41**
	4	.15*	.46**	—	.48**

[a] Mean training level minus mean baseline level.
[b] Estimated from systolic and diastolic pressures according to the formula mean pressure = diastolic pressure + ⅓ pulse pressure.
* $p < .05$.
** $p < .01$.

learned heart rate control are different early in training from those late in training.

I want to document these two conclusions with another set of data from the same study. Fig. 6 shows the heart rate and blood pressure responses emitted by these animals during escape-avoidance responding. That is, during a 30 sec interval while they escape from shock and avoid subsequent shocks. Note that the mean blood pressure responses are similar during slowing and speeding while heart rate responses differ markedly. From a hemodynamic point of view, therefore, slowing and speeding are mediated differently.

Note also that during early training blood pressure rises in the post-shock period while during the late training period blood pressure does not change. Note further that the heart rate speeding response during early training is about twice as great as it is during late training. What all this means is that the hemodynamic mechanisms associated with escape-avoidance performance are different early and late in training.

There is one last set of data I would like to cite before I leave hemodynamics. Forsyth (1971) has recently reported that monkeys which are maintained on a 20 sec, Sidman avoidance schedule for 72 consecutive hours, initially respond to this stress with a hemodynamic response pattern which includes high heart rates, elevated mean arterial pressures, and small increments in total peripheral resist-

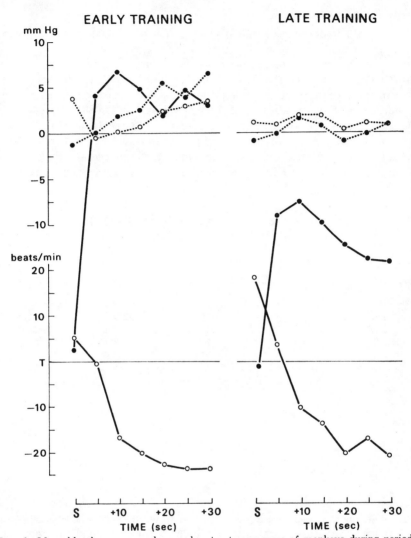

Fig. 6. Mean blood pressure and mean heart rate responses of monkeys during periods of escape from shock and avoidance of subsequent shocks. Symbols are: heart rates during speeding sessions, ●——●; heart rates during slowing sessions, ○——○; mean blood pressures during speeding sessions, ●-----●; mean blood pressures during slowing sessions, ○-----○. See text and Table 1 for further explanation.

ance. As the stress continues, however, heart rates and cardiac outputs return toward pre-stress levels while pressures remain elevated and total peripheral resistance increases. While these data do not tell us anything about hemodynamic mechanisms during operant cardiac conditioning, they do illustrate vividly that cardiovascular response patterns change systematically in the course of an experiment.

Anyone who is seriously interested in learning about cardiovascular control mechanisms must be willing to study the system chronically as well as acutely.

Neural Regulations

The central nervous system (CNS) mechanisms which mediate operant cardiac control are virtually unknown. There are a few observations, however, which are relevant. In the non-human primate studies I cited earlier, we also included measurements of non-specific EEG changes. What we did was to implant electrodes in the parietal cortexes of the monkeys, and measure the percentage of time during which the monkey emitted "high voltage" in the 8–14 Hz band of the EEG. The purpose of this study was, obviously, to try to determine whether heart rate speeding and heart rate slowing are associated with different levels of "cortical arousal." As you can see from Table 2, the percentage time that high voltage is present in the EEG tends to be greater during slowing sessions than

TABLE 2

Percentage time high voltage present in the EEG (8–14 Hz)

Animal	Early Training		Late Training	
	Slow	Speed	Slow	Speed
Mean Responses[a]				
1	2.8**	−2.8**	2.0**	−4.4**
2	3.6**	2.5**	0	−2.4**
3	6.5**	−7.7**	−0.7*	−1.0**
4	3.9**	2.2**	—	0.5**
Correlations (with Heart Rate)				
1	−.09	−.12	−.23**	−.26**
2	−.06	−.19**	−.31**	−.22**
3	−.06	−.11	−.26**	.02
4	−.10	−.46**	—	−.15*

[a] Mean training level minus mean baseline level.
* $p < .05$.
** $p < .01$.

during speeding sessions despite the marked differences between early training and late training; and heart rate and high voltage in the EEG tend to be negatively correlated: increases in heart rate tend to be associated with low voltage, high-frequency responses in the EEG. Furthermore, this relationship tends to become more consistent during late training.

The response to shock is also interesting. As you can see from Fig. 7, escape-avoidance behavior is associated with an absence of high voltage, synchronized activity in the EEG when the heart rate speeds. During slowing there is an increment in high voltage early in training and very little high voltage late in training. However, there is more high voltage during slowing than speeding.

In general then, I would conclude from this study that there is a small but reliable negative correlation between heart rate and high voltage in the EEG; and that there is more synchronized, high voltage activity in the EEG during heart rate slowing sessions than during heart rate speeding sessions. The evidence

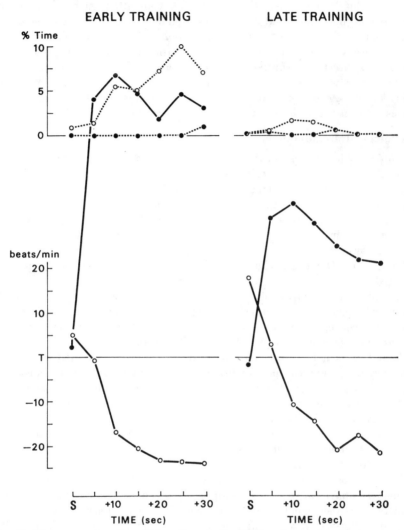

Fig. 7. Mean percentage time high voltage in the EEG and mean heart rate responses of monkeys during periods of escape from shock and avoidance of subsequent shocks. Symbols are: heart rates during speeding sessions, ●——●; heart rates during slowing sessions, ○——○; percentage times high voltage in the EEG during speeding sessions, ●----●; percentage times higher voltage in the EEG during slowing sessions, ○----○. See text and Table 2 for further explanation.

thus suggests that the monkey's brain is somewhat more "aroused" when it is speeding its heart than when it is slowing its heart. The relationships are thin, however, and a more precise functional analysis is needed.

I have focused my presentation of mechanisms on operant conditioning. I suspect that the future will show that distinctions between operant and classical conditioning are arbitrary, and that concepts about mechanism will apply equally to the two technologies. Thus, while I will not review the work of others as part of this lecture, I feel the need to call your attention to the fact that a good deal of important and careful research has already been done on CNS mechanisms during classical cardiac conditioning. Russian physiology is almost all concerned with questions of CNS mechanisms of conditioning, and much of that work is directed at cardiac function (Chernigovski, 1967). Among American investigators I would call your attention to the pioneering work of W. Horsley Gannt (1966). Among contemporary studies the work by Cohen and his colleagues with the pigeon is especially noteworthy (Cohen, 1969).

Psychological Regulations

Before considering some of the psychological variables we have worked with, I would like to intrude a few, brief, critical notes. A number of psychologists have proposed that it has not yet been shown that autonomic responses can be operantly conditioned because these responses may be mediated by somatic, respiratory, or cognitive processes. First, it should be absolutely clear that the definition of operant conditioning has no such contingencies in it. As a matter of fact one of the most powerful features of operant conditioning is the way in which responses can be chained. Obviously, chained responses are both operantly conditioned and mediated.

In addition, many of the assertions about mediation are merely *non sequiturs*. Anyone who has spent any time as an experimental scientist knows that ubiquitous, global concepts such as cognitive mediation add nothing to scientific knowledge, and may even detract from it by inhibiting research.

All of this is not to say that there are no physiological and psychological correlates of operant cardiac conditioning. There must be. But the task of the proper scientist is to identify specifically what these correlates are, and to specify the conditions under which these relationships may be changed.

It is my personal view that there must be some degree of integration among cardiovascular behavior, somatic behavior, and central nervous system behavior. Specificity is biologically useful since it supports adaptation. However, specificity to the point of physiological disintegration would be biologically disastrous.

There is another issue which is related to the mediation question which some psychologists have raised which also deserves a comment in passing: namely, the distinction between conditioning and control. The argument is that there are many behaviors, such as running around the block, which will increase heart rate, and which one would not say were conditional responses if they were emitted. The relationship of the conditioning/control question to the mediation question derives from the obvious fact that any variable which can control cardiac function is also a potential mediator of cardiac function.

I think the distinction between conditioning and control is a glib one. First of all, it is not clear how one differentiates between a conditioned response and a controlled response: After all, operant conditioners do refer to their techniques as ones which bring behavior under stimulus control. Presumably, the distinction is based upon the number of trials one requires to reach a criterion. If the behavior is being conditioned, learning takes time; if the behavior is being controlled, there would be a high level of performance on the first trial. It should be obvious that a distinction between conditioning and control based upon trials to a criterion is logically identical to the distinction between learning and insight which I thought Harlow had dealt with a number of years ago when he showed that insight, *qua* control, was an index of the experimenter's ignorance about the subject's learning history (Harlow, 1949).

I want to emphasize my point. I believe that the distinction between conditioning and control is a glib one since it really says that one should reject all performance which the subject learned before the experimenter decided to carry out his research. As a matter of fact, I could make a strong case for seeking out such subjects since they may prove to be ideal for some studies of mechanism. An example of such a subject would be a Yogi.

I hope it is quite clear that in rejecting the distinction between conditioning and control I have in no way denied the importance of understanding the mechanisms which a subject uses to emit his cardiac response. Mediation, in terms of specific experimental variables, is a valid question.

There is another question which has never been made explicit, but which I think should be raised. The question is, "of what importance is it to the science of psychology that autonomic responses can be operantly conditioned?" And my answer is, "probably none whatsoever." Whether or not autonomic responses can be operantly conditioned in an empirical question which adds nothing conceptually to our knowledge about the principles of learning. In my opinion the great importance of operant conditioning of autonomic responses comes from what it tells us about the autonomic nervous system. It is physiology, not psychology, which is going to have to revise some of its principles.[1] The *psychological* unconscious has in no way been affected by the research on operant autonomic conditioning, but the *physiological* unconscious will never, ever be the same again.

Our research on psychological variables has been quite limited. Since the main thrust of our human studies has been toward clinical questions, we have tended to concentrate on a few factors which might be useful in expediting that work. Specifically, we have tried to determine if the behavior which was learned in the laboratory can transfer to other environments; we have tried to learn if there was any consistency among the strategies patients used to gain control over their hearts; we have looked to see if training cardiac patients to control their hearts makes these patients overly anxious about their hearts; and we have fretted about techniques which might optimize learning.

Unfortunately, I have little to say about optimizing procedures. Except for a few ideas about pretesting patients with drugs we have done virtually nothing about optimization studies. I do wish that someone who was especially interested in schedules of reinforcement would carry out the appropriate studies to define the best ways to optimize training.

The question of extinction is one we have merged with the question of transfer to outside environments. Since most of our human research has been with patients, and since we have often had clinical as well as scientific end points in

---

[1] I believe Skinner made this same point when he said: "Not only are laws of behavior independent of neurological support they actually impose certain limiting conditions upon any science which undertakes to study the internal economy of the organism. The contribution that a science of behavior makes to neurology is a rigorous and quantitative statement of the program before it [Skinner, 1938, p. 432]."

mind, extinction is something we have not wanted to see. Usually we have added a procedure to our studies which progressively reduces feedback. I can illustrate that technique best in terms of a specific example: suppose we want to teach a subject to slow his heart rate. After he has been trained to slow with continuous reinforcement, we carry out a series of sessions in which the feedback is alternately available and unavailable on a one to one schedule; then it is available on a one to three schedule; and finally it is available on a one to seven schedule. Fig. 8, taken from a study with a patient who was diagnosed as having sinus tachycardia, illustrates the procedure and the results. Feedback was progressively reduced during the later stages of training, and performance was equally good with and without feedback. In a number of different kinds of patients that has been a consistent outcome, performance tends to be equally good with and without feedback.

A derivative of the question about feedback has to do with performance away from the laboratory. In our studies with patients we encourage the subjects to practice their techniques away from the laboratory. In some of these studies we have actually monitored cardiac function with telemetry devices while the patients were on the ward. The results show that when the patients perform successfully in the laboratory, they show similar changes in cardiac function on the

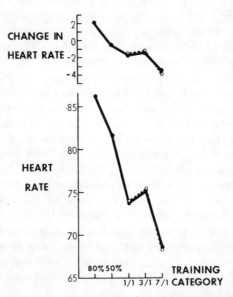

Fig. 8. Mean heart rates and mean changes in heart rate from baseline levels in a patient with sinus tachycardia being trained to slow her heart rate. Numbers on the abscissa refer to various stages of training: 80%, continuous feedback with the trigger set so that maintenance of baseline heart rate would result in successful performance 80% of the time; 50%, continuous feedback with the trigger set so that maintenance of baseline heart rate would result in successful performance 50% of the time; 1/1, alternate 1 min periods of feedback and no feedback during which the patient was instructed to slow her heart rate; 3/1, alternate 3 min periods of no feedback and 1 min periods of feedback; 7/1, alternate 7 min periods of no feedback and 1 min periods of feedback. Performance is equally good with (●——●) and without (○-----○) feedback.

ward. What is especially significant is that this effect occurs in both directions; i.e., sometimes successful performance in the laboratory is associated with an exacerbation of clinical signs as for example in the patients with PVCs that Dr. Weiss and I studied. When PVCs increased in the laboratory they increased on the ward at times when the patients said they were practicing. When PVCs decreased in the laboratory, they likewise tended to decrease on the ward.

We do have other, anecdotal examples of instances where patients demonstrated their skills away from the laboratory. An example is one patient with atrial fibrillation whom Dr. Bleecker and I studied. The patient was hospitalized for studies preceding valve surgery. He reported that on occasions the nurse would withhold his digoxin because his ventricular rate was too slow. On several of those occasions he would increase his rate and he reported that he was successful in getting his digoxin.

We also have somewhat "harder" reports from physicians who have told us that the patients have demonstrated their skills during examinations. Finally, we have retested a few patients in the laboratory after 6 months or one year during which they received no intervening training sessions. These patients performed successfully. I must also point out here that some of the patients who performed successfully after a lapse of some months did not receive the feedback/no feedback training I just described. These data suggest that the additional training with and without feedback may not be necessary.

The question of iatrogenic anxiety is one to which the answer seems quite clear. I can think of no instance in which a patient whom we trained showed any evidence of increased anxiety about his cardiac function. As a matter of fact, in almost every instance the patients seemed less anxious. The impression one gains from talking to patients who have learned to control their hearts is that the patients have a sense of mastery over themselves that they lacked earlier.

The last behavioral variable about which I want to comment is the question of consistency of strategies among patients. I have a confession to make. I began this research about 1962, and I have asked every subject whom I ever tested the same question: "What did you do?" "How did you do it?" And after ten years of this nonsense I finally recognized this year how silly those questions are.

This year I decided to take up golf. Like most new golfers I hit a few balls well, most balls badly. And I asked myself on those occasions when I hit the ball well, "What did you do?" "How did you do it?" And the answer came back loud and clear: "I don't know."

If learned cardiac control is a form of motor learning as I believe is the case, then why should someone be able to describe the details of his performance during the early stages of learning? I think that the day may come when I will be able to describe how I hit a golf ball, but I assure you, that day is far off. I am much too concerned with ritualistic and superstitious acts—I always hold my breath at end-expiration, except on those occasions when I hold my breath at end-inspiration, before I hit the ball. Intellectually I know that this act must be irrelevant to hitting the ball, however, if I do not perform it, the realization that I have not done so preoccupies me throughout all of my swing and I inevitably duff the ball.

It is small wonder that I have not been able to find any consistency among the stories the subjects have told me. I am certain that they do not know what they are doing, and that they are just making up stories to please me. I think what we should try, is to withhold our questions until some months have passed and the subject is highly skilled. Maybe then his answer will be meaningful. In the meantime a more sensible scientific strategy would be to institute experimental controls over some variables to learn whether these are really relevant. Experimental science is always better than free association, even when it is the scientist who is on the couch.

REFERENCES

Chernigovskiv, V. N. *Interoceptors* (Trans. by G. Onischenko). Washington: American Psychological Association, 1967.

Cohen, D. H. Development of a vertebrate experimental model for cellular neurophysiologic studies of learning. *Conditional Reflex*, 1969, *4*, 61–80.

Cohen, D. H., & Pitts, L. H. Vagal and sympathetic components of conditioned cardio-acceleration in the pigeon. *Brain Research*, 1968, *9*, 15–31.

DiCara, L. V., & Miller, N. E. Instrumental learning of systolic blood pressure responses by curarized rats: Dissociation of cardiac and vascular changes. *Psychosomatic Medicine*, 1968, *30*, 489–494.

Dykman, R. A., & Gantt, W. H. The parasympathetic component of unlearned and acquired cardiac responses. *Journal of Comparative & Physiological Psychology*, 1959, *52*, 163–167.

Engel, B. T., & Bickford, A. F. Response specificity: Stimulus-response and individual-response specificity in essential hypertensives. *Archives of General Psychiatry*, 1961, *5*, 479–484.

Engel, B. T., & Gottlieb, S. H. Differential operant conditioning of heart rate in the restrained monkey. *Journal of Comparative & Physiological Psychology*, 1970, *73*, 217–225.

Engel, B. T., & Moos, R. H. The generality of specificity. *Archives of General Psychiatry*, 1967, *16*, 574–581.

Flynn, J. P. Discussion of papers by Reese and Gantt. *Physiological Reviews*, 1960, *40* (Suppl. 4), 292–294.

Forsyth, R. P. Regional blood flow changes during 72-hour avoidance schedules in the monkey. *Science*, 1971, *173*, 546–548.

Gantt, W. H. The meaning of the cardiac conditional reflex. *Conditional Reflex*, 1966, *1*, 139–143.

Harlow, H. F. The formation of learning sets. *Psychological Review*, 1949, *56*, 51–65.

Moos, R. H., & Engel, B. T. Psychophysiological reactions in hypertensive and arthritic patients. *Journal of Psychosomatic Research*, 1962, *6*, 227–241.

Obrist, P. A., Wood, D. M., & Perez-Reyes, M. Heart rate during conditioning in humans: Effects of VCS intensity, vagal blockade, and adrenergic block of vasomotor activity. *Journal of Experimental Psychology*, 1965, *70*, 32–42.

Schwartz, G. E. Operant conditioning of human cardiovascular integration and differentiation. *Psychophysiology*, 1971, *8*, 245. (Abstract)

Skinner, B. F. *The behavior of organisms.* New York: Appleton-Century-Crofts, 1938.

Weiss, T., & Engel, B. T. Operant conditioning of heart rate in patients with premature ventricular contractions. *Psychosomatic Medicine*, 1971, *33*, 301–321.

Voluntary Control of Human Cardiovascular Integration and Differentiation through Feedback and Reward

Gary E. Schwartz

Abstract. *Human subjects can learn to control the relation between their systolic blood pressure and heart rate when they are given feedback and reward for the desired pattern of blood pressure and heart rate. They can learn to integrate these functions (increase or decrease both jointly), or to a lesser degree, differentiate them (raise one and simultaneously lower the other). The extent of this learning is predicted by a behavioral and biological model that explains specificity of learning in the autonomic nervous system.*

The traditional notion of the autonomic nervous system is of a tightly controlled homeostatic network capable of little response separation or voluntary control. Contrary to this belief, recent research on instrumental learning of visceral responses has shown that individual functions can be brought under voluntary control and can show specificity of learning similar to that found for skeletal responses (*1*). For example, research on systolic blood pressure (BP) and heart rate (HR), two closely related autonomic responses, has shown that each can be separately controlled. (i) Providing human subjects with feedback and reward for increases or decreases in systolic BP leads to learned control of BP without corresponding changes in HR (*2, 3*). (ii) If increases or decreases in HR are reinforced, the subjects learn to control HR without similarly changing BP (*4*). Although this phenomenon has both theoretical and clinical importance (for example, as a potential treatment for decreasing specific symptoms in psychosomatic disorders), a suitable explanation for it has not yet appeared.

The purpose of the proposed integration-differentiation (ID) model (*5*) is to help provide a behavioral and biological framework for understanding and predicting learned patterns of physiological activity. The model takes into account (i) the behavioral relations between responses as defined by the operant (feedback) procedure (*6*) and (ii) natural physiological changes or constraints that occur over time. In this context, the term integration is reserved for the response pattern in which two functions simultaneously change in the same direction (both increasing and decreasing together in a sympathetic-like pattern), and the term differentiation refers to the response pattern in which two functions simultaneously change in opposite directions (*7*). The first part of the model is an attempt to assess the activity of other

Reprinted by permission from *Science*, Vol. 175, 90-93. Copyright 1972 by the American Association for the Advancement of Science.

physiological functions at the instant when a given physiological function is being reinforced, and the second part is an attempt to relate this information to naturally occurring changes in physiological activity due to stimulation, adaptation, and other factors not related to the contingency of reinforcement per se (8).

The general approach can be illustrated through a behavioral analysis of systolic BP and HR, two functions showing discrete bursts of activity that can be easily reduced to binary response units. With every heart contraction, BP rises and reaches a systolic peak, its magnitude determined by a complex interaction of HR, stroke volume, and peripheral resistance (9). If average (tonic) levels are specified for HR (in beats per minute) and for BP (in millimeters of mercury), each heart beat can be classified according to whether HR and BP values are above (^{up}) or below ($_{down}$) their tonic levels. At each heart beat only four coincidence patterns are possible: BP^{up} HR^{up}, BP_{down} HR_{down}, BP^{up} HR_{down}, and BP_{down} HR^{up}.

If the two functions naturally rose and fell together all of the time (BP^{up} HR^{up} and BP_{down} HR_{down} only), then even though an experimenter might select only one system for reward, he would unwittingly provide the same reinforcement for parallel changes in the other system. Accordingly, both functions would show simultaneous learning in the same direction (integration). At the other extreme, if the two functions always changed in opposite directions (BP^{up} HR_{down} and BP_{down} HR^{up} only), then when changes in one function were selected for reward, opposite changes in the other function would also receive contingent (100 percent correct) reinforcement. Again, both functions would show simultaneous learning, this time in opposite directions (differentiation). However, if the two functions were unrelated to the point of actually producing equal numbers of the four coincidence patterns, then when changes in one function were reinforced, the other function would receive a sequence of random reinforcement. (For example, if every BP^{up} was rewarded, HR would simultaneously be rewarded half the time for HR^{up} and half for HR_{down}.) Therefore, only the response receiving contingent reinforcement would be learned. I refer to this latter situation (when only the response chosen for reinforcement is learned) as specificity of learning.

My experiment was designed with two major purposes. One was to empirically determine the naturally occurring behavioral relationship between BP and HR, to test the hypothesis that, in order for specificity of BP and HR learning to occur, these two functions vary independently and thus go up and down together only about 50 percent of the time. The second aim was to answer the following question: if HR and BP (or any two functions) are randomly related, then is it possible to make them change together, or change in opposite directions? One implication of the ID model is that it should be possible to control a combination of functions, for example BP and HR, by rewarding the subject only when he shows the desired coincidence pattern of simultaneous changes in both functions (10). This procedure, although it reduces the percentage of correct reinforcement each function can receive (since neither function alone controls the reward), eliminates incorrect (undesired) rewards. A behavioral analysis would predict that if the frequency of the four coincidence patterns of BP and HR were equal and randomly distributed, then each pattern would show learning to the same degree. However, this prediction would fail if physiological constraints were preventing HR and BP from changing independently. In this case, the discrepancy between the predicted and measured extents of learning would uncover the ways in which the constraints were operating.

This approach requires a method for measuring on-line the phasic (beat-by-beat) and tonic (median) relationships between autonomic responses. Instrumentation was developed which determines at each heart beat the four coincidence patterns for BP and HR and also provides tonic value for each function during each trial (*11*) (Fig. 1).

Forty normotensive males (paid volunteers), 21 to 30 years old, were seated in a sound- and temperature-controlled room and connected to the physiological recording devices. Systolic BP, HR, and respiration were recorded on an Offner type R dynagraph. The electrocardiogram was measured with standard plate electrodes and displayed on one channel of the dynagraph. An electronic switch triggered Grason-Stadler (model 1200) solid-state programming equipment at each heart cycle (R spike of the electrocardiogram). The equipment automatically detected the four coincidence patterns for BP and HR and also presented all stimuli to subjects. Respiration was recorded by a strain gauge placed around the waist.

For each trial, median systolic BP was defined as the constant cuff pressure at which 50 percent of possible Korotkoff sounds occurred. Korotkoff sounds were displayed on one channel of the dynagraph in conjunction with a second electronic switch. With the use of this switch and of appropriate logic modules it was possible to count heart beats accompanied (within 300 msec) by a Korotkoff sound and heart beats not followed by a Korotkoff sound. Similarly, median HR was obtained by the use of a third electronic switch calibrated in beats per minute. The HR for each beat was displayed on one channel of the dynagraph through a cardiotachometer (Lexington Instruments model 107). With the use of appropriate logic modules in conjunction with the third (cardiotachometer) electronic switch, it was possible to count heart beats that were faster or slower than the value set by the cardiotachometer electronic switch.

If Korotkoff sounds (BP^{up}) or fast heart beats (HR^{up}) or both exceeded the median levels by 35 in a given trial (50 heart beats in length), on the next trial the applied cuff pressure was raised 2 mm-Hg, or the cardiotachometer electronic switch was raised 2 beats per minute, or both changes were made. If the number of BP^{up} or HR^{up} or both exceeded the medians by less than 15 in a given trial, on the next trial the applied cuff pressure was lowered 2 mm-Hg, or the cardiotachometer electronic switch was lowered 2 beats per minute, or both changes were made. This shaping procedure made it possible to accurately track both BP and HR independently and simultaneously, and at the same time to obtain comparable information about relative changes in BP and HR at each beat.

Subjects were told that the purpose of the experiment was to determine whether they could learn to control certain physiological responses that are generally considered involuntary. They were instructed to refrain from moving and to breathe regularly. They were not told the nature of the specific responses to be controlled or the required directions of change. Twenty subjects were studied in two integration conditions. Half of these subjects received a 100-msec light and tone as feedback for each pulse cycle in which BP and HR simultaneously increased (BP^{up} HR^{up}); the other half received the same feedback each time their BP and HR simultaneously decreased (BP_{down} HR_{down}). Another 20 subjects were studied in two differentiation conditions. Half re-

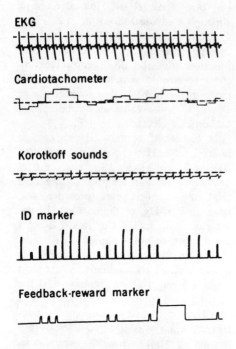

Fig. 1. Representative portion of a polygraph record of the integration-differentiation (ID) system in operation. Shown are the electrocardiogram (EKG), heart rate (HR) displayed through a cardiotachometer, Korotkoff sounds measured at a constant cuff pressure, and two marker channels. Dashed lines represent the approximate levels of the three electronic switches. The presence or absence of a Korotkoff sound relative to the constant pressure in the cuff indicates whether blood pressure (BP) is up or down, while HR is rated up or down relative to the median HR. After each heart cycle (except during a reward) one of four possible marks appears on the ID marker channel. The longest and shortest marks indicate integration, with $BP^{up} HR^{up}$ producing the longest mark and $BP_{down} HR_{down}$ producing the shortest mark. The other two marks indicate differentiation, with $BP^{up} HR_{down}$ producing the third longest mark, and $BP_{down} HR^{up}$ producing the second shortest mark. The bottom channel indicates which one of the four possible combinations is eliciting feedback (short mark) and reward (long mark). In this example, feedback is occurring for $BP_{down} HR^{up}$ differentiation.

ceived the feedback only when their BP increased and their HR simultaneously decreased ($BP^{up} HR_{down}$), and the other half received the feedback for $BP_{down} HR^{up}$ responses.

After giving every 12 correct responses, subjects were rewarded with a 3-second view of slides showing landscapes, attractive nude females, or cumulative bonuses earned (each slide was worth 5 cents). No feedback was given during the slides. All subjects receive 5 adaptation, 5 random reinforcement, and 35 conditioning trials. A blue light signaled the onset of random reinforcement and conditioning trials and remained on during the trials. Each trial was 50 heart beats long and was preceded by 10 seconds of cuff inflation. Intervals between trials ranged from 20 to 30 seconds; during this time the cuff was deflated. Subjects were matched according to their resting median BP's and randomly placed in one of the four conditions (*12*).

Figure 2 shows the average conditioning results for the four groups, based on the median cuff pressures and HR's obtained in each trial. Results are summarized as follows.

1) Subjects learned to directly integrate their BP and HR in a single session. Separate analyses of variance revealed highly significant group-by-trial interactions for BP ($P < .0001$) and HR ($P < .0001$), a result which indicates that the divergence of the two BP curves and of the two HR curves is reliable (*13*). Note that when $BP^{up} HR^{up}$ is rewarded, both BP and HR increase somewhat and then return to baseline, while $BP_{down} HR_{down}$ reinforcement yields sustained decreases in both. This result is very similar to earlier data obtained when each was separately rewarded (*2, 4*), except that in the present experiment control of both BP and HR is learned, and the learning effect is greater.

2) Integration was learned more easily than was differentiation. In fact, each of the integration groups earned more slides than either differentiation group. The divergence of the two BP curves in Fig. 2 (right) is reliable ($P < .0001$) and is very similar to the results for integration. In contrast, the divergence of the two HR curves in Fig. 2 (right) is not reliable, primarily because of the lack of sustained HR decreases in the $BP^{up}\ HR_{down}$ group (13).

3) If the curves in Fig. 2 (left and right) are compared, the $BP^{up}\ HR_{down}$ group is similar to the $BP^{up}\ HR^{up}$ group, while the $BP_{down}\ HR^{up}$ group is more similar to the $BP_{down}\ HR_{down}$ group.

4) Clear evidence for learning of differentiation was obtained only for the $BP_{down}\ HR^{up}$ group ($P < .0001$); the reverse pattern ($BP^{up}\ HR_{down}$) proved difficult to learn (14).

These results were anticipated from the ID model, by an analysis of the natural beat-by-beat and tonic relation of BP to HR during rest and during random stimulation in the present experiment and by a comparison of these data with earlier data on BP and HR control (2–4). A summary of the predictions from the model follows.

1) The $BP^{up}\ HR^{up}$ pattern would look as if it were learned less effectively than the reverse $BP_{down}\ HR_{down}$ pattern, when the levels found at random stimulation were used as the standard. This prediction was based on data suggesting that systolic BP and HR tend to normally adapt (decrease relative to initial rest or random stimulation levels) during this type of experiment (2). Adaptation would therefore act to lower both the increase and decrease curves for BP and HR.

2) Tonic integration would be more readily learned than tonic differentiation. This prediction was based on the observation that median levels of BP and HR tend to increase (or decrease) together ($r = +.36$, $P < .05$) during the initial period of random stimulation. (For example, if a subject reacts to the beginning reinforcement trials with a large tonic increase in BP, his HR will tend to show a comparable rise in level.) In other words, to the extent that the systems are tonically integrated in this experimental situation (other situations might heighten, eliminate, or possibly reverse this relationship), it should be easier to make them change together than in opposite directions.

3) Subjects reinforced for the $BP^{up}\ HR_{down}$ pattern would raise both functions (and thus appear similar to the $BP^{up}\ HR^{up}$ group); while subjects reinforced for the $BP_{down}\ HR^{up}$ pattern would tend to lower both functions. This hypothesis is based on the following analysis of the BP and HR coincidence patterns. As predicted from the specificity findings (2–4), BP and HR change in the same direction only 53 percent of the time, and the frequency of the four coincidence patterns is approximately equal. However, close inspection of the coincidence patterns reveals that the relation between BP and HR over time is not truly random. Changes in HR precede those of BP; this relation produces a constant sequence of coincidence patterns (15). For example, the data indicate that when a $BP^{up}\ HR^{up}$ pattern changes, 70 percent of the changes will be to a $BP^{up}\ HR_{down}$ pattern. Therefore, to the extent that the $BP^{up}\ HR^{up}$ pattern consistently precedes the $BP^{up}\ HR_{down}$ pattern, then reinforcement for the latter will produce consistent (although delayed) reinforcement of the former as well. This is an example of a response chaining factor. Similarly, the data indicate that when a $BP_{down}\ HR_{down}$ pattern changes, it will change with a 70 percent frequency to a $BP_{down}\ HR^{up}$ pattern; this observation explains the similarity to $BP_{down}\ HR_{down}$ learning when $BP_{down}\ HR^{up}$ responses are re-

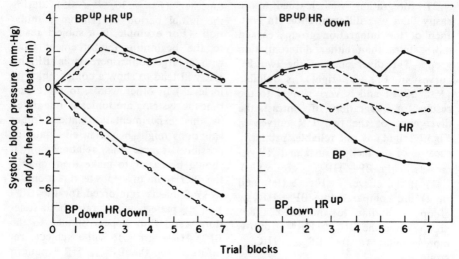

Fig. 2. Average systolic blood pressure (BP) and heart rate (HR) for the subjects being conditioned in the four coincidence patterns. On the left are data for the two integration conditions, on the right are data for the two differentiation conditions. Solid lines are BP, dashed lines are HR. Each point is the mean of five trials, set to zero by the last random trial. Beats per minute and millimeters of mercury are therefore on the same axis.

inforced. However, with natural tonic integration acting as a constraint, the response chaining factor may have little influence when integration patterns for BP and HR (both increasing and decreasing) are directly reinforced.

4) The BP_{down} HR^{up} pattern would be more readily learned than the opposite BP^{up} HR_{down} pattern. This was predicted from the interaction of tonic and phasic factors in the present reinforcement procedure. To the extent that BP and HR normally adapt within a trial (but the electronic switches remain constant), the frequency of BP_{down} HR_{down} patterns would tend to increase toward the end of a trial. The BP_{down} HR^{up} pattern naturally occurs after a BP_{down} HR_{down} pattern, whereas the BP^{up} HR_{down} pattern rarely occurs after a BP_{down} HR_{down} pattern. In other words, it appears more feasible to change from a BP_{down} HR_{down} pattern by raising HR than by raising BP (16).

To test for conditioning of respiration during the experiment, analyses were made of respiration rate, variation of respiration rate, and variation of respiration amplitude. No differences were found between the four groups of subjects (17). Also, cognitive activity was assessed at the end of the experiment by a comprehensive questionnaire. Subjects in the BP_{down} HR_{down} group tended to check more items associated with relaxation than did the BP^{up} HR^{up} group, a result suggesting that some kind of arousal variable may have been operating. However, the two differentiation groups were indistinguishable from the BP^{up} HR^{up} group. Apparently, when a person is required to decrease the activity of more than one function simultaneously, subjective relaxation may occur.

Altogether, these data demonstrate that it is possible for human subjects to develop some control over the relation between their systolic BP and HR when they are provided with feedback and reward for the desired pattern of BP and HR. When the behavioral relation of these functions as "seen through the eyes of feedback and reward" is analyzed, and when natural biological constraints are taken into account, it appears possible to understand and predict the resulting pattern of learned control (*18*). To this extent, these results clarify earlier specificity findings and provide a new framework for research and theory in the control of multiautonomic functions. It is suggested that the present technique may be a tool for studying and controlling not only the relationship between visceral responses but the interaction of visceral responses with somatic and central behavior as well (*19*). This could be accomplished by assessing the extent and ease with which specific patterns of activity can be learned. The importance of considering feedback and reward in biological perspective is stressed, since natural physiological relationships and constraints do occur (*20*). It may be possible to apply these techniques to the treatment of specific clinical disorders; for example, to condition decreases in BP and HR to reduce the pain of angina pectoris (*21*).

GARY E. SCHWARTZ

Department of Social Relations,
Harvard University,
Cambridge, Massachusetts 02138

References and Notes

1. For reviews of human research, see H. D. Kimmel, *Psychol. Bull.* **67**, 337 (1967); E. S. Katkin and E. N. Murray, *ibid.* **70**, 52 (1968). For reviews of animal research see N. E. Miller, *Science* **163**, 434 (1969); L. V. DiCara, *Sci. Amer.* **222**, 30 (January 1970).
2. D. Shapiro, B. Tursky, E. Gershon, M. Stern, *Science* **163**, 588 (1969); D. Shapiro, B. Tursky, G. E. Schwartz, *Circ. Res.* **26-27** (Suppl. 1), I-27 (1970).
3. J. Brener and R. Kleinman, *Nature* **226**, 1063 (1970).
4. D. Shapiro, B. Tursky, G. E. Schwartz, *Psychosom. Med.* **32**, 417 (1970).
5. A complete description of the ID model and the present experiment can be found in G. E. Schwartz, thesis, Harvard University (1971).
6. For this report, no formal distinction is made between feedback (implying response information) and reinforcement (implying response contingency). Since both terms require experimental procedures for systematically producing changes in the environment which closely follow changes in behavior, their similarities (in obtained results) rather than differences (in theoretical underpinnings) are emphasized. Although I have primarily used the terminology developed in operant conditioning, feedback terminology can be easily substituted; the major implications and conclusions remain the same.
7. These definitions depart somewhat from those generally used in biology, where the term integration refers to any consistent pattern of unified activity, regardless of direction, while differentiation refers to a separation of one response from others (here called specificity).
8. For example, if a response decreases naturally from the beginning to the end of an experiment this changing operant baseline must be taken into account to measure the direction and extent of learning. See A. Crider, G. E. Schwartz, S. R. Shnidman, *Psychol. Bull.* **71**, 455 (1969).
9. R. F. Rushmer, *Cardiovascular Dynamics* (Saunders, Philadelphia, 1961).
10. A similar approach has been used to train rats to differentially control blood volume between the two ears. See L. V. DiCara and N. E. Miller, *Science* **159**, 1485 (1968).
11. A preliminary experiment on learned cardiovascular integration in which the procedure was tested can be found in G. E. Schwartz, D. Shapiro, B. Tursky, *Psychosom. Med.* **33**, 57 (1971).
12. No subjects showing initial systolic blood pressures of 135 mm-Hg or more during the last adaptation or random reinforcement trial were included in this experiment.
13. Analyses of variance were performed on an IBM 360 computer with the Biomed 08V program. Groups was the between factor and trials was the within factor. The degrees of freedom for the group by trial interaction was 34/612. For the BP data, the analyses of variance (comparison of two groups at a time; six possible combinations) revealed that over trials the two BP^{up} conditions were each significantly higher than each of the two BP_{down} conditions ($P < .0001$), but were not different from each other. For HR, all two-group comparison trials were significant ($P < .01$ to .0001) except BP^{up} HR^{up} with BP^{up} HR_{down}, and BP^{up} HR_{down} with BP_{down} HR^{up}. The corresponding group main effects for these comparisons were also significant for BP ($P < .025$ to .005) and HR ($P < .10$ to .0001).
14. For example, separate analyses of variance on each of four groups, with measures (2) (change in BP versus change in HR) and trials (35) as within factors, indicated that BP_{down} HR^{up} reinforcement was the only condition that produced a reliable divergence between BP and HR during trials (d.f. = 34/306).
15. Sinus arrhythmia, a condition in which HR leads BP, is discussed by A. M. Scher, in *Physiology and Biophysics*, T. C. Ruch and H. D. Patton, Eds. (Saunders, Philadelphia, 1965), p. 660.
16. This pattern of results is also consistent with physiological theory suggesting that the parasympathetic system is capable of finer differentiation than is the sympathetic system. Raising the HR while lowering the BP may constitute a parasympathetic pattern, since HR may be increased by a decrease in vagal

tone. Unlike elevation of HR, elevation of systolic BP requires sympathetic activity, hence the observed difficulty in lowering HR at the same time.
17. Finer analysis procedures (for example, coincidence measures of BP, HR, and respiration) may be necessary to assess such effects.
18. Predictive power of the ID model will, by definition, be limited to the extent that (i) operant (feedback) theory adequately handles the learning of individual responses (the interaction of other variables such as cognitive set and motivation is yet little understood); and (ii) physiological mechanisms and constraints can be empirically assessed in the given situation.
19. The recently published report by E. E. Fetz and D. V. Finocchio [*Science* 174, 431 (1971)], which demonstrates operant conditioning of specific patterns of neural and muscular activity in the monkey, strongly supports this view.
20. For example, when changes at each beat in diastolic (as opposed to systolic) BP are reinforced, some conditioning of HR also takes place (D. Shapiro, G. E. Schwartz, B. Tursky, *Psychophysiology*, in press). This implies that diastolic BP and HR are partially (but not completely) integrated with respect to phase. Analysis of $BP^{up} HR^{up}$ and $BP_{down} HR_{down}$ coincidence responses has confirmed this prediction (average phasic integration is 70 percent). Research is required to determine the extent to which these two integrated functions can be separated through operant differentiation reinforcement (for example, requiring the subject to decrease diastolic BP by reducing peripheral resistance, while at the same time increasing HR).
21. The circumstances under which cardiac oxygen requirements in angina pectoris can be reduced if both HR and BP are lowered are discussed by E. Braunewald, S. E. Epstein, G. Glick, A. A. Wechsler, N. H. Wechsler, *N. Engl. J. Med.* 277, 1278 (1967); E. H. Sonnenblick, J. Ross, E. Braunewald, *Amer. J. Cardiol.* 22, 328 (1968).
22. Supported by NIMH research grant MH-08853; research scientist award K5-MH-20,476; ONR contract N00014-67-A-0298-0024; and the Milton Fund of Harvard. I especially thank D. Shapiro and B. Tursky for guidance and encouragement, and J. D. Higgins for comments on the manuscript.

4 April 1971; revised 8 July 1971

Control of Diastolic Blood Pressure in Man by Feedback and Reinforcement

15

David Shapiro, Gary E. Schwartz, and Bernard Tursky

ABSTRACT

When provided with external feedback of their diastolic blood pressure and incentives to respond appropriately, normal male Ss learned to raise or lower their diastolic pressure in a 35-min training session. The difference between increase and decrease groups at the end of conditioning was 7.0 mm Hg or 10% of baseline. This difference was augmented to 10.4 mm Hg or 15% of baseline during extinction when half the Ss were asked to maintain continuing "voluntary control" even though feedback and incentives were withdrawn. Heart rate was also influenced when diastolic blood pressure was reinforced, although less markedly. Further analysis indicated that when diastolic pressure is reinforced, heart rate is partially reinforced in the same direction, accounting for the coincidental conditioning of the related cardiovascular measure. No consistent changes in respiration or post-session verbal reports were obtained. These results lend support to the possibility of therapeutic application of the techniques in patients with essential hypertension.

DESCRIPTORS: Operant autonomic conditioning, Visceral learning, Biofeedback, Diastolic blood pressure, Heart rate, Cardiovascular functions, Blood pressure measurement, Integration-differentiation.

One area of potential application of operant conditioning and physiological feedback techniques is to the control of symptoms in psycho-physiological disorders (Miller, 1969; Shapiro, Tursky, Gershon, & Stern, 1969). Essential hypertension is a common disorder in which the primary symptom is an elevation of blood pressure, the etiology of which is not well understood (Merrill, 1966). If operant-feedback techniques can be used effectively to reduce pressure levels in hypertensive patients, this may provide a significant means of treatment in addition to others currently in practice.

It has been shown that systolic blood pressure can be relatively raised or lowered in man by the use of feedback and operant reinforcement (Shapiro et al., 1969). These effects were found to be independent of heart rate (Shapiro, Tursky, & Schwartz, 1970a) and were replicated (Shapiro, Tursky, & Schwartz,

This research was supported by: NIMH Research Scientist Award K5-MH-20,476; NIMH Research Grants MH-08853 and 04172; Office of Naval Research Contract N00014-67-A-0298-0024; and the Milton Fund of Harvard University.

Address requests for reprints to: Dr. David Shapiro, 74 Fenwood Road, Boston, Massachusetts 02115.

Reprinted with permission of the publisher and author from *Psychophysiology*, Vol. 9, 296-304. Copyright 1972, The Society for Psychophysiological Research.

1970b). Comparable results were obtained for systolic blood pressure in rats (DiCara & Miller, 1968) and for mean arterial blood pressure in squirrel monkeys (Benson, Herd, Morse, & Kelleher, 1969). More recently, we have obtained evidence that coincident patterns of change in systolic blood pressure and heart rate, both integration (Schwartz, Shapiro, & Tursky, 1971) and differentiation (Schwartz, 1972), can be effectively conditioned by operant procedures.

In hypertension, the physician is usually concerned with an increase in peripheral resistance to blood flow which is more closely represented by diastolic rather than systolic blood pressure (Merrill, 1966). In addition to its relation to peripheral resistance, diastolic blood pressure can also be affected by changes in heart rate from heartbeat to heartbeat, decreasing as the time between one heartbeat and the next increases.

In view of the physiological and medical significance of diastolic blood pressure, the present study was carried out to determine the degree to which it could be modified in normal human subjects in a single experimental session, with the same feedback and operant procedures used previously in our laboratory to control systolic pressure. To assess the role of heart rate when diastolic blood pressure is reinforced, simultaneous changes in heart rate were also evaluated using an on-line tracking procedure for assessing tonic and phasic interrelationships of heart rate and blood pressure (Schwartz et al., 1971; Schwartz, 1972).

Method

Subjects

The Ss were 20 volunteer healthy (normotensive) college students.

Apparatus

Instrumentation previously used to measure and reinforce systolic blood pressure (Shapiro et al., 1969) was adapted in this study to provide similar information about diastolic blood pressure on each heart cycle.

A conventional blood pressure cuff was wrapped around the upper arm and pumped up approximately to the S's diastolic blood pressure. The pressure applied to the cuff could be held constant for any period of time, and could be increased or decreased in increments of 2 mm Hg by use of a regulated low-pressure compressed air source. A Statham pressure transducer (P23AC) monitored the pressure in the cuff and recorded it on one channel of an Offner Type R Dynograph. A crystal microphone was mounted under the cuff over the brachial artery to detect Korotkoff sounds which are displayed on a second channel of the Dynograph. The EKG was recorded on a third Dynograph channel from a pair of standard limb electrodes, and a cardiotachometric record of heart rate was displayed on a fourth channel of the recorder. The R-wave of each successive heart cycle was detected by an electronic switch, and a second switch was used to detect the presence of criterion amplitude Korotkoff sounds. Detection of the R-wave opened an electronic gate for 300 msec, and the coincidence of this event with the presence, or absence, of a Korotkoff sound was used as the criterion of feedback, as described below. Solid-state logic modules (Grason-Stadler Series 1200) were used to control the programming of experimental events. Respiration was recorded on a fifth Dynograph channel by means of a strain gauge belt strapped around the waist.

Median Diastolic Pressure

In the casual measurement of clinical diastolic blood pressure, either the muffled Korotkoff sound or the absence of the Korotkoff sound is used as the criterion of diastolic pressure. Since the procedure for electronically isolating the muffled sound is more difficult, the present study employed the absence of Korotkoff sound as the criterion of diastolic pressure. The criterion Korotkoff amplitude was determined empirically prior to conditioning by slowly bleeding down the cuff pressure from above systolic to below diastolic pressure and finding the exact amplitude prior to disappearance of the sounds. In a few Ss in whom the disappearance of sound was not clear cut, an arbitrary Korotkoff amplitude had to be selected that could be differentiated from other sounds in this range of pressures. In all Ss, the criterion Korotkoff amplitude appeared to be consistent over trials.

Median diastolic blood pressure was determined for each S in the following manner: When a cuff is inflated and held at a constant pressure, a Korotkoff sound will be present or absent as a function of the variation in blood pressure in the brachial artery at each heart cycle. When cuff pressure is lower than diastolic pressure, the cuff does not impede the flow of blood in the artery, and no Korotkoff sound can be detected. If the applied cuff pressure is greater than diastolic and below systolic pressure, the flow of blood is constantly impeded and the resultant turbulence produces a Korotkoff sound on every heart cycle. Then the pressure is adjusted to a constant level where a Korotkoff sound is present on 50% of the heart cycles; that pressure is by definition the median diastolic pressure. This principle was used to establish each S's baseline median diastolic pressure and to track median pressure from trial to trial during conditioning. At the median diastolic pressure, when a Korotkoff sound is present, the S's diastolic level is at or below the applied constant cuff pressure; when the sound is absent, the S's diastolic level is above the constant pressure in the cuff. This makes it possible to provide feedback to Ss on relative increases or decreases in diastolic pressure on each successive heartbeat.[1]

Median Heart Rate

Median heart rate was determined and tracked over time for each experimental trial using a method described in detail elsewhere (Shapiro et al., 1970a). Median heart rate is defined as that heart rate for which half the beat-to-beat intervals during a trial are faster and half are slower. This was determined by using logic modules and a cardiotachometer level detector. At the conjunction of each R-wave and the tachometer level, each beat-to-beat interval could be designated as fast or slow relative to median heart rate.

Relationship of Blood Pressure and Heart Rate

The phasic interrelationships of heart rate and diastolic blood pressure were measured on line using a tracking procedure developed in prior experiments

---

[1] Blood pressure recording and conditioning apparatus based on the system described in this paper can be purchased from Lexington Instruments, Inc., Waltham, Massachusetts.

(Schwartz et al., 1971; Schwartz, 1972) and were displayed on a sixth channel of the Dynograph. Briefly, the procedure determines at each heart beat whether blood pressure and heart rate are integrated, that is, both going in the same direction; or differentiated, that is, both going in opposite directions, as measured from median levels. Using this procedure, it is possible to assess the extent of blood pressure-heart rate integration for a given S as defined by the present feedback-reward system. For example, 100% integration means that for every heartbeat, blood pressure and heart rate are either both high or both low, relative to median levels.

Experimental Procedure

Upon entering the laboratory, Ss were seated in a semi-reclining position in a sound-light controlled room isolated from the apparatus. They were told that the research concerned the ability of individuals to control certain physiological responses usually considered involuntary, and that they would be given feedback about their own responses and rewards to help them achieve control. Ss were asked not to try to influence these responses by physical manipulations such as moving about or tensing muscles, and they were told to keep their breathing regular. All Ss were given exactly the same instructions and were not informed about the particular physiological activity being studied. They were paid $3.00 per hour plus bonuses earned during the session.

Ss were given 5 resting, 5 random reinforcement, 35 conditioning, and 10 extinction trials, each trial consisting of 50 heart beats preceded by 5 sec for cuff inflation. After each trial, there was a 20 to 25-sec period of cuff deflation. The resting trials were calibration (blank) trials for purposes of determining initial median pressure. During random reinforcement, feedback on the correct response was randomly programmed at 50%; this procedure served to familiarize Ss with the feedback stimuli and slides used as reinforcers (see below) and provided uniform rates of feedback and reinforcement to all Ss regardless of initial reaction to the task.

Two conditions were studied. Ten Ss were reinforced for increasing (*up*) and 10 for decreasing (*down*) diastolic pressure. During each conditioning trial, a success was defined by the absence of a Korotkoff sound for *up* Ss and the presence of a sound for *down* Ss on each successive heart beat. This information was fed back to Ss by programming apparatus which produced a 100-msec flash of white light and a simultaneous 100-msec tone of moderate intensity for each success. Ss earned rewards consisting of slides of landscapes, nude girls, and monetary bonuses after every 20 successes (feedback). The slides were arranged randomly and projected for 3 sec during which time no feedback was given. Ss had been instructed about the feedback stimuli and incentive slides, and told to look at the flashing light and slides when they appeared, and to keep their eyes open throughout the session. A steady blue light signalled each experimental trial (random reinforcement and conditioning) and was turned off during initial cuff inflation and intertrial periods.

In previous research (Shapiro et al., 1969; Tursky, Shapiro, & Schwartz, in press) it was determined that a change in the percentage of coincidences of R-wave and criterion Korotkoff sound of more than 25%, approximately, from median diastolic pressure (50%), represented a change in pressure of 2 mm Hg.

This enabled the shaping and tracking of a S's pressure from trial to trial. In this study, whenever the coincidence rate was 72% or 28% on any one trial, a 2 mm Hg adjustment was made in the constant pressure applied to the cuff on the next trial. Thus, if a *down* S succeeded in reducing pressure on a trial by increasing the coincidence rate to 72% or greater, the cuff pressure was reduced by 2 mm Hg on the next trial, making the task more difficult. If he failed by showing an increase in pressure (coincidence rate < 28%), the applied pressure was increased by 2 mm Hg, making the task easier. Opposite adjustments were made for *up* Ss.

The same percentage limits (28%–72%) were used in tracking median heart rate. If the percentage of fast beats on a given trial was greater than 72%, the tachometer level detector was modified so that the median heart rate in the next trial was 2 beats higher. An opposite adjustment was made for trials on which the percentage of fast beats was less than 28%.

After the 35 conditioning trials were completed, there was a 2-min pause while further instructions were given. Five Ss in each group (*up* and *down*) were asked to continue trying to make correct responses during the succeeding trials but without feedback or slides. They were told that any additional bonus actually earned during this period would be added to the total accumulated. This extinction condition is called *voluntary control (vol)*. The remaining 5 Ss in each group were told that the task proper was terminated, and they were asked to sit quietly and stay awake as they had done up to that time while additional measurements were taken. This extinction condition is called *rest*.

Thus, there were 5 Ss in each of the four conditions: *up rest, up vol, down rest, down vol*. These four groups were equated for initial median diastolic blood pressure by a matching procedure in which Ss having approximately the same initial pressure were assigned at random to the four different conditions.

Finally, a questionaire was administered to Ss at the end of the session concerning subjective and physical reactions to the experiment.

Results and Discussion

The basic data are the constant cuff pressures on each experimental trial, indicating median diastolic pressure over time. An analysis of variance of diastolic blood pressure during the 35 conditioning trials was computed for two variables: condition (*up* vs *down*) and trials (35 repeated trials). A highly significant interaction, $F\ (34/612) = 3.872$, $p < .001$, was obtained between condition and trials.

Fig. 1 shows the differential trend in blood pressure for the two groups. From beginning to end of conditioning, 7 *up* Ss raised their pressure from 2 to 18 mm Hg while 3 showed no change, and 8 *down* Ss lowered their pressure from 2 to 10 mm, 1 showed no change, and 1 an increase. On the last conditioning trial (trial 35), the two groups showed an average difference of 7.0 mm Hg or 10% of a baseline of 70 mm Hg (*up* increased 4.0 mm Hg, *down* decreased 3.0 mm Hg). This difference was greater than previously obtained for systolic pressure (Shapiro et al., 1969) (4 mm Hg or 3.3% of a baseline of 120 mm Hg). Also, it appeared somewhat easier to raise diastolic pressure than to lower it, a pattern opposite to that found for systolic pressure. Main effects for trials and for conditions were not significant.

Fig. 1 suggests that during extinction *rest* Ss tended to return to baseline pressures while *vol* Ss tended to exert continuing control over their pressure or even to augment the change in the appropriate direction, the latter especially in *up vol* Ss. The major effect, however, occurred immediately after the different instructions for extinction were read to the Ss. This is seen by comparing pressures on conditioning trial 35 with extinction trial 1 for the four groups of Ss. An analysis of variance showed a significant, $F\ (1/16) = 8.862$, $p < .01$, interaction between condition (*up* vs *down*) and extinction procedure (*rest* vs *vol*). In *up*

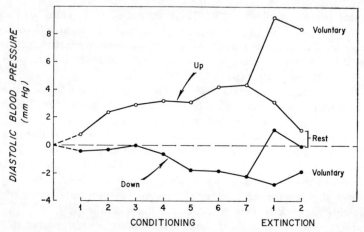

Fig. 1. Average diastolic blood pressure in groups reinforced for increasing (*up*) and decreasing (*down*) diastolic blood pressure. Each point is a mean of 10 Ss, 5 trials each. The curves were adjusted for slight pre-conditioning differences in diastolic blood pressure between the two groups and set to zero. The two extinction procedures were Rest and Voluntary Control. Changes during extinction were plotted with respect to final conditioning values, separately for Rest and Voluntary Control within each condition.

rest, 1 out of 5 Ss raised his pressure compared to 4 out of 5 *up vol* Ss. In *down rest*, 4 out of 5 Ss raised their pressure compared to 1 out of 5 *down vol* Ss. On the first extinction trial, diastolic pressure for the *up vol* group was 7.2 mm Hg higher than their pre-conditioning baseline; for the *down vol* group, diastolic pressure was 3.2 mm Hg lower than their pre-conditioning baseline. The difference between these two *vol* groups was 10.4 mm Hg or about 15% of baseline.

While Fig. 1 suggests that the *rest* conditions tended to converge to baseline more than the *vol* conditions, an analysis of variance of the diastolic pressures during the 10 extinction trials revealed no significant interactions between direction (*up* vs *down*), extinction method (*rest* vs *vol*), and trials. Neither the three-way nor the two-way interactions were significant. Changes in pressure during extinction itself were approximately the same in all four conditions.

Median heart rate was analyzed in the same fashion as diastolic pressure (Fig. 2). The analysis of variance of the conditioning trials (1–35) revealed a highly significant main effect for trials, $F\ (34/612) = 7.125$, $p < .001$. *Both up* and *down* groups showed a decline in median heart rate. From conditioning trial 1 to 35, the *up* group dropped 2.4 bpm and the *down* group 7.0 bpm. This difference in amount of decline in heart rate is seen in a significant Trials × Condition interaction, $F\ (34/612) = 2.120, p < .01$, an effect parallel to that obtained for diastolic pressure.

In contrast to the finding for diastolic pressure, there was no immediate maintenance of the heart rate effect after the extinction instructions were read to the Ss. While heart rate in the *vol* conditions appeared to continue in the appropriate direction as compared to the *rest* conditions, no significant effects were obtained in the analysis of variance comparing conditioning trial 35 with extinction trial 1 or in the analysis of the extinction data *per se*. The heart rate differences brought about during conditioning did not carry over to a significant degree in the extinction period.

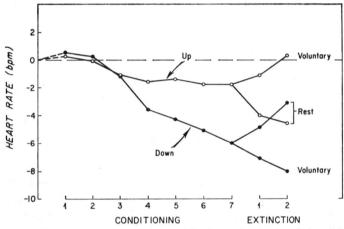

Fig. 2. Average heart rate in groups reinforced for increasing (*up*) and decreasing (*down*) heart rate. Each point is a mean of 10 Ss, 5 trials each. The curves were adjusted for slight preconditioning differences in heart rate between the two groups and set to zero. The two extinction procedures were Rest and Voluntary Control. Changes during extinction were plotted with respect to final conditioning values, separately for Rest and Voluntary Control within each condition.

In summary of the findings, Ss were able to increase or decrease diastolic pressure with the aid of feedback and rewards. These differences were maintained when Ss were asked to continue controlling their physiological activity without external feedback or reward. Although heart rate was also influenced by the reinforcement of diastolic pressure, the effects did not persist significantly during the trials without feedback and rewards. Moreover, correlations computed between changes in diastolic blood pressure and heart rate from the beginning to the end of conditioning were positive but small and not statistically significant.

These data are in contrast to earlier findings demonstrating that systolic blood pressure and heart rate can be controlled independently, that is, reinforcing systolic blood pressure leads to learned blood pressure changes without corresponding changes in heart rate, while reinforcing heart rate produces learned heart rate control without changing blood pressure (Shapiro et al., 1970a). An explanation of these findings is that systolic blood pressure and heart rate must be naturally randomly integrated (move in the same direction 50% of the time) as defined by the present reward system in order for the above results to be obtained, because if the two systems were substantially integrated, then reinforcing one would also result in the simultaneous reinforcement of the other as well, and both should have shown learning in the same direction. (See Schwartz, 1972.) The heart rate data in the present experiment, therefore, suggests that heart rate is more closely integrated with diastolic than with systolic pressure. This is borne out in the analysis of coincident changes in both functions during a preconditioning rest trial. Counts were made of the instances in which heart rate and diastolic pressure were simultaneously integrated (both above or both below their respective medians) or simultaneously differentiated (one above and the other below the medians). On the average, heart rate and diastolic pressure are integrated 70% of the time. In effect, in reinforcing diastolic blood pressure, heart rate is partially reinforced in the same direction most but not all of the time. This coincidental reinforcement probably accounts for the parallel though less effective control of heart rate when diastolic pressure is conditioned. In contrast, systolic pressure and heart rate are only weakly integrated and thereby can be more independently controlled (Schwartz, 1972).

The interpretation may therefore be offered that changes in heart rate observed in the conditioning of diastolic pressure were probably secondary in nature and not substantial. This interpretation is supported by the above analysis of coincidental changes in both functions and by the failure to obtain statistically significant heart rate effects during extinction. It should also be noted that the learned changes in diastolic pressure appear to occur prior to differences in heart rate. Comparing the *up* vs *down* curves in Figs. 1 and 2, the divergence in pressure precedes the divergence in heart rate. Moreover, obtained differences in heart rate in the present study were smaller than those observed when heart rate itself was directly reinforced (Shapiro et al., 1970a). Altogether, the data suggest that factors other than heart rate (e.g., peripheral resistance) were probably being modified to produce the observed changes in diastolic pressure.

Finally, respiration records were examined closely on a trial at the peak of conditioning. *Down* Ss showed slightly faster and more shallow patterns of breathing than *up* Ss, but these differences were not significant. Irregularities in rate and amplitude of respiration occurred about equally often in both groups. In view of the observed association between heart rate and diastolic pressure, coincident changes in respiration might also be expected. Other approaches to the measurement of respiration may be needed to tease out these relationships. Whether measures of other skeletal motor activity were affected by the conditioning procedures is not known.

It must be emphasized that Ss were not differentially instructed about increases or decreases in blood pressure. In post-session questionnaires, almost all Ss reported that the feedback meant an *increase* in arousal and blood pressure. Aside from more frequent mention of alertness and attentiveness by *up* Ss, on a variety of questions no consistent differences in reports of subjective or physical state were found between *up* vs *down* or *rest* vs *vol* conditions.

Conclusion

When provided with external feedback of their diastolic pressure and incentives to respond appropriately, normal human Ss can learn to raise or to lower their diastolic blood pressure. Reductions ranged from 2 to 10 mm Hg during conditioning, and up to 14 mm Hg during an additional set of voluntary control trials without feedback. Considering the use of minimal instructions and relatively brief (35 min) period of training, the results lend additional support to the possibility of application of the techniques to the lowering of diastolic blood pressure in patients with essential hypertension (Benson, Shapiro, Tursky, & Schwartz, 1972). The ability of Ss to exert some voluntary control over their blood pressure after training, without external feedback, reinforcement, or specific knowledge of what they are controlling, is also therapeutically significant. The apparent facilitation of the conditioning during the "voluntary" extinction trials needs further study.

The findings suggest that heart rate is partially conditioned when diastolic blood pressure is reinforced. The extent of the interrelationships, though not extensive, can be predicted by the tendency for heart rate and diastolic pressure to be phasically integrated around their median levels (Schwartz, 1972). This normal integration of the two functions may also account for the apparent larger differences obtained for diastolic compared to systolic pressure in the previous research (Shapiro et al., 1969). When integrated systolic pressure and heart rate changes are reinforced, larger effects were also obtained for systolic pressure (Schwartz et al., 1971; Schwartz, 1972).

Additional research is needed to determine the degree to which two closely integrated functions such as diastolic pressure and heart rate can be dissociated by operant techniques. It may be necessary, for example, to reinforce decreases in diastolic blood pressure on a partial schedule whereby the probability of a simultaneous decrease in heart rate is 50%. This will make it possible to establish whether other determinants of diastolic pressure such as peripheral resistance can be modified independently of heart rate. Evidence on the operant control of blood pressure in relation to other physiological functions may help further elucidate some of the mechanisms of the physiological and behavioral regulation of blood pressure and related physiological processes.

REFERENCES

Benson, H., Herd, J. A., Morse, W. H., & Kelleher, R. T. Behavioral induction of arterial hypertension and its reversal. *American Journal of Physiology*, 1969, *217*, 30–34.

Benson, H., Shapiro, D., Tursky, B., & Schwartz, G. E. Decreased systolic blood pressure through operant conditioning techniques in patients with essential hypertension. *Science*, 1971, *173*, 740–742.

DiCara, L. V., & Miller, N. E. Instrumental learning of systolic blood pressure responses by curarized rats: Dissociation of cardiac and vascular changes. *Psychosomatic Medicine*, 1968, *30*, 489–494.

Merrill, J. P. Hypertensive vascular disease. In J. V. Harrison, R. D. Adams, I. L. Bennett, W. H. Resnik, G. W. Thorn, & M. M. Wintrobe (Eds.), *Principles of internal medicine*. New York: McGraw Hill, 1966. Pp. 702–712.

Miller, N. E. Learning of visceral and glandular responses. *Science*, 1969, *163*, 434–445.

Schwartz, G. E. Voluntary control of human cardiovascular integration and differentiation through feedback and reward. *Science*, 1972, *175*, 90–93.

Schwartz, G. E., Shapiro, D., & Tursky, B. Learned control of cardiovascular integration in man through operant conditioning. *Psychosomatic Medicine*, 1971, *33*, 57–62.

Shapiro, D., Tursky, B., Gershon, E., & Stern, M. Effects of feedback and reinforcement on the control of human systolic blood pressure. *Science,* 1969, *163,* 588–590.

Shapiro, D., Tursky, B., & Schwartz, G. E. Differentiation of heart rate and systolic blood pressure in man by operant conditioning. *Psychosomatic Medicine,* 1970, *32,* 417–423. (a)

Shapiro, D., Tursky, B., & Schwartz, G. E. Control of blood pressure in man by operant conditioning. *Circulation Research,* 1970, *26* and *27* (Suppl. 1), I-27–41. (b)

Tursky, B., Shapiro, D., & Schwartz, G. E. Automated constant cuff pressure system to measure average systolic and diastolic blood pressure in man. *IEEE Transactions on Bio-medical Engineering,* in press.

Sources of Information which Affect Training and Raising of Heart Rate

Joel S. Bergman and Harold J. Johnson

ABSTRACT

This study was concerned with the effects of cardiac information and reinforcement on raising heart rate (HR). The experiment consisted of a 3×2 design with three types of cardiac information and two reinforcement conditions. The cardiac information given to Ss consisted of instructions to control an internal response (no specific HR information), instructions to increase HR (specific HR information), or instructions to increase HR while hearing heart beats through earphones (augmented HR information). External reinforcement was given to one-half of these Ss, while the remaining Ss received no external reinforcement. Sixty female undergraduates were randomly assigned to one of these six experimental groups.

Analyses of these data indicate that Ss in both the augmented and specific HR information groups were able to increase their HR. The no specific HR information groups showed no increases in HR, suggesting that awareness of the criterion response plays an important role in raising HR. No differences were found between the reinforcement conditions.

DESCRIPTORS: Heart rate, Feedback. (J. Bergman)

Within the past few years, results from some studies suggest that feedback is an important factor contributing to cardiovascular control in human Ss. The term feedback simply refers to knowledge of the results of a behavior which facilitates or modifies further behavior of the S. Feedback, however, is an ambiguous term and subsumes different types of information. Because of this ambiguity and the variety of information used in past studies on cardiac control, the present study is an attempt to analyze the types of information which contribute to raising heart rate (HR).

The types of information used to facilitate changes in cardiac responses in previous studies can be classified into two categories, specific information about cardiac responses and the sources of reinforcement of these cardiac responses. The first category consists of the kinds of information about cardiac responses which are introduced to Ss. The introduction of specific cardiac information can be conceptualized in three different forms: no specific HR information, specific HR information, and augmented HR information.

No specific HR information instructions often consist of asking Ss to try to make a light on a panel go on as often as possible. A light goes on whenever the

---

This research was supported in part by dissertation research funds awarded by the Graduate School and Department of Psychology of Bowling Green University to the first author, and by NIMH Grant MH 13373 awarded to the second author.

Address requests for reprints to: Joel S. Bergman, Ph.D., Department of Psychology, Smith College, Northampton, Massachusetts 01060.

Reprinted with permission of the publisher and author from *Psychophysiology*, Vol. 9, 30-39. Copyright 1972, The Society for Psychophysiological Research.

criterion HR response occurs. Some studies have demonstrated changes in HR under these conditions while the Ss were *unaware* that HR was the response of interest (Ascough & Sipprelle, 1968; Brener & Hothersall, 1966, 1967; Engel & Chism, 1967; Engel & Hansen, 1966; Frazier, 1966; Levene, Engel, & Pearson, 1968).

Specific HR information consists of asking Ss to attend to and concentrate on their HR. Here Ss are aware that changes in HR are the desired criterion responses. Under these conditions, giving specific requests to Ss to control HR has been shown to facilitate increases and decreases in HR (Bergman & Johnson, 1971; Brener, Kleinman, & Goesling, 1969).

Subjects receiving augmented cardiac information have also been asked to attend to and concentrate on their heart. In addition, these Ss have received cardiac information in some augmented form. For example, Ss have heard their heart beat through a loudspeaker (Shearn, 1962), viewed their heart beats on an oscilloscope (Donelson, 1966), or viewed their HR variability on a meter display (Lang, Sroufe, & Hastings, 1967).

In summary, the first category of specific types of information about cardiovascular responses deals with the kind of information about the response of interest which is given to Ss through instructional sets.

The second category, or sources of reinforcement of cardiac responses, has to do with the manner in which cardiac responses are interpreted and reinforced. Under certain conditions a particular portion of the cardiac response is selected by the E as the criterion response and the E rewards the S with some external form of reinforcement. In contrast to the first category where Ss receive all the feedback there is for a particular experimental condition, in the reinforcement category Ss receive only part of the feedback which the E has selected and has made the criterion response.

The use of this form of external reinforcement has been shown to facilitate increases and decreases in HR. Such external reinforcement has appeared in the form of tones of different frequency (Brener & Hothersall, 1966, 1967), lights (Brener et al., 1969; Engel & Chism, 1967; Engel & Hansen, 1966; Levene et al., 1968), electrical shock (Frazier, 1966; Shearn, 1962), and positive verbal statements (Ascough & Sipprelle, 1968).

By reducing the concept of feedback into components of specific kinds of cardiac information and forms of reinforcement, perhaps one could clarify the results of the above mentioned studies which employed different combinations of information and reinforcement. Because of the different types of feedback used in past experiments, the studies are not directly comparable, and therefore create difficulty in determining what information was most effective in facilitating the greatest degrees of HR control.

The experimental hypotheses in this study predicted that external reinforcement would enhance increases in HR more than no external reinforcement; augmented cardiac information would enhance increases in HR more than specific HR information which in turn would enhance increases more than no specific HR information. It was further hypothesized that more than one combination of these sources of information would enhance increases in HR more than sources of information introduced separately, and that the largest increases in

HR would be facilitated by augmented information combined with external reinforcement.

Method

Design

The experiment used a 3 × 2 design consisting of three types of cardiac information and two reinforcement conditions. The first variable dealt with Ss receiving specific HR information, augmented HR information, or no specific HR information.

Subjects in the specific HR information group were asked to try very hard to make their HR go faster whenever an amber light went on. This instructional set has been used successfully in the past to have Ss concentrate on their HR and produce cardiac changes (Bergman & Johnson, 1971).

In addition to the above instructions, Ss in the augmented cardiac information group were told that their task would be facilitated by listening to their heart beat which they could hear through earphones which they were wearing.

The no specific HR information group was unaware of the learning contingency of the experiment, and Ss in this group were asked to try and control *an internal response* whenever they saw an amber light illuminated. These instructions are almost identical to those used previously by Brener.[1]

The second variable dealt with attempts to reinforce cardiac increases. Subjects in the external reinforcement groups saw a blue light go on whenever HR increased above a certain criterion. These Ss were told that whenever the blue light went on, they were successfully increasing HR (or controlling the internal response). It was assumed that the blue light had reinforcing values since its flashing indicated to Ss that they were being successful in their task, and that such success would please both Ss and E. Subjects not receiving external reinforcers did not see the blue light.

Subjects

The Ss in this experiment were 80 female undergraduates who received credit in their introductory psychology course for participating in the study. Twenty Ss were eliminated either because of equipment failures (12), erratic respiration patterns (6), or because the no specific HR information Ss correctly guessed that HR increase was the criterion response (2). The remaining 60 experimental Ss were randomly assigned to one of the six groups by the order in which they appeared for the study.

Procedure and Training

Subjects were tested individually in a sound- and electrostatically-shielded room, and had physiological recording devices attached. A 10 min adaptation period served to obtain baseline readings.

Subjects received general introductory instructions about experiments in psychophysiology, such as being told not to move, or change their natural breathing

---

[1] J. Brener, personal communication, 1970.

rate during the study. Specific instructions were also given to Ss based on the experimental group to which the Ss were assigned. All Ss were told that whenever the amber light on the panel went on, they were to try to increase their HR (or control an internal response).

A trial consisted of a 20 sec presentation of the amber light. The inter-trial interval randomly ranged from 20–30 sec. Ss received 30 trials which were divided into three blocks of 10 trials each. Blocks of trials were separated by rest periods of 60 sec.

Half the Ss from the three cardiac information groups received external reinforcement in the form of a blue light illuminated for 0.5 sec whenever their HR increased 10% above their baseline HR. Reinforcement was presented with each heart beat whenever the HR at that beat was above the 10% criterion. When HR was considerably above the 10% criterion for more than 2 successive trials, the criterion was increased to a rate where Ss received 15 reinforcements during each amber light period. When the 10% criterion was initially too difficult to reach, Ss were shaped to the 10% level by first reinforcing 5% increases in HR above base level, and then raising the criterion to the 10% level.

Ss receiving augmented cardiac information heard their own heart beats through earphones. They heard the beats while the amber light was both on and off.

Testing

After Ss received the 30 trials, the training period was considered over. Ss were then told they did very well, and that they were able to increase their HR (control the internal response). They were then asked to try to increase their HR (or control the internal response) in the absence of the blue light or augmented information, whichever was applicable. Those Ss not receiving external reinforcement during training and given the no HR information or specific HR information instructions were told that they did so well during the training period that we wanted to check out our equipment. These Ss were then asked to continue doing what they had been (successfully) doing during the training period.

Following the administration of these instructional sets, the amber light was introduced for an additional 10 trials of 20 sec duration with an inter-trial interval randomly ranging from 20–30 sec. During the testing period Ss in the augmented cardiac information or external reinforcement groups did not hear their heart beats or see the blue light.

At the end of the testing period, Ss were interviewed extensively to determine the techniques they used when they tried to increase HR. A questionnaire with specific and open-ended questions was used to obtain information about Ss' attempts to increase HR.

Apparatus

Heart rate was measured with recording electrodes attached to S's right forearm and above the left ankle and recorded on a Grass Model 7D polygraph. Respiration amplitude and rate were measured with a strain gauge transducer attached to S's chest and recorded on the polygraph. Respiration measurements were taken to ensure that the predicted changes in HR were not due to changes in respiration. Exosomatic measurements of galvanic skin response (GSR) and skin resistance level were obtained with Beckman Biopotential Skin Electrodes placed on the S's palm. A constant 50 μa current was used. These skin resistance measures were taken to determine if the changes in HR were associated with changes in general arousal level.

Reinforcement of HR at or above the 10% criterion HR was automatically administered through the use of digital logic circuits manufactured by BRS Electronics. Ss who received augmented cardiac information heard their heart beats through headphones in the form of a 1000 Hz tone of 40 db intensity lasting 100 msec for each heart beat. Each heart beat triggered a circuit which delivered the tone to the Ss.

Results

Heart Rate

Heart rate was recorded on the cardiotachograph and was therefore in rate form of bpm. Heart rate scores were calculated by taking the mean HR from as many HRs occurring within each 1 sec period during the stimulus interval. Thus 20 HRs were obtained for each trial. A mean HR was also taken from 5 sec of HR which preceded the stimulus interval. This prestimulus mean HR was then subtracted from each of the 20 mean HRs, which resulted in 20 difference scores.

The first analysis included HR changes occurring during three blocks of trials. Mean HR changes for the first 10 trials were obtained for the 20 difference scores, and then a mean score was calculated from these 20 difference scores. The same procedure was used to calculate a mean HR score for the second and third blocks of 10 trials. The three mean scores were included in a 3 × 2 × 3 Winer (1962) Case II design where the variables represented types of cardiac information, reinforcement conditions, and blocks of trials, respectively.

Results of the analysis showed that the cardiac information variable was significant, $F(2/54) = 9.07$, $p < .01$. Ss who received augmented information ($\bar{X} = 1.30$) showed higher HRs than Ss receiving no specific HR information ($\bar{X} = -1.07$), $t(54) = 4.02$, $p < .001$. The groups receiving specific HR information ($\bar{X} = 0.86$) also showed higher HRs than the no specific HR information groups, $t(54) = 3.27$, $p < .005$. No differences in HR were found between the augmented and specific HR information groups, $t(54) = 0.74$, $p > .5$. It is of interest that no significant differences were found between the two reinforcement conditions, $F(1/54) = 0.02$, $p > .5$. Also noteworthy is the absence of HR increases as a function of blocks of trials. No statistically significant interaction between the cardiac information and reinforcement variables was found, $F(2/54) = 0.53$, $p > .5$.

A second analysis, similar to the first, was performed where the 20 mean HRs representing the 20 sec stimulus interval were obtained for all 30 trials. From these 20 scores, four mean scores were calculated which represented the first, second, third, and fourth 5 sec intervals of the 20 sec stimulus period. The four means were then included as the repeated measure in a 3 × 2 × 4 Case II design.

The results of this analysis are shown in Fig. 1. The cardiac information variable was found to be significant, $F(2/54) = 8.81$, $p < .001$. The augmented information groups ($\bar{X} = 1.28$) showed higher HRs than the no specific HR information groups ($\bar{X} = -1.05$), $t(54) = 3.01$, $p < .001$, and the specific HR

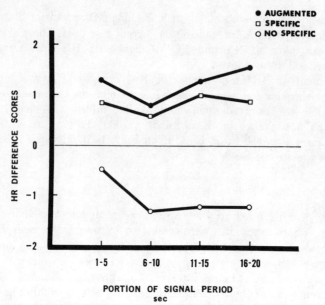

Fig. 1. Changes in heart rate for the cardiac information groups within the signal period during training.

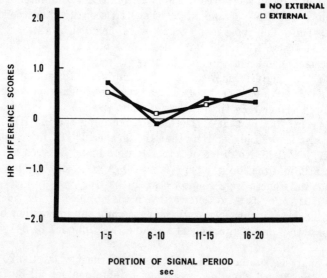

Fig. 2. Changes in heart rate for the reinforcement groups within the signal period during training.

information groups ($\bar{X} = 0.86$) showed greater increases in HR than the no specific HR information groups, $t(54) = 3.21$, $p < .005$. No differences were found between the specific HR and augmented HR information groups, $t(54) = 0.70$, $p > .5$. The interaction between cardiac information and the reinforcement conditions was not statistically significant, $F(2/54) = .51, p > .5$. Illustrated in Fig. 2 are the mean scores for the external compared with the no reinforcement conditions. No differences in HR were found between these two groups, $F(1/54) = 0.13, p > .5$.

Scoring of HR for the testing period was identical to the scoring of HR for the training period. One analysis compared HR increases during the training period with HR increases displayed during testing, and another analysis viewed the testing period alone. Results of the analyses indicated that the same increases in HRs evidenced during the training period for the various groups were displayed during the testing period. Relative to the training period, there were no significant increases in HR due to possible learning, or decreases in HR due to the removal of external reinforcement or augmented information.

Respiration

Respiration amplitude was measured by taking the peak to trough distance of an inspiration cycle. For scoring purposes, the 20 sec stimulus interval was divided into four intervals of 5 sec each. Mean amplitude scores for each interval were obtained by averaging whatever number of amplitude peaks fell within a 5 sec interval. Typically, one respiration cycle occurred during each 5 sec interval, although sometimes a mean was obtained from two respiration cycles falling within one 5 sec interval.

A prestimulus respiration amplitude mean was obtained by averaging respiration amplitudes occurring during a 10 sec period preceding each stimulus interval. This prestimulus mean was then subtracted from the stimulus means, resulting in four difference scores for each trial. These four difference scores were the scores from which all of the analyses to be discussed were derived.

Amplitude means for the first, second, third, and fourth 5 sec intervals of the 20 sec stimulus interval were averaged over the 30 trials and included in a $3 \times 2 \times 4$ analysis. Results of this analysis showed no difference in amplitude for the cardiac information variables, $F(2/54) = 1.12, p > .5$. The absence of respiration changes here suggests that the HR differences observed between these information groups were not due to changes in respiration amplitude. Significant differences were found for the reinforcement variable. The external reinforcement Ss ($\bar{X} = 0.43$) showed larger increases in amplitude than the no reinforcement Ss ($\bar{X} = -0.67$), $F(1/54) = 3.51, p < .05$. Respiration differences found between these reinforcement groups are interesting in light of finding no differences between these groups in HR increases. The mean amplitude scores for these reinforcement groups are illustrated in Fig. 3.

In the above analysis, a significant interaction was found between the reinforcement conditions and the 5 sec intervals, $F(3/162) = 4.03, p < .01$. No differences were found for the reinforcement variable during the first and second 5 sec intervals. However, during the third interval, the external reinforcement scores ($\bar{X} = 0.61$) were higher than the no reinforcement scores ($\bar{X} = -0.76$), $q(3/162) = 3.17, p < .05$, and the former group ($\bar{X} = 0.55$) again showed higher scores than the latter group ($\bar{X} = -0.85$) during the fourth interval, $q(3/162) = 3.23, p < .05$.

Mean scores for the four 5 sec intervals of the 20 sec stimulus interval for the

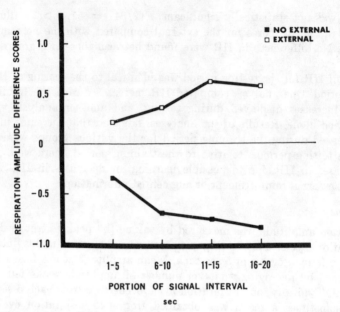

Fig. 3. Changes in respiration amplitude for the reinforcement groups within the signal period during training.

10 testing trials were included in a 3 × 2 × 4 analysis. Results of this analysis indicated significant differences found for respiration amplitude across the 5 sec intervals, $F(3/162) = 4.64$, $p < .01$. Amplitude scores during the first interval ($\bar{X} = 0.22$) were greater than during the third interval ($\bar{X} = -0.35$), $q(4/162) = 4.51$, $p < .01$, and greater than during the fourth interval ($\bar{X} = -0.33$), $q(4/162) = 4.35$, $p < .05$. Of interest here are the significant decrements in respiration amplitude across time within the stimulus interval, while HR increases were maintained in a constant fashion for the groups which showed HR increases during the testing period.

Skin Resistance Level

Skin resistance level scores in ohms were obtained by taking the differences in level between the first and last trial from each of the four blocks of trials. Analyses using level difference scores were performed for the training and testing periods separately and taken together.

The only analysis showing significant differences was for the testing period for the reinforcement variable, $F(1/54) = 5.56$, $p < .05$. The no reinforcement Ss ($\bar{X} = 6.59$) showed larger resistance changes than the external reinforcement Ss ($\bar{X} = 2.88$). In other words, the external reinforcement Ss showed less adaptation or were more aroused than no reinforcement Ss. Of interest in these analyses was the absence of significant differences in skin resistance level for the various types of cardiac information groups, $F(2/54) = 1.38$, $p > .5$.

Number and Amplitude of GSRs

A GSR was operationally defined as a 1000 ohm or greater decrease in skin resistance level occurring within the 20 sec stimulus interval, and the frequency of these responses was included in analyses similar to those analyses used for HR and respiration.

The only significant difference found in these analyses was the progressive decrease in the number of GSRs across blocks of trials which can be attributed to adaptation, $F(2/108) = 22.89, p < .01$.

Analyses of GSR amplitude also showed adaptation changes with progressive decreases in amplitude found across blocks of training and testing trials, $F(2/108) = 11.70, p < .01$.

Post-Experimental Questionnaire

Questionnaire responses from Ss in the augmented and specific HR information groups were recorded, and a frequency count was tabulated for each item on the questionnaire. Responses to the questionnaire indicated that Ss tried to increase their HR in different ways. When Ss attempted to increase their HR by thinking certain thoughts, the most frequently reported thoughts were categorized as exciting (40%), frightening (58%), enjoyable (39%), and physically exerting (40%). Ss showing HR increases also more often employed covert verbalizations such as cheering or rooting for increases in HR (80%) than Ss showing no changes in HR (50%), but these differences were not statistically significant.

Discussion

Results from this study suggest that augmented and specific HR information facilitates the raising of HR while the no specific HR information situations do not contribute to raising HR. External reinforcement also did not enhance raising HR under the present experimental conditions. These findings do not discount other types of information which might also influence HR changes, nor are they necessarily relevant to the control of other types of autonomic responses.

Analysis of HR across blocks of training and testing trials indicated that the increases in HR when displayed remained about the same as a function of trials. Perhaps these results are due to using less than the sufficient number of trials which may be necessary to produce progressive increases in HR.

The inability of Ss in the no specific HR information group to increase HR supports the experimental hypothesis that awareness of the learning contingency is an important variable in raising HR. The suggestion by Engel and Hansen (1966) that knowledge of the learning contingency interferes with changing HR appears applicable only to HR decreases, since in the present study awareness of the criterion response *enhanced* HR increases. All four of the experimental groups aware of the criterion response showed increases in HR, while both of the unaware groups displayed no increases.

The difficulty in raising HR displayed by Ss receiving no specific HR information is particularly interesting for the Ss who received external reinforcement for increases in HR. This group is similar to groups used by Brener and Hothersall (1966, 1967) who report that Ss were capable of raising HR without being aware of the response of interest. Several hypotheses might account for the disparity in results found between the latter and present study.

First the blue light, which indicated to the Ss that they were being successful in their task, did not appear to have reinforcing value. Perhaps the strength of this reinforcement was not sufficient to enhance raising HR, and some other form of reinforcement might have been more effective.

For some Ss 5% increases in HR were due to HR increases occurring as part of sinus arrhythmia. Consequently, although 5% increases in HR were reinforced, so were increases in inspiration amplitude simultaneously reinforced, which might account for the increases in respiration amplitude for the external reinforcement groups.

A third hypothesis accounting for the differences between the reinforcement groups deals with the nature of the instructional sets. In this study Ss in the reinforcement groups had to concentrate on their HR (internal responses) and also attend to the blue light. Perhaps having to attend to both internal and external cues made the processing of both sets of information too difficult a task for Ss.

Finally, some of the subjects in the external reinforcement groups were aware of their inability to increase HR by virtue of the observed decrement in the number of times the blue light was displayed. Consequently these Ss may have reverted to using other means of increasing HR which altered level of arousal or changed their breathing pattern. Since the blue light was displayed when HR increased, independently of the manner in which HR increased, Ss continued to use other means which consequently led to the reported differences in respiration amplitude and skin resistance level between the two reinforcement groups.

The precise way in which Ss raised their HR remains an open question. Subjects tried various methods to increase HR, none of which systematically differentiated successful from unsuccessful attempts to raise HR. Methods of increasing HR by altering general arousal level, by using covert verbalizations, or by producing muscular changes of small magnitude have been discussed previously in detail (Bergman & Johnson, 1971).

The failure to find systematic ways in which Ss successfully raised HR suggests that many approaches may have been employed. Some Ss may have been successful with some methods, while other Ss used different methods.[2] If Ss did successfully use a variety of different methods, then future studies should be directed toward finding the most effective and unique ways in which a S can raise HR. Once these subject-specific methods are determined, then the introduction of types of augmented and specific HR information, and forms of reinforcement which are relevant to the S, will probably facilitate increasing HR.

REFERENCES

Ascough, J. C., & Sipprelle, C. N. Operant verbal conditioning of autonomic responses. *Behavior Research & Therapy*, 1968, *6*, 363–370.

Bergman, J. S., & Johnson, H. J. The effects of instructional set and autonomic perception on cardiac control. *Psychophysiology*, 1971, *8*, 180–190.

Brandt, K., & Fenz, W. D. Specificity in verbal and physiological indicants of anxiety. *Perceptual & Motor Skills*, 1969, *29*, 663–675.

---

[2] A variable thought to be related to increasing HR was the way in which a S responded on questionnaires reported to measure autonomic perception (Mandler & Kremen, 1958) and autonomic and somatic anxiety (Brandt & Fenz, 1969). No relationship was found between degree of increasing HR and high, middle, or low scores on these questionnaires.

Brener, J., & Hothersall, D. Heart rate control under conditions of augmented sensory feedback. *Psychophysiology*, 1966, *3*, 23-28.

Brener, J., & Hothersall, D. Paced respiration and heart rate control. *Psychophysiology*, 1967, *4*, 1-6.

Brener, J., Kleinman, R. A., & Goesling, W. J. Effects of different exposures to augmented sensory feedback on the control of heart rate. *Psychophysiology*, 1969, *5*, 510-516.

Donelson, F. E. Discrimination and control of human heart rate. Unpublished doctoral dissertation, Cornell University, 1966.

Engel, B. T., & Chism, R. A. Operant conditioning of heart rate speeding. *Psychophysiology*, 1967, *3*, 418-426.

Engel, B. T., & Hansen, S. P. Operant conditioning of heart rate slowing. *Psychophysiology*, 1966, *3*, 176-187.

Frazier, T. W. Avoidance conditioning of heart rate in humans. *Psychophysiology*, 1966, *3*, 188-202.

Lang, P. J., Sroufe, L. A., & Hastings, J. E. Effects of feedback and instructional set on the control of cardiac-rate variability. *Journal of Experimental Psychology*, 1967, *75*, 425-431.

Levene, H. I., Engel, B. T., & Pearson, J. A. Differential operant conditioning of heart rate. *Psychosomatic Medicine*, 1968, *30*, 837-845.

Mandler, G., & Kremen, I. Autonomic feedback: A correlational study. *Journal of Personality*, 1958, *26*, 388-399.

Shearn, D. W. Operant conditioning of heart rate. *Science*, 1962, *137*, 530-531.

Winer, B. J. *Statistical principles in experimental design*. New York: McGraw-Hill, 1962.

17
Some Experiments on Instrumental Modification of Autonomic Responses

Keiichi Hamano and Tsunetaka Okita

In this paper we attempted to make a summary of some experiments on the instrumental modification of autonomic responses to date made by the present authors. Studies were done on heart rate (HR) and GSR. In the rewarding situation, the reinforcer such as tone, light or slide of a nude female in color was presented, while the noxios electric stimulus as UCS was applied in the avoidance situation. In addition, in order to examine the effect of drive strength on the modifiability, the subject was motivated by water deprivation in two instrumental rewarding experiments. Finally some questions to be determined by future studies were pointed out.

The purpose of this paper is to summarize experiments by the present authors on the instrumental modification of autonomic responses in humans. This paper is divided into four parts. The first part presents an introduction to the problem. The second part reviews some experiments on rewarding situation, and the third part those on avoidance situation. The final part presents some summarizing remarks regarding the work by the present authors.

INTRODUCTION

One of the persistent problems in the area of conditioning and learning has been the question of whether or not an autonomic response can be modified through the instrumental as well as the classical paradigm. The question has been in circulation since the 1930's, but the experimental evidence for a long time has been limited to results of two studies (Mowrer, 1938; Skinner, 1938). Skinner, in collaboration with Delabare, attempted to condition the vasoconstriction of the arm in humans by making a positive reinforcement depending upon the vasoconstriction. Mowrer also tried to modify the GSR instrumentally in a shock-avoidance situation. The results of these two pioneering studies failed to demonstrate the instrumental modification of autonomic responses. But for some reason, since the early evidence, investigators gave strangely little attention to this question during the next twenty years, merely accepting their seniors' conclusions. Traditionally many learning theorists have differentiated between these two types of conditioning, not only on the basis of experimental operations, but on the assumption that dif-

---

* Notre Dame Seishin University, 700 Okayama, Ifuku-cho 2–16–9, Japan.
** Department of Psychology, Kwansei Gakuin University, 662 Nishinomiya, Uegahara, Japan.
*** The critical comments of Prof. Y. Kotake and Prof. Y. Miyata were invaluable in preparing the final version. The authors desire to express here both their indebtedness and their thanks for their generous services. Also they are grateful to Prof. S. Miyake for helpful advices. Finally they wish to thank Prof. M. Frances for lending her talents to their English manuscript.

Reprinted with permission of the Psychologia Society and the author from *Psychologia*, 1972, Vol. 15, 101-109.

ferent fundamental psychological and physiological mechanisms are involved. They have believed that because an autonomic response does not act directly on the external environment, the response is incapable of being modified instrumentally.

In the early 1960's, though, the situation greatly changed. Some positive reports on the instrumental modification of autonomic responses began to appear. Kimmel and Hill's study which was made public in 1960 was the very first successful one on humans, while the first successful experiment on animals was Fromer's in 1963. In the past decade quite a few successful experiments using a variety of autonomic responses have been published, and their results have become a center of attraction in recent years (Hamano et al., 1970; Katkin & Murray, 1968; Kimmel, 1967; Miller, 1966; Miyata & Hamano, 1967).

Such attempts at modifying autonomic system instrumentally also have had deep implications for the clinical as well as the experimental and theoretical levels. The findings that the autonomic responses may be influenced by something like rewards in the external environment have documented the remarkable sensitivity of autonomic function to stimuli of which it is remained ignorant, and provided a basis for finding better explanations of the mechanism of occurrence of psychosomatic disorders which are generally believed to be related to the theories of emotionality.

But the study on this topic in Japan did not begin until 1968. Then it was planned and executed the series of studies which are now to be described. Most of them were carried out at the Osaka Prefectural Institute of Public Health.

Rewarding Situation

The first effort was an historical review of this topic by Miyata and Hamano (1967) tracing back the efforts to date made chiefly by Western psychologists. The entrance into the experiment came in the spring of 1968 when Hamano conducted a preliminary study which was attempted to modify instrumentally as preparatory steps, to examine some attributes of spontaneous autonomic response itself. The experiment was designed to investigate the habituation of spontaneous autonomic response over time and the effect of stimulation on it. As a result, it was found that the frequency of spontaneous response is influenced by the passage of time or stimulation (Hamano, 1969).

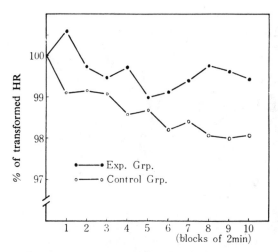

Fig. 1. Percentage of mean transformed HR during 20 min. reinforcement period.

Based upon these fundamental data, Hamano and Miyake started in the same year to determine the possibility of instrumental modification of spontaneous HR acceleration using a visual reinforcer. Twenty Ss were divided into the two groups. After a 15-min. rest, the experimental group received a dim light following each instantaneous HR acceleration during a 20-min. reinforcement, then 10-min. of extinction. The control group received the same number of lights per min. during reinforcement when no HR acceleration was occuring. The results were somewhat encouraging, but it was difficult to be completely sure that the autonomic response could be modified through the instrumental paradigm. Although the group differences were highly reliable statistically, the HR of the experimental group did not increase so remarkably as to go over the resting level as shown in Fig. 1. In this respect it was suggested that the strength of the reinforcer may be a determinant of the instrumental modifiability of autonomic response by reinforcer. The main reason for making this statement was that a dim light as a reinforcer used in this experiment was supposedly not very effective in modifying the autonomic response (Hamano & Miyake, 1971).

In order to investigate the assumption, Hamano and Miyake in 1969 proceeded to modify spontaneous GSR instrumentally using slides of a nude female in color as more effective reinforcers. Twenty-eight Ss were divided into 3 groups. Each S was subjected to two different processes of reinforcement within the same individual. In the case of Group I, the reinforcer was given when spontaneous GSR occurred during the initial period of reinforcement of 16-min., while, during the latter period of reinforcement of 16-min., on the contrary, the slide was given when the GSR did not occur. Group II was reinforced in an order inverse to that of Group I. Group III was left under no stimulation through the entire process of the experiment. As evident in Fig. 2, the results obtained showed that, in the initial period of reinforcement, GSR had the greatest frequency of occurrences in the case of Group I, followed by Group II and Group III. In the latter period, on the other hand, Group I was seen to have the decreased frequency of occurrences as compared with that of the initial period. The findings were interpreted as supporting the assumption stated above (Hamano & Miyake, 1972 a). Moreover, in order to confirm

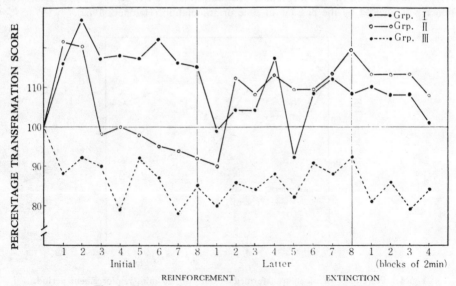

Fig. 2. Percentage of mean transformed GSR frequencies.

this more clearly, Tanaka (1970) conducted a supplementary experiment. He employed GSR and divided 24 Ss into 4 groups, namely the group to which some slides of a nude female were used as reinforcers (SE group), the group to which light was presented as reinforcer (LE group), and the control group to each of the above, and then tried to modify GSR instrumentally. As a result, he found that SE group had a tendency to show more frequent occurrences of GSR than LE group as shown in Fig. 3, although the difference between SE and LE groups was not statistically significant (Tanaka, 1970). And also Okita is attempting to modify the HR, using money as a reinforcer at present. In the experiment, 2 Ss were given some money by changing their HRs toward appropriate bidirection during a CS presentation. The result suggests that the modification is larger than the previous findings in HR.

Taking a step forward in order to obtain more detailed consideration of the evidence, Hamano and Miyake (1971) tried to modify the HR instrumentally with 3 Ss under three different conditions of water deprivation in order to investigate any difference in the modifiability as a function of time of water deprivation. Each of 3 Ss was put under water deprivation for 24 hours, 12 hours or 6 hours, and the experiment was intermittently performed over a period of 2 days. On each of the two days of the experiment, training in HR speeding and HR slowing were run separately from one another. A procedure similar to a shaping schedule in the conventional instrumental conditioning study was employed. A weak tone of 1000 cps instead of water was presented for 1 sec. as a reinforcer in the experiment. The tone was explained as having the meaning of a reward in the pre-experimental instruction in which every S was told that he would be forced to have water deprivation for another 4 hours even after the experiment, that the tone might be presented during the experiment, and that a presentation of tone would mean reduction of the post-experimental 4-hour water deprivation by 1 min. In order to estimate physiological conditions of water deprivation in Ss, the experimenters collected 2 cc of blood and salivation from Ss before and after the experment. The blood was analysed for water changes in blood by means of the hematocrit method.

The experiment showed that there was an instrumental modification of HR toward

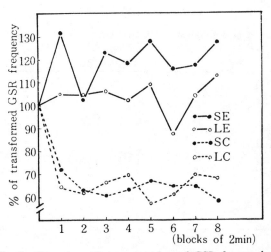

Fig. 3. Percentage of mean transformed GSR frequencies.

bidirection in all Ss, even though the degree of modifiability in the slowing training was inferior to that in the speeding one. The difference in the modifiability at the drive level was not markedly observed in the process of training. As a whole, however, there occurred differences in the modifiability as a function of the drive level as shown in Fig. 4 (Hamano & Miyake, 1971).

In the same year Hamano and Miyake further attempted to modify the HR over a period of 2 days under the same conditions by means of the discrimination training. An extremely weak white light of 25 cps or 5 cps was used as a CS. Both lights were presented for 1 min. The former was employed as a CS for the speeding trial, and the latter for the slowing trial. Each of the lights was presented at random 40 times up to a total of 80 times. It was found that the most favorable discriminative modification was observed in the Ss under the 24-hour water deprivation, followed by the 12-hour Ss and lastly by the 6-hour S as in Fig. 5. These findings are only inconclusive ones, but they seem to indicate the energizing effect of drive (Hamano & Miyake, 1972 b).

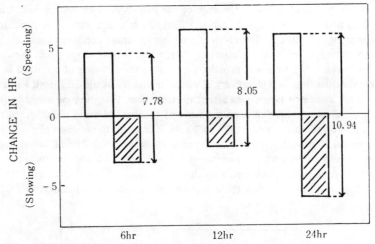

Fig. 4. Comparison of instrumental modifiability.

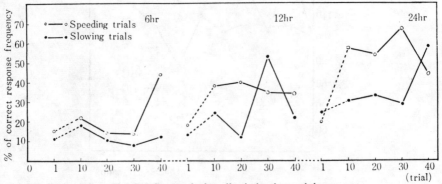

Fig. 5. Curves during discriminative training.

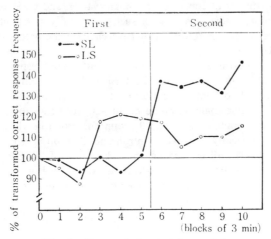

Fig. 6. Curves during training period.

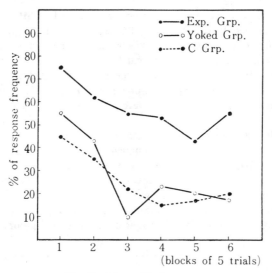

Fig. 7. Mean OR frequency.

Then, in an attempt to study how much the modifiability might be affected by variation of the amount of reinforcer used as a reward, Kitano (1971) picked up the problem of intensity variables. The experiment was based on a within-Ss design. He used a light as a reinforcer, and examined the modifiability of the HR by varying the amount of the reinforcer in the first half and second half of the training period. The results are shown in Fig. 6. The LS group was reinforced with a large amount of reinforcer during the first half of the period, and with a small amount of reinforcer during the second half. The SC group was reinforced in the reversed order. As evident in Fig. 6, there were differences noted in the modifiability of the HR in accordance with the quantitative change in the reinforcer. This is, of course, a preliminary study, and there are several questions to be solved by future studies. As one of such questions, a possibility remains that the quantitative change of the reinforcer might have changed the nature of the reinforcer itself (Kitano, 1971).

A study on the instrumental modification of autonomic responses was made from a different point of view. Using the HR, Hyoda (1971) investigated the instrumental modification of the OR to light. The purpose of investigation was to see whether such a modification of the autonomic response might actually be attributable to a contingent effect. Twenty-four Ss were divided into 3 groups; E group in which the reinforcer was given to the OR contingently, its yoked control, and C group in which the reinforcer was not given to the OR. In all the groups, the habituation of the OR over time occurred gradually, but the habituation was of the least degree in the E group, with a statistically significant difference from the other two groups as shown in Fig. 7 (Hyoda, 1971).

Avoidance Situation

Studies in an avoidance situation have been undertaken and continued mainly by Okita since 1970. The results in this situation reported by other investigators were contradictory, so he started to search out the factors causing such a discrepancy.

Okita (1971) began to see whether or not the elicited GSR can be influenced by reinforcement in an avoidance situation. The experimental design was replicated the previous study, with some minor procedural changes; (a) systematic biasing effects in the yoked control design were controlled by matching the members of yoked pairs in conditionability, (b) the criterion GSR avoiding the UCS was suited to each S. The results showed that the experimental subjects' response frequency was significantly higher than the control one as shown in Fig. 8. However, he could not obtain clear differences in magnitude.

Next, after due consideration of the general argument on the control treatment in this sort of experimental design, Okita conducted the experiment with the bidirectional control design using HR as a bidirectional response. He divided 14 Ss into 2 groups to which the UCS was omitted, namely those whose HR increased, and those whose HR decreased. Ten adaptation trials of CS were followed by 30 acquisition trials and then 10 extinction trials of CS alone. As indicated in Fig. 9, HR of the decreased group had a tendency to decrease more rapidly than that of the increased group. Those findings in

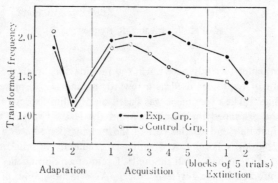

Fig. 8. Mean response frequency.

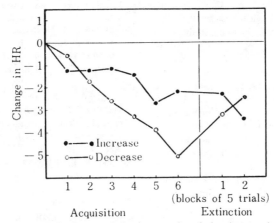

Fig. 9. Mean HR change of the increased and the decreased group.

the avoidance situation were somewhat ambiguous and not a clear-cut evidence of the autonomic avoidance conditioning. However, he is now planning to design a further experiment that will be more rigorously unequivocal (Okita, 1971).

SUMMARIZING REMARKS

Some experiments on the instrumental modification of autonomic responses to date made by us have been summarized and interpreted leading to the conclusion that (a) an autonomic response can be modified by instrumental paradigm, although much still remains to be done about this topic and (b) the instrumental modifiability of autonomic response may be influenced in some degree by the strength of reinforcer. Therefore, the drive variables, the investigation of the analogy among such studies and the more conventional instrumental conditioning of skeletal response, and the better theoretical structure regarding its mechanism will be pointed out as the main problems toward which future studies should be directed.

REFERENCES

Hamano, K. 1969. Studies on modification of spontaneous autonomic activity: I. Preliminary report on habituation of spontaneous heart rate with concomitant measurement of three autonomic variables. *Jap. psychol. Res.*, 11, 110–116.

Hamano, K., & Miyake, S. 1971. Studies on modification of spontaneous autonomic activity: II. Modification of heart rate change through training procedure of instrumental conditioning. *Jap. psychol. Res.*, 13, 19–25.

Hamano, K., & Miyake, S. 1971. Modification of autonomic response through training procedure of instrumental conditioning (III). The paper read at the 35th Annual Meeting of Jap. Psychol. Association.

Hamano, K., & Miyake, S. 1972 a. Studies on modification of spontaneous autonomic activity: III. The relation of strength of reinforcer to instrumental modifiability of autonomic response. *Jap. psychol. Res.*, 14, 38–42.

Hamano, K., & Miyake, S. 1972 b. Modification of autonomic response through training procedure of instrumental conditioning (IV). The paper prepared for the 36 th Annual Meeting of Jap. Psychol. Association.

Hamano, K., Okita, T., & Miyata, Y. 1970. Instrumental conditioning of autonomic responses. *Jap. psychol. Rev.*, **13**, (In Japanese Pp. 244–263 with English abstract P. 264).

Hyoda, M. 1971. A preliminary study on instrumental conditioning of autonomic response. Bachelor thesis, Kwansei Gakuin University (In Japanese).

Katkin, E. S., & Murray, E. N. 1968. Instrumental conditioning of autonomically mediated behavior: Theoretical and methodological issues. *Psychol. Bull.*, **70**, 52–68.

Kimmel, H. D. 1967. Instrumental conditioning of autonomically mediated behavior. *Psychol. Bull.*, **67**, 337–345.

Kitano, Y. 1971. The effect of the amount of reward on instrumental conditioning of autonomic response. Bachelor thesis, Kwansei Gakuin University (In Japanese).

Miller, N. E. 1966. Experiments relevant to learning theory and psychopathology. Moscow.

Miyata, Y., & Hamano, K. 1967. Can the autonomic response be trained through the operant paradigm? *Humanities Review: Kwansei Gakuin University*, **18**, 1–18.

Mowrer, O. H. 1938. Preparatory set (expectancy)—A determinant in motivation and learning. *Psychol. Rev.*, **45**, 62–91.

Okita, T. 1971. Avoidance conditioning of autonomic responses in man. *Jap. psychol. Res.*, **13**, 131–138.

Skinner, B. F. 1938. *The behavior of organism: An experimental analysis.* New York: Appleton.

Tanaka, R. 1970. A preliminary study on instrumental conditioning of GSR. Bachelor thesis, Kwansei Gakuin University (In Japanese).

Large Magnitude Heart Rate Changes in Subjects Instructed to Change their Heart Rates and Given Exteroceptive Feedback

Joseph H. Stephens, Alan H. Harris, and Joseph V. Brady

ABSTRACT

Four subjects who were given exteroceptive, auditory and visual feedback, and were asked to raise and to lower their heart rate on signal, were able to produce large magnitude changes in both directions. The fact that some of these changes occurred immediately suggests that feedback may not be as important as some authors have suggested. Respiratory changes and changes in muscle tension did not appear to be mediators.

DESCRIPTORS: Voluntary control, Heart rate, Exteroceptive feedback. (J. H. Stephens)

This paper reports on 4 subjects who were able to produce large magnitude heart rate (HR) changes consequent to a feedback procedure that will be described. The 4 subjects were drawn from a sample of 25 used in a study designed to investigate the interaction between ability to change HR during the performance of concurrent motor tasks superimposed on this procedure. Details on the other 21 subjects relevant to the present study will be found in a forthcoming paper.

In the majority of studies on conditioned HR control reported in the literature, the subjects have not been informed that HR was being measured. In our study, we have not attempted to rule out cognitive mediation of HR change but instead have tried to maximize it. Murray (1968) has suggested that voluntary control through conscious mental activity may be a more crucial mediator of HR change than some of the reinforcement contingencies utilized to this end. Our subjects have accordingly been asked to try to change their HR on signal by "purely mental means." They have also been given visual and auditory feedback of their performance and been paid according to their success.

This research was supported by the Office of Naval Research, Sub-contract No. N0014-70-C-0350.
Address requests for reprints to: Joseph H. Stephens, M.D., Associate Professor of Psychiatry, The Henry Phipps Psychiatric Clinic, Johns Hopkins Hospital, Baltimore, Maryland 21205.

We are aware of only two studies reporting large magnitude HR changes comparable to those we obtained on our 4 subjects. In one of these (Frazier, 1966), avoidance conditioning with an electric shock given randomly during a visual display task was employed. The visual display without the shock later produced increases in HR of up to 35 bpm which were sustained for up to 30 min. In the other study (Headrick, Feather, & Wells, 1971), a single subject was asked to change his HR and was given visual and auditory feedback. With this procedure, which was similar to ours, HR increases of as great as 44 bpm were obtained in the subject, although decreases of only 5 bpm from the baseline were obtained.

Procedure

All Ss were tested in a reclining position in a sound-proof room. Attached to them were physiological recording devices leading to a Beckman Offner polygraph in an adjoining room. Electrocardiogram (EKG), HR, respiration, and palmar skin potential were recorded. Visual feedback to Ss was provided by a meter giving continuous HR in bpm. Auditory feedback consisted of a moderately soft tone which sounded with each heart beat and varied in pitch as a function of instantaneous HR. As HR increased, so did the pitch of the tone.

Ss were asked to raise their heart rate above a criterion number whenever a red light, displayed adjacent to the heart rate meter, was illuminated. When a green light was displayed, Ss were instructed to lower their heart rate below the criterion number. For each second during which the S met the criterion, a yellow light was also displayed, and a counter accumulated the number of seconds the yellow light was illuminated. Ss were paid ½ cent for each second during which the yellow light was on. Parts of this procedure were modifications of the one described by Engel and Chism (1967).

All Ss were measured during a 15 min base period (before any lights were displayed), as well as during a 10 min post period. The number and duration of "Red" and "Green" sessions varied for each of the 4 Ss as described below.

Results

Subject A

This 24-year-old male medical student took part in the study 17 times at weekly intervals. For the first 4 sessions, the red light was presented 4 times for 5 min, and the green light 4 times, also for 5 min. The intertrial intervals were also 5 min each, and the sequence of lights was random. During the red light periods the HR rose an average of 23 bpm. The resting rate and intertrial rate mean was 91 bpm. This ability to increase the HR was noted at the very first presentation of the red light, and instantaneous increases in HR remained unchanged throughout these first 4 sessions. Furthermore, the increased HR was sustained throughout the red light period. By contrast, the HR during the green light periods did not differ from the intertrial intervals during the first 2 sessions. However, there was a mean decrease of 2 beats during the third session, and 5 beats during the fourth session.

Subsequent to these 4 sessions, the trial periods were increased until red or green lights were presented for up to 30 minutes at a time. For the 30 min period, the HR was increased from the intertrial interval by a mean of 17 bpm with little variability throughout the period. The mean decrease during the 30 min period was 8 bpm. At the 17th and last session, the mean decrease in HR from the intertrial period was 22 bpm and the increase 24 bpm.

Subject B

This 20-year-old male subject took part in the study 18 times at weekly intervals. Because his resting mean HR was 48, no attempt was made at lowering it. For the first 13 sessions, the red light was kept on for 60 min at a time. During the first session there was an average decrease in HR during the red light of 1 bpm. After this first session the S was able to maintain elevated increases to the red light with means of from 12 to 23 bpm for the entire 60 min period.

During the last 4 sessions, the S was presented alternately with 5 min intervals of red light and 5 min of no light for a total of 7 red light periods in a session. During these 4 sessions, the mean HR during the red light intervals was 24 bpm higher than during the intertrial periods with a maximum increase of 34 bpm on Day 18, the last session.

Subject C

This 23-year-old male student was the brother of Subject B. His resting heart rate was 69, and an attempt was made both to raise and lower his HR. The S took part in the study 10 times at weekly intervals. He was given 5 min presentations of red light, 5 min presentations of green, and 5 min intertrial periods between the colored lights. The red and green lights were each presented 4 times at each session in a random order. At the first session, the S showed a mean increase over the intertrial rate of 7 bpm to the red lights and a mean decrease of 4 bpm to the green lights. All changes were immediate and were sustained. After 9 more similar sessions, the mean increase was 15 bpm for the red light with a 6 bpm decrease for the green light.

Subject D

This 20-year-old male student had a mean resting rate of 60 and no attempt at lowering his HR was made. The S took part in the study 12 times at weekly intervals. For the first 6 sessions, the red light was kept on for 60 min at a time. On the first day of the study, the S's HR rose immediately on the presentation of the red light and sustained an average increase of 16 bpm. In subsequent 60 min sessions, the greatest maintained increase was 21 bpm.

During the last 6 sessions, the red light was presented 8 times for 5 min at a time. Intertrial periods were also 5 min. Under this schedule the S achieved a mean increase of 29 bpm on the 9th day.

Mediating Factors

All Ss were instructed to breathe as normally as possible and to avoid muscle tension or movement during the study. Respiratory changes during attempts at lowering and raising HR were variable and followed no discernible pattern. There were no noticeable changes in muscle tension.

All Ss were questioned as to how they were able to bring about changes in HR. All agreed that they relaxed as much as possible when trying to lower their HR. To increase HR, all said they at first thought of exciting things but later simply concentrated on the red light and the actual HR. They denied voluntarily tensing their muscles, but no myographic determinations were made. Neither was any attempt made to pace respiration as had been done in previous studies (Brenner & Hothersall, 1967).

Commentary

This preliminary report suggests that certain subjects can voluntarily bring about rather large magnitude increases and decreases in their HR. The immediate, first day, response in several subjects suggests that feedback may not be as important as some authors have suggested. Brenner, Kleinman, and Goesling (1969) have suggested that subjects can control HR in the absence of feedback although they are more successful if feedback is provided. However, they report very small magnitude changes. Bergman and Johnson (1971) reported that subjects could increase or decrease their HR in the absence of externalized feedback although these changes were of only 2 or 3 bpm. It would appear that in this preliminary study of 4 subjects, the control of HR changes may have in part occurred independently of the feedback procedure employed to facilitate HR changes. Another paper in preparation will deal with the many variables related to this seemingly voluntary ability to change resting HR in the absence of motor activity and systematic variations in respiration.

JOSEPH H. STEPHENS,
ALAN H. HARRIS, AND
JOSEPH V. BRADY
Department of Psychiatry and Behavioral Science
Johns Hopkins University School of Medicine

REFERENCES

Bergman, J. S., & Johnson, H. J. The effects of instructional set and autonomic perception on cardiac control. *Psychophysiology*, 1971, *8*, 180-190.

Brener, J., & Hothersall, D. Paced respiration and heart rate control. *Psychophysiology*, 1967, *4*, 1-6.

Brener, J., Kleinman, R. A., & Goesling, W. J. The effects of different exposures to augmented sensory feedback on the control of heart rate. *Psychophysiology*, 1969, *5*, 510-516.

Engel, B. T., & Chism, R. A. Operant conditioning of heart rate speeding. *Psychophysiology*, 1967, *3*, 418-426.

Frazier, T. W. Avoidance conditioning of heart rate in humans. *Psychophysiology*, 1966, *3*, 188-202.

Headrick, M. W., Feather, B. W., & Wells, D. T. Unidirectional and large magnitude heart rate changes with augmented sensory feedback. *Psychophysiology*, 1971, *8*, 132-142.

Murray, E. N. Comment on two recent reports of operant heart rate conditioning. *Psychophysiology*, 1968, *5*, 192-195.

V

ELECTROENCEPHALOGRAPHIC CONTROL

Similar Effects of Feedback Signals and Instructional Information on EEG Activity

Jackson Beatty

BEATTY, J. *Similar effects of feedback signals and instructional information on EEG activity.* PHYSIOL. BEHAV. **9** (2) 151–154, 1972.—Normal human subjects may systematically modify the proportions of alpha and beta frequency activity in the occipital EEG in either of two conditions. They may be provided with information giving appropriate strategies for producing such changes or with only second by second feedback about their success in achieving the response criteria. Subjects provided with both did no better than subjects given prior information or feedback alone. Subjects who received no treatment, or who were given information inappropriately matched to the feedback signal, showed no systematic changes. The similarity in both the development and final magnitude of the differential responsiveness in the groups which learned emphasizes the difficulty in ascribing learned changes to the effects of feedback if subjects are also informed about the nature of the task.

EEG   Alpha rhythm   Instructions   Operant methods   Biofeedback

IT HAS BEEN argued that people can selectively control the frequency spectrum of their EEG if they are given a signal indicating their success in producing the desired EEG pattern [2, 3, 4, 10]. This procedure, often termed bio-feedback, has aroused much interest both within and outside the scientific community, since it seems to offer voluntary control of central neural activity. Various EEG patterns have been associated with psychologically defined states of consciousness [1, 6, 8, 12]. The promise of a non-behavioral index of mental state which may be voluntarily controlled makes bio-feedback techniques attractive tools for research on consciousness as well as for clinical applications. Similar shifts of EEG spectra, however, result from simple changes in cognitive set. Orienting to novel stimuli, solving mental problems, frustration and tension tend to produce electroencephalographic desynchronization or beta activity [8]. Conversely, alpha band activity is often reported to increase during states of calmness and meditation [1, 6, 12], although sometimes it may be triggered by cognitive activity [9]. Therefore, when evaluating the effects of feedback on autoregulation of the

---

[1]This research was supported by the Advanced Research Projects Agency of the Department of Defense and was monitored by the Office of Naval Research under Contract N0001-70-C-0350 to the San Diego State College Foundation. Thanks are expressed to K. Roberts for helpful conversations on the general role of instructions in psychological experiments, and to Linda Weekley, Karen Sabovich, and Carl Figueroa who participated in various phases of this study.

Reprinted with permission of Microforms International Marketing Corp. from *Physiology and Behavior*, 1972, Vol. 9.

EEG, care must be taken to dissociate the effects of informational feedback from possible effects of the expectations and the cognitive set of the subject. Subjects drawn from the university community may now be especially aware of the relationship between alpha wave activity and mediational procedures. Thus, if subjects are told that the EEG patterns under study include alpha band activity, the possible influence of the subject's set cannot be excluded as an explanation of any observed changes. Beatty [2] tried to minimize this problem by not identifying the EEG patterns as alpha and beta band activity, falsely informing his subjects instead that the responses selected for modification were arbitrarily chosen from a large population of possible patterns. Nonetheless, significant changes in the probabilities of alpha and beta frequency activity were found as a function of reinforcement contingency.

The present experiment attempted to investigate explicitly the roles of prior information about the psychological correlates of EEG spectra and of informational feedback on the autoregulation of the spectra of the occipital EEG. By comparing the performance of subjects who did and did not receive such instruction, and the performance of subjects who were and were not provided with informational feedback, taken in all combinations, the effects of prior information and EEG based feedback can be systematically evaluated.

METHOD

Forty-five UCLA undergraduate students served as subjects, thereby partially satisfying the requirements of an introductory psychology course. Subjects were assigned to this experiment without detailed prior knowledge of its nature, minimizing problems of subject self-selection.

Occipital EEG was recorded from position 0_z of the 10–20 system [5] referred to the right earlobe. The subject's left earlobe was grounded. The frequency response of the EEG amplifying system was essentially flat between 2 and 20 Hz, with 1/2 amplitude attenuation at 0.6 and 32 Hz. This signal was monitored on an oscilloscope and was available at the analog/digital converter of the computer.

Subjects were first told that they were participating in a study of brain wave activity. After the recording electrodes were attached, they were seated in an electrically shielded room with a low level of ambient lighting. They were asked to keep their eyes open and refrain from moving for a 300 sec period, during which the baseline spectra of their EEG was calculated.

Subjects were then assigned to 1 of 5 experimental groups where they were given an opportunity to show discriminative control of EEG spectra. All groups were approximately matched for baseline level of alpha frequency activity. The first 4 groups (Information and Feedback, Information Only, Feedback Only, and Control, i.e., no information and no feedback) constituted a factorial experimental design and were so analyzed. The fifth group, like the first, received both information and feedback, but the instructions for alpha production were coupled with feedback for beta activity and vice versa. Thus in these subjects prior information and experimental feedback were pitted against each other.

At this point the subjects in the various groups were instructed about the nature of their tasks. Those in the feedback conditions were instructed in words similar to the following: "While you have been sitting here, the computer has made a number of measurements of your EEG activity. You now have the opportunity to learn to control your own brain waves. The EEG is a complex waveform which may show many different patterns. From all these patterns we have arbitrarily selected two for today's study. Each second the computer will sample your EEG, looking for one of the two selected wave patterns. Before each testing period I will tell you which of the two patterns will be reinforced. During each trial either the "1" or "2" signal light will be on. Between trials the door will be opened, the "3" signal will be displayed and you may rest. During each trial when the appropriate pattern occurs, the loudness of the background tone will increase for 1 sec. Your job is to learn to produce the 2 kinds of wave patterns which will keep the tone louder in the 2 conditions. As before, keep your eyes open and refrain from moving during trials."

Subjects in the Information Only group were not provided with the feedback signal (the tone), but they were told that the responses being measured were alpha and beta frequency activity. In addition they received task instructions drawn from the literature which made the following points:

(a) Beta activity is related to the orienting response to novel stimuli, sensory or mental attentiveness, tension, aggravation, and frustration.

(b) Alpha rhythms are often said to be associated with feelings of calmness, pleasant relaxation, and increased inner awareness. However, some naive subjects report very different feeling states, such as sadness, while producing alpha rhythms.

(c) Relaxation and regular breathing sometimes facilitate production of alpha waves.

Subjects in the information and feedback groups received the combined set of instructions.

For the Feedback Only and the Control groups, no mention was ever made of alpha or beta wave patterns or the psychological activities associated with them. No subject in any group was ever given any information about his performance other than the feedback signal, if present. All subjects were told they would receive an extra hour of experimental credit if they successfully doubled their baseline level on either index.

Subjects were then run for 8 experimental trials of 200 sec duration, which were divided into 4 blocks of 2 trials each. On one trial in a block indicated to the subject by a signal light displaying a 1, subjects in the Feedback groups were reinforced with an auditory signal for alpha band activity. Subjects in the Information groups were told to produce alpha wave activity, except for subjects in the mismatched condition, who were told to produce beta activity but were given alpha criterion reinforcement. On the other trial in the block an illuminated 2 was presented and beta frequency activity was reinforced. The ordering of each of the 2 trials within each block was randomized for each subject. Between blocks, subjects were forced to get up and walk around for a period of not less than 1 min. This procedure was used to counteract a growing drowsiness which subjects in psychological experiments often experience and to maximize baseline stability against which discriminative control of EEG spectra could be shown. The experimenter would converse with the subject if the subject desired, but no discussion of the experiment or related issues was permitted. All subjects were free to examine their own instructions if they chose to during these breaks between trial blocks.

The entire experiment was run under digital computer control. During a trial or baseline period, the EEG was randomly sampled each second for one complete wave, referenced to 0 potential and beginning with a positive deflection. The period of this wave was then measured and classified by the method of Legewie and Probst [7] as a single wave at X Hz. This probability measure converges upon the estimate of the power spectrum as computed by standard spectral methods [11]. For feedback groups, if the wave was within the criterion frequency band (8-12 Hz for alpha, 13 Hz or more for beta) the intensity of a quiet 400 Hz tone was immediately augmented for 1 sec. During that second, the random onset sampling and measurement procedure was repeated and if the sampled wave was again within the criterion bounds the tone intensity remained high for another second. During the baseline trial and for groups receiving no feedback, the tone was totally suppressed.

RESULTS AND DISCUSSION

For each subject the probabilities of obtaining an alpha and a beta wave were computed for the baseline period and each of the 8 experimental trials. Figure 1 presents the mean probabilities for each group by trial block and experimental condition.

Subjects may modify the probability of alpha wave activity (Fig. 1, top row) if they are provided with either appropriate prior information, appropriate feedback, or with both. If they are given neither treatment, or both treatments inappropriately matched, no significant changes appear. These conclusions are supported by analyses of variance (ANOVAs) for the probabilities of alpha wave in the initial and final blocks of the experiment. In the main ANOVA, the alpha probabilities in block 1 and block 4 were analyzed for the first four groups by trial block (1 or 4) and contingency (discriminative stimulus 1 or 2), prior information (present or absent), and feedback (available or unavailable). The main effect of contingency was significant ($F = 13.75$, 1 and 32 df, $p < 0.001$), indicating greater production of alpha wave activity during the trials in which animals attempted to produce alpha activity. A significant effect of trial ($F = 6.78$, 1 and 32 df, $p < 0.025$) reflected the greater probability of alpha waves at the end of training. The significant interaction of contingency and trial ($F = 15.57$, 1 and 32 df, $p < 0.001$) means the differential effects of treatment increased with trials. A significant triple interaction of prior information, contingency and trials ($F = 7.12$ with 1 and 32 df, $p < 0.025$) and a quadruple interaction of feedback, information, contingency, and trials ($F = 5.55$, 1 and 32 df, $p < 0.05$) suggest support for the conclusion that all three of the treatment groups (Feedback Only, Information Only, Feedback and Information) show a similar pattern of effects, which differs substantially from the no treatment control group.

To test for significant effects in the mismatched feedback and information group upon the probability of alpha, this group's performance in blocks 1 and 4 was compared with that of the no treatment control group in a second ANOVA. No significant main effect or interaction was found in that analysis. An examination of Fig. 1 shows that neither group increases either the absolute probability of alpha or the discrimination between conditions as a function of trials.

Parallel ANOVAs were carried out on the probabilities of beta activity (Fig. 1, bottom row). Desynchronization, like occipital alpha, may be modified by either appropriate prior information or feedback. From the main beta ANOVA, the

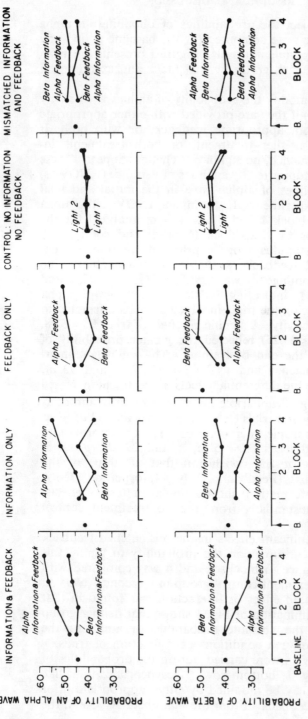

FIG. 1. Mean probabilities of obtaining an EEG wave of alpha or beta frequencies, by experimental conditions and trial. The top row of graphs shows the probability of obtaining an alpha wave. The initial point is the baseline probability. The remaining points plot the mean probability in each block under each of the two states of the discriminative stimulus. The groups which received either feedback, information or both properly matched show a discrimination between the two trials of the block. The discrimination improves with trials. This pattern differs from that shown by subjects who receive no treatment or conflicting prior information and task feedback. The bottom row of graphs plot the probability of observing a beta wave in the same subjects and the same trials. The same groups show discrimination and learning. Within each group which learned, the probability of beta activity is higher during beta contingency than alpha contingency trials.

only main effects reaching significance were Contingency (F = 18.26, 1 and 32 df, $p < 0.001$) and Trials (F = 4.31, 1 and 32 df, $p < 0.05$). As was the case with the probability of alpha activities, the interaction of Contingency and Trial was also significant (F = 8.03, 1 and 32 df, $p < 0.01$). The effects of Feedback and Information are shown in the significant triple interactions of Feedback, Information and Contingency (F = 4.90, 1 and 32 df, $p < 0.05$) and Feedback Information and Trial (F = 4.26, 1 and 32 df, $p < 0.05$). As with alpha probabilities, the second ANOVA comparing Mismatched Information and Feedback with control revealed no significant main effects or interactions.

The spectra of the occipital EEG may be systematically modified by subjects who either are provided with appropriate strategies for producing the desired changes but denied any information about their success in the task, or are given no prior information about appropriate strategies, but are given second by second information about their success.

Both procedures appeared to work equally well for the sorts of changes demanded of our subjects. The performance of the three groups with Information, Feedback and both Information and Feedback is very similar, not only in the absolute magnitude of the effect at the end of training, but also in the time course of the developing differences in responding to the discriminative stimulus. For all groups, the differences are minimal in the first trial block and increase with practice. This is somewhat surprising. One might have expected the group which received Information to have been provided with a relatively well articulated response strategy which would lead to large first trial differences. Further it was not expected that this group should show increasing differential responsiveness as a function of practice.

These similarities, plus the fact that the group receiving both Information and Feedback did not do strikingly better than the groups receiving only one or the other, suggest that all three groups were in fact performing similar tasks. Since the response strategies which were provided for the Information groups were simple and relatively dominant in a subject's behavioral repertoire, Feedback Only subjects could be expected to utilize similar strategies early in a period of trial and error learning. If the psychological maneuvers required of the subject were more unusual or bizarre than controlling alertness, frustration, emotionality, etc., then one might expect the Information groups to show a superiority in the early trials. Similarly it is reasonable to believe that Information Only subjects are able to judge their own performance in the absence of an external feedback system by monitoring their own internal states as suggested in their instructions. Subjects know what it feels like to be calm or aroused. Internally generated knowledge of results could account for the improvement of the Instruction Only group with practice.

The similarity in both the development and the final magnitude of differential responsiveness in the groups which learned, either with or without feedback, emphasizes the difficulty in ascribing learned changes to the effects of feedback when subjects are also informed about the nature of the task. Further, these data suggest that, while proper feedback may be used by naive subjects to produce significant changes in the EEG spectra, it is not the only method by which such changes may be quickly and easily produced. Feedback about alpha or beta frequency activity does not provide unique access to these possible physiological substrates of conscious states.

In the present study a short (1 hr) period of feedback training or instructed behavior was used, during which subjects in both conditions behaved similarly. What would happen if testing were to continue for a very much longer period? While there are no carefully controlled data on this point, an examination of the literature is instructive. Both procedures with sufficient practice produce strikingly large changes in the abundance of the alpha rhythm. The effects of continued feedback reported by Green, Green and Walters [4] appear similar to the EEG changes reported by Wallace [12] in his study of practiced transcendental meditators. It therefore seems unlikely that the initial similarities between the Instruction and Feedback groups in the present experiment would disappear if the period of training were extended. In fact it is this first stage of training that might be expected to show between-group differences if any exist.

Further, the fact that very simple and general instructions can effectively induce alpha wave activity argues against the popular notion that "the alpha state" is quite unique, attainable only by meditation or EEG contingent reinforcement. This conclusion is also informally supported by the lack of uniformity of subjective reports gathered from subjects in the Feedback Only condition. Subjects in the Information conditions, however, presumably because of their initial biases, reported the typical correlates of brain alpha rhythms—relaxation, calmness, inner awareness, etc. The apparent psychological specificity of the alpha state may be primarily attributable to cognitive components in the experimental setting, and not to the brain process itself.

Finally, the Mismatched Information and Feedback group was included to obtain an estimate of the relative power of the two methods of modifying EEG activity. The absence of effect in this condition suggests that neither clearly predominates. This conclusion is congruent with the similarity of response in the three groups which learned. Subjects in the Mismatched group reported spending a great deal of effort seeking appropriate response strategies, which would be both compatible with their instructions and effective in controlling the feedback signal, a very difficult task all things considered.

REFERENCES

1. Anand, B. K., G. S. Chhina and B. Singh. Some aspects of electroencephalographic studies in yogis. *Electroenceph. clin. Neurophysiol.* **13**: 452–456, 1961.
2. Beatty, J. T. Effects of initial alpha wave abundance and operant training procedures on occipital alpha and beta activity. *Psychonom. Sci.* **23**: 197–199, 1971.
3. Brown, B. B. Recognition of aspects of consciousness through association with EEG alpha activity represented by a light signal. *Psychophysiology* **6**: 442–452, 1970.
4. Green, E. E., A. M. Green and E. D. Walters. Self-regulation of internal states. In *Proceedings of the International Congress of Cybernetics*, edited by J. Rose, London, 1969, London: Gordon and Breach, 1970.
5. Jasper, H. H. The ten-twenty electrode system of the international federation. *Electroenceph. clin. Neurophysiol.* **10**: 371–375, 1958.
6. Kasamatsu, A. and H. Tomio. An electroencephalographic study on the zen meditation (Zazen). *Folia Psyciatrica et Neurologica Japonica* **20**: 315–336, 1966.
7. Legewie, H. and W. Probst. On-line analysis of EEG with a small computer (period-amplitude analysis). *Electroenceph. clin. Neurophysiol.* **27**: 533–536, 1969.
8. Lindsley, D. B. Attention, consciousness, sleep and wakefulness. In: *Handbook of Physiology*, Section I, Vol. 3, Edited by J. Field 1553–1593, 1960.
9. Morrell, L. K. Some characteristics of stimulus-provoked alpha activity. *Electroenceph. clin. Neurophysiol.* **21**: 552–561, 1966.
10. Nowlis, D. P. and J. Kamiya. The control of electroencephalographic alpha rhythms through auditory feedback and the associated mental activity. *Psychophysiology* **6**: 476–484, 1970.
11. Stark, L. Comments in C. R. Evans and T. B. Mulholland (Eds.) *Attention in Neurophysiology.* New York: Appleton-Century–Crofts, 1969, 162–163.
12. Wallace, R. K. Physiological effects of transcendental meditation. *Science* **167**: 1751–1754, 1970.

Localized EEG Alpha Feedback Training: A Possible Technique for Mapping Subjective, Conscious, and Behavioral Experiences

Erik Peper

Abstract

Subjects who received EEG alpha feedback recorded from two homologous scalp areas (central-temporal) were trained to have ON-OFF control over the left and right sides. The partial success in demonstrating localized control suggests that subjects may be trained for very specific control. Localized training may be used to partition the subjective, conscious and behavioral experiences associated with selected EEG patterns and to develop an independent subjective physiological language. Applications to medicine and altered states of consciousness are discussed.

Introduction

While receiving feedback, subjects have learned to control their alpha, beta and theta electroencephalographic (EEG) patterns (Kamiya, 1968; Peper and Mulholland, 1970; Green et al., 1970; Brown, 1971); in addition, some epileptic patients have learned to control (inhibit) their seizures by enhancing their sensory motor rhythm (Sterman, 1972). Eventhough subjects have learned to control their EEG, the mechanism, modes of control, and mapping of subjective conscious experience associated with certain

---

* Presented in part at the Biofeedback Research Society Meeting, St. Louis, Mo. (1971).

** Reprint requests: Erik Peper, Perception Laboratory, Veterans Administration Hospital, Bedford, Mass, 01730, U.S.A.

EEG patterns are not understood. The EEG varies spontaneously in amplitude, frequency, and spatial location (the surface of the scalp) as the person is learning to control his EEG with feedback or shifts levels of alertness. In many cases, the subjective experience associated with EEG training (such as alpha) results from constraining the EEG within a narrow boundary and the subject's effort to stay awake and refraining from orienting commands while being in a sensory limited environment (Peper, 1971b).

The following experiment presents a methodology to explore the subjective experience and hopefully the mechanisms associated with localized EEG control. Instead of learning control over a certain EEG pattern, the subject controls one pattern at one scalp location and suppresses the pattern at another scalp location.

In human studies, the localized control of the EEG should tease out those subjective experiences associated with the varying and self-controlled EEG pattern. This paradigm may partially map the physiological language for internal states of consciousness that Kamiya (1968) has advocated. Moreover, if the strategies by which the subjects develop localized EEG control were known, they could be used to enhance the training of subjects with abnormal EEG's and the associated behavior aberrations – possibly offering treatment through self-control. In addition, if localized training is feasible, then the differential control would imply a greater learned specificity and not a general change in the level of arousal, alertness, or orienting. This localized control would be similar with human subjects to the results of DiCara and Miller (1968) who showed that curarized and artificially respirated rats could learn vasomotor dilation in one ear independent of the other ear. This demonstration of localized control self-control suggests applications for the understanding and treatment of psychosomatic illness.

A previous exploration (Peper, 1971a) indicated that localized EEG control may be feasible. In that initial experiment two out of six subjects were successfully trained to produce occipital asymmetry when given feedback. However, the feedback indicated only absolute asymmetry and the paradigm could not resolve subjective strategies or asymmetry because of an overall increase in percent time occipital alpha during the asymmetry trials.

When subjects were trained for absolute occipital alpha asymmetry, the extent of control was surprising: 31 sec out of 120 sec of alpha asymmetry for the asymmetry trial; contrasted with 3 sec out of 120 sec of alpha asymmetry for the symmetry trial; moreover, the seconds of asymmetry increased above the two baseline conditions.

With a different experiment, Fehmi (1971) has reported that: after a 20-minute exposure to a feedback tone which signaled when the right and left occipital rhythms were within 15 degrees of being in phase, the subjects demonstrated an ability to increase and decrease phase synchrony with respect to the baseline values. He reported that these results are statistically significant; moreover, yoked controls who received non-contingent feedback, failed to demonstrate a difference. The verbal reports from these subjects and 87 others associated attentional focussing upon a relatively stable mental image with the occurrence of the feedback signal. When such a mental focus was disrupted, the phase synchrony of occipital waves was lost.

In addition, psychological and behavioral correlates have been associated with localized brain disorders and epileptic foci. The localization of cerebral function has been repeatedly demonstrated in the brain injured and split brain studies (Sperry, 1964; Gazzaniga, 1970). This data indicates that the left hemisphere is dominant in processing language functions, mathematical and analytic tasks; while the right hemisphere is dominant in processing spatial relations and gestalt tasks. Although the data comes from split brain and brain-damaged subjects; data from normal subjects also suggests that the two sides of the brain act differently.

Recently in an ingenious study, Galin and Ornstein (1972) looked at the power of the EEG spectrum in normal subjects when those subjects where presented with predominantly right (spatial) or left (verbal) hemisphere tasks. EEG recording were made from the left and right temporal-parietal areas and the rations of average power (1-35 Hz) in homologous T4/T3 and P4/P3 were computed. This ratio, average power of the homologous areas (right over left), was greater in the verbal tasks than in the spatial tasks. Using this measure from scalp recordings, they have been able to distinguish between these two cognitive modes as they occur in normal subjects. Generally, verbal behavior is mediated in the left hemisphere; a finding that may explain why a large number of subjects show alpha rhythms of higher mean voltage and of somewhat wider distribution over the right hemisphere (non-dominant) than the left hemisphere (Kiloh and Osselton, 1966). Subjects when tested are somewhat nervous, anxious and continuously analyzing the situation; that behavior would decrease the alpha amplitude on the left side since that side is giving more efferent commands.

One hypothesis suggests that alpha EEG activity may attenuate when efferent commands are given. Specifically, occipital alpha will attenuate when the person is no longer "passively observing" but giving visual motor commands – accommodation and convergence – while looking and orienting (Mulholland and Peper, 1971; Peper, 1971b).

Procedure for Localized Central-Temporal Alpha EEG Training

The EEG was recorded on a Grass model 5 and the two feedback paths were identical. The feedback system for each channel has been described previously (Peper, 1971a); Except that the electrodes were placed at the mid-points between P4–T4 and F4–T4 on the right side of the scalp and P3–T3 and F3–T3 on the left side of the scalp; the right mastoid was ground. The bandpass filter for each channel was set ± 1 Hz for each individual's resting alpha frequency. The criterion to define alpha was set and measured by an optical meter relay at the output of the bandpass filters. The relay was set to close when the alpha amplitude reached 30% of the resting amplitude and this system has been described in detail (Boudrot, 1972). The relays operated a timer and closed the circuits to present the feedback tones. A high tone was presented when ever alpha was defined in one of the channels and a low tone was present when alpha was defined in the other channel. The sound was 50 Db and presented by a speaker behind the subject.

Eight unpaid volunteers were trained for 2 sessions, four of the subjects received the high feedback tone when alpha occurred in the right hemisphere while the other 4 subjects received the high feedback tone when alpha appeared in the left hemisphere. The low tone was present when alpha occurred in the other hemisphere. The subject sat in a comfortable reclining chair in a sound attenuated light-proof room. An intercom linked the experimental chamber with the experimenter.

For each session, the subjects were instructed that after the two baseline conditions (eyes open and closed) with no feedback, two tones would sporadically go on. These tones were produced by their own brain waves and that during the session they would attempt to gain voluntary control over these tones (their EEG). The subjects were then given a 10 min practice session with feedback which was followed by four or six 180 sec training trials in which the subjects would attempt in the first trial to keep the high tone on and suppress the low tone; in the second trial, the directions were reversed, the subject would attempt to keep the low tone on and suppress the high tone. This pattern was repeated for the following trials. After the feedback trials, 2 trials were given without feedback to check whether the subject had any method of control. Final baselines were recorded and the subjects were interviewed. Since this study was an exploratory approach, in some cases the subjects were asked to keep their eyes open while in other cases the light was on in the experimental room.

Results

Even though localized control of the EEG is extremely difficult at least one subject showed control, see figure 1. Although the subject did not have control over the per cent alpha in the left central-temporal areas – it stayed relatively constant; the subject demonstrated independent control over her right

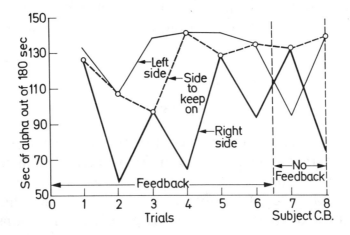

Fig. 1. Seconds of EEG alpha recorded from the homologous areas of the scalp (central-temporal) are shown, while the subject alternately enhances alpha on the left or right side by maintaining a feedback tone on one side (designated by small circles) and suppressing the tone on the other side. The subject demonstrates right hemisphere control when feedback tone is given (trials 1–6) and demonstrates both right and left hemisphere control when no feedback is given (trials 7 and 8)

central-temporal alpha EEG. Ironically, once the feedback was removed the subject had some control over the alpha EEG on the left side. This subject (C. B.) reported that to keep the low tone on (left central-temporal alpha), she quietly sang to herself, felt the rhythm of dancing and noticed (felt) how relaxed she was; to keep the high tone on (right central-temporal alpha) she spelled things named the states of the U. S. and multiplied high figures.

Only one other subject (T. M.) reported that the tones made sense to him although he could not control them. To keep the high tone on (right central-temporal alpha) he talked and issued verbal commands; to keep the low tone on (left central-temporal alpha) he relaxed his tongue and felt like a zombie. He reported that his main difficulty was that he could not stop self-observing and analyzing his own behavior. The low tone kept going off as he inadvertently observed, gave strong verbal commands and fixated visually.

Discussion

Localized EEG alpha control appears feasible and was demonstrated by C. B. It seems likely that many more training sessions are necessary (two sessions are not enough) for subjects to learn extremely fine control of their EEG patterns at selected locations, especially if shaping procedures are used. In order to "learn" control, subjects have to "learn to forget about trying" and just "let it occur".

The limited subjective reports corroborated Galin and Ornstein's (1972) results, who found that alpha amplitude is less in the hemisphere which is performing a cognitive task. For example, subject, C. B., kept the left central-temporal alpha on by singing and feeling the rhythm of dancing which would activate the EEG and therefore suppress alpha on the right side (non analytic – spatial analyses). She kept alpha on on the right side by spelling things, naming the states and multiplying high figures which would activate the left side, allowing relatived more alpha on the right side.

Perhaps her inability to change the percent alpha on the left side during feedback conditions meant that she continued to subverbalize at a certain level, analytically observed herself while attempting to keep the tones on and off. Only after she no longer "heard" the tones did she relax and modulate the alpha on the left side. The other subject's (T. M.) subjective reports fit a similar hypothesis. His constant level of self-awareness or mentation prevented any control which supports the localization of function.

Though this is an exploratory study, learned localized EEG control may have applications to the study of altered states and medicine. Localized training may help partition the unique altered states and subjective experiences associated with a particular EEG pattern. It might be used to enhance differential experience by training relative suppression or enhancement of left hemisphere orienting (analytic functions) and right hemisphere responses (spatial and global tasks). Furthermore, this technique, used in longitudinal study of children may help to explore how language functions consolidate in the left hemisphere.

In the clinical context this method of partitioning the EEG may be used to train people to recognize and abort epileptic discharges and other localized EEG disturbances. Finally, it may be used to develop diagnostic tests which would identify the conditions under which EEG asymmetry is pathological.

References

Boudrot, R.: An alpha detection and feedback control system. Psychophysiology **9**, 461–466 (1972).

Brown, B.: Awareness of EEG-subjective activity relationships detected within a closed feedback system. Psychophysiology **7**, 451–464 (1971).

DiCara, L. V., and Miller, N. E.: Instrumental learning of vasomotor responses by rats: learning to respond differentially in the two ears. Science **159**, 1485–1486 (1968).

Fehmi, L. G.: Bio-feedback of electroencephalographic parameters and related states of consciousness. Paper presented at the Annual American Psychological Association Convention. Washington, D. C. (1971).

Galin, D., Ornstein, R.: Lateral Specialization of cognitive mode: an EEG study. Psychophysiology **9**, 412–418 (1972).

Gazzaniga, M.S.: The bisected brain. New York: Appleton-Century-Crofts, 1970.

Green, E. E., Green, A. M., Walters, E. D.: Voluntary control of internal states: Psychological and physiological. Journal of Transpersonal. Psychology **2**, 1–26 (1970).

Kamiya, J.: Conscious control of brain waves. Psychology Today 1968 I, 57–60.

Kiloh, L. G., Osseton, J. W.: Clinical electroencephalography, p. 19. London, Butterworths, 1966.

Mulholland, T. B., Peper, E.: Occipital alpha and accommodative vergence, pursit tracking, and fast eye movements. Psychophysiology **8**, 556–575 (1971).

Peper, E.: Comment on feedback training of partietal-occipital alpha asymmetry in normal human subjects. Kybernetik **9**, 156–158 (1971 a).

Peper, E.: Reduction of efferent motor commands during alpha feedback as a facilitator of EEG alpha and a precondition for changes in consciousness. Kybernetik **9**, 226–231 (1971 b).

— Mulholland, T. B.: Methodological and theoretical problems in the voluntary control of electroencephalographic occipital alpha by the subject. Kybernetik **7**, 10–13 (1970).

Sperry, R. W.: The great cerebral commissure. Sci. Amer. (1964).

Sterman, M. B.: Paper presented at the Veterans Administration Hospital Bedford, Mass., May 10, 1972.

E. Peper
Perception Laboratory
Veterans Administration Hospital
Bedford, Mass. 01730, U.S.A.

VI
ELECTROMYOGRAPHIC CONTROL

Electromyography Comes of Age   21

John V. Basmajian

Man has shown a perpetual curiosity about the function and control of skeletal muscles. Aristotle, Galen, Leonardo da Vinci, and Vesalius, the "father of modern anatomy," all were fascinated by the organs of locomotion and power (1). Galvani may be considered the originator of electromyography, or the study of electrical activity in muscle. He believed that muscles stored and discharged electricity received from the nerves, very much like Leiden jars (2). While many uses were made of Galvani's other findings, muscle electricity remained a scientific curiosity until the 20th century when improved methods of detecting and recording minute electrical discharges became widely available.

Much of the credit for launching modern electromyography about 40 years ago goes to Adrian and Bronk (3) and their colleagues. Being neurophysiologists, they did not attempt to determine the functions of individual muscles and groups of muscles; instead they improved electrical methods of studying the nervous system. It was not until the end of World War II, when there was a marked improvement in the technology and availability of electronic apparatus, that anatomists, kinesiologists, and clinical scientists began to make increasing use of electromyography.

The first study to gain wide acceptance was that of Inman et al. who investigated the movements of the human shoulder region (4). During the 1950's electromyography was used frequently in kinesiological studies, and it is now used in many different fields of biology, in studies ranging from the activity of the middle ear muscles of bats to the psychophysiology of human relaxation (5).

The Basis of Electromyography

The cellular unit of contraction in skeletal muscle is the muscle cell or muscle fiber (Fig. 1). Best described as a very fine thread, this muscle fiber has a length of up to 30 centimeters or more but is less than 100 micrometers (0.1 millimeter) wide. On contracting it will shorten to about 57 percent of its resting length. By looking at the intact normal muscle during contraction one could easily believe that all the muscle fibers were in some stage of continuous smooth shortening. But in fact, there is a widespread, rapid series of twitches which occur asyn-

---

The author is director of the Regional Rehabilitation Research and Training Center, and professor of anatomy, physical medicine, and psychiatry in the School of Medicine at Emory University, Atlanta, Georgia 30322.

chronously among the fibers. The apparently smooth contraction is a summation of the asynchronous twitches.

In normal mammalian skeletal muscle, single fibers probably never contract individually. Instead, small groups of them contract at the same moment because they are supplied by the terminal branches of one nerve fiber or axon. The cell body from which the axon arises is in the anterior horn of the spinal gray matter or related brainstem areas. Collectively, the nerve cell body, its axon, and its terminal branches, and all the muscle fibers supplied by these branches, constitute a motor unit; this is the functional unit of striated muscle (Fig. 2). An impulse descending the nerve axon causes all the muscle fibers in one motor unit to contract almost simultaneously. In man, impulses are generated at various frequencies, usually below 50 per second.

The number of muscle fibers that are served by one axon, that is, the number of fibers in a motor unit, varies widely. Generally, muscles controlling fine or delicate movements and adjustments have the smallest number of muscle fibers per motor unit. For example, the muscles that move the eyes have fewer than ten fibers per unit, as do the muscles of the middle ear and the larynx (5). On the other hand, large muscles in the limbs have larger motor units. One of the muscles of the thigh has motor units with about 2000 muscle fibers supplied by each axon. Individual muscles of the body consist of many hundreds of such motor units and it is their summated activity that develops the tension in the whole muscle.

The amount of work produced by a single motor unit is quite small. In a living human being it is usually insufficient to show any external movement of a joint spanned by the whole muscle of which it is a part. Even in small joints, such as those of the thumb, at least two or three motor units are needed to give a visible movement. Under normal conditions small motor units are recruited early, and, as the force is automatically or consciously increased, larger and larger motor units are recruited (6), while all the motor units also increase their frequency of twitching. There is no single, set frequency; individual motor units can fire very slowly and will increase their frequency of response on demand (7).

Motor Unit Potentials

The motor unit potential has a brief duration (with a median of 9 milliseconds) and a total amplitude measured in microvolts or millivolts. When displayed on the cathode-ray oscilloscope, most potentials are sharp triphasic or biphasic spikes (Fig. 3). Generally, the larger the spike the larger is the motor unit that produced it. However, the distance from the electrode, the type of electrodes used, and the equipment are among the many factors which influence the final size and shape of the oscilloscopic display.

Much disagreement in electromyography has centered around the different types of electrodes used. Because kinesiological studies are often made by investigators who are not medically qualified, surface electrodes have been popular. However, for discrete recordings without "cross talk" between muscles, there are many difficulties with surface electrodes. Electrodes inserted through the skin are no longer as forbidding as they once were. They are made from a nylon-insulated Nichrome alloy or any other similar material and are only 25 or 75 micrometers in diameter (Fig. 4) (5). Because of their size they are painless and are easily injected and withdrawn. They can collect as much data from a specific muscle as can the best surface electrodes, and give excellent recordings on the oscilloscope. With 1 millimeter of an inserted electrode left exposed, the voltage from a muscle can be recorded much better than with a surface electrode. Bipolar electrodes restrict their pickup either to the whole muscle being studied or to the confines of the compartment within a muscle if it has a multipennate structure. Barriers of fibrous connective tissue within a muscle or around it act as insulation. Thus, one can record all the activity as far as such a barrier without interference from beyond it.

Electromyograms are obtained by means of amplifiers with most sensitive to frequencies from about 10 to 2000 hertz. Amplifiers sensitive to frequencies of no more than 1000 hertz are satisfactory. Ideally, the results should be recorded either photographically, or by means of electromagnetic tape. In recent years multitrack tape recorders have provided a relatively cheap means of storing electromyogram signals, especially for computer analysis later.

As with any electric signals, motor unit potentials can be transmitted for long distances either by telephone lines or by FM radio. The latter promises to be useful in field studies of wild animals or in experiments in which the subjects must be completely unfettered by dragging cables (as in running and jumping, for example). A number of excellent telemetering systems have been described and good standard equipment is available from commercial sources. Single-channel telemetering in electromyography is simple; multichannel telemetry is more difficult but will be used more frequently as techniques and equipment improve.

Muscle Tone and Fatigue

Most neurophysiologists agree that the complete relaxation of normal human striated muscle at rest can be demonstrated conclusively by electromyography. That is, a normal human being can abolish neuromuscular activity in a muscle by relaxing it. This

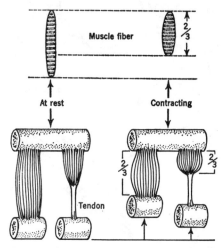

Fig. 1. The structural unit of contraction in skeletal muscle is the muscle fiber which can shorten to approximately two thirds of its resting length. Thus, the excursion of the ends of a whole muscle is limited by this constraint.

does not mean that there is no tone (or tonus) in skeletal muscle, as some investigators have claimed. It does mean, however, that the usual definition of tone should be modified to state that the general tone of a muscle is determined both by the passive elasticity or turgor of muscular (and fibrous) tissues and by the active (though not continuous) contraction of muscle in response to the reaction of the nervous system to stimuli. Thus, at complete rest, a muscle has not lost its tone even though there is no neuromuscular activity (5).

In the clinical appreciation of tone, the more important of the above two elements is the reactivity of the nervous system. One cannot lay hands on a normal limb without causing such a

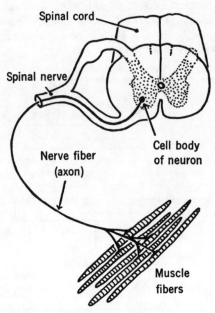

Fig. 2. Diagram of a single motor unit, the functional unit of contraction.

reaction. Therefore, the clinician soon learns to evaluate the level of tone and it may seem of little consequence to him that the muscle he is feeling is, in fact, capable of complete neuromuscular inactivity. In spite of this, he would be surprised to learn that an experienced subject can simulate hypotonia or even the atonia of complete paralysis and so deceive—if only for a brief period—the most astute physician (8).

Electromyograms from normal human muscles generally show that relaxation occurs completely and almost instantaneously when subjects are ordered to relax. However, a small number of normal subjects do have great difficulty in relaxing quickly. In no normal muscle at complete rest is there any sign of neuromuscular activity, even with multiple electrodes. As Stolov (9) has said: "We can therefore conclude that no alpha motor neuron discharge is present in normal muscle at rest, but may be present during a stretch that is rapid enough to initiate a reflex response."

"Complete rest" requires some qualification. A normal person does not completely relax all his muscles at once. Reacting to multiple internal and external stimuli, various groups of muscles show increasing and decreasing amounts of activity. Goldstein (10) found that certain muscle groups at rest showed electromyogram activity related to a general muscular tension: these muscles were mostly in the limbs and neck. Anxious subjects were not greatly different from nonanxious ones —except that the very anxious subjects showed marked reaction to new stimuli.

Electromyography has been used by some workers in the investigation of fatigue. The fatigue of strenuous effort may be quite different from the weariness felt after a long day's routine sedentary work and undoubtedly the following types of fatigue exist: emotional fatigue, central nervous system fatigue, "general" fatigue, and peripheral neuromuscular fatigue of special kinds.

Investigations in my laboratory have shown that postural fatigue is caused largely by painful strain not on muscles (which often have been quite quiescent) but on ligaments, capsules, and other inert structures. This has led to the concept that muscles are spared when ligaments suffice. This is important not only in understanding the sources of postural fatigue in man but also because it underlies another important principle of man's posture: There should be a minimal expenditure of energy consistent with the ends to be achieved. Thus, MacConaill and I (11) proposed two laws which express the operation of this principle: (i) The law of minimal spurt action: no more muscle fibers are brought into action than are both necessary and sufficient to stabilize or move a bone against gravity or other resistant forces, and none are used insofar as gravity can supply the motive force for movement. (ii) The law of minimal shunt action: only such muscle fibers are used as are necessary and sufficient to ensure that the transarticular force directed toward a joint is equal to the weight of the stabilized or moving part together with such additional centripetal force as may be required because of the velocity of that part when it is in motion.

That these two laws are valid has been demonstrated clearly by electromyography. In studies of the biomechanics of the shoulder (glenohumeral) joint, it was shown that the whole weight of an unloaded arm is counteracted by the upper part of its capsule alone, no muscles being required (5). In this area a previously puzzling ligament (the coracohumeral) is very strong; it appears to reinforce the action of the joint capsule in preventing the head of the humerus from sliding downward on the inclined plane formed by the glenoid cavity or shoulder socket. With moderate or heavy loads, a muscle adhering to the capsule (the supraspinatus) is brought into action to reinforce the original tension in the capsule.

Everyone knows that carrying a heavy weight (say, a suitcase) can become a painful experience. It is often assumed that the reactive component which becomes fatigued and exhibits pain is muscular; in fact, the fatigue is chiefly ligamentous. This demonstrates the importance of the inert structures.

Another example of the minimal principle is that of arch support in the human foot. Generations of surgeons have stressed the importance of the muscular tie-beams in the plantigrade foot, yet electromyography has repeatedly pointed to the fundamental importance of plantar ligaments. Indeed, even very heavy loads do not recruit muscular activity in the leg and foot (5).

A variety of phenomena that occur with the progressive fatigue of continuous voluntary activity have been described. Some are probably of spinal cord origin and are not well understood. Other phenomena include the synchronization of potentials, an augmentation of the amplitude and duration of potentials, and an increase in polyphasic potentials. An excellent review of these observations has been written by Scherrer and Monod (12).

Coordination of Muscle Function

The statement that the brain does not order a muscle to contract, but orders movements of a joint, is true only in part. Under certain circumstances the movement is the result of contraction in only one or two muscles. For example, pronation of the forearm (turning the hand into the palm-down position) is usually produced by one muscle alone, the pronator quadratus, unless there is added resistance to the movement, in which case more muscles are used (5). In elbow flexion, the brachialis (not biceps) often suffices (5). On the other hand, there are complex movements (such as rotation of the scapula on the chest wall during elevation of the limb) which obviously require groups of cooperating muscles.

It has generally been believed that during the movement of a joint in one direction, muscles that move it in the opposite direction show some sort of antagonism. It has been shown by Sherrington (13), however, that the so-called antagonist relaxes completely except perhaps with one exception, at the end of a whiplike motion of a hinge joint (5). Sherrington referred to the relaxation of the antagonist as re-

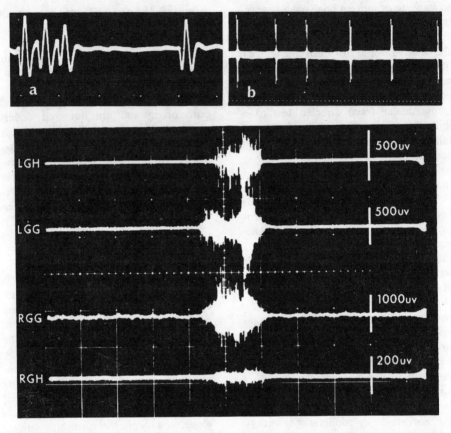

Fig. 3. Potentials demonstrated by electromyography. (a and b) Single motor unit potentials with varying time bases (time scale in each, 10 msec). The lower panel shows simultaneous recordings obtained at slow speeds from muscles in the tongue while the subject was swallowing. (*LGG* and *RGG*, left and right genioglossi; *LGH* and *RGH*, left and right geniohyoids; time scale is 0.5 second per major division.)

ciprocal inhibition; the term antagonist should thus be replaced by synergist. The activity of muscles in the position of antagonists during a movement may also be a sign of nervous abnormality (for example, in the spasticity of paraplegia), or, in the case of fine movements requiring training, a sign of ineptitude. Indeed, the athlete's continued drill to perfect a skilled movement exhibits a large element of progressively more successful repression of undesired contractions.

Conscious Control and Training of Single Motor Units

A recently developed technique in electromyography, known as motor unit training, offers a useful approach to studies of the conscious control over spinal motoneurons, of the neurophysiological processes underlying proprioception and feedback, and of many related psychophysiological phenomena. Although in the 1930's Smith (*14*) and Lindsley (*15*) gave brief accounts of man's ability to discharge single motor units and to vary their rates, electromyographers tended to take this phenomenon for granted and performed no systematic studies of it until the early 1960's (7, 16). Advances in the development of intramuscular electrodes then permitted rapid progress. In man it is much easier to place an electrode in or near the muscle fibers rather than in the spinal cord; thus, electromyography provides a simple way to record the activity of the motoneurons.

Human subjects undergoing motor unit training are given auditory and visual displays of their individual myoelectric potentials recorded by means of intramuscular electrodes. The cues provide the subjects with an awareness of the twitching of individual motor units. They learn in a few minutes to control this activity and can give many bizarre responses with only the feedback cues as a guide (7).

Subjects are invariably amazed at the responsiveness of the loudspeaker and cathode-ray tube to their slightest efforts, and they accept these as a new form of proprioception without difficulty. It is not necessary for them to have any knowledge of electromyography. After being given a general explanation of the procedure they need only to concentrate their attention on the response of the electromyograph. Even the most naive subject is soon able to maintain various levels of activity in a muscle on the sensory basis provided by the monitors. Indeed, the actions that are required of him generally demand such gentle contractions that his only awareness of them is through the apparatus. After a period of orientation, the subject can be put through increasingly difficult tasks for many hours.

After acquiring good control of the first motor unit—that is, being able to fire it in isolation, speed it up, slow it down, and turn it "off" and "on" in various set patterns, most subjects can

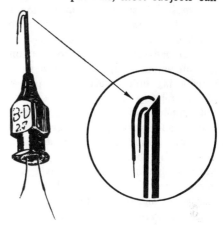

Fig. 4. Bipolar fine-wire electrode (top) greatly magnified in the carrier needle (bottom). The carrier needle will be withdrawn after insertion, leaving the hooked wires in the muscle.

isolate a second unit with which they then learn the same tasks; then a third, and so on. In a serial procedure, the next task may be to recruit, unerringly and in isolation, the several units over which he has gained the best control. The best subjects (about 1 in 20) can learn to maintain the activity of specific motor units in the absence of either one or both of the visual and auditory feedbacks. That is, as the monitors are gradually turned off the subjects must try to maintain or recall a well-learned unit without the artificial "proprioception" provided earlier.

Any skeletal muscle may be selected for motor unit training. In my laboratory we usually select the abductor pollicis brevis, abductor digiti minimi, tibialis anterior, biceps brachii and brachialis, and the extensors of the arm and forearm. But we have also successfully trained motor units in back muscles, shoulder and neck muscles, tongue muscles, and others; there appears to be no limit.

Factors Influencing Motor Unit Training

Almost all subjects (among many hundreds tested) have been able to produce well-isolated contractions of at least one motor unit, turning it off and on without any interference from neighboring units. Only a few people fail completely to perform this basic task. Analysis of subjects that perform poorly and very poorly reveals no common characteristic that separates them from subjects that perform well. Most people are able to isolate and master one or two units readily; some can isolate and master three units, four units, and even six units or many more. This last level of control is of the highest order, for the subject must be able to give an instant response to an order to produce contractions of a specified unit without the interference of neighboring units; he also must be able to turn the unit off and on at will.

Once a person has gained control of a spinal motoneuron, it is possible for him to learn to vary its rate of firing. This rate can be deliberately changed in immediate response to a command. The lowest limit of the range of frequencies is zero—that is, one can start from neuromuscular silence and then give single isolated contractions at regular rates as low as one per second and at increasingly faster rates. When the more able subjects are asked to produce special repetitive rhythms and imitations of drum beats, almost all are successful (some strikingly so) in producing subtle shades and coloring of internal rhythms.

Some persons can be trained to gain control of isolated motor units to such a degree that, with both visual and aural cues shut off, they can recall any one of three favorite units on command and in any sequence. They can keep such units firing without any conscious awareness other than the assurance (after the fact) that they have succeeded. In spite of considerable introspection, they cannot explain their success except to state they thought about a motor unit as though they had seen and heard it personally. This type of training may underlie some facets of the development of ordinary motor skills.

We have found no distinct personal characteristics that reveal reasons for the quality of performance. The best subjects are of various ages, either sex, and may be either manually skilled or unskilled, educated or uneducated, bright or dull personalities [see (17)]. Some nervous persons do not perform well—but neither do some very calm persons.

In a recent study (18) we set out to discover whether previous training in special skills is a factor in the time it takes to achieve control over the first motor unit. We used the time required for training the motor units in one of the hand muscles as the criterion.

The time required to train most of the manually skilled subjects was above the median, although one might expect the opposite. Henderson (19) has suggested that the constant repetition of a specific motor skill increases the probability of its correct recurrence by the learning and consolidation of an optimal anticipatory tension. Perhaps this depends on an increase in the background activity of the gamma motoneurons regulating the sensitivity of the muscle spindles of the muscles used in performing that skill. Wilkins (20) postulated that the acquisition of a new motor skill leads to the learning of a certain "position memory" for it. If anticipatory tensions and position memory, or both, are learned, regardless of whether they are integrated at the cerebral level, the spinal level, or both, these or some other cerebral or spinal mechanisms may be acting temporarily to block the initial learning of new skills. It may be that some neuromuscular pathways acquire a habit of responding in certain ways, and it is not until that habit is broken that a new skill can be learned.

When a large number of subjects were studied on two occasions using a different hand each time, Powers (21) found that they always isolated a unit more quickly in the second hand. Isolation was twice as rapid when the second hand was the preferred (dominant) hand; it was almost five times as rapid when the second hand was the nonpreferred one. The time required to train a subject to control a previously isolated unit was shortened significantly only when the preferred hand was the second hand.

Any changes in the action potentials of trained motor units as a result of electrical stimulation of the motor nerve supplying the whole muscle must reflect neurophysiologic changes of the single neuron supplying the motor unit. Therefore, Scully and I investigated the influence of causing strong contractions in a muscle to compete with a discrete motor unit in it which was being driven consciously (22). Each of a series of subjects sat with his forearm resting comfortably on a table top. The stimulator cathode was applied to the region of the ulnar nerve above the elbow. The effective stimuli were 0.1-millisecond square-wave pulses of 70 to 100 volts, delivered at a frequency of 90 per minute. Because stimuli of this order are not maximal, all axons in the ulnar nerve were not shocked and slight variation must have existed in axons actually stimulated by each successive shock (Fig. 5).

Contrary to expectation, when the massive contraction of a muscle was superimposed on the contraction of only one of its motor units, the regular conscious firing of that motor unit was not significantly changed. Our experiments leave little if any doubt that well-trained motor units are not blocked in most persons. Even the coinciding of the motor unit potential with elements of the electrically induced massive contraction would not abolish the motor unit potential.

Subjects asked to move a neighboring joint while a motor unit is firing find such a request distracting, but can usually achieve the task in spite of the distraction (23). Carlsöö and Edfeldt (24) concluded that proprioception can be assisted greatly by exteroceptive auxiliary stimuli in achieving motor precision. Some persons can be trained to maintain the isolated activity of a motor unit regardless of changes in the position of various joints of a limb. In order to maintain or recall a motor unit at different positions, the subject must keep the motor unit active during the performance of the movements; and, therefore, preliminary training is necessary.

Applications of Motor Unit Training

In the search for new models of the learning process, motor unit training emerges as an important technique demanding widespread exploration. In addition to providing a model for neuromuscular training and operant conditioning, it also has practical applications in such fields as rehabilitation and the design of myoelectric artificial aids for the physically handicapped. Normal persons can adequately control several motor units connected electronically to electric motors. With motors having several steps or rates, the strength of contraction can be made variable. Many practical problems in designing such equipment must

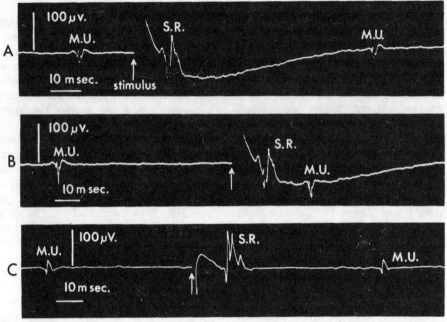

Fig. 5. Undisturbed firing of trained isolated motor units in the abductor digiti minimi of three different subjects, A, B, and C. Arrows indicate application of stimuli (0.1-msec square-wave pulses of 70 to 100 volts) to other axons in the same muscle. Each such stimulus is followed by a massive response (S.R.) while the isolated motor units continue to fire uninterruptedly (M.U.).

be overcome, but there is no question of the feasibility of using individual motor units to act as switches for separate channels of mechanical devices (25).

Motor unit training has also been used in the retraining of muscles in physical therapy and in teaching patients to relax specific muscles. Jacobs and Felton (26) have shown patients with painful traumatic neck spasms how to relax the affected muscle, thus effecting relief from the pain. Hardyk et al. (27) have used similar techniques to abolish laryngeal muscular activity in certain types of reading problems where the speed of silent reading is impeded by subvocalization. There is no limit to the novel applications in research and technology that are possible from the basic knowledge that, given electronic feedbacks, man can consciously control individual motoneurons with extreme precision.

Electromyographic Kinesiology

Much literature has already accumulated on the role of various muscles as revealed by electromyography. Most of it is only of interest to specialists with a clinical interest and to groups such as athletes, musicians, and dancers that are directly concerned with human physical performance. Most investigations have been centered on man, but comparative studies with other mammals are attracting growing interest among biologists. Here I shall describe some of the studies that have the greatest potential for stimulating increased interest.

Many of the widely held beliefs on human posture—and to a lesser extent, human gait—are based on teleology and metaphysics. Many are not borne out by electromyography. It has long been thought that man has had a complicated evolution and that his posture and locomotion are second-rate adaptations. Adaptations they certainly must be, but electromyography has demonstrated their superb functional efficiencies. Man's upright posture is extremely economical in energy expenditure, and there is no clear evidence that his back and lower limbs are hurt more often than those of quadrupeds. The large "antigravity" muscles in man permit the powerful movements necessary for the major changes from lying, to sitting, to standing, and thus cannot be equated with the antigravity muscles of animals that habitually stand on flexed joints.

To stand erect, most human subjects require very slight activity and some intermittent reflex activity of the intrinsic muscles of the back (5). During forward flexion there is marked activity until flexion is extreme, at which time the ligamentous structures assume the load and the muscles become silent. In the extreme-flexed position of the back, the erector spinae remain relaxed in the initial stages of lifting heavy weights (28).

Asmussen and Klausen (29) concluded that the force of gravity is counteracted by one set of muscles only, usually the back muscles, but in 20 to 25 percent of subjects, by means of the abdominal muscles. The line of gravity passes very close to the axis of movement of the fourth lumbar vertebra and does not intersect with the curves of the spine as is often postulated. Klausen (30), investigating the effect of changes in the curve of the spine, the line of gravity in relation to the fourth lumbar vertebra and ankle joints, and the activity of the muscles of the trunk, concluded that the short, deep intrinsic muscles of the back must play an important role in stabilizing the individual intervertebral joints. The long intrinsic muscles and the abdominal muscles stabilize the spine as a whole.

A load placed high on the back automatically causes the trunk to lean slightly forward. The increased pull of gravity is counteracted by an increased activity in the lower back muscles. A load placed low on the back reduces the activity of the back muscles (31). There is increased activity when a load is held in front of the thighs. Thus, the position of the load—either back or front—either aids the muscles or by reflex action calls upon their activity to prevent forward imbalance.

While some investigators believe that the vertebral part of the psoas major (a large hip flexor which lies on the side of the spine) acts on the spine, I have found with Greenlaw (32) that this muscle shows only slight activity during standing. Even strong attempts to increase the natural lumbar lordosis (the hollow of the back) which is said to be a function of psoas in man, recruits little activity in the muscle.

Multifactorial studies are difficult and time consuming, and only recently has equipment improved to the point where electromyography gives especially useful results. In walking there is a very fine sequence of activity in various groups of muscles in the lower limb. As the heel strikes the ground the hamstrings and pretibial muscles reach their peak of activity. Thereafter the quadriceps, the large muscle mass which extends the knee, increases in activity as the torso is carried forward over the limb, apparently to maintain stability of the knee. As the heel lifts off the ground, the calf group of muscles build up a crescendo of activity which ceases as the toe leaves the ground, although the quadriceps and sometimes the hamstrings reach another (but smaller) peak of activity (5).

Conclusions

Among the many studies being made of groups of muscles in human beings, there is one that I shall describe briefly because of its novelty. With a group of music professors, I have been recording the activity of the lip and cheek muscles of wind-instrument players possessing various levels of proficiency. We have found evidence of very clear differences between performers. We have demonstrated a relation between the skill of players and specific electromyographic patterns which requires extensive exploration. Since one purpose of these studies is to establish methods for improving performance, some efforts are being made to apply feedback techniques based on motor unit training. Subjects with multiple electrodes in the muscle of the cheek (buccinator) and lips can selectively activate local areas of these muscles quite easily. This finding provides a firm beginning for many applied studies of performance and its direct modification.

Another study that Tuttle and I began recently (33) is concerned with the posture and locomotion of apes with an emphasis on evolutionary theory. Initially, we are examining the "knuckle-walk" of gorillas by means of videotapes and multichannel electromyography from muscles of the hand and forearm. We hope to determine whether man's direct ancestry included a knuckle-walking stage. As the reader can imagine, electromyography in apes is more difficult—and much more hazardous—for the investigator. Our techniques have allowed extensive electromyographic exploration of muscles throughout the forelimb in one gorilla and limited studies in others. In recent months we have also succeeded in recording the concurrent muscular activity in muscles of the hip and thigh regions, of great interest to anthropologists attempting to explain the erect posture of man. The synchronized videotape record displays both the spontaneous behavior of the unrestricted animal and five channels of electromyography from intramuscular electrodes. To extend this work to a larger population without further increasing the risk to the investigators, future work will make increasing use of telemetering by means of multiple transmitters implanted surgically.

Electromyography promises to make significant contributions to a host of biological sciences. Zoologists and psychophysiologists have not taken full advantage of the technique until recently, but this picture seems to be changing. The electronic and recording equipment now available is both better and cheaper than it was formerly, and is improving rapidly. Almost 200 years since Galvani described animal electricity, this branch of science is finally coming into its own.

References

1. C. Singer, *The Evolution of Anatomy* (Trench, Trubner, London, 1925).
2. L. Galvani, *De Viribus Electricitatis* (1791), quoted by S. Licht, *Electrodiagnosis and Electromyography* (Licht, New Haven, Conn., 1961).
3. E. D. Adrian and D. W. Bronk, *J. Physiol.* 67, 119 (1929).
4. V. T. Inman, J. B. DeC. M. Saunders, L. C. Abbott, *J. Bone Joint Surg.* 26, 1 (1944).
5. J. V. Basmajian, *Muscles Alive: Their Functions Revealed by Electromyography* (Williams & Wilkins, Baltimore, ed. 2, 1967).
6. E. Henneman, G. Somjen, D. O. Carpenter, *J. Neurophysiol.* 28, 599 (1965).
7. J. V. Basmajian, *Science* 141, 440 (1963).
8. ———, *Can. Med. Assoc. J.* 77, 203 (1957).
9. W. C. Stolov, *Arch. Phys. Med. Rehabil.* 47, 156 (1966).
10. I. B. Goldstein, *Psychosom. Med.* 27, 39 (1965).
11. M. A. MacConaill and J. V. Basmajian,

Muscles and Movements: A Basis for Human Kinesiology (Williams & Wilkins, Baltimore, 1969).
12. J. Scherrer and H. Monod, *J. Physiol. Paris* **52**, 419 (1960).
13. C. S. Sherrington, *Proc. Roy. Soc. Ser. B* **105**, 332 (1929).
14. O. C. Smith, *Amer. J. Physiol.* **108**, 629 (1934).
15. D. B. Lindsley, *ibid.* **114**, 90 (1935).
16. V. F. Harrison and O. A. Mortensen, *Anat. Rec.* **144**, 109 (1962).
17. J. V. Basmajian, M. Baeza, C. Fabrigar, *J. Clin. Pharmacol. J. New Drugs* **5**, 78 (1965).
18. H. E. Scully and J. V. Basmajian, *Psychophysiology* **5**, 625 (1969).
19. R. L. Henderson, *J. Exp. Psychol.* **44**, 238 (1952).
20. B. R. Wilkins, *J. Theor. Biol.* **7**, 374 (1964).
21. W. R. Powers, thesis, Queen's University, Kingston, Ontario, Canada (1969).
22. H. E. Scully and J. V. Basmajian, *Arch. Phys. Med. Rehabil.* **50**, 32 (1969).
23. J. V. Basmajian and T. G. Simard, *Amer. J. Phys. Med.* **46**, 1427 (1967); I. H. Wagman, D. S. Pierce, R. E. Burger, *Nature* **207**, 957 (1965).
24. S. Carlsöö and A. Edfeldt, *Scand. J. Psychol.* **4**, 231 (1963).
25. T. G. Simard and H. W. Ladd, *Develop. Med. Child Neurol.* **11**, 743 (1969); *Arch. Phys. Med. Rehabil.* **52**, 447 (1971).
26. A. Jacobs and G. S. Felton, *Arch. Phys. Med. Rehabil.* **50**, 34 (1969).
27. C. D. Hardyk, L. F. Petrinovich, D. W. Ellsworth, *Science* **154**, 1467 (1966).
28. W. F. Floyd and P. H. S. Silver, *J. Physiol. London* **129**, 184 (1955).
29. E. Asmussen and K. Klausen, *Clin. Orthop. Related Res.* **25**, 55 (1962).
30. K. Klausen, *Acta Physiol. Scand.* **65**, 176 (1965).
31. S. Carlsöö, *Acta Orthop. Scand.* **34**, 299 (1964).
32. J. V. Basmajian and R. K. Greenlaw, *Anat. Rec.* **160**, 310 (1968).
33. R. Tuttle and J. V. Basmajian, in preparation.

22 The Basic Principles Underlying Neuromuscular Re-education

Alberto A. Marinacci

VOLUNTARY motor function depends upon the transmission of nerve impulses from the cortical centers through the upper motor neuron to reach the anterior horn cells. The axons of these cells (the lower motor neurons) extend to the group of muscle fibers which they innervate. The electrical waves resulting from such impulses are commonly referred to as motor unit potentials. The motor unit is therefore the ultimate unit of voluntary function. Under normal conditions this function depends upon the activity of two basic structures, the nerve and the muscle.

Any alteration in either the nerve or the muscle results in an impairment of voluntary function. Any disease which affects either the upper or lower motor neuron pathway impedes the nervous impulse and results in muscular weakness or paralysis. Any method which can restore the transmission of neuromuscular impulses is of vital importance in the restoration of voluntary function. One reason for failure of transmission of the nerve impulses is that the pathways have not as yet been fully established, such as in infants (Marinacci, 1959). When the neuromuscular mechanism is first established, the ability of the patient to transmit impulses makes voluntary function possible.

The established transmission of the neuromuscular impulses may then be disrupted by a physiological or pathological interruption of either the upper or lower motor neuron pathways. In reversible physiological block, interruption may be temporary, lasting from a few seconds to weeks or months, or even years, depending upon the severity of the insult. This type of block occurs in cases of temporary edema or vascular changes of the brain, spinal cord, nerve roots, or peripheral nerves. On the other hand, pathological blocks may result in prolonged interruption of transmission for months or years as seen in nerve injury. Impairment of nerve function then persists until full nerve regeneration

Reprinted with permission of the publisher from *Applied Electromyography*. Copyright 1968 by Lea & Febiger.

has taken place. Permanent paralysis also occurs in degenerative diseases of the brain, anterior poliomyelitis, progressive spinal muscular atrophy, tumors of the spinal cord, and in those cases wherein nerve regeneration has not taken place.

With the aid of the EMG it is found that transmission of nerve impulses may be impaired by numerous factors: (1) The inability of the patient to initiate the nerve impulses such as in sensory aphasia. (2) In situations in which a certain number of nerve cells have escaped destruction remaining in a state of temporarily impaired function. This is exemplified in lesions of the upper motor neuron as in cerebral thrombosis, in which many of the cortical nerve cells have been destroyed, while those which survive are not able to transmit nerve impulses because they are in a crippled state. (3) Temporary blocking of the pathways (reversible physiological block) due to local edema, as in cellulitis of one limb.

(4) In cases wherein the anatomical pathways have re-established themselves but the impulses are not being transmitted because the cerebral motor engrams have become disorganized. This is especially true in cases wherein the patient has had severe pain (sympathetic causalgia) in a limb during which time the patient has not used the affected member. Following recovery from the pain, it is found that the patient has forgotten how to transmit nervous impulses to the muscles of the limb with a resulting atrophy of disuse. (5) A similar mechanism occurs in peripheral nerve lesions in which the nerve has regenerated. Although the anatomical neural pathways have been re-established, the engrams here again have been disorganized and the patient does not remember how to use these re-established pathways. (6) Another phenomenon is that of substitution. Although present to a certain degree in all these conditions, substitution mechanism implies that when a group of muscles become paralyzed due to nerve injury, either the proximal, adjacent, or distal muscle groups take over the function of the paralyzed ones. This substitution mechanism is characteristically seen in axillary nerve injury when the muscle group of the shoulder cuff compensates for the paralyzed deltoid. The muscles most frequently called upon for substitution in the event of injury to the axillary nerve are the supraspinatus, teres major, trapezius, coracobrachialis and the pectoralis muscles. By the time the function of the axillary nerve returns, the patient has forgotten how to use the deltoid and continues to use the substituting muscle group. This mechanism can best be described as one of "detour of function." Another outstanding example is in the case of paralysis of the biceps muscle by injury to the musculocutaneous nerve. The brachioradialis and extensor carpi radialis take over completely the function of the biceps.

Again when the musculocutaneous nerve regenerates, it is found that the patient has forgotten how to use the re-established pathways. Cases of old Bell's palsy may show a similar situation.

(7) There are other cases such as anterior poliomyelitis wherein the patient has been completely paralyzed for months and a considerable amount of destruction of the lower motor neurons has taken place. The pathways for certain numbers of cells that have not been destroyed but which were under a state of physiological block are functionally normal although the patient has forgotten how to transmit the impulses over them. (8) Functional paralysis has also been observed in which the patient has forgotten how to transmit the impulses to an extremity. This may result in complete uniform atrophy of disuse. Occasionally such cases have misled the physician to believe that the paralysis is of true neurogenic origin to the extent that tendon repair has been done.

(9) The EMG is also now proving useful in those conditions in which an excessive neuromuscular transmission of impulses is due to local muscular hypertonia. This is especially true of patients who have neuromuscular tension of the muscles of the neck producing tension headaches. After the electrode is inserted into these muscles, the patient hears a constant sound from the loud speaker produced by the contraction of the muscles of the neck and is able to learn how to relax. When the muscles are completely relaxed, the sound from the speaker stops. Thus, by listening to the speaker, the patient is able to learn when the excessive tension in the muscles of the neck can return to his voluntary control for his relaxation (Marinacci and Horande, 1960).

ROLE OF THE ELECTROMYOGRAM IN FACILITATION OF TRANSMISSION OF NEUROMUSCULAR IMPULSES

The basic problem in neuromuscular re-education is the transmission of the nerve impulses on attempted voluntary movement on the part of the patient. Thus any method which will facilitate the transmission of these impulses will be of definite value in obtaining voluntary muscle activity. I have already made certain observations referable to this mechanism. The first important clinical observations, however, were made by Borsook, Billig and Golseth (1952). They observed that the EMG facilitated the transmission of nerve impulses and as a consequence the motor units increased in frequency as well as in amplitude (power).

The use of electromyography in neuromuscular re-education is a simple procedure. In order for the patient to understand the purpose of this method, it is advisable for him to hear and recognize the sound from the loud speaker produced by contracting his own muscles. This is done by simply inserting the active electrode into a normal muscle such as the deltoid (Marinacci, 1955). The examiner then presses his thumb against the outer aspect of the elbow. The patient is then instructed to exert slight pressure against the examiner's thumb, that is, to abduct the arm slightly. This slight exertion, produces a repetitive, sharp, clear, knocking sound from the loud speaker. The patient is allowed to relax the deltoid and the repetitive motor unit potential will cease. Again the patient is instructed to abduct the arm slightly, and the transmission of the neuromuscular impulses will again be heard from the loud speaker. This procedure is repeated a few times until the patient is able to reproduce the sound. This method is similar to that of driving an automobile and learning how one can control the motor by the amount of pressure exerted on the accelerator. If the patient, however, is severely paralyzed, and there is no normal muscle present, in order to demonstrate what is expected of him it is advisable for the examiner to demonstrate the technique on himself.

The next step consists of inserting the needle electrode into the paralyzed muscle and instructing the patient to exert voluntary effort to determine whether there exists any latent muscular activity. If such is found, it is brought to the attention of the patient by directing him to the sound emitted from the speaker. With the patient's repeated efforts, the sound from the speaker guides him to exert the proper degree of voluntary effort and thus facilitates the transmission of neuromuscular impulses. It is apparent therefore that the auditory impulses can be a useful aid in facilitating transmission of the motor impulses.

In the following paragraphs an attempt will be made to explain the different types of impulses that can be detected in this process with the aid of the EMG. The range of progressive responses varies according to the following pattern: (1) a complete failure to transmit impulses, (2) only reflex nerve impulses are elicited, (3) the occurrence of partially reflex and partially voluntary control, (4) only transient voluntary control, (5) faltering impulses not under complete voluntary control, and (6) impulses under full voluntary control.

ABSENCE OF NERVE IMPULSES. In the examination of weakened and atrophic muscles incident to a neurogenic disorder, one may find at first no evidence of motor impulses. This does not prove, however, that no nerve impulses can be transmitted to the muscle examined. Experience has shown that by repeated searching on multiple trials one may find a few isolated impulses (motor units).

REFLEX NEUROMUSCULAR IMPULSES. This type of impulse frequently occurs during the examination of patients with long-standing neuromuscular atrophy. As the needle is being inserted into the atrophic muscle, one may detect no motor units at all. By repeated voluntary efforts, however, one may hear the responsive activity of a motor unit which may fire only once or, at most a few times. Occasionally the unit may keep firing during the entire period of observation without any evidence that it is under voluntary control. This reflex neuromuscular impulse appears to be induced essentially by the insertion of the needle electrode.

PARTIAL REFLEX AND PARTIAL VOLUNTARY CONTROL. These impulses constitute a step which takes place during the period of change from a reflex status to voluntary control of the nervous impulses. When reflex activity is being elicited, the patient is instructed to listen to it from the loud speaker and to attempt to bring the musical note under voluntary control. When the patient learns to control the motor unit by listening to the speaker, transition from the reflex to the voluntary state can usually be encouraged. Thus, the reflex phase can be progressively transferred to the voluntary phase. This does not mean that in this phase the patient will have full control of the transmission of the nerve impulse, for he often becomes easily fatigued. Moreover, if the patient becomes too anxious the firing may cease. Too much effort on the part of the patient may also cause the motor unit to cease firing. Only with continued persistent but gentle effort on the part of both examiner and the patient can the ultimate facilitation of transmission of the neuromuscular impulses be accomplished.

TRANSIENT MOTOR UNITS. These are commonly observed. Many nerve impulses under transitory voluntary control may cease because the neural pathways are not yet fully re-established. Then it usually takes a considerable amount of patience to reactivate these impulses. These transient motor units come and go but eventually with persevering yet cautious effort, can be brought once more under voluntary control. In this phase, there is usually considerable amount of hesitation. This may be compared to an attempt to start the motor of an automobile on a cold morning; it falters and dies at first but finally the motor "turns over" and begins to function.

FALTERING IMPULSES NOT UNDER COMPLETE VOLUNTARY CONTROL. Motor units which show hesitation, not being under complete voluntary control, will at first start gradually and then increase in frequency. Some, however, will become fatigued and disappear. Only by listening to the loud speaker and using a little strength can the patient assist these faltering motor units. The transmission becomes progressively easier and full voluntary transmission of the nerve impulses can ultimately be achieved.

FULL VOLUNTARY CONTROL. In some cases, full voluntary control of the neuromuscular mechanism can be accomplished. When this has been achieved, there is at first an increase in frequency followed by an increase in magnitude of the waves. Next "use" hypertrophy of the muscle fibers will take place. This phase is marked by a further increase in the frequency and the voltage of the motor units. This process leads to the development of giant motor units which serve to compensate in part for the loss of nerve tissue.

In searching for latent neuromuscular potentials, it has been found that in lesions of long standing resulting in groups of paralytic muscles with evidence of fibrosis, there is no hope of salvaging any neuromuscular function. If one finds areas of fibrillation in muscles that have undergone fibrosis, this very activity suggests the presence of an "oasis" of a few latent motor units. When after repeated trials, one can elicit either a reflex potential or a faltering motor unit, these units can usually be brought under voluntary control. This phenomenon of areas of survival of denervated muscle fibers is well demonstrated by a partially paralyzed group of muscles wherein the denervation activity had survived for long periods. This was recorded in a man of 74 years who had apparently suffered a brachial plexus injury at birth (Erb's paralysis). In this case the presence of partial neuromuscular activity helped the denervated muscle fibers to survive. However, this is not true when the lesion is complete, for then there is no help from antagonistic groups of muscles. In other words, in cases wherein there is a complete lesion of the radial, median and ulnar nerves in the forearm, no denervation activity can be recorded. On the other hand, if there is a radial nerve lesion but the median and ulnar nerves are intact, this very function of flexion of the muscles by two healthy nerves helps the survival of the extensors supplied by the impaired radial nerve for considerable intervals.

There are two specific groups of patients in whom audioneuromuscular re-education has been found to be of considerable value. The first is the large group of individuals in whom a considerable amount of latent function is detected. For them, the unused neuromuscular function is brought to the patient's attention and within a short time, the apparently completely paralyzed limb is brought under voluntary control and function is re-established. Usually only one trial is necessary to start the patient toward a degree of voluntary function from the heretofore paralyzed muscles. The neuromuscular activity found in this group consists mostly of simple and complex motor units; as a rule no giant motor units are found.

In the second group, the latent function is minimal in degree. With this group, it is a problem of conversion of very minute nonproductive potentials to the production of giant motor units.

Giant motor unit activity is the result of repair of the neuromuscular mechanism and represents a process compensating for the deficit of the motor nerve cells. Nevertheless, the salvaged power in the giant motor units is not as efficient as in the normal motor units. One giant motor unit can generate as much power as fifty normal ones, but this power is not as useful as a normal one since it fatigues readily, and lacks the kind of potency which is necessary to perform finer movements such as those of the fingers. Moreover, the response of these units to voluntary command is somewhat slow. During the regular process of aging, these giant units also lose their power much faster than normal motor units. This loss of power incident to aging gives rise to a progressive clinical picture which may simulate various neuromuscular disorders (Marinacci, 1960). In order to develop these giant motor units by means of audio-neuromuscular re-education, a considerable amount of effort, patience, time and mutual comprehension on the part of the examiner and patient is required.

The most important factors in neuromuscular re-education or reactivation of the motor unit are an increase in the voltage, the duration, the frequency and the promptness of the motor unit's response to voluntary efforts. These results are best obtained from the development of physiological compensatory factors as seen in the formation of the giant motor units (Marinacci, 1955). Motor units have been observed to *increase in voltage* from about $0.5\mu V$. to as much as $25\mu V$. They may *increase in duration* from 5 to 30 msec. within a period of 12 to 18 months. This increase in voltage and duration is most likely due to a hypertrophy of the muscles from overuse, and the acquisition of additional muscle fibers by the surviving motor units. This addition of muscle fibers is the result of sprouting or budding of the terminals of functional axons which assume control of adjoining denervated elements. This process of sprouting is a natural tendency of healthy motor nerve fibers to re-innervate the neighboring denervated muscle fibers. By this process, therefore, the motor units acquire additional muscle fibers which add to the size and increase the duration of action of the giant motor units. It is presumed that the denervated fibrillating muscle somehow stimulates sprouting for its own re-innervation, possibly as a means for self-survival. The terminal axons of those lower motor neuron cells indicate that the process of sprouting must be quite common. These terminal axons find themselves enveloped by numerous denervated and fibrillating muscle fibers which appear to be fertile soil for the sprouting process. This process is, therefore, common in diseases which produce a large amount of denervated muscle fibers such as those resulting from severe anterior poliomyelitis, severe neuropathies and instances in which the nerve cells have survived severe nerve injury, and/or in which a few nerve cells have managed to regenerate from severe nerve root lesions such as affections of the cauda equina.

Still another ideal situation for the sprouting process presents itself in a patient who has had an end-to-end suture for nerve repair. In examining a large number of patients who have had end-to-end suture it was found that the maximum amount of return of motor units (axon regeneration detected by the EMG) does not exceed 10 per cent. In this group of cases, the apparently good functional return is the result of the production of the giant motor units. It is assumed therefore that giant motor units represent an expression of increased biological efficiency as a result of hypertrophy of the re-innervated muscle fibers.

OBJECT OF NEUROMUSCULAR RE-EDUCATION

The ultimate objective of neuromuscular re-education is to re-establish functionally the neural pathways. Thus, the frequency of neuromuscular potentials is increased. The increase in frequency also facilitates overuse of the muscle fibers resulting in their hypertrophy. These two elements result in production of giant motor units which are responsible for the increase in voluntary power.

By listening to the sound the patient is able to increase the frequency of the nerve impulses resulting in hypertrophy of the muscle fibers from overuse. Sprouting of the terminals of the axons to re-innervate the adjacent fibrillating muscle fibers is not entirely adequate to accomplish this end. It appears that all the muscle fibers that have been re-innervated by this process of sprouting do not come under voluntary control, for not all of the new-formed terminals transmit motor impulses. The increase in frequency and the overuse of the muscles are therefore mainly responsible for the increase in voltage and appear to be the most important basis for any increase of voluntary neuromuscular power.

The promptness of response of muscular contraction to voluntary effort is dependent upon the physiological status of the neural pathways as well as on the ability of the patient to transmit neuromuscular impulses. If the anterior horn cells and their axons are relatively normal, they will eventually transmit the nerve impulses as voluntary effort demands. In those cases, however, in which the nerve cells are injured or "sick," the response to voluntary effort is faulty and may never be completely accomplished. In severe lesions, the transmission of the neuromuscular impulses to voluntary demand is usually poor.

SUMMARY

The purpose of audio-neuromuscular re-education in the process of restoration of the motor function is to detect by means of the EMG the latent microphysiological neuromuscular potentials and to bring these potentials once more under useful voluntary control.

References

Borsook, M. E., Billig, H. K. and Golseth, J. G.: Betaine and glycocyamine in the treatment of disability resulting from acute anterior poliomyelitis. Ann. West. Med. & Surg. 6:423, 1952.

Marinacci, A. A.: Horse serum neuropathy. Bull. Los Angeles Neurol. Soc. 23:149, 1958.

———: Embryological development of neuromuscular mechanism. Preliminary report of its electromyographic evaluation and clinical significance. Bull. Los Angeles Neurol. Soc. 24:36, 1959.

———: Clinical Electromyography. A brief review of the electrophysiology of the motor neuron disease, peripheral neuropathy and the myopathies. Los Angeles, San Lucas Press, 1955.

———: Lower motor neuron disorders superimposed on the residuals of poliomyelitis. Value of the electromyogram in differential diagnosis. Bull. Los Angeles Neurol. Soc. 25:18, 1960.

———: Diagnosis of paralysis following surgical procedures. Indus. Med. & Surg. 29:137, 1960.

Marinacci, A. A.: Neurological aspects of complications of spinal anesthesia. With medicolegal implications. Bull. Los Angeles Neurol. Soc. 25:170, 1960.

———: Some unusual causes of pressure neuropathies. Bull. Los Angeles Neurol. Soc. 25:223, 1960.

Marinacci, A. A. and Horande, M.: Electromyogram in neuromuscular reeducation. Bull. Los Angeles Neurol. Soc. 25:57, 1960.

VII

ELECTRODERMAL CONTROL

23
Effects of Exteroceptive Feedback and Instructions on Control of Spontaneous Galvanic Skin Response

Valerie Klinge

ABSTRACT

To more fully explore the possibility of control of the autonomic nervous system (ANS) in the human S, this study investigated the effects of instructions and exteroceptive feedback on the control of spontaneous galvanic skin responsivity (GSR). Two sets of instructions ("Relax" and "Think") were alternately presented to Ss under four types of meter feedback: accurate, positive, negative, and control. In addition to GSR activity, respiration and cardiac rate were recorded. The results indicated that Ss receiving accurate feedback were significantly better able to comply with the "Relax"–"Think" instructions; next most effective was positive feedback, while negative and control feedback were least effective. No within-subject relationships were found between control of spontaneous GSRs and respiration or cardiac activity. The results suggest that Ss are responsive to the consequences of their autonomic behavior in much the same manner as they are responsive to the consequences of other learned behaviors.

DESCRIPTORS: GSR, Exteroceptive feedback, Instructions, Control.

The autonomic nervous system (ANS) has long been considered an index of the individual's activation level, emotional arousal, intensity of excitation, etc. Traditionally, this system has been regarded as involuntary or self-modifying, and not amenable to volitional regulation. While studies using classical conditioning procedures to modify ANS functions have a long history in the literature, the more recent studies on ANS conditioning fall into two general categories: those employing operant procedures (reward and punishment in the more traditional sense), and those employing externalization procedures (exteroceptive or augmented sensory feedback). Although the effects of exteroceptive feedback have been studied on a number of autonomic responses, to date only three studies (Edelman, 1970; Stern & Kaplan, 1967; Stern & Lewis, 1968) have attempted to assess the effects of feedback on voluntary control of the galvanic skin response (GSR).

This investigation was conducted in partial fulfillment for the Ph.D. degree in Psychology at the State University of New York at Stony Brook.

Appreciation is expressed to Dr. James H. Geer for his encouragement of this study and his consultation as Chairman of the Dissertation Committee.

Address requests for reprints to: Valerie Klinge, Ph.D., Department of Psychology, The Lafayette Clinic, 951 East Lafayette, Detroit, Michigan 48207.

Reprinted with permission of the publisher and author from *Psychophysiology*, Vol. 9, 305-317. Copyright 1972, The Society for Psychophysiological Research.

The purpose of the present research was to more fully explore the possibility of ANS control by examining the effects of feedback and instructions on the ability of human Ss to control their levels of sympathetic activation. Since level of emotional arousal is generally considered a function of sympathetic nervous system involvement, it was an index of this sympathetic involvement with which the present studies were concerned. Specifically, it was predicted that sympathetic control, as indexed by galvanic skin responsivity, would be, in part, a function of two independent factors: the type of instructions S received and the type of feedback he received. In the below experiments, each S received both "Relax" and "Think" instructions; it was predicted that GSR responsivity would be significantly lower under the "Relax" instruction than under the "Think" instruction. It was also predicted that the type of GSR feedback S receives, whether accurate, positive, or negative, would influence his control of GSR responsivity. More specifically, it was hypothesized that while the effect of the "Relax"–"Think" instructions would lead to increased control, this control would be heightened when accurate feedback was also present. In other words, a significant interaction was predicted between type of feedback and type of instructions.

Study I

This study was conducted to determine if there was a general feedback or instructional effect on the production of spontaneous GSRs. A spontaneous GSR was defined as a fluctuation in electrical resistance which exceeded 550 ohms and which occurred in the apparent absence of external stimulation.

Method

Subjects. The Ss in this study were 14 Caucasian college males, ranging in age from 17 to 21.

Apparatus. The GSR data were collected on a multi-channel Beckman Type R Dynograph, which was situated in a room adjacent to S's soundproof room. The input circuitry used in the reactive GSR channel consisted of output from the standard GSR coupler which was put through a 6 sec time constant capacitor. The electrodes used were Beckman bio-potential 2 cm^2 silver-silver chloride, and were fastened to S by adhesive collars attached with sodium chloride paste. The GSR data were used as the feedback data for the Ss in the Feedback group; Ss in the control group received no feedback.

The "Relax" and "Think" instructions were presented in alternating 1-min blocks on 2 in. × 2 in. transparent slides which were flashed on a screen directly in front of S. The "Relax" and "Think" conditions were begun as soon as the instructions (described below) had been given. The slide projector itself was in the adjacent polygraph room, and the slides were projected through the window connecting E's polygraph room with S's soundproof room.

Procedure. Feedback Ss received feedback data from their own physiological systems. For these Ss, the centerline feedback meter was wired in parallel with the reactive GSR channel, thus enabling S to view a continuous display of his own GSR responsivity as indexed by the needle deflections on the meter.

Upon entering the laboratory each S was told to remove his wristwatch and any bulky sweaters or shirts which might interfere with recording, and asked to sit in the lounge chair in the S room. Every S was tested individually in this soundproof, temperature-controlled room. After being seated, the S was asked to remain as quiet as possible for the duration of the experiment. The GSR electrodes were then attached, and S was told he would receive further directions over the intercom. During the 10-min period, when base data were collected, the feedback meter was not in operation. After the base data had been collected, the No-Feedback Ss were told:

"During the session I will flash either the word 'Relax' or the words 'Think arousing thoughts' on the screen. What I want you to do is this: when you see the word 'Relax' on the screen, think peaceful, relaxing, non-emotional thoughts; when you see the words 'Think arousing thoughts,' think arousing, emotional thoughts. Keep thinking these thoughts until the alternate word reappears on the screen. Do you understand? Any questions? The entire session will take about half an hour. We are ready to begin now."

Subjects in the Feedback group were then given these directions:

"The meter to your left records the activity level of your thoughts as you are thinking them. If you think strongly emotional thoughts, the needle on the meter will move back and forth at a fairly rapid rate; if you think restful, non-emotional thoughts, the needle will remain in a relatively stationary position. Try it and see." (At this point, the meter was made operative so S could demonstrate the validity of this statement.)

"During the session, I will flash either the word 'Relax' or the words 'Think arousing thoughts' on the screen. What I want you to do is this: when you see the word 'Relax' on the screen, think peaceful, relaxing, non-emotional thoughts; when you see the words 'Think arousing thoughts,' think arousing, emotional thoughts. Keep thinking these thoughts until the alternate word reappears on the screen.

"While you are thinking these thoughts, continue to look at the meter which will tell you the activity level of your thoughts. Remember, when you see the word 'Relax' make the needle remain as stationary as possible by thinking quiet, relaxing thoughts. When you see the words 'Think arousing thoughts,' make the needle move back and forth by thinking arousing, emotional thoughts. Do you understand? Any questions? The entire session will take about half an hour. We are ready to begin now."

The only dependent variable of concern in this study was spontaneous GSR frequency. The method of counting spontaneous responses was identical to that described by Edelberg (1967). According to this method, waves occurring on the downward slope of a preceding inflection were not considered in the frequency count unless they reached the assigned (550 ohm) amplitude from their own inflection to peak. GSRs given to slide onset or offset were not included in the frequency count.

Results

Study I produced 15 min of continuously recorded data for each of the Ss: the last 5 min of base-rate data and 5 min each of data obtained under the "Relax" and "Think" instructions. Analysis of the frequency data produced during the 5-min base period revealed that the two groups, Feedback and No-Feedback, did not differ significantly from each other ($t < 1$). To determine the significance of the effects of feedback and instruction, the mean number of spontaneous responses per minute were then analyzed in a 2 × 2 (Feedback vs No-Feedback ×

"Relax" vs "Think") analysis of variance. The results of this analysis indicated that Ss produced significantly ($F = 56.01$, $df = 1/12$, $p < .0001$) fewer spontaneous responses under "Relax" than under "Think" instructions. This analysis further revealed a significant ($F = 9.69$, $df = 1/12$, $p < .01$) Feedback × Instructions interaction; t tests computed on this interaction indicated that Ss who receive feedback are significantly ($p < .005$) better able to modify their GSRs in accordance with the instructions than are the No-Feedback Ss. The means (in GSRs/min) of this interaction are: "Relax" with feedback, 3.60; "Relax" without feedback, 4.43; "Think" with feedback, 7.34; "Think" without feedback, 5.71.

Since it was possible that the meter needle deflections themselves acted as stimuli in eliciting GSRs, the percentage of the total GSR frequency in each minute block in which two GSRs occurred within a 1.5 sec interval were compared for the two groups. (The 1.5 sec interval was used since the literature [Edelberg, 1967] describes this as being the average GSR latency time.) If this percentage were higher for the Feedback than for the No-Feedback Ss it might be assumed that the meter was eliciting GSRs; naturally, however, the 1.5 sec proximity of two GSRs does not necessarily indicate that the second was elicited by feedback from the first. Analysis of these data (Feedback vs No-Feedback × "Relax" vs "Think" analysis of variance) produced no significant main effects or interactions.

Study II

The results of Study I suggested that GSR control might be related to cardiac or respiratory activity. Inspection of the data and questioning of the Ss indicated that 5 of the 14 Ss (3 in the Feedback group and 2 in the No-Feedback group) seemed to be regulating their breathing and heart rates to better comply with the "Relax"–"Think" instructions. To more fully explore this possibility, it was decided to incorporate respiration and cardiac measures in a subsequent study. Since the findings of Study I indicated that feedback did facilitate GSR control, Study II was designed to investigate the effects of various types of feedback on control of the GSR.

Method

Subjects. The Ss in Study II were 80 Caucasian college males, ranging in age from 17 to 21.

Apparatus. All data were collected on the multi-channel Dynograph described in Study I. Respiration data were collected on one polygraph channel, heart rate on another, and GSR tonic and GSR reactive data on two others. The GSR reactive data were used as the feedback data as explained below, and were presented to the Ss on a centerline meter. The heart and GSR electrodes used were Beckman bio-potential 2 cm^2 silver-silver chloride and were fastened to the S by adhesive collars attached with sodium chloride electrode paste. The heart electrodes were placed on the volar surface of S's left and right lower arms; GSR electrodes were placed on the palmar surface of S's non-preferred hand. A strain gauge, similar to that described by Ackner (1956), was used to measure respiration, and was placed around S's chest just underneath his armpits.

Procedure. The "Relax" and "Think" instructions were presented to S as described for Study I. In this study, half of the Ss in each group started the "Relax"–"Think" series with the "Relax" slide, while the other half were presented first with the "Think" slide. Each slide was presented 5 times for 1 min each time in ABAB order.

In order to assess the relative effectiveness of various types of feedback, four groups of 20 Ss each were used in this study. One group received accurate feedback, one positive, one negative, and one control feedback. Accurate feedback Ss received feedback data from their own physiological systems, as in Study I. As will be recalled, for these Ss the feedback meter was wired in parallel with the reactive GSR polygraph channel, thus enabling the S to view an accurate, continuous, and simultaneous display of his own GSR responsivity as indexed by the needle deflections on the meter. Because data from the Accurate Ss were used in compiling the feedback for the other three groups, it was, of course, necessary to test the Accurate Ss prior to those in the other three groups; S assignment to these latter groups was random.

Analysis of the GSR frequency data produced by Accurate feedback Ss revealed that the mean number of spontaneous responses emitted during the last 5 min of the base period was 3/min. Since the distribution of these spontaneous responses was approximately normal, Ss who produced 3 or more per minute were termed High spontaneous responders while those who produced fewer than 3 per minute were Low responders. There was a significant positive correlation between the number of spontaneous responses emitted in the base period and the number emitted in both the "Relax" ($r = +.685$, $p < .01$) and the "Think" ($r = +.537$, $p < .02$) periods for the Accurate Ss. This suggested that Ss in the other three feedback goups should be categorized as High or Low responders on the basis of the number of responses they emitted during their base periods. This categorization would increase the probability that the number of responses the Positive, Negative, and Control Ss received during the "Relax" and "Think" periods would be related to the number they actually produced during the base periods.

The GSR data emitted by the Accurate Ss during the 5 "Relax" and 5 "Think" periods were used in determining the feedback for the other three groups. The average number of spontaneous GSRs produced by Low-responding Accurate Ss under the "Relax" instruction was computed and presented as feedback to Low Positive Ss during the "Relax" periods and to Low Negative Ss during the "Think" periods. The average number of spontaneous responses produced by the Low-responding Accurate Ss under the "Think" periods was presented to Low Positive Ss under "Think" and to Low Negative Ss under "Relax." Similarly, the average number of spontaneous responses produced by High-responding Accurate Ss was presented to High-responding Positive and Negative Ss as shown in Table 1. The feedback presented to the Positive, Negative, and Control Ss, in addition to being based on the GSR frequency data produced by the Accurate Ss, was also related to the amplitude and temporal spacing of the spontaneous GSRs produced by the Accurate Ss. The feedback was presented to the Ss via a meter needle deflection produced by the output of a tape recorder whose recordings had been made by manually operating a polygraph pen so that its tracings met the criteria noted above. That is, the feedback recordings were based on the minute

TABLE 1
Feedback presented to groups – Study II

Groups	Feedback	
	With "Think" Instructions	With "Relax" Instructions
Low Accurate (N = 10)	Self-produced feedback.	Self-produced feedback.
High Accurate (N = 10)	Self-produced feedback.	Self-produced feedback.
Low Positive (N = 10)	Feedback based on GSRs produced by Low Accurate Ss during their "Think" period.	Feedback based on GSRs produced by Low Accurate Ss during their "Relax" period.
High Positive (N = 10)	Feedback based on GSRs produced by High Accurate Ss during their "Think" period.	Feedback based on GSRs produced by High Accurate Ss during their "Relax" period.
Low Negative (N = 10)	Feedback based on GSRs produced by Low Accurate Ss during their "Relax" period.	Feedback based on GSRs produced by Low Accurate Ss during their "Think" period.
High Negative (N = 10)	Feedback based on GSRs produced by High Accurate Ss during their "Relax" period.	Feedback based on GSRs produced by High Accurate Ss during their "Think" period.
Low Control (N = 10)	Feedback based on Low Accurate Ss' mean number of GSRs produced under "Relax" and "Think."	Feedback based on Low Accurate Ss' mean number of GSRs produced under "Relax" and "Think."
High Control (N = 10)	Feedback based on High Accurate Ss' mean number of GSRs produced under "Relax" and "Think."	Feedback based on High Accurate Ss' mean number of GSRs produced under "Relax" and "Think."

by minute average frequency, amplitude, and spacing of the GSRs produced by the Accurate Ss.

Positive feedback Ss received prerecorded meter read-out which had been set to coincide with the "Relax"–"Think" instructions they received. For example, when the instruction was "Think," the meter indicated to the Positive feedback S that he was producing a high frequency of responses, regardless of his actual GSR performance. Thus, Ss in the Positive feedback group received feedback which was non-contingent on their performance but which was positively related to instructional compliance.

Negative feedback Ss received prerecorded meter read-out which had been set in opposition to the "Relax"–"Think" instructions. Here, for example, if the instruction was to "Think," the meter needle remained relatively stationary regardless of S's actual performance, indicating to him that he was not complying with the instructions. Thus, Ss in the Negative feedback group received feedback which was non-contingent on their performance and which was negatively related to instructional compliance.

Control feedback Ss received prerecorded meter read-out which was non-contingent on their performance and which was also unrelated to the "Relax"–"Think" instructions. Low responding Control Ss received feedback based on the Low responding Accurate Ss' mean number of "Relax" and "Think" responses; these Control Ss received the same number of needle deflections in the "Relax" as in the "Think" instruction. High responding Control Ss received feedback based on the High responding Accurate Ss' mean number of "Relax" and "Think" responses; again, these Control Ss received the same amount of feedback under the "Relax" instruction as they did under the "Think" instruction. Thus, all four feedback groups received the same total number of needle deflections from the meter.

The procedure for testing these Ss was similar to that employed in Study I. Accurate, Positive, and Negative Ss in Study II received the identical directions that the feedback Ss in Study I had been given. However, the Control Ss in Study II were told:

"Throughout the experiment, I want you to pay attention to the meter on your left. All you need do is watch the needle closely. I will explain later why this is necessary." (At this point, the meter was made operative.)

"During the session, I will flash either the word 'Relax' or the words 'Think arousing thoughts' on the screen. What I want you to do is this: when you see the word 'Relax' on the screen, think peaceful, relaxing, non-emotional thoughts; when you see the words 'Think arousing thoughts,' think arousing, emotional thoughts. Keep thinking these thoughts until the alternate word reappears on the screen. While you are thinking these thoughts, continue to look at the meter. Do you understand? Any questions? The entire session will take about half an hour. We are ready to begin now."

Thus, Control Ss were not informed that the meter deflections were relevant to the activity level of their emotional thoughts; the purpose of instructing these Ss to attend to the meter was to control for attention factors. Subjects in the other three groups were not advised that the meter was monitoring their GSR activity; they were told only that the meter was indexing their "emotional activity."

After the Ss had been given their directions, the "Relax" and "Think" slides were presented; each slide presentation lasted for 1 min and each slide was shown 5 times. There was no further communication between E and S until the end of the experiment.

Results

Study II produced 15 min of continuously recorded data for each of the 80 Ss: the last 5 min of base rate data, and 5 min each of data obtained under the "Relax" and "Think" instructions. The dependent variables analyzed in this experiment were spontaneous GSR frequency, GSR amplitude, tonic GSR level, respiration frequency, and heart rate.

As mentioned earlier, all Accurate feedback Ss were tested prior to Ss in the other three feedback groups. To ascertain if there were significant experimenter changes in, for example, familiarity in attaching electrodes, calibrating the polygraph, etc., several internal checks were made comparing the first 5 Accurate Ss tested with the final 5. Tests were calculated comparing the means and variances of the 5 first-run and 5 last-run Accurate Ss on spontaneous GSR frequency, amplitude, and tonic level, initial resistance level, heart rate, and respiration rate. On none of these six variables were there any differences between the first and last run Ss in means or in variances. On the basis of the finding that no significant internal differences were found within the Accurate S data, the primary dependent variables from all four feedback groups were then analyzed.

Spontaneous GSR Frequency Data. The dependent variable of greatest interest in Study II was spontaneous GSR frequency since it was these data which had been presented as feedback. The method of determining spontaneous GSR frequency was identical to that described in Study I. The number of spontaneous responses produced during the 5 min base period were analyzed and it was determined that none of the four feedback groups differed among themselves in the average number of responses emitted during the base period (all t values <1). The data for the "Relax" and "Think" periods were then cast in a Lindquist modified Type VI design (Lindquist, 1953).

The significant ($F = 29.371$, $df = 1/72$, $p < .0001$) Low-High responsivity effect obtained in this analysis indicated that Low and High spontaneous responders during the base period maintained their relative level of producing spontaneous GSRs throughout the experimental periods. The intercorrelations of base spontaneous responding with "Relax" and "Think" responding substantiate this finding. The correlation of base with "Relax" ($r = +.653$) and with "Think" ($r = +.619$) and the correlation of "Relax" with "Think" responding ($r = +.756$) are all significant at $p < .0001$.

Another significant effect of the GSR frequency analysis was that of the significant ($F = 64.925$, $df = 1/72$, $p < .0001$) "Relax"–"Think" main effect, a finding also obtained in Study I. The mean number of responses emitted under

TABLE 2

Mean numbers of spontaneous GSRs per minute and mean GSR amplitudes for the four feedback groups in Study II

Variables	Means							
	Accurate Ss		Positive Ss		Negative Ss		Control Ss	
	"Relax"	"Think"	"Relax"	"Think"	"Relax"	"Think"	"Relax"	"Think"
Spontaneous GSRs (frequency per min)	2.89	5.86	3.71	5.60	4.09	4.82	2.69	3.64
Amplitude GSRs (square root conductance)	.478	.664	.667	.809	.733	.776	.927	.818

Note.—See text for significance levels of all interactions and t tests.

the "Relax" instruction was 3.35/min, while the mean under "Think" was 4.49. This ability to control frequency of GSR output in the presence of differential instructions occurs regardless of the type of feedback the S receives. It will be recalled the present investigation hypothesized that while the instructions alone would lead to increased control of spontaneous responding, this control would be augmented when accurate feedback was also present. This hypothesis was confirmed by the findings of Study I and also by the significant Feedback × Instructions interaction ($F = 6.290$, $df = 3/72$, $p < .005$) of Study II. T tests computed on this interaction reveal that while all four feedback groups emit significantly more responses under the "Think" instruction than they do under the "Relax" instruction, this increment in responsivity is differential among the groups. (See Table 2.) Under the "Relax" instruction, the mean number of responses emitted per minute by the Accurate Ss was significantly fewer than the number emitted by the Positive ($p < .05$) or the Negative ($p < .01$) Ss, but was not significantly different from the Control Ss. The number emitted by the Positive Ss was not significantly different from the number emitted by the Negative Ss, but was different ($p < .02$) from the Control Ss. The number emitted by the Negative Ss was significantly ($p < .01$) greater than the Control Ss. Under the "Think" instruction, the number of responses emitted by the Accurate Ss was not significantly different from the Positive Ss, but was significantly more than the number emitted by the Negative Ss ($p < .02$) and the Control Ss ($p < .01$). Positive Ss emitted more responses than either Negative Ss ($p < .05$) or Control Ss ($p < .01$). Negative Ss produced significantly more ($p < .01$) responses than Control Ss.

Two-tailed t tests calculated on the "Relax"–"Think" difference scores show that those Ss who receive Accurate feedback are significantly better able to comply with the instructions than are Ss receiving Positive, Negative, or Control feedback (in each comparison, $p < .01$). Further t tests reveal that Ss receiving Positive feedback are superior to Negative ($p < .01$) and Control ($p < .05$) feedback Ss in complying with the instructions. Subjects receiving Negative feedback do not differ significantly ($t < 1$) from the Controls.

Other significant findings of the GSR frequency analysis were the Feedback × Trials ($F = 3.527$, $df = 12/288$, $p < .0001$) and the Feedback × Trials × Instructions ($F = 2.000$, $df = 12/288$, $p < .025$) interactions. Essentially what these rather complex interactions show is that the Accurate and Positive feedback Ss decrease most in responsivity over the course of the "Relax" trials and maintain a higher level of responsivity over the "Think" trials than do the Negative or Control Ss. That is, throughout the experiment Ss receiving Accurate or Positive feedback are better able to comply with the "Relax" and "Think" instructions than are Ss receiving Negative or Control feedback.

To determine if the effects of feedback and instructions were obtained on GSR components other than frequency, similar analyses were computed on the GSR amplitude and tonic level data.

Spontaneous GSR Amplitude Data. GSR amplitude was calculated by converting each amplitude resistance to square root conductance change (Montagu & Coles, 1966). The amplitude analysis revealed that the main effect of Low-High level of responsivity was significant ($F = 9.602$, $df = 1/72$, $p < .005$), indicating that Ss who are Low spontaneous frequency responders also produce amplitudes of less magnitude than do High frequency responders.

An interesting finding which approached significance in the GSR amplitude analysis of variance was the Feedback × Instruction interaction ($F = 2.463$, $df = 3/288$, $p < .10$). (See Table 2.) T tests computed on this interaction demonstrated that for Accurate feedback Ss there was a significant increase ($p < .05$) in amplitude between the "Relax" and "Think" instructions: for Positive feedback Ss this increase approached significance ($p < .10$), while for Negative and Control Ss the increment between "Relax" and "Think" instructions was not significant. It will be recalled that this Feedback × Instruction interaction which approached significance in the amplitude analysis of variance parallels that reported above for the frequency analysis of variance in both the preliminary and major study. Further, this finding suggests that the differential feedback, which reflected GSR frequency, also had a similar effect on the amplitude component of the GSR.

Another noteworthy finding in the amplitude data was the very stable intercorrelations: the correlation of base period amplitude with "Relax" amplitude = +.944; base period with "Think" = +.940; and "Relax" with "Think" = +.944. It is likely that this very stability of the amplitude component of the GSR makes it less amenable than the frequency component to the modifying effects of feedback and instructions.

GSR Tonic Data. To determine if the feedback or instructions had similar effects on the tonic component of the GSR, analyses were also compared on this variable using as the basic datum mean level of conductance (mhos) per minute. Analyses of these data indicated that the main effect of Low-High responsivity approached significance ($F = 3.421$, $df = 1/72$, $p < .10$); Low GSR frequency responders demonstrated a tendency to have lower tonic conductance (higher resistance) levels than High frequency responders. This tendency is consistent with the more definitive findings reported above for the GSR frequency and amplitude analyses, and suggests that the levels of responsivity for the various components of the GSR are within-subject related.

Another finding of interest in the tonic analysis is the significant main effect of Instruction. The result here is that Ss' tonic conductance levels are significantly ($F = 4.857$, $df = 1/72$, $p < .05$) lower (resistance is higher) during the "Relax" instructions than during the "Think." Again, this finding is consonant with results reported for the GSR frequency analysis where it was found that significantly more spontaneous GSRs were emitted under the "Think" than the "Relax" instruction.

Respiration Data. The data on respiration were analyzed to determine if the effects of feedback and instructions on differential GSR responding were in any way related to changes in respiration. Respiration rate (number of inspiration-expiration cycles per minute) was used as the dependent variable. Here, a finding of particular interest was the significant "Relax"–"Think" instruction effect ($F = 18.024$, $df = 1/72$, $p < .0005$), which demonstrated that under the "Relax" instruction, respiration rate was lower than under the "Think" instruction. It will be recalled that the main effect of instruction was also significant in the GSR frequency analysis. To investigate any possible relationship between respiration rate and frequency of spontaneous GSR responding, several sets of correlations were computed. Because the issue of mediation of ANS activity has reached such proportions in the literature (Black, 1969; Crider, Schwartz, & Shnidman, 1969; Katkin & Murray, 1968; Katkin, Murray, & Lachman, 1969), correlations between GSR frequency and respiration rate for every possible different subject grouping were computed. The obtained correlations did not substantiate any relationship between respiration rate and number of spontaneous GSRs. Correlations were also computed between respiration rate and GSR amplitude for all possible subject groups during the "Relax" and "Think" instructions. Again, none of these values reached significance.

Heart Rate Data. To determine if the GSR feedback or "Relax"–"Think" instructions had effects on modifying cardiac function, analyses were calculated on heart rate (HR) activity (bpm).

Analyses of the HR data produced during the 5 min of base period revealed no significant differences among the four feedback groups. The HR data from the "Relax" and "Think" periods were then analyzed in a Type VI design.

The noteworthy finding in this analysis was the significant ($F = 54.390$, $df = 1/72$, $p < .0001$) Instruction main effect which revealed that the mean bpm was 78.33 during the "Relax" periods and 82.28 during the "Think" periods. To determine if this difference in HR between instructions was related to the similar difference in GSR frequency, amplitude, or tonic components reported above, several series of correlations were calculated on different subject groupings. None of these correlations approached significance.

A serendipitous finding of Study II was that both the frequency and amplitude of orienting responses (ORs) were related to slide content. Subjects gave significantly more ORs and ORs of greater amplitude to the "Think" than to the "Relax" slides.

Discussion

Study I revealed that the ability to comply with the "Relax"–"Think" instructions occurred for both feedback and no-feedback Ss, but that this ability was heightened when S was also provided with accurate, exteroceptive feedback on his own GSR responsivity. A second investigation revealed that the kind of feedback S received, whether Accurate, Positive, or Negative, had differential effects on his ability to conform to the "Relax"–"Think" instructions. In this latter study, a major finding was that providing S with accurate contingent information most augmented his compliance with the instructions. Subjects who received positive, non-contingent feedback, while not performing as well as the accurate Ss, nonetheless were superior in instructional compliance to Ss in either the negative or control feedback groups. This finding raised an interesting issue. Since positive, negative, and control Ss all received information which was non-contingent on their actual GSR responsivity, it appeared that the kind of feedback, rather than the contingency of this feedback, was the important factor involved. Merely informing the S non-contingently that his performance was in congruence with the instructions increased his ability to comply with these instructions.

There are at least two explanations for this outcome. It is possible that providing positive, "successful," feedback to the S increased his general level of arousal and caused him to "try harder" to comply with the instructions. The S who is informed he is successfully complying with the instructions might, for example, attend to the needle deflections of the meter more closely than negative or control group Ss. To assess the possibility of differential group attention to the meter, the study had incorporated a control group; further, each S in all groups was constantly observed by the E through the laboratory window to ascertain that he was looking at the meter. However, since the present study provided no psychophysiological check on Ss' attention to the meter, as might be indexed, for example, by eye muscle movement, it is not possible to assess differential group attention.

Another possible, although somewhat less likely, explanation for positive S superiority in instructional compliance might be that, by chance, not all of the feedback the positive S received was non-contingent. That is, the positive S might have emitted GSR behavior which was inadvertently positively reinforced by the needle deflections, increasing the likelihood that he would engage in "superstitious" GSR producing behavior. It will be recalled that regardless of what GSR behavior the positive S produced, he was given positive feedback, thus allowing opportunity for what the S might think was a contingency between his superstitious behavior and the needle deflections to develop. While this explanation appears reasonable theoretically, it becomes less tenable when one considers that the development of an inadvertent contingency would probably require the positive S to emit more responses under the "Think" and fewer under the "Relax" instruction than he did, in fact, emit. While the data indicate that the "Relax"–"Think" difference for the positive Ss was statistically larger than the comparable difference for negative or control Ss, this difference does not seem sufficiently large in absolute terms to definitely conclude that a contingency was in operation for the positive Ss.

The explanation of either general arousal or "superstitious" behavior, which might account for the superior instructional compliance of the positive feedback Ss over the negative and control Ss, is consistent with the finding that these latter two groups are not significantly different from one another in the frequency of spontaneous responses they produce. This study has yielded no evidence suggesting that the negative S, who receives negative, "unsuccessful," feedback, increases his general level of arousal or "tries harder" to comply with the instructions. Furthermore, for the negative S, there is no opportunity for an inadvertent contingency to develop between "superstitious" behavior and feedback, since regardless of what the negative S actually does, the feedback tells him he is wrong, i.e., not complying with the instructions.

The results also indicated that the feedback had effects on only the response system it was modifying. Modification of GSR responsivity occurred independent of identifiable mediation or involvement of systematic changes in respiration or heart rate. While both these variables and GSR control show similar changes over the course of the experiment there is no evidence indicating that these three systems are in any way within-subject related; they appear to be functioning independently of one another. Naturally, though, there is still the possibility that some other unmeasured, somatic mediator is responsible for GSR control.

The conclusion of specificity of feedback effects drawn from the present research has particularly interesting implications for psychosomatic medicine where it has been observed that each psychosomatic disturbance has its own specific physiological concomitants. Further, the knowledge that autonomic functioning obeys many of the laws of learning obviates the well-established assumption that psychosomatic disorders are fundamentally different from other learned problems. The applicability of learning principles suggests that if one could present accurate feedback to the psychosomatic patient on the individual response system involved in his particular disturbance, it would be possible to modify the response system implicated without altering other systems which are functioning adequately. This outcome on the specific effects of feedback has been earlier concluded by Shapiro and Crider (1967) who report that, "... conditioned variations in response rate are specific to the reinforced responses and do not necessarily depend on corresponding variations in other autonomic functions [p. 173]." To investigate the possibility that some unmeasured skeletal mediator is involved with or responsible for GSR control, it would be desirable to duplicate the Miller (1969) research using curare on human Ss. (One study, Birk, Crider, Shapiro, & Tursky, 1966, has attempted to do this.) However, as Black (1969) contends, the issue of mediation is more relevant to theoretical concerns on instrumental conditioning than it is to the more practical issues involved in gaining control of ANS functions.

The findings of the present study which suggest that instructions and feedback do influence the ability of subjects to modify their autonomic activity have particularly relevant implications for the area of behavior therapy. It appears that relaxation training, for example, could be greatly facilitated by providing the client with simultaneous exteroceptive feedback on decreases of his autonomic arousal level. Currently, feedback on arousal level is usually available to the individual only proprioceptively or from observations the therapist may relay to him. Providing the client with feedback in another modality might speed the process of relaxation and offer more convincing evidence to him that he is, in fact, successful at controlling his level of relaxation. One would predict that such a procedure would be particularly useful with retarded or chronologically young individuals who are often considered to be externally oriented and generally less sensitive to internal cues related to their own effectiveness.

The present research has provided evidence suggesting that autonomic activity, as indexed by the GSR, can be acquired, maintained, and modified by the same principles which govern other learned behavior. It appears that individuals are responsive to the consequences of their autonomic behavior in much the same way as they are responsive to the consequences of other learned behavior.

REFERENCES

Ackner, B. A simple method of recording respiration. *Journal of Psychosomatic Research,* 1956, *1,* 144–146.

Birk, L., Crider, A., Shapiro, D., & Tursky, B. Operant electrodermal conditioning under partial curarization. *Journal of Comparative & Physiological Psychology,* 1966, *62,* 165–166.

Black, A. H. Mediating mechanisms of conditioning. Paper presented at the meeting of the Pavlovian Society of North America, Princeton, November 1969.

Crider, A., Schwartz, G. E., & Shnidman, S. On the criteria for instrumental conditioning. *Psychological Bulletin,* 1969, *71,* 455–461.

Edelberg, R. Electrical properties of the skin. In C. C. Brown (Ed.), *Methods in psychophysiology.* Baltimore: Williams & Wilkins Co., 1967. Pp. 1–53.

Edelman, R. I. Effects of differential afferent feedback on instrumental GSR conditioning. *Journal of Psychology,* 1970, *74,* 3–14.

Katkin, E. S., & Murray, E. N. Instrumental conditioning of autonomically mediated behavior. *Psychological Bulletin,* 1968, *70,* 52–68.

Katkin, E. S., Murray, E. N., & Lachman, R. Concerning instrumental autonomic conditioning: A rejoinder. *Psychological Bulletin,* 1969, *71,* 462–466.

Lindquist, E. F. *Design and analysis of experiments in psychology and education.* New York: Houghton Mifflin Co., 1953.

Miller, N. E. Learning of visceral and glandular responses. *Science,* 1969, *163*, 434–445.

Montagu, J. D., & Coles, E. M. Mechanism and measurement of the galvanic skin response. *Psychological Bulletin,* 1966, *65*, 261–279.

Shapiro, D., & Crider, A. Operant electrodermal conditioning under multiple schedules of reinforcement. *Psychophysiology,* 1967, *4*, 168–175.

Stern, R. M., & Kaplan, B. GSR: Voluntary control and externalization. *Psychosomatic Medicine,* 1967, *10*, 349–353.

Stern, R. M., & Lewis, N. L. Ability of actors to control their GSRs and express emotions. *Psychophysiology,* 1968, *4*, 294–299.

24
Timing Characteristics of Operant Electrodermal Modification: Fixed-Interval Effects

David Shapiro and Takami Watanabe

Palmar skin potential responses in human subjects were reinforced under several fixed-interval schedules of reinforcement varying from 50 to 150 seconds. With the exception of an initial response burst immediately after reinforcement, the response characteristics appear to be comparable to those observed for skeletal operants. Data are also presented showing that efficiency of performance is improved when subjects are provided an external clock. Concomitant changes in other physiological variables recorded do not show any consistent evidence for somatic mediation of the learned patterns of skin potential activity.

Accumulating evidence in the last decade has dispelled doubt about the appropriateness of an operant scheme in the modification of autonomic activity (see reviews by Katkin & Murray, 1968; Kimmel, 1967; Miller, 1969). One of the next steps is to extend the analogy between an autonomically mediated response and a somatic-motor behavior like a lever press. Chief among the unexplored questions is that of the effect of different reinforcement schedules: whether autonomic activity can be subjected to the schedules worked out for overt somatic behavior and whether the effects are comparable. Autonomic responding is suggested to vary depending on the schedule of reinforcement correlated with an external stimulus (Shapiro & Crider, 1967). Almost all studies in man, however, have utilized continuous reinforcement or short fixed-interval (FI) schedules, the latter to avoid reinforcing responses elicited by the reinforcer. On the other hand, little study has been done on the influence of intermittent reinforcement upon autonomic responses. Greene (1966) demonstrated that fixed ratio reinforcement influences the spontaneous electrodermal response in a fashion only remotely similar to its influence on somatically mediated responses. Shapiro and Crider (1967) showed that a variable ratio schedule yielded more skin potential responses than an intermittent schedule in which long pauses or periods of nonresponse before responding are reinforced.

The exploratory studies reported here are concerned with two issues concerning the timing characteristics of an autonomic operant, the spontaneous skin potential response (SPR): one is the effects of a rather long FI schedule of reinforcement, and the other the possible role of a behavioral clock with respect to re-

[1] This research was supported by National Institute of Mental Health Research Grant MH-08853 and Research Scientist Award K5-MH-20,476; also by Office of Naval Research Contract N00014-67A-0298-0024.

inforcement under a FI schedule. In the FI procedure, the first occurrence of a SPR is reinforced, and each subsequent SPR occurring at least 50 seconds following a reinforcement is reinforced. In the case of skeletal operants, a fixed-interval schedule generates a stable pattern of behavior in which a pause (no response) follows the presentation of the reinforcer, after which the rate of response accelerates to a terminal moderate level, as the interval of time is concluded (Ferster & Skinner, 1957). This gives the characteristic appearance of "scalloping" in cumulative recordings of such responses. In this study, several autonomic responses simultaneously measured were subjected to analysis in addition to SPR modification in order to check on the role of somatic changes in the "shaping" process. This paper reports individual data of subjects studied in a variety of conditions.

Method

Skin potentials were measured using non-polarizing silver-silver chloride electrodes (O'Connell & Tursky, 1960), one on the thenar eminence of the right palm and the other on the dorsal surface of the right forearm. SPRs were recorded via an Offner R-C coupled amplifier having an input time constant of 1 second. The autonomic operant was a fluctuation of ± 0.25 mV or greater. These criterion SPRs were automatically detected by an electronic switch. When each response occurred, subjects were given immediate external sensory feedback, a 200-msec tone of moderate intensity.

Each reinforcer was worth 10 cents on the short interval schedules and proportionately more for longer intervals and was administered by advancing a slide projector when a criterion SPR occurred after the appropriate time period had elapsed. A screen, placed directly in front of the subject, indicated the accumulated monetary bonus to be paid the subject at the end of the session. Because the time course of a SPR is at least 3 seconds and its wave form is often diphasic, after each criterion response there was a 3-second dead period to avoid triggering the feedback or reinforcement program on both positive and negative waves of the same response complex.

The experiments were run completely automatically, using timing and logic devices, to be free from "experimenter effects" (Rosenthal & Rosnow, 1969) in determining responses and deciding whether or not to reinforce them. Criterion responses and reinforcers were marked on the polygraph record when they occurred, as shown in Fig. 1, a segment of an actual record. Fig. 1 shows the other recordings taken at the same time. Skin potential level is the DC measure of potential obtained from the same electrodes as the SPR. Respiration was recorded by means of a strain gauge belt fixed around the lower rib cage. A relative measure of gross bodily movements was recorded as the integrated output of a transducer attached to the springs of the subject's chair. A cardiotachometer measured beat-by-beat variations in heart rate. For additional details on the apparatus, see Crider, Shapiro and Tursky, 1966.

The subjects were college students who were selected because of their electrodermal lability. They had been used in previous experiments in which SPRs were reinforced on different schedules from the present. These were very cooperative subjects who could sit still, breathe regularly, and refrain from tensing their muscles. The subjects were informed that the tone feedback indicated their body was making some physiological response. They were also informed of the exact contingencies in each schedule.

S were seated in a lounge chair in a sound and light controlled room, isolated from the experimental apparatus. Physiological recordings were made on an Offner Type R dynagraph.

Results

Effects of FI schedules of reinforcement

Fig. 2 shows a cumulative recording of one subject run for two sessions on FI 100″. Criterion SPRs are cumulated, and the pip marks indicate when the reinforcer was presented. Note the early

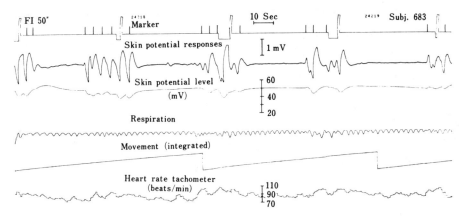

FIG. 1. Segment of polygraph record of a subject reinforced on Fixed-Interval 50″. The marker channel indicates occurrence of all criterion responses (small spikes) and the responses which were reinforced (large deflections).

development of a stable pattern of scalloping in Session 1. This pattern was maintained throughout Session 2 (45 minutes). A slight response burst following reinforcement was also observed. The portion of data in Fig. 1 was selected to show these same early and late peaks of activity during the interval. This subject was also run under the longer FI schedule with similar results. Under a variable interval schedule (average 30″), this subject was reinforced for comparison and showed a steadily maintained, fairly high response rate, as seen in the bottom panel of Fig. 2.

Three other subjects were studied under FI 50″, FI 100″, and FI 150″ in varying orders, three sessions for each interval, or nine sessions in all per subject. The bonus value of the reinforcer was 10, 20, and 25 cents, respectively. To portray the timing characteristics of the SPR, response distributions were prepared as follows: For each criterion SPR, the time was measured from the previous reinforcer. With each new presentation of the reinforcer, the measurement started again. The data for each interval within a session were superimposed to give a picture of the response patterning for each session under each schedule. Fig. 3 shows the response distributions for one subject. Note two peaks in each of the nine response curves, one immediately after reinforcement and one prior to reinforcement. The first peak is followed by a sudden drop, and then a gradual acceleration is observed to the end of the time period of the fixed interval.

Fig. 4 shows data for another subject, studied for three sessions each on FI 100″, FI 150″, and FI 50″, in that order. The response patterns are similar to the ones observed in Fig. 3, with somewhat less peaking immediately after reinforcement.

Fig. 5 is a third subject run in three sessions each of FI 50″, and FI 100″, in that order. His curves appear more sharply differentiated than those of the previous two subjects.

All three subjects (Fig. 3, 4, and 5) reveal some response properties in common under FI schedules. First, with the exception of the immediate and short-lived response burst following the reinforcer, the response curves follow the same pattern of " scalloping " obtained in skeletal operants. Second, at the longer fixed intervals, especially 150″, the curves seem to be less sharply differentiated, more erratic. That is, the stability of timing of the SPR appears to break down at longer intervals. Third, there are occasional periods of apparent cessation in SPR activity, even when reinforcement is already available at the end of the interval.

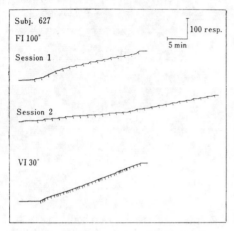

FIG. 2. Cumulated criterion skin potential responses. Top panel shows two sessions on Fixed-Interval 50″. For comparison, bottom panel presents a cumulative recording of the same subject on Variable-Interval 30″ for a single session.

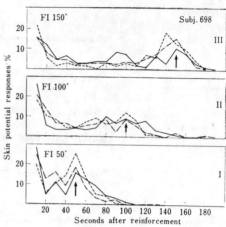

FIG. 3. Skin potential response distributions for three sessions each on Fixed-Interval 50″, 100″, and 150″, in that order. The order of sessions is indicated by dot-dash, dash, and solid lines for sessions 1, 2, and 3, respectively, for each schedule.

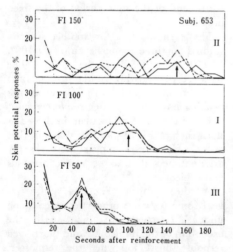

FIG. 4. Skin potential response distributions for three sessions each on Fixed-Interval 100″, 150″, and 50″, in that order (see Fig. 3).

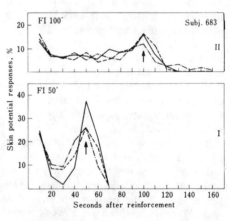

FIG. 5. Skin potential response distributions for three sessions each on Fixed-Interval 50″ and 100″, in that order (see Fig. 3).

For example, under FI 50″, in two of the three subjects (Fig. 3, 4), there were some delays in response up to 140″. These periods of delay tended to decrease from session to session, however. In all but one of the eight sets of three sessions each (Fig. 3, 4, 5), a larger number of reinforcements was achieved during the same overall training period, with repeated practice from session to session. Fourth, as to effects of one schedule upon a succeeding one, in going from FI 50″ to FI 100″ (Fig. 3, 5), the peaks under FI 50″ do not seem to carry over to FI

100″, in the initial session with the latter schedule. In going from FI 100″ to FI 150″, however, there is a suggestion of some peaking of response at 100″ in the FI 150″ curves.

To examine possible correlated changes in the other variables recorded, a detailed analysis was made of illustrative examples in which SPR rate was highly differentiated, particularly under FI 50″. Average curves for respiration rate and gross bodily movement were computed for each interval within a session. These curves were almost completely flat during the time interval, unrelated to the SPR distributions. Skin potential level tended to show a slight peak at the time of reinforcement but a flat pattern during the remainder of the time preceding each reinforcement. Heart rate data were more interesting. Specifically, these were analyzed by sampling the tachometer record every 4 seconds and averaging over successive periods of the fixed interval during a session.

Fig. 6 gives data for the third session under FI 50″ of the subjects presented in Fig. 3, 4, and 5. Only one of these shows the U-shape pattern in heart rate.

Although there were indications of variations in breathing rate and amplitude correlated with SPR activity in some subjects, these relationships were not consistent within or between subjects.

Effects of providing an external clock

The behavior under fixed-interval schedules is thought of as a clock by which the organism paces itself with respect to reinforcement (Ferster & Skinner, 1957). This behavioral clock is not a perfect timepiece, otherwise a single response would occur when reinforcement becomes available.

Three additional subjects were studied over four sessions each to compare their behavior with and without an external clock. In two sessions, they were provided with an external stimulus correlated with time, a large electric clock placed directly in front of the subjects. In the other two sessions, there was no external clock. Response distributions were prepared as described above, and averaged over sessions with and without the clock, separately (Fig. 7). The external clock was present in Sessions 1 and 4 for subject 608 and in Sessions 3 and 4 for the other two subjects.

The general form of the curves is similar to those shown in Fig. 3, 4, and 5. Although the differences are not large, the availability of the clock appears to have improved the efficiency of performance in all three instances: that is, the curves are more peaked at the terminal point when

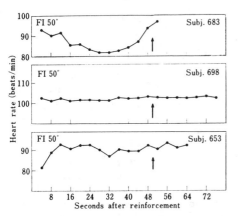

Fig. 6. Heart rate averaged over successive intevals on Fixed-Interval 50″.

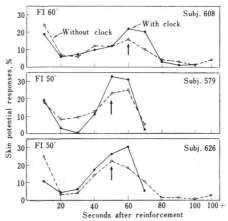

Fig. 7. Skin potential response distributions for three subjects, two on Fixed-Interval 50″, one on Fixed-Interval 60″. Each subject was run for two sessions with an external clock and two without a clock.

the external clock is present. Analogous results for skeletal operants are described in Ferster and Skinner (1957).

Concomitant heart rate data for the three subjects studied with and without the external clock are shown in Fig. 8. To some degree, the heart rate pattern in these cases seems to resemble the SPR behavioral pattern. A further look at the actual momentary changes in heart rate accompanying each criterion SPR was undertaken in a subject (608) during one session in which he showed the U-shaped SPR and heart rate pattern (clock present). To a degree, SPRs at the beginning and end of the time period were accompanied by heart rate *speeding* and during the middle portion by heart rate *slowing*. This apparent fractionation of heart rate and skin potential change and the inconsistent patterns shown in Fig. 6 and 8 indicate that there is no simple relationship of these two systems.

Informal verbal reports obtained from subjects following some of these sessions and those reported previously suggested

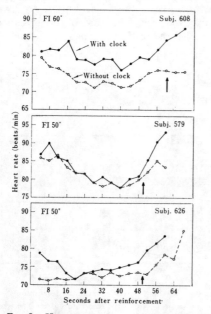

FIG. 8. Heart rate averaged over successive fixed intervals for two sessions with an external clock and two with out a clock (see Fig. 7).

that idiosyncratic patterns of imagery or physical sensation may occur in subjects such as these who are capable of exerting fairly effective control of their electrodermal activity. The subjects report these sensations as accompanying rather than mediating the feedback and rewards. However, relationships between the verbal reports and specific patterns of SPR activity were not consistent from subject to subject or session to session.

DISCUSSION

The timing characteristics of the skin potential response as an autonomic operant are similar in some respects to those observed for skeletal operants under fixed interval schedules. The pause following reinforcement is not immediate, however. There is an initial peak in response rate in the first few seconds following reinforcement which then falls off sharply. Responsivity in this initial peak would appear to be due to reinforcer-elicited responses having a time course of excitability longer than the 3-second dead period employed, rather than due to responses dependent on the particular FI schedules employed or on repeated practice in each schedule. Evidence supporting this possibility is the maintenance of such elicited electrodermal response noted in operant training (Gavalas, 1967; Shapiro & Crider, 1967). Gavalas, using FI 15″, reported only a slight decrement in GSRs elicited by the reinforcer during the first three days of operant reinforcement, but she did note a larger drop in a fourth and final session. Shapiro and Crider observed the maintenance of elicited SPRs to bonus slides under several intermittent schedules of reinforcement.

The persistence of elicited SPRs suggests that the same autonomic function has both respondent as well as operant properties. Kimmel (1966), making this same distinction, pointed out that the classically conditioned GSR tends to show response decrement or inhibition during conditioning in contrast to persistence of the instrumentally reinforced GSR in an avoidance conditioning procedure. A reward such as money is probably highly salient for an individual, and it appears to maintain its response-eliciting power. The increase in heart rate for some subjects prior to reinforcement may also reflect this prior conditioning. Whether autonomic responses to potent or salient stimuli such as rewards can be influenced by operant procedures is a problem for further research. Evidence supporting this possibility is available for responses elicited by moderate stimuli (Shnidman, 1970; Shnidman & Shapiro, 1970).

It is perhaps useful to consider whether the effects of fixed-interval reinforcement on skin potential responses are a function of temporal conditioning in which the unconditioned stimulus is presented on an irregular timed basis. Effective temporal conditioning with a monetary reinforcer presented at varying intervals does not seem very likely. Moreover, whenever noncontingent reinforcements are used in related experimentation (Shapiro, Crider & Tursky, 1964; Crider, Shapiro & Tursky, 1966; Watanabe & Shapiro, 1971) skin potential activity is quick to die out altogether. Further studies with the same reinforcer presented at a constant interval of time or a yoked-control FI procedure are needed to clarify this issue.

It is encouraging to find maintained and regulated patterns of spontaneous autonomic activity in an operant procedure because of the rather rapid habituation usually noted for such fluctuations. The coordination is far from perfect, however. For example, there were periods of nonresponse observed at times when reinforcement was available. Under long fixed intervals (FI 150''), response patterning appeared unstable. The occurrence of volleys of closely spaced skin potential responses interspersed with long periods of nonresponse was also observed under multiple reinforcement schedules (Shapiro & Crider, 1967). More information is needed on the flow characteristics of spontaneous autonomic responses, on the mechanisms of excitation and inhibition controlling these responses, and on the degree to which still finer control can be obtained through operant techniques.

How autonomic responses interact with one another to make a behavioral clock also warrants attention. In the present studies, auditory feedback was provided to subjects for each criterion response. External sensory feedback provides S with immediately externalized information concerning his interoceptive activity which is not perceived under normal conditions. The role of this feedback is a close analogue to that of natural feedback of proprioception in skeletal-muscular responses. Lisina (1965) has shown that avoidance conditioning of vasomotor responses is only possible when Ss are provided with exteroceptive feedback from the relevant smooth muscles. External sensory feedback may be necessary to facilitate autonomic learning in a situation involving expectancy responses such as in avoidance and FI reinforcement. This possibility might be supported by the poor results obtained in avoidance autonomic training research in which no external feedback has been used (Katkin & Murrary, 1968). Whether autonomic responses can cue one another without external sensory feedback is a problem for further research.

Acceleration in heart rate for some subjects prior to reinforcement may reflect their internal attention to produce responses, thereby getting reward, although relevant verbal reports were not obtained. According to Lacey (1967), internal attention leads to "environmental rejection" associated with heart rate acceleration. Heart rate acceleration during heart rate estimation has been reported by Porges and Raskin (1969) as evidence for Lacey's theory. However, from concomitant changes in the other variables recorded in this study, in conjunction with previous work on SPR (Crider et al., 1966; Birk, Crider, Shapiro & Tursky, 1966) and on deeply curarized animals (Miller, 1969), operant modifications of the SPR do not seem of necessity to be mediated by or correlated with gross bodily activity, breathing, skin potential level or heart rate.

The data support the conclusion that autonomic fluctuations have some of the timing properties akin to skeletal operants although not in all respects. Operant conditioning procedures offer an effective means of comparing behavioral processes mediated by different nervous pathways.

REFERENCES

BIRK, L., CRIDER, A., SHAPIRO, D., & TURSKY, B. 1966 Operant conditioning of electrodermal activity under partial curarization. *J. comp. physiol. Psychol.*, **62**, 165–166.

CRIDER, A., SHAPIRO, D., & TURSKY, B. 1966 Reinforcement of spontaneous electrodermal activity. *J. comp. physiol. Psychol.*, **61**, 20–27.

FERSTER, C. B., & SKINNER, B. F. 1957 *Schedules of reinforcement.* New York: Appleton-Century-Crofts.

GAVALAS, R. J. 1967 Operant reinforcement of an autonomic response: Two studies. *J. exp. Anal. Behav.*, **10**, 119–130.

GREENE, W. A. 1966 Operant conditioning of the GSR using partial reinforcement. *Psychol. Rep.*, **19**, 571–578.

KATKIN, E. S., & MURRAY, E. N. 1968 Instrumental conditioning of autonomically mediated behavior: Theoretical and methodologic issues. *Psychol. Bull.*, **70**, 52–68.

KIMMEL, H. D. 1966 Inhibition of the unconditioned response in classical conditioning. *Psychol. Rev.*, **73**, 232–240.

KIMMEL, H. D. 1967 Instrumental conditioning of autonomically mediated behavior. *Psychol. Bull.*, **67**, 337–345.

LACEY, J. I. 1967 Somatic response patterning and stress: Some revisions of activation theory. In M. H. Appley & R. Trumbull (Eds.), *Psychological stress: Issues in research.* New York: Appleton-Century-Crofts.

LISINA, M. I. 1965 The role of orientation in the transformation of involuntary reactions into voluntary ones. In L. G. Voronin, A. N. Leontiev, A. R. Luria, E. N. Sokolov, & O. S. Vinogradova (Eds.), *Orienting reflex and exploratory behavior.* Washington, D. C.: American Institute of Biological Sciences.

MILLER, N. E. 1969 Learning of visceral and glandular responses. *Science*, **163**, 434–445.

O'CONNELL, D. N., & TURSKY, B. 1960 Silver-silver chloride sponge electrodes for skin potential recording. *Amer. J. Psychol.*, **73**, 302.

PORGES, S., & RASKIN, D. 1969 Respiratory and heart rate components of attention. *J. exp. Psychol.*, **81**, 497–503.

ROSENTHAL, R., & ROSNOW, R. L. (Eds.). 1969 *Artifact in behavioral research.* New York: Academic Press.

SHAPIRO, D., & CRIDER, A. 1967 Operant electrodermal conditioning under multiple schedules of reinforcement. *Psychophysiology*, **4**, 168–175.

SHAPIRO, D., CRIDER, A., & TURSKY, B. 1964 Differentiation of an autonomic response through operant reinforcement. *Psychon. Sci.*, **1**, 147–148.

SHNIDMAN, SUSAN R. 1970 Instrumental conditioning of orienting responses using positive reinforcement. *J. exp. Psychol.*, **83**, 491–494.

SHNIDMAN, SUSAN R., & SHAPIRO, D. 1970 Instrumental modification of elicited autonomic responses. *Psychophysiology*, **7**, 395–401.

WATANABE, T., & SHAPIRO, D. 1971 Operant control of spontaneous skin potential responses in Japanese and American subjects: A comparative study. *Jap. J. Psychol.*, **42**, 79–86. (Japanese).

(Received Aug. 16, 1971)

VIII
ANIMAL STUDIES

Effects of Activity and Immobility Conditioning Upon Subsequent Heart-Rate Conditioning in Curarized Rats

Wendall J. Goesling and Jasper Brener

Prior to heart-rate conditioning under curare, rats were submitted to one of four pretraining procedures in the noncurarized state. Forty female rats served either as experimental or yoked-control subjects in the conditioning of activity in a running wheel or conditioning of immobility in a small open field. Following 10 days under these procedures, all rats were curarized and punished either for the emission of high or low heart rates. It was observed that the pretraining procedures contributed significantly more to the heart-rate changes observed under curare than did the reinforcement contingencies imposed under this latter condition. The implications of this finding are discussed with respect to the specificity of operant cardiovascular conditioning and the general issue of somatocardiovascular relations.

That somatomotor and cardiovascular activities are related in a lawful manner is a well-documented fact. The manner in which they are related is, however, a rather critical theoretical issue in the area of operant autonomic conditioning. Two clearly delineated schools of thought have offered their interpretations of this relationship. DiCara and Miller and their associates have repeatedly asserted that evidence of cardiovascular conditioning under curare provides evidence of the independence of somatomotor and cardiovascular control. Black (1967), Obrist, Webb, Sutterer, and Howard (1970), and the authors of the present article (Brener & Goesling, 1968) have, on the other hand, proposed that cardiovascular and somatomotor activities, rather than being autonomously controlled activities, represent two components of a general response process.

The published data relevant to this issue are extremely ambiguous. DiCara and Miller (1969b) conditioned heart-rate increases in one group of curarized rats and heart-rate decreases in another group. Retention of the conditioned heart-rate responses was then tested in the noncurarized state. It was found that rats that had been conditioned to increase their heart rates displayed substantially higher activity levels and respiratory rates than rats that had been conditioned to decrease their heart rates. These data strongly support the existence of a central linkage between somatomotor and cardiovascular activities since under curare variations of striate muscle and respiratory activities were not possible and therefore could not have been fortuitously reinforced. However, in the same year, DiCara and Weiss (1969) using a procedure very similar to that employed in the experiment just described obtained contradictory transfer data. In this experiment, rats that previously had been reinforced for heart-rate increases under curare displayed lower respiration rates and activity levels than rats that had been reinforced for heart-rate decreases. Another experiment by DiCara and Miller (1969a) provided additional data relevant to this issue. Here rats received heart-rate training in the noncurarized state prior to being tested in the curarized state. It was observed that rats that were rein-

[1] This research was supported by Grant 17061 from the National Institute of Mental Health.
[2] Now at the University of Santa Clara.
[3] Requests for reprints should be sent to Jasper Brener, Department of Psychology, University of Tennessee, Knoxville, Tennessee 37916.

forced for high heart rates displayed significantly higher activity levels and similar respiration rates to rats that were reinforced for heart-rate decreases.

All three studies indicate a linkage between somatomotor and cardiovascular activities. In the two that investigated transfer from the curarized to the noncurarized state, the evidence strongly suggests a central linkage. In the investigation of heart-rate conditioning in the noncurarized state, the data are again consonant with this interpretation, but could possibly be interpreted as evidence of peripheral mediation of the cardiovascular effect. Although in normal functioning a strong case can be made for the central linkage of somatomotor and cardiovascular activities, this linkage should not be seen as immutable. Thus it may be observed in the DiCara and Miller (1969b) and DiCara and Weiss (1969) studies that further training or testing in the noncurarized state following heart-rate conditioning under curare led to a decrease in the somatomotor components of the conditioned response and an increase in the heart-rate component.

Since curare inhibits only the most peripheral manifestations of the somatomotor act, it is possible that somatic processes proximal to the myoneural junction retain an influence over cardiovascular activity even in the curarized animal. The purpose of the present experiment was to investigate whether such influences may be observed in the curarized rat. The basic procedure was to condition antagonistic somatocardiovascular response sets in two groups of rats and then compare the effects of heart-rate conditioning under curare in these two groups and in rats that had acted as yoked controls in the pretraining procedure. If the two response systems are independently controlled, then we might expect the experimental contingencies introduced under curare to serve as the primary determinants of heart-rate change. On the other hand, if we accept that the two systems are served in parallel by a common central process, we would expect the effects of the different pretraining procedures to be manifested in the curarized state.

METHOD

Subjects

Forty naive female hooded rats of the Long-Evans strain weighing between 200 and 250 gm. at the commencement of the experiment were arbitrarily divided into two groups of 20 each. Within each group they were paired on a random basis with one subject in each pair serving as the experimental subject and the other as a yoked control. Yoked control subjects received the same temporal sequence of experimental stimuli as the experimental subjects to which they were yoked, but could not influence the occurrence of these stimuli. The sequence of stimuli received by each pair of rats was therefore controlled by the behavior of the experimental subject in that pair. One group of 10 experimental and 10 control rats served as subjects in the activity-conditioning experiment and the other in the immobility-conditioning experiment.

Apparatus

Activity conditioning. Conditioning of activity was performed in two similar running wheels, 15 in. in diameter and 5 in. wide. The floors of the wheels were made of $1/8$-in. brass rods spaced $5/8$ in. apart. Electric shock (.75 ma.) was delivered through these grids via a Grason-Stadler grid scrambler (Model E1064GSP) from an Applegate stimulator (Model 250). A Midland Model 60-172 8-ohm speaker was mounted on the back wall of a sound-attenuating enclosure that housed the running wheel for the presentation of a 1,000-Hz. tone at 80 db., and a 20-w. lamp was mounted in the center of the ceiling of the enclosure. Wheel rotation was recorded from a microswitch operated from a cam attached to the hub of the wheel. A contact closure occurred each time the wheel traveled $1/4$ of its rotation. The experimental contingencies were programmed using standard electromechanical relay equipment so that the activity of the experimental rat determined the temporal sequence of the experimental stimuli received both by itself and its yoked control.

Immobility conditioning. Conditioning of immobility was carried out in two cages measuring 12 × 12 × 12 in. The floors of these cages were constructed of $1/4$-in. brass rods mounted $3/4$ in. apart and the walls of $1/8$-in. brass rods all mounted in a Plexiglas frame. Shock (.75 ma.) was applied by a constant-wattage shock generator scrambler (BRS SGS-001) through the grid walls and floor. Each activity cage was housed in a larger sound- and light-attenuating box measuring 2 × 2 × 2 ft. A 4-in. Midland Model 60-172 8-ohm speaker was mounted on the back wall of the box for the delivery of a 1,000-Hz. tone at 80 db., and a 20-w. bulb (houselight) was mounted in the center of the ceiling. Mounted in 6-in. stands on the side of the sound-attenuating boxes were the two transducers of an ultrasonic activity detector (Alton Electronics Model M4A). These devices were calibrated to provide one output pulse on any movement of 1 in/sec in any direction. This output from the cage housing the experimental subject was fed to a solid-state programming circuit (BRS electronics) that controlled the experimental contingencies and presented the same temporal sequence of experimental stimuli to both cages.

Heart-rate conditioning under curare. During heart-rate conditioning the subjects lay in a small hammock and were fitted with a balloon mask similar to that described by DiCara (1970) through which they were respirated by an E & M Instrument Co. small-animal respirator (Model V5KG). Heart rate was recorded from chronically implanted electrodes on a Model 7 Grass polygraph. The output from J6 of the polygraph was connected to a pulse-shaping circuit that provided a square-wave output on each R wave of the EKG. This signal constituted the input to a solid-state (BRS) programming circuit that analyzed the subject's heart rate and controlled the reinforcement contingencies. Facilities similar to those described for the activity- and immobility-conditioning apparatus were available in the animal enclosure for the presentation of auditory and visual stimuli. The aversive stimulus consisted of a brief electric shock (.1 sec.) at an intensity of .5 ma. The shock was delivered through Grass earclip electrodes attached to the base of the tail about 1 in. apart.

Procedure

Activity conditioning. Activity conditioning was carried out over 10 40-min. sessions, following an initial adaptation session. Sessions were 48 hr. apart. During adaptation, the houselights were off for the first 20 min. and on for the second 20 min. No experimental contingencies were programmed during this session. Heart rate and number of activity responses (¼ rotations of the wheel) were monitored continuously during the adaptation and final conditioning sessions. During the first nine conditioning sessions, however, heart rate was not monitored.

Each conditioning session commenced with a 20-min. S^Δ period during which the houselights were out. During this period, if the subject rotated the wheel through ¼ of its rotation, it received a .5-sec. shock. This procedure was implemented because during pilot studies it had been noted that the rats tended to be active during the S^Δ periods unless a contingency of this sort were introduced. In view of the design and goals of the experiment, it was deemed desirable to maximize the activity differences between S^D and S^Δ. During the second 20-min. period of each session, the houselights came on (S^D) and the experimental contingencies were in effect. If the rat failed to emit an activity response (turning the wheel through ¼ of its rotation) within a given period of time (the response-tone interval), a tone (1,000 Hz. at 80 db.) came on. In the absence of an activity response, the tone remained on and 8 sec. after its onset an electric shock of .75 ma. was delivered to the rat's feet through the grid floor of the running wheel. If, however, the rat emitted an activity response, the tone or tone and shock were terminated and a new response-tone interval commenced. During the first two sessions the response-tone interval was set at 20 sec. and in all subsequent sessions, at 5 sec. Thus, during the last eight conditioning sessions, the rat could avoid the shock continuously by emitting activity responses at intervals less than 13 sec. and avoid the tone by emitting activity responses at intervals of less than 5 sec. Yoked control subjects received the same temporal sequence of experimental stimuli as did the experimental subjects to which they were yoked but their activity was ineffective in controlling these stimuli.

Immobility conditioning. Immobility conditioning was carried out over 10 40-min. conditioning sessions following an adaptation session which was also 40 min. in duration. During the first 20 min. of the adaptation session, the houselights were off (S^Δ), and during the second 20 min., they were on (S^D). Each conditioning session began with a 20-min. S^Δ period during which the houselights were off. Thereafter, the houselights came on and remained on for 20 min. (S^D) during which the experimental contingencies were in effect. During S^D any movement detected by the ultrasonic motion detector resulted in the occurrence of a 2-sec. tone (1,000 Hz. at 80 db.). Each activity unit emitted by the rats in the presence of the tone resulted in a brief (.5 sec.) shock (.75 ma.) and programmed an additional 2 sec. of the warning stimulus. The tone always terminated 2 sec. following the last recorded activity unit. Experimental and yoked control rats in each pair were run simultaneously with the activity of the experimental rat determining the temporal sequence of experimental stimuli received by both subjects. The activity of the control subjects had no influence on the experimental stimuli. Sessions were 48 hr. apart.

Heart-rate conditioning under curare. Two days following the last session of the pretraining procedure each rat was submitted to heart-rate conditioning under curare. In this procedure half of the rats in each group could avoid electric shock by increasing their heart rates and half by decreasing their heart rates. Thus, five experimental and five control subjects that had received the activity pretraining procedure were punished for high heart rates under curare and five subjects from each of these groups were punished for decreasing their heart rates. Similarly five experimental and five control subjects that had been submitted to the immobility pretraining procedure were punished for high heart rates and five subjects from each of these groups were punished for decreases in heart rates.

At the beginning of the conditioning session, the subject received 8 U/kg of d-tubocurarine chloride administered ip in two injections separated by 10 min. Following the first signs of respiratory difficulty, the subject was attached to the respirator via a balloon face mask. The rats were respirated at a rate of 40 cycles/min at a peak pressure of 16-cm. H_2O with a 1:1 inspiration: expiration ratio. The houselights were out during the first 20 min. of the session and no experimental stimuli were presented. During this adaptation period, a heart-rate reinforcement criterion was established. This was done by selecting an interbeat interval (IBI) that approximated the median IBI of the subject's distribution. Any IBI longer than this criterion was classified as a low heart rate and any IBI shorter than the criterion, as a high heart rate.

Punishment of low heart rates. The procedure for the punishment of low heart rates is represented in Figure 1. Whenever the subject emitted an IBI that was longer than the criterion for five consecutive IBIs, a tone (1,000 Hz. at 80 db.) was presented. If, in the presence of this tone, the rat

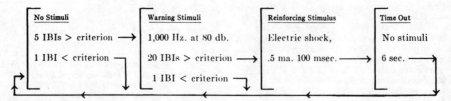

FIG. 1. Procedure employed in the punishment of low heart rates. (A diagrammatic representation of the procedure employed in the punishment of high heart rates may be obtained by reversing the < and > signs.)

continued to emit IBIs longer than the criterion for an additional 20 consecutive IBIs, an electric shock (.5 ma. for 100 msec.) was delivered. The shock was followed by a 6-sec. time out (TO) during which no stimuli were presented, following which the procedure recycled. When the rat emitted an IBI that was shorter than the criterion, a safe period commenced during which no stimuli were presented. This safe period only terminated when the rat emitted five consecutive IBIs longer than the criterion. If the rat succeeded in avoiding the warning stimulus continuously for a period of 3 min., the criterion was adjusted so as to make avoidance more difficult. In the present procedure this was accomplished by decreasing the IBI criterion by 10 msec.

Punishment of high heart rates. The procedure used in the punishment of high heart rates was identical to that just described except that in the present case, the warning stimulus and shocks were made contingent on the IBIs shorter than the criterion. Adjustment of the criterion in this procedure followed 3 min. of continuous avoidance of the warning stimulus and involved increasing the value of the criterion by 10 msec.

RESULTS

Since the primary purpose of the present experiment was to assess the effects of different pretraining procedures upon subsequent heart-rate conditioning under curare, the description of results of the pretraining procedures will be limited to the terminal performances exhibited by the rats submitted to the activity- and immobility-conditioning procedures.

Activity Conditioning

The mean heart rates and activity scores (number of ¼ rotations of the activity wheel) recorded during the S^Δ and S^D segments of the tenth conditioning session are presented in Table 1. Experimental and control subjects displayed significantly more activity in S^D than S^Δ ($F = 36.16$, $df = 1/18$, $p < .01$). In addition, it will be observed that experimental subjects displayed substantially more activity during S^D and substantially less activity during S^Δ than did their yoked controls. These differences were found to be significant in an analysis of variance which yielded a significant Groups $\times$ Periods (S^D/S^Δ) interaction ($F = 26.43$, $df = 1/18$, $p < .01$).

Overall, heart rate was higher in S^D than S^Δ ($F = 35.93$, $df = 1/18$, $p < .01$). A significant Groups $\times$ Periods interaction ($F = 18.21$, $df = 1/18$, $p < .01$) was associated with experimental subjects displaying higher heart rates in S^D and lower heart rates in S^Δ than their yoked controls.

Immobility Conditioning

The corresponding data for rats conditioned to be immobile are presented in Table 2. It will be seen that both groups of rats displayed substantially lower levels of activity during S^D than during S^Δ and that these activity differences are paralleled by differences in heart rate. Although analysis of variance did not yield a significant Groups $\times$ Periods interaction for activity or heart rate in this experiment, t tests did indicate that during S^D, experimental subjects displayed significantly less activity than controls ($t = 2.54$, $df = 18$, $p < .05$). Although the heart-rate differences paralleled these activity differences, they were not of sufficient magnitude to achieve statistical significance.

TABLE 1

MEAN ACTIVITY AND HEART-RATE SCORES FOR EXPERIMENTAL AND YOKED CONTROL SUBJECTS DURING THE TENTH DAY OF CONDITIONING IN THE RUNNING WHEEL

Group	S^Δ		S^D	
	Activity	Heart rate	Activity	Heart rate
Experimental	47.2	386	600.5	461
Control	159.0	396	202.1	409

When the heart-rate data for all four groups were submitted to analysis of variance, it was observed that none of the main variables accounted for a significant proportion of the total variance. However, it was found that all interactions with the period variable (S^D/S^Δ) proved to be significant. The Procedure (active/immobile) × Periods interaction ($F = 53.09$, $df = 1/36$, $p < .01$) was attributable to active subjects displaying higher heart rates (435 bpm) in S^D and lower heart rates (392 bpm) in S^Δ and immobile subjects displaying lower heart rates in S^D (410 bpm) and higher heart rates in S^Δ (429 bpm). The Groups (experimental/yoked control) × Periods interaction ($F = 7.90$, $df = 1/36$, $p < .01$) was due to the experimental subjects displaying higher heart rates in S^D (433 bpm) than they displayed in S^Δ (410 bpm) or than the yoked control subjects displayed in S^D (407 bpm) or S^Δ (410 bpm). The basis of the significant Procedure × Group × Period interaction ($F = 14.59$, $df = 1/36$, $p < .01$) may be understood by reference to Tables 1 and 2.

Heart-Rate Conditioning under Curare

The data of primary interest here were the heart-rate changes evidenced during the 40-min. conditioning session. These data were derived for each group of subjects by subtracting the mean heart rate recorded during the last 8 min. of adaptation (see Table 3) from the successive 8-min. heart-rate averages recorded over the 40-min. conditioning session. Figure 2 illustrates the heart-rate changes exhibited by each of the eight groups of subjects during the heart-rate conditioning session. It will be noted that experimental rats that had learned to be immobile in the pretraining procedure displayed the greatest decrements in heart rate under curare when punished for high heart rate, and experimental rats that had learned to be active in the pretraining procedure displayed the greatest increments in heart rate when punished for low heart rates.

Although there was a tendency among rats that had served as experimental subjects during the pretraining procedure to display heart-rate changes under curare in the direction that one might anticipate on the basis of the law of effect, it should be noted that these differences were small when compared to the influence of the pretraining

TABLE 2
MEAN ACTIVITY AND HEART-RATE SCORES FOR EXPERIMENTAL AND YOKED CONTROL SUBJECTS DURING THE TENTH DAY OF CONDITIONING IN THE ULTRASONIC APPARATUS

Group	S^Δ		S^D	
	Activity	Heart rate	Activity	Heart rate
Experimental	32.85	428	2.60	404
Control	34.30	424	15.60	415

procedure. The mean terminal heart-rate changes from adaptation were computed by subtracting the heart rates recorded during the final 8 min. of adaptation from the heart rates recorded during the final 8 min. of conditioning for experimental and yoked-control subjects trained under the activity and immobility conditioning procedures. The adaptation heart rates and mean terminal heart-rate changes are presented in Table 3. It will be seen here that the active experimental subjects displayed a mean increment in heart rate regardless of the contingencies in effect under curare and that all other groups displayed a mean decrement in heart rate with the immobile experimental subjects displaying the greatest mean decrement.

The terminal heart-rate differences for all subjects were then submitted to an analysis of variance to determine the relative contributions of pretraining and contingency

TABLE 3
GROUP MEAN HEART RATES RECORDED DURING FINAL 8 MIN. OF ADAPTATION AND CHANGES IN HEART RATE FROM ADAPTATION PERIOD TO FINAL 8 MIN. OF CONDITIONING UNDER CURARE

Precurare treatment	Curare treatment	M adaptation heart rates (bpm)	Terminal heart-rate change (bpm)
Active group			
Experimental	Punish high HR	443	7.0
	Punish low HR	423	15.4
Control	Punish high HR	406	−19.4
	Punish low HR	429	−18.4
Immobile group			
Experimental	Punish high HR	435	−36.7
	Punish low HR	417	−14.9
Control	Punish high HR	408	−4.2
	Punish low HR	413	−17.2

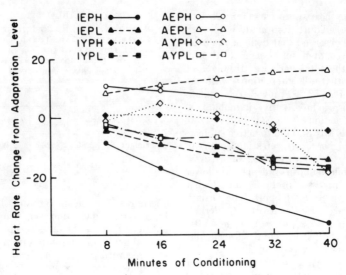

Fig. 2. Heart-rate change from adaptation level as a function of minutes of conditioning under curare. (The first two letters in the legend designate the pretraining procedure [I = immobile; A = active; E = experimental; Y = yoked control] and the last two letters, the curare treatment [PH = punish high heart rate; PL = punish low heart rate].)

under curare to the observed effects.

This analysis reveals that pretraining factors alone and not the curare treatment factor accounted for a significant proportion of the heart-rate change variability under curare. In particular, the analysis of variance yielded a significant active/immobile pretraining effect ($F = 6.40$, $df = 1/33$, $p < .05$) and a significant Active/Immobile × Experimental/Control interaction ($F = 15.04$, $df = 1/33$, $p < .01$). In other words, the magnitude and direction of heart-rate change under curare was a function of whether rats had been submitted to the active or immobile pretraining procedures as experimental or yoked-control subjects. Particularly germane to the primary thrust of this article is the observation that the curare treatment factor (punish high or low heart rates) did not approach significance.

DISCUSSION

Although the data of primary interest in this experiment relate to the effects of pretraining different response patterns in the noncurarized state upon subsequent heart-rate conditioning under curare, the results of the pretraining procedure warrant some consideration. Traditional notions of avoidance behavior (e.g., Solomon & Wynne, 1954) suggest that the motivational substrate of this class of behavior is manifested in increased sympathetic activity. In terms of this hypothesis we would expect experimental rats in the immobile- and active-avoidance-conditioning procedures to exhibit higher heart rates in S^D (when the avoidance contingencies were in effect) than in S^Δ (when they were not in effect). Clearly the data do not support this hypothesis. Immobile experimental rats displayed lower heart rates in S^D than in S^Δ and active experimental rats displayed higher heart rates in S^D than in S^Δ. In other words, heart rate, rather than reflecting the motivational substrate of behavior, reflected the general levels of activity exhibited by the rats under the various experimental conditions employed in this experiment.

The major results of this experiment indicate that the effects of somatomotor conditioning may be clearly evidenced in the cardiovascular activity of curarized rats. Within the constraints of the particular design employed here, these pretraining effects were more significant determinants of heart-rate change under curare than the heart-rate reinforcement contingencies imposed.

This finding has important implications for the issue of specificity of operant cardiovascular conditioning. When these data are considered together with those mentioned in the introduction to this article and those of other investigators such as Obrist et al. (1970) and Black (1967), the case for independent central control of somatomotor and cardiovascular activities becomes extremely tenuous. This case is further hindered by consideration of the neurophysiological evidence on somatocardiovascular integration (Germana, 1969), which indicates that the same central structures are involved in the control of both processes.

With respect to clarifying the relationship between cardiovascular and somatomotor activity, the procedure of curarization has had a negative utility. Its use has fostered the erroneous idea of conditioning effects that are specific to arbitrarily designated response units. This is most clearly exemplified in the results of DiCara and Miller's (1969b) study of the transfer of conditioned heart-rate responses from the curarized to the noncurarized states. During the initial transfer tests in the noncurarized state, subjects displayed unmistakable evidence of somatic responses that are normally associated with the heart-rate responses that had been conditioned under curare. In other words, the use of curare during the initial heart-rate conditioning phase of the experiment did not prevent somatomotor activity from occurring—it simply prevented the experimenter from observing it by blocking its most peripheral manifestations. This is not at all to deny the evidence of the development of somatocardiovascular dissociation provided by DiCara and his associates. Their data do indicate that following heart-rate conditioning under curare, further training or testing in the noncurarized state leads to a magnification of the cardiovascular effects and a diminution of the somatomotor correlates. A somewhat overzealous treatment of the topic of somatocardiovascular independence has led to the neglect of this potentially very significant dissociation process.

REFERENCES

BLACK, A. H. *Operant conditioning of heart rate under curare.* (Tech. Rep. No. 12) Hamilton, Ontario: McMaster University, Department of Psychology, 1967.

BRENER, J. M., & GOESLING, W. J. Heart rate and conditioned activity, 1968. Paper presented at the meetings of the Society for Psychophysiological Research, Washington, D.C., October 1968.

DICARA, L. V. Analysis of arterial blood gases in the curarized artificially respirated rat. *Behavioral Research Methods and Instruction*, 1970, 2, 67–69.

DICARA, L. V., & MILLER, N. E. Heart-rate learning in the non-curarized state, transfer to the curarized state and subsequent retraining in the curarized state. *Physiology and Behavior*, 1969, 4, 612–624. (a)

DICARA, L. V., & MILLER, N. E. Transfer of instrumentally learned heart-rate changes from curarized to noncurarized state: Implications for a mediation hypothesis. *Journal of Comparative and Physiological Psychology*, 1969, 68, 159–162. (b)

DICARA, L. V., & WEISS, J. M. Effect of heart-rate learning under curare on subsequent non-curarized avoidance learning. *Journal of Comparative and Physiological Psychology*, 1969, 69, 368–374.

GERMANA, J. Central efferent processes and autonomic-behavioral integration. *Psychophysiology*, 1969, 6, 78–90.

OBRIST, P. A., WEBB, R. A., SUTTERER, J. R., & HOWARD, J. L. The cardiac-somatic relationship: Some reformulations. *Psychophysiology*, 1970, 6, 569–587.

SOLOMON, R. L., & WYNNE, L. C. Traumatic avoidance learning: The principles of anxiety conservation and partial irreversibility. *Psychological Review*, 1954, 61, 353–385.

(Received November 26, 1971)

26 Discriminative Shock Avoidance Learning of an Autonomic Response Under Curare

Ali Banuazizi

In Experiment 1, 16 curarized, artificially respirated rats were assigned to two groups of 8 subjects each, rewarded for intestinal contraction (C) or relaxation (R), respectively. Each animal received a predetermined series of CS+, CS−, and blank (B) trials, the latter involving no stimulation. On CS+ trials, rats in Group C received a noxious electric shock to the tail for episodes of intestinal relaxation, while those in Group R received the shock for episodes of intestinal contraction. The results indicated that both groups spent significantly greater amounts of time, both toward the end of acquisition and during extinction, in the intestinal state for which they were rewarded, manifesting stimulus-specific escape and avoidance learning. In Experiment 2, using four curarized rats as subjects, no consistent unconditioned effects on intestinal motility were observed as a result of shock application or shock offset.

Reversing a widely held and authoritative conclusion of a decade ago (e.g., Kimble, 1961), the recent evidence seems to point overwhelmingly to the inference that autonomically mediated responses can be modified by instrumental learning methods.[3] In an earlier experiment (Miller & Banuazizi, 1968) it was shown that the spontaneous motility of the intestine in deeply curarized rats can be increased or decreased by making a reward, direct stimulation of the rewarding areas of the brain (ESB), contingent upon intestinal contraction or relaxation, respectively. The present study was undertaken in an attempt to determine further the similarity of the operant conditioning of intestinal contraction and relaxation states to that of skeletal responses by the use of different parameters and conditions of learning. A subsidiary goal was a systematic assessment of the unconditioned effects of the reinforcing event employed, since such effects are potentially important to alternative interpretations of apparent instrumental conditioning of intestinal states.

EXPERIMENT 1

While most experiments in instrumental learning of autonomic responses in curarized animals have used ESB as the rewarding event, comparable results have also been obtained with aversive, peripheral stimuli as reinforcers (Black, 1967; DiCara & Miller, 1968a, 1968b, 1969; DiCara & Weiss, 1969). In the case of intestinal motility, the question of whether certain properties of ESB render it uniquely capable of producing the observed changes in the response is particularly cogent because of the involvement of the hypothalamus in both the reward (ESB) and the response (feed-

[1] This research was based on a dissertation submitted to the Graduate School of Yale University in partial fulfillment of the requirements for the PhD degree (Banuazizi, 1968). The author gratefully acknowledges his indebtedness to Allan R. Wagner and Neal E. Miller for their encouragement and guidance at various stages of this investigation. This research was in part supported by Grant GB-6534 from the National Science Foundation to Allan R. Wagner. A portion of this paper was presented at the Eighth Annual Meeting of the Psychonomic Society, Chicago, October 1967.

[2] Requests for reprints should be sent to Ali Banuazizi, who is now at the Department of Psychology, Boston College, Chestnut Hill, Massachusetts 02167.

[3] For summaries of the evidence and some methodological and theoretical issues raised by it, see Kimmel (1967), Katkin and Murray (1968), Crider, Schwartz, and Shnidman (1969), Katkin, Murray, and Lachman (1969), Miller (1969), and Black (1971).

Copyright 1972 by the American Psychological Association and reproduced by permission. From the *Journal of Comparative and Physiological Psychology*, November 1972, 236-246.

ing) systems. A recent study by Folkow and Rubinstein (1965), for instance, has shown that topical electrical stimulation of the hypothalamic area ("feeding center") in the cat elicits a relatively specific autonomic adjustment, involving increased intestinal motility and characteristic cardiovascular changes, in addition to the frequently observed "feeding response." These considerations provided the first reason for conducting the present investigation, which attempts to demonstrate that a different, more conventional reinforcer could successfully be employed in the operant modification of intestinal motility.

A second reason had to do with certain questions arising from the study by Miller and Banuazizi (1968) referred to above. While the use of a bidirectional control procedure by these investigators ensured that the response changes obtained could not be explained on the basis of some lingering, cumulative effects of the reward (ESB), it is still possible that the different number and temporal pattern of reinforcement received could partially account for the observed differences between the two experimental groups rewarded for contraction and relaxation of the intestine. To circumvent this difficulty in the present experiment, the bidirectional control procedure was used in conjunction with a discrimination learning paradigm which required that the learned changes in the response occur more frequently in the presence of the stimulus situation associated with the reward. Thus, the purpose of Experiment 1 was to determine whether states of intestinal contraction or relaxation in deeply curarized rats could be brought under the control of a specific stimulus in an instrumental learning procedure involving escape and/or avoidance of an electric shock as reinforcing events.

Method

Subjects

The subjects were 16 male albino Sprague-Dawley rats weighing 450–600 gm.

Apparatus

Throughout the experimental session, the deeply curarized rat lay on a sponge pad in a 42 × 24 × 34 in. sound-attenuated and ventilated chamber.

The animal was artificially respirated through a mask made from the mouthpiece of a rubber balloon. The balloon piece was secured by placing the lower lip behind the upper incisors and by pulling the upper portion tightly over the top of the rat's mouth. The other end of the balloon piece was connected to a rubber stopper which held a short Y junction to the inspiration and expiration tubes of a small animal respirator (E & M Instruments Co., Model V5KG). The inspiration/expiration ratio was maintained at 1:1, with 70 respiratory cycles per minute and a peak reading of 20 cm. of water pressure.

The conditioned stimuli used in the experiment were a flickering light flash and a continuous "white" noise. The light flash was produced by a Grass photostimulator (Model PS-2) with nominal frequency and intensity settings of 10/sec and 8, respectively, and the lamp situated approximately 20 in. above the position of the rat's head. The auditory stimulus was produced by a Foringer (Model 1291) white-noise generator and was delivered through a loudspeaker inside the enclosure at a loudness of 80 db., as measured by a sound-level meter (General Radio, Model 1551-C) with a weighting of 20 kHz. at the position of the rat's head. The reinforcement stimulus was a discontinuous electric shock of .45-ma. intensity, comprised of 1 .1-sec. pps, delivered through two electrodes firmly attached to the base of the rat's tail. The shock source was a voltage divider on the secondary of a 500-v. ac transformer, delivering the current to the animal through a fixed resistance of 220 K ohms.

To monitor intestinal motility, a small balloon tied to the end of a 7-cm. dull-point gauge No. 19 hypodermic needle was inserted through the rat's rectum into the large intestine approximately 4 cm. beyond the anal sphincter. Figure 1 shows an X-ray picture of a rat with the balloon inserted in its intestine after a barium solution was injected into the gastrointestinal tract. The balloon was filled with 2 cc of water; the needle was connected via a short length of polyethylene tubing to a pressure transducer (Statham Laboratories, Model P23Dc, calibration factor = 52.03 μv/v/cm Hg). The signal from the pressure transducer was amplified at a sensitivity of 200 μv/cm and fed in parallel to two recording pens of an Offner Dynograph. At this sensitivity, both pens showed a deflection of 3.9 cm. for an intraluminal pressure equivalent to 15 cm. of water. The first pen traced a graphic record of the intestinal motility. The second "pen" had a ball point, which made contact with the surface of a plate constructed of two strips of brass inlaid in a piece of Plexiglas and separated from each other by approximately 1 mm. By appropriately connecting each brass strip to electronic relays and counters, the amount of time that the rat's intestinal state placed the pen (to which a ground pulse was applied) on one or the other strip of the plate could be monitored. The line of transition between the two strips of this commutator plate was designated as the *contraction criterion*. Hence, the lower strip was defined as the "relaxation region," while the upper strip was defined as the "contraction region." In short, the function of the commutator plate was to provide a means for objectively measuring the amount of time that the rat's intestine was in the state of contraction or relaxation and, furthermore, to

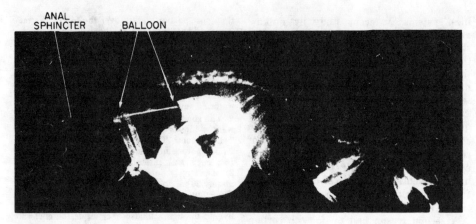

FIG. 1. X-ray of a rat showing the intestinal balloon in relation to the gastrointestinal tract and anal sphincter, taken after the injection of a barium solution.

serve as an automatic device for determining the administration of reinforcement.

Electromyographic (EMG) activity was recorded through two stainless-steel electrodes, one inserted in the lateral and the other in the medial muscles of the thigh of the rat's right hind limb. The EMG signal was amplified, integrated, and traced by the polygraph at a sensitivity of 20 μv/cm.

Procedure

At the end of a 3-hr. period of food deprivation, the subject was initially injected intraperitoneally with 3 mg/kg of d-tubocurarine chloride in a solution containing 3 mg/cc, and every hour subsequently until the end of the training session with a supplementary dose of 1 mg/kg. After the initial curare injection, the animal was carefully observed for the first signs of motor paralysis and difficulty in breathing, which appeared within 3–5 min. after the injection. Once immobilization occurred, artificial respiration was begun and the measurement and stimulation devices were attached.

Representative samples of both EMG activity and intestinal motility recorded from a typical subject are presented in Figure 2. The EMG recordings were made on half of the subjects, selected randomly, from the beginning of the experimental session to the point of recovery from curarization, usually occurring 2–3 hr. after the termination of training. The top section of the record, taken from an early phase in training, and showing essentially no EMG activity, is quite representative of the level of activity observed throughout the duration of training in all animals from which EMG recordings were taken. The lower sections of the figure show that some muscle potentials appeared about 60 min. after the termination of training, increasing considerably by about 120 min. after training. And, finally, twitches and overt skeletal movements occurred at about 2–3 hr. after the end of the experiment.

The pattern of intestinal motility depicted in Figure 2 is also typical of all rats. Rather regular and frequent contraction episodes were invariably followed by periods of relaxation, which could easily be distinguished by their falling on the same base line. For each subject the polygraph balancing circuit was adjusted so that the typical relaxation base-line level of intestinal motility placed the ball-point pen at approximately 2.5 mm. below the contraction criterion on the commutator plate. In this position the ball-point pen would be in the contraction region when the graphic record, as seen in Figure 2, was above the contraction criterion indicated, and in the relaxation region when the graphic record was below that criterion.

The 16 rats in this experiment were randomly assigned to two groups of 8 subjects each, one group (C) rewarded for intestinal contraction and the other group (R) rewarded for intestinal relaxation. After a 15-min. habituation period, in which intestinal and EMG responses were simply monitored, the training session consisted of four successive stages as follows.

Stage 1. During this stage all animals received three types of trials: noise, light flash, and blank (B), the last type involving no stimulation. No reinforcement was administered in this stage on any of the trials, but for half of the rats in each experimental group the light flash was designated as the subsequently positive stimulus (CS+) and the noise as the negative stimulus (CS−), while for the other half of the rats this assignment was reversed.

A total of 15 trials, 5 of each kind, were presented to each animal based on the following predetermined, balanced sequence: CS+, CS−, CS+, CS−, B, CS−, CS−, CS+, CS+, B, CS+, B, CS−, CS−, B, B, CS+.

In this and all subsequent stages the intertrial intervals between the end of one trial and the beginning of the next were irregularly 15, 30, and 45 sec., with an average of 30 sec., and each trial was 1 min. in duration.

Stage 2. In this stage, the noxious electric shock was administered at the onset and anytime during the CS+ trials that the rat's intestine was in the

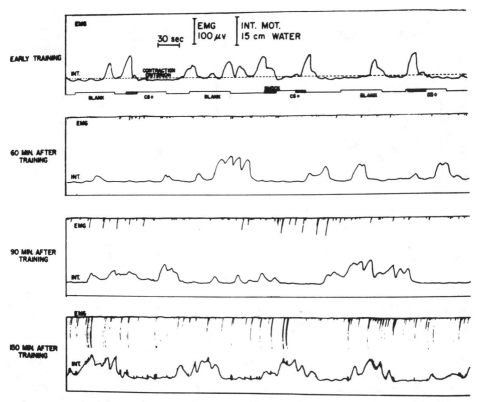

FIG. 2. Typical records of electromyographic and intestinal responses of a curarized rat during and after training for intestinal relaxation.

state of relaxation (Group C) or contraction (Group R). While CS+ stayed on for the entire length of each CS+ trial, shock was given during this time only when the rat's intestinal state was *other* than that for which it was being rewarded. In this way the subject could completely avoid the shock by remaining in the reinforcement region throughout the trial; could escape the shock, once it was initiated, by making the criterion response; or could experience a number of shock onsets and escapes within the same trial by making repeated transitions between the two intestinal states.

Each subject received 40 CS+ and 40 B trials according to repetitions of the following predetermined sequence: CS+, B, CS+, B, CS+, B, B, CS+, B, B, CS+, CS+, B, CS+, CS+, B.

Stage 3. In this stage rats continued to receive CS+ and B trials as in Stage 2, but, in addition, they received interspersed CS− trials during which no shock was delivered.

Each animal received 15 CS+, 15 CS−, and 15 B trials based on the same sequence as in Stage 1.

Stage 4. This stage was an exact replica of Stage 1. Each rat received five CS+, five CS−, and five B trials, with no shock administered on any of the trials.

In each of the four stages the total amount of time that the rat's intestine was in the state of contraction was measured on every trial. After the completion of the training session, the subject was maintained under artificial respiration until the time of recovery from the curare-induced paralysis, normally a minimum of 2–3 hr.

Results and Discussion

Figure 3 presents changes in the amount of time spent in the state of intestinal *contraction* during the course of training for both experimental groups. Each point on the graphs represents the mean number of seconds per trial spent above the contraction criterion, averaged over blocks of five trials of the same type (i.e., CS+, CS−, or B). It can be seen that in Stage 1, responding to the two different types of stimuli and B trials was essentially the same in either experimental group. In Stage 2, the amount of time spent in the state of contraction

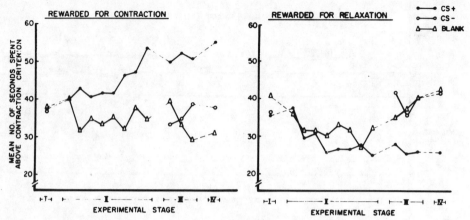

Fig. 3. Mean number of seconds per trial spent above the contraction criterion on the three different types of trials in experimental groups rewarded for intestinal contraction and relaxation.

during CS+ trials, in comparison with B trials, increased for Group C and decreased for Group R. In Stage 3, when CS− was reintroduced, a separation between responding during CS+ trials on the one hand, and CS− and B trials on the other, may be seen for both experimental groups. Similarly, during Stage 4, subjects in Groups C and R spent a greater amount of time in states of contraction and relaxation, respectively, on CS+ trials as compared with CS− and B trials.

The statistical reliabilities of the above findings were tested initially by conducting appropriate analyses of variance on the data from each of the four stages of the experiment for Groups C and R, separately. The number of seconds per trial spent in the intestinal state for which the rat was rewarded constituted the scores on which these analyses were performed. Individual comparisons among the means of the different groups were made subsequently by using the Newman-Keuls method.

In Stage 1, two 2 × 3 (CS Modality × Type of Trial) analyses of variance revealed no significant differences among the three types of trials ($F < 1$ and $F = 1.92$ for Groups C and R, respectively) or the two CS modalities ($F < 1$ for both groups) in either of the two experimental groups.

In Stage 2, a separate 2 × 8 × 2 (CS Modality × Block of Trials × Type of Trial) analysis was performed for each of the two experimental groups. For Group C, there was a significant difference between CS+ and B trials ($F = 13.67$, $df = 1/6$, $p < .05$) and also a significant Block of Trials × Type of Trial interaction ($F = 2.64$, $df = 7/24$, $p < .05$). In the case of Group R, however, the difference between CS+ and B trials was not statistically reliable ($F = 2.08$, $df = 1/6$, $p > .05$), nor was there a significant Block of Trials × Type of Trial interaction.

In Stage 3, highly reliable differences appeared among the three different types of trials for both Group C ($F = 36.50$, $df = 2/12$, $p < .001$) and for Group R ($F = 17.39$, $df = 2/12$, $p < .001$) in separate analyses of variance. In addition, as in Stage 2, a significant Block of Trials × Type of Trial interaction ($F = 2.87$, $df = 4/24$, $p < .05$) was found for Group C only. Individual comparisons among the three different types of trials, averaged over the three blocks in Group R, showed highly significant differences between CS+ trials and either CS− or B trials ($p < .01$ in each case). For this same group the difference between CS− and B trials was not statistically reliable. For Group C, due to the appearance of a significant Block of Trials × Type of Trial interaction, individual comparisons of the three different types of trials were made separately for each of the three blocks of trials. The difference between CS+ and CS− trials was statistically highly reliable ($p < .01$) for all three blocks. Also, CS+ trials differed significantly from B trials ($p < .05, < .01$, and $< .01$ on Blocks 1, 2, and 3, respectively). Only in the last block of trials was there a significant difference between CS− and B trials in this stage ($p < .05$), with a greater amount of time spent in the state of contraction on CS− trials than on B trials.

In Stage 4, separate 2 × 3 (CS Modality × Type of Trial) analyses of variance showed highly reliable differences among the three types of trials for both Group C ($F = 20.02, df = 2/12, p < .001$) and Group R ($F = 9.80, df = 2/12, p < .01$). Individual comparisons among the means of the three different types of trials revealed statistically significant differences between CS+ trials on the one hand and CS− and B trials on the other ($p < .01$ in all cases) for both experimental groups. The CS− and B trials did not differ significantly from each other during this stage in either group.

The results of this experiment, therefore, clearly show that an increase in time of intestinal contraction and relaxation states can be produced by making electric shock contingent upon intestinal relaxation or contraction, respectively. Furthermore, for both experimental groups the discrimination training appears to have been successful in increasing the likelihood of responding (changes in duration and/or frequency of intestinal contraction) in the presence of CS+ as compared to CS− and B trials. Such differences in Stages 2 and 3, where shock was administered on CS+ trials, would potentially have to be attributed to shock-escape or shock-avoidance learning, or some combination of the two. During Stage 4, however, since shock was never given, the greater amount of time spent in the rewarded state by both Groups C and R on CS+ trials, as compared to CS− and B trials, can only be described as the result of stimulus-specific avoidance learning.

The above increases in the amount of time spent in the rewarded intestinal states on CS+ trials, compared to CS− and B trials, might have been produced in a variety of ways. For example, a prolongation of intestinal contraction or relaxation episodes could have occurred only on those trials that were initiated while the rat's intestine was already in the rewarded state; alternatively, subjects could have made shifts from the nonrewarded intestinal state to the rewarded one in order to escape and/or avoid the noxious electric shock. If the first alternative were solely the case, one would expect to find significant differences between CS+ trials, on the one hand, and CS− and B trials, on the other, only when these were initiated while the subject was in the rewarded intestinal state.

In order to determine the degree to which differences in responding on the three types of trials were a function of rats' intestinal states at the onset of the trials, the data from Stage 4 were analyzed in further detail. For both experimental groups, trials from this stage were divided into two subgroups. The first subgroup included trials which initiated when the animal was in the state of intestinal contraction, while the second subgroup consisted of trials which were initiated when the rat's intestine was in the state of relaxation.

Figure 4 presents frequency distributions for trials with different amounts of time spent above the contraction criterion for both experimental groups. The distributions are plotted separately for the three types of trials (CS+, CS−, and B) and for the intestinal states in which each trial was initiated (contraction or relaxation). The small arrows on the graphs point to the mean number of seconds per trial spent above the contraction criterion averaged over the trials represented in each distribution. The numerals in each box indicate the total number of trials upon which the distribution is based. As might readily be seen, for both experimental groups, the differences between CS+ trials, on the one hand, and CS− and B trials, on the other, are present in the expected directions irrespective of rats' intestinal states at the onset of the trials. These differences appear to be somewhat more striking for the group rewarded for intestinal contraction.

Therefore, it may be concluded that, at least during the avoidance phase of this experiment, rats' discriminative responding on the three types of trials could not be solely attributed to the differential prolongations

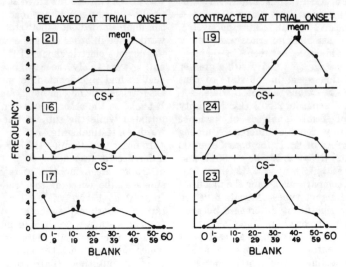

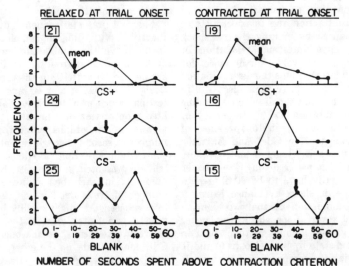

NUMBER OF SECONDS SPENT ABOVE CONTRACTION CRITERION

FIG. 4. Frequency distributions of trials based on the different amounts of time spent above the contraction criterion on each trial.

of the rewarded intestinal states in which trials were initiated; that is, subjects seemed to be also capable of avoiding the noxious stimulus by making shifts from the nonrewarded intestinal state to the rewarded one. A gross examination of the records from Stages 2 and 3 of the experiment revealed similar trends for all subjects.

An inspection of Figure 4 indicates also that, in both experimental groups, animals seldom spent the entire length of a trial (60 sec.) in the intestinal state in which the trial was initiated. In the case of CS+ trials, for example, there were very few occasions in which rats totally stayed out of the region in which the electric shock had been administered earlier in training.

A related question concerning the qualitative aspects of the obtained changes in the intestinal response is whether or not the increases in the amount of time spent in the rewarded intestinal states were accompanied by changes in the *frequency* of contraction and relaxation episodes. The reinforcement criterion in the present experiment was not sensitive to the above distinction, i.e., animals could escape and/or avoid the electric shock through an increase in the duration *and/or* the frequency of the rewarded intestinal state. Thus, changes in the frequency of contraction and relaxation episodes on different trials might have been determined by one or more of the following variables: (a) type of trial, (b) intestinal state rewarded, and (c) intestinal state at the onset of the trial. To evaluate the degree to which the above factors influenced the frequencies of intestinal contractions and relaxations, the trials in Stage 4 were grouped along the three dimensions suggested by these variables. On each trial the number of transitions between the two intestinal states was recorded.

Table 1 presents the mean number of transitions made on CS+, CS−, and B trials, initiated in states of relaxation or contraction, for both experimental groups. It can be seen that, irrespective of rats' intestinal states at points of trial onset, the mean numbers of transitions made during different types of trials are not appreciably different in the group rewarded for contraction. When intestinal relaxation was rewarded, a smaller number of transitions occurred during CS+ trials initiated in the state of relaxation than on CS+ trials initiated in the state of contraction. The magnitude of this difference, however, was quite small. Furthermore, no consistent differences in the mean numbers of transitions appeared among the three types of trials in this group. Qualitative changes occurring in the intestinal response during Stages 2 and 3 of this experiment were similar to those reported for Stage 4. Thus, the increases in

TABLE 1
MEAN FREQUENCIES OF CHANGES IN INTESTINAL STATES DURING TRIALS INITIATED IN STATES OF CONTRACTION AND RELAXATION

Intestinal state rewarded	Type of trial	Intestinal state at trial onset	
		Contraction	Relaxation
Contraction	CS+	2.32 (19)	2.48 (21)
	CS−	2.00 (24)	2.44 (16)
	B	2.87 (23)	2.35 (17)
Relaxation	CS+	2.84 (19)	1.90 (21)
	CS−	1.88 (16)	2.62 (24)
	B	1.87 (15)	1.84 (25)

Note.—Numbers inside parentheses indicate the number of trials on which each mean is based.

the amount of time spent in the rewarded intestinal states appeared to have primarily resulted from changes in the duration, rather than in the frequency, of intestinal contraction and relaxation episodes.

EXPERIMENT 2

Although the stimulus-specific modification of an autonomic reaction in two opposing directions, as was reported in the previous experiment, cannot be explained on the basis of some invariant unconditioned effect associated with the reinforcing event, it is conceivable that a reinforcer may have different unconditioned effects depending upon the particular state of autonomic activity during which it is administered. Thus, for example, electric shock delivered to the tail might elicit intestinal contraction if it is administered during a state of relaxation, while the same shock delivered during a contraction cycle might cause intestinal relaxation. This type of unconditioned effect, which may be referred to as "state dependent," has been observed with several behavior systems (e.g., shock applied to an extended limb may produce flexion, while shock applied to a flexed limb may produce extension) and has an important history in the literature comparing classical and instrumental learning (e.g., Sheffield, 1948).

If present, such a state-dependent effect could enable one to explain an apparently instrumental, bidirectional learning of a visceral response on the basis of classical conditioning effects arising from the repeated pairings of the reinforcer and the specific states of autonomic activity.

To illustrate this point in more detail, we may consider the performance of rats in Group C of Experiment 1. The greater amount of time spent in the contraction region on CS+ trials during Stages 2 and 3 might presumably be attributed to elicited contraction responses by the application of the electric shock when and only when the rat's intestine was in the state of relaxation at the onset or in the course of the shock signal. Furthermore, the separation between CS+ trials, on the one hand, and CS− and B trials, on the other, during Stage 4, i.e., in the absence of the shock, might be described as a classical conditioning effect resulting from the contiguous occurrence of the shock signal and the elicitation of relaxation-to-contraction changes in the rat's intestinal state during Stages 2 and 3 of the experiment. The same line of reasoning could, of course, be applied also to Group R in Experiment 1, where the administration of the shock whenever the rat's intestine was in the state of contraction might presumably have produced intestinal relaxation.

Therefore, the interpretation of the findings of the previous experiment on the basis of the dependency between response and reinforcer may not be completely unequivocal unless it is demonstrated that a state-dependent unconditioned response is not associated with the reinforcing agent employed. Relevant to this problem, Experiment 2 was designed to provide a direct and systematic assessment of the unconditioned effect(s) of an electric shock to the tail on the intestinal state of deeply curarized rats.

Method

Subjects

The subjects were four male albino Sprague-Dawley rats weighing 450–600 gm.

Apparatus

The apparatus for artificial respiration, recording of intestinal motility, and delivery of electric shock was the same as that used in Experiment 1. Records of intestinal motility were traced by the polygraph pen only, and no use was made of the commutator plate. Also, no conditioned stimuli were used in this experiment.

Procedure

The procedures for food deprivation and curarization were identical to those used in the previous experiment. After a 15-min. habituation period which started with the onset of muscular paralysis, each animal received 54 shock (Sh) trials, divided equally among three durations of 5, 10, and 25 sec. These Sh trials were unsignaled and totally noncontingent with respect to the rat's intestinal state. In order to provide a basis of comparison for assessing the effect of the shock, 54 blank (B) trials involving no stimulation were interspersed among the Sh trials. The B trials were of the same duration (5, 10, and 25 sec.) as the Sh trials. A total of 108 Sh and B trials was presented to each rat according to repetitions of the following predetermined balanced sequence: Sh-5, Sh-10, Sh-10, B-25, B-5, B-10, Sh-25, Sh-5, B-5, B-5, Sh-10, B-10, B-25, Sh-5, Sh-5, Sh-25, B-5, Sh-25, B-10, Sh-10, B-5, B-25, Sh-25, Sh-10, Sh-5, B-10, B-5, Sh-5, B-25, Sh-10, Sh-25, Sh-25, B-25, B-10. The intertrial intervals were irregularly 30, 60, or 90 sec., with an average of 60 sec. After the administration of these 108 trials, the subject was maintained on artificial respiration until the time of recovery from curare, normally a minimum of 2–3 hr.

For each rat, a "contraction criterion" was designated at the same position with respect to the relaxation base line as that of Experiment 1. Individual Sh and B trials were then scored at both the point of trial onset and the point of trial offset as contraction (c) or relaxation (r), depending upon whether the graphic record of the rat's intestinal motility was above or below the contraction criterion, respectively. Hence, for every trial the rat received two scores indicating its intestinal state at the beginning and at the termination of the trial.

Results and Discussion

Frequencies of change in the intestinal states of each rat, as measured at points of trial onset and trial offset, are presented in Table 2. The frequencies are arranged separately for trials initiated in states of contraction and relaxation, making possible comparisons between Sh and B trials initiated in either intestinal state. Since no consistent differences in the intestinal response could be noted among the three durations of shock trials (5, 10, and 25 sec.), the data for the three durations were combined. Hence, the entries in the 2 × 2 contingency tables in Table 2 represent, in each case, totals for the three different trial durations. Chi-square analyses of these data, as seen in Table 2, did not reveal a significant association between the type of trial (Sh vs. B)

TABLE 2

FREQUENCIES OF CHANGE IN INTESTINAL STATES AS MEASURED AT POINTS OF TRIAL ONSET AND TRIAL OFFSET

Subject	Intestinal state at trial onset	Type of preceding trial	Intestinal state at trial offset		x^2
			Changed	Not changed	
1	c	Sh	3	16	1.01
		B	7	13	
	r	Sh	7	28	.06
		B	7	27	
2	c	Sh	10	15	2.22
		B	4	20	
	r	Sh	8	21	.01
		B	17	13	
3	c	Sh	8	20	.08
		B	9	16	
	r	Sh	7	19	.10
		B	10	19	
4	c	Sh	11	17	.74
		B	7	21	
	r	Sh	7	19	2.03
		B	13	13	

Note.—Abbreviations: c = contraction; r = relaxation; Sh = shock; B = blank.

and frequency of change in the intestinal state for any of the four animals in the experiment. This was the case irrespective of the initial state of the intestine at the time of trial onset.

Subsequently, in order to determine whether or not there is a dependency between the effects of *shock offset* and the probability of intestinal change, the records were scored, according to the same criteria as specified above, at points of trial offset and 10 sec. after trial offset. Since, as in the case of shock application, no consistent differences in rats' intestinal response in the 10-sec. intervals following the three durations were observed, trial frequencies for the three trial durations were combined. These frequencies are presented separately for each subject in Table 3. Chi-square analyses of these data, also, failed to show any significant association between the type of preceding trial (Sh vs. B) and the frequencies of change in intestinal state at the end of a 10-sec. interval following a trial.

This was the case for both trials terminated in a state of contraction and those terminated in a state of relaxation.

In summary, the findings of this experiment fail to demonstrate that the application of an unsignaled, noncontingent electric shock to the tail could significantly influence the course of the spontaneous activity of the intestine by either sustaining or changing the particular intestinal state in which shock application is initiated. Similarly, the cessation of an electric shock does not appear to have such a systematic effect on this response.

GENERAL DISCUSSION

The findings of Experiment 1 clearly show that stimulus control of intestinal contraction and relaxation can be obtained in deeply curarized rats as a result of discriminated escape and avoidance training with electric shock. The use of the bidirectional procedure in that experiment and the absence of any significant shock effects that

TABLE 3

FREQUENCIES OF CHANGE IN INTESTINAL STATES AS MEASURED AT POINTS OF TRIAL OFFSET AND 10 SEC. AFTER TRIAL OFFSET

Subject	Intestinal state at trial offset	Type of trial	Intestinal state at 10 sec. after trial offset		x^2
			Changed	Not changed	
1	c	Sh	6	17	.09
		B	7	13	
	r	Sh	6	25	.82
		B	11	23	
2	c	Sh	9	14	1.36
		B	8	29	
	r	Sh	10	21	1.19
		B	9	8	
3	c	Sh	5	22	.08
		B	5	21	
	r	Sh	7	20	.01
		B	6	22	
4	c	Sh	9	15	.37
		B	9	25	
	r	Sh	9	21	.21
		B	4	16	

Note.—Abbreviations: c = contraction; r = relaxation; Sh = shock; B = blank.

would be dependent on the intestinal state in which shock is administered, as evidenced in Experiment 2, provide safeguards against the possibility that the obtained changes in responding might be explained on the basis of any unconditioned effects associated with the reinforcing agent and/or as a result of classical conditioning.

These findings corroborate those reported by Miller and Banuazizi (1968), where increases and decreases in intestinal motility were produced through an instrumental training procedure using ESB as reward. Therefore, the possible range of events capable of reinforcing operant conditioning of intestinal motility may now be extended to include both positive and negative stimuli.

It should be noted, however, that changes in responding obtained in this study appear to be qualitatively somewhat different from those reported by Miller and Banuazizi (1968, see Figure 1). Specifically, subjects in their experiment appeared to have met the criteria of learning through a considerable disruption of the regular pattern of the intestinal response; i.e., contraction episodes became more or less frequent as a result of reinforcing contraction or relaxation, respectively. As noted earlier, an examination of the records of animals in the present experiment revealed that the basic rhythmicity of the response persisted throughout the training session and that learning was manifested more as a result of changes in *duration*, rather than in frequency, of contraction or relaxation episodes.

There are numerous differences in experimental procedure in the two studies that might account for the above discrepancy. But, one especially intriguing possibility is that it may have resulted from differences in the reinforcement schedule such that the more efficient satisfaction of the reward criteria in each study produced the different patterns of change. Thus, it might be feasible to employ different instrumental procedures in order to produce relatively specific *patterns* of activity in autonomically mediated response systems.

REFERENCES

BANUAZIZI, A. *Modification of an autonomic response by instrumental learning.* (Doctoral dissertation, Yale University) Ann Arbor, Mich.: University Microfilms, 1968, No. 69-8311.

BLACK, A. H. Heart rate conditioning in curarized animals: What is operantly conditioned. Invited address at the 75th Annual Meeting of the American Psychological Association, Washington, D.C., September 1967.

BLACK, A. H. Autonomic aversive conditioning in infrahuman subjects. In F. R. Brush (Ed.), *Aversive conditioning and learning.* New York and London: Academic Press, 1971.

CRIDER, A., SCHWARTZ, G. E., & SHNIDMAN, S. On the criteria for instrumental autonomic conditioning: A reply to Katkin and Murray. *Psychological Bulletin*, 1969, **71**, 455-461.

DICARA, L. V., & MILLER, N. E. Changes in heart rate instrumentally learned by curarized rats as avoidance responses. *Journal of Comparative and Physiological Psychology*, 1968, **65**, 8-12. (a)

DICARA, L. V., & MILLER, N. E. Instrumental learning of systolic blood pressure responses by curarized rats: Dissociation of cardiac and vascular changes. *Psychosomatic Medicine*, 1968, **30**, 489-494. (b)

DICARA, L. V., & MILLER, N. E. Transfer of heart-rate changes from curarized to noncurarized state: Implications for a mediational hypothesis. *Journal of Comparative and Physiological Psychology*, 1969, **68**, 159-162.

DICARA, L. V., & WEISS, J. M. Effect of heart-rate learning under curare on subsequent noncurarized avoidance learning. *Journal of Comparative and Physiological Psychology*, 1969, **69**, 368-374.

FOLKOW, B., & RUBINSTEIN, E. H. Behavioural and autonomic patterns evoked by stimulation of the lateral hypothalamic area in the cat. *Acta Physiologica Scandinavica*, 1965, **65**, 292-299.

KATKIN, E. S., & MURRAY, E. N. Instrumental conditioning of autonomically mediated behavior: Theoretical and methodological issues. *Psychological Bulletin*, 1968, **70**, 52-68.

KATKIN, E. S., MURRAY, E. N., & LACHMAN, R. Concerning instrumental autonomic conditioning: A rejoinder. *Psychological Bulletin*, 1969, **71**, 462-466.

KIMBLE, G. A. *Hilgard and Marquis' conditioning and learning.* (2nd ed.) New York: Appleton-Century-Crofts, 1961.

KIMMEL, H. D. Instrumental conditioning of autonomically mediated behavior. *Psychological Bulletin*, 1967, **67**, 337-345.

MILLER, N. E. Learning of visceral and glandular responses. *Science*, 1969, **163**, 434-445.

MILLER, N. E., & BANUAZIZI, A. Instrumental learning by curarized rats of a specific visceral response, intestinal or cardiac. *Journal of Comparative and Physiological Psychology*, 1968, **65**, 1-7.

SHEFFIELD, F. D. Avoidance training and the contiguity principle. *Journal of Comparative and Physiological Psychology*, 1948, **41**, 165-177.

(Received December 22, 1971)

27
Sequential Representation of Voluntary Movement in Cortical Macro-Potentials: Direct Control of Behavior by Operant Conditioning of Wave Amplitude

Joel P. Rosenfeld and Stephen S. Fox

WE HAVE RECENTLY reported that a stereotyped, voluntary reaching movement in the cat is associated with an EEG potential in the contralateral sensorimotor cortex (6). In that report we noted that the slope and peak latency of a late component in the movement-associated potential (MEP) covaried with the duration of the animal's reaching movement. In this report, we have extended the earlier findings by demonstrating 1) that during the excursion of the cat's limb, the sequential, instantaneous values of limb displacement covary systematically with sequential values of movement-evoked potential amplitude; and 2) that instrumental conditioning of the amplitude of selected MEP components directly is associated with modifications in the details of movement. Such findings suggest cortical waveform as a candidate code for voluntary movement.

METHODS

Five adult cats were trained to reach from a 1.5-inch white circle on the floor of a running chamber to a hole in the wall before them. In the elevated rings surrounding the floor circle and wall hole were photocells whose outputs delimited the start and finish of the reaching movement. The cats were rewarded for these self-paced movements with .6 ml milk pumped directly to the mouth through a chronically implanted cannula extending from skull to palate.

Attached to the shaved distal forelimb was a grain-of-wheat light bulb. This was mounted on a tape bracelet which the animals wore around the limb just proximal to the wrist joint. A line drawn on the dorsal-lateral aspect of the forelimb in gention violet was aligned each day with a line drawn on the bracelet so as to assure that the position of the bulb on the limb remained invariant over repeated sessions. The bulb was powered through two fine wires which extended from the bracelet to a light wire harness with which the animals were fitted each day. The power wires, leaving the harness, traveled with the conventional chronic recording cable to a remote battery-power source. Two modified television cameras, permanently mounted outside the running chamber, tracked the excursion of the light bulb in the horizontal and vertical dimensions. Their output, after on-line computer processing, corresponded to vertical and horizontal displacement. The PDP8 computer controlling the experiment stored on magnetic tape the digitized movement values for each dimension for each trial. Sessions consisted of 200–400 trials.

Simultaneously stored on tape were the digitized values of the EEG potential recorded from 200 msec before the movement began to 200 msec after the movement ended. The potential was recorded from the contralateral cortex at Jasper and Ajmone-Marsan coordinates anterior 23, lateral 9. The electrodes were intended to be at the lateral extent of the region just posterior to the postcruciate sulcus where units sensitive to joint rotation have been reported (3, 4). A bipolar electrode pair (of .008-inch nichrome wires cemented together) was driven into cortex until the upper element of the pair was heard to just penetrate dura; intertip distance was 1 mm. The recording wires were soldered to a standard chronic recording plug fastened to the skull with dental acrylic. We recorded differentially and transcortically (sur-

---

Received for publication February 22, 1972.

Reprinted by permission of The American Physiological Society from the *Journal of Neurophysiology*, Vol. 35, 879-891, 1972.

face to depths) with a screw in frontal sinus attached to system ground. The potentials were transmitted first through an FET preamplifier-follower mounted on the recording plug. By vigorously shaking the recording cable with no load and observing no voltage deflections, we verified that this arrangement obviated cable or movement artifact. The potentials were then amplified with a Grass P511 amplifier with filters set to pass frequencies from 1.5 to 100 Hz. The output was led to the A-D converter of our on-line computer for further processing.

There were two phases of our experiments. In the first phase we did extensive linear cross-correlational analyses between movement and wave variables. The values at a given latency in a movement variable (e.g., vertical displacement) would be correlated over trials with wave amplitude at a specified latency. We would then have the correlation coefficient, r, between, say, the vertical displacement of the limb at time t_1 and the amplitude of the associated EEG potential at time t_2. We would then find r between vertical displacement again at t_1 and wave amplitude at a different lag, t_3. By continuing this process for several t values for wave, we obtained a plot of r between displacement at t_1 and wave at t_x as a function of x or lag between movement point and wave point. We obtained several such plots for two time locations in various movement-describing variables. For each cat, the two time loci in any given movement variable which we examined corresponded to an early and middle part of that subject's typical reaching behavior. The movement variables examined included vertical and horizontal displacement, instantaneous vertical and horizontal speed, and "absolute" displacement and its derivative. If x and y are vertical and horizontal displacement, absolute displacement $= z = \sqrt{x^2 + y^2}$; absolute speed $= dz/dt$). The movement point cross-correlated with wave points always preceded all wave points in time since we were concerned with afferent wave activity due to feedback from the moving limb.

In the second, experimental, phase of our experiments we rewarded the cats only when the MEP value at a specified latency exceeded a specified amplitude criterion. The time location or "point" chosen for conditioning was actually a 28-msec segment centered at a given latency from movement onset. The correlations described earlier between time loci in movement and time loci in wave, likewise, were between means of 28-msec segments (wave) and 8-msec segments (movement). The cats in the second phase of the experiments were rewarded for changing the mean value of amplitude during the specified 28-msec segment of the MEP for that trial. The segment of the wave chosen for instrumental conditioning was, for each cat, the segment showing the best relationship (highest correlation) with any of the movement variables in the correlation analysis of the first phase of our experiment. We reasoned that if a given movement variable at a given time correlated highly with wave amplitude at a given time, then if the correlation was not spurious, conditioned changes in the time-lagged wave segment should be paralleled by time-localized changes in the correlated movement variable, and the direction of these latter changes should be predictable from the sign of the previously obtained correlation coefficient.

The method of instrumentally training brain waves has been described (1, 2, 5): briefly, an evoked potential is sampled repeatedly so as to obtain the mean and standard deviation (SD) of the amplitude of the "critical" component centered at specified latency. The subject is then rewarded for producing samples of that component (a 28-msec segment in the present research) falling 1 SD from the pretraining mean. As the chance probability of such values is .16, it is assumed that learning occurs if subjects achieve performance levels upward of 32%. The procedures used in this report were similar, except that since the potentials here were generated by voluntary movements, it was necessary to gradually shape the behavior of the cats so as to avoid a potentially disruptive, sudden shift in reinforcement contingency from continuous (100%) reinforcement to the partial reinforcement schedule in force as the animal is introduced to the neural reinforcement contingency. Thus, the cats were gradually shifted from 100 to 90% reinforcement, to 80%, and so on in steps of 10% a day until they were performing the reaching behavior for reinforcement of a random 20% of their reaching responses; (e.g., the probability of reinforcement was .2). After performing at this level for 2 weeks, cats were run under the neural reinforcement contingency that reward, .6 ml of milk, be delivered if the criterion 28-msec MEP component amplitude was $-.85$ SD from the pretraining mean. After 2 weeks of this training, the reinforcement contingency was reversed: cats were rewarded for criterion component values $+.85$ SD from their pretraining means. The chance probability of either criterion response is .2.

We collected base-line data on each cat: during the last week of performance on 20% random partial reinforcement, the collected tapes (400 trials/day) were analyzed off-line to determine the percent of responses which would have reached the $-.85$ SD and $+.85$ SD criteria if they had been imposed; 19.96 and 21.49% were the respectively obtained percentages. These are near the theoretical 20% level.

RESULTS

Cross correlations between movement and wave

It will be recalled that for each cat, values of each movement variable studied at two time loci were individually correlated over trials with subsequent sequential temporal loci in the MEP. Figure 1A is a plot of the correlation over 250 trials (in one cat) of absolute displacement at 8 msec following movement onset with movement-evoked potential (MEP) amplitude at various lags from 0 to 320 msec (e.g., from 8 msec following movement onset to 328 msec following movement onset). It is seen that the Pearson correlation coefficient, r, reaches a distinct minimum at a lag of 40 msec, after which it stays near 0. This means that absolute displacement during the early portion of the movement, at 8 msec, is maximally related to wave amplitude at a lag of 40 msec (or $40 + 8 = 48$ msec following movement onset) and that there is no relation between absolute displacement at 8 msec and wave for wave latencies after 48 msec. (The negative or positive signs of peak r values depend upon recording polarity.)

In Fig. 1B, two replications (taken on 2 separate days separated by an intervening week) of another correlation function for the same animal are shown. The same movement variable, absolute displacement, is correlated with sequential evoked potential-amplitude values, however the movement variable has been sampled at 156 msec following movement onset, and the sampled wave points are thus lagged from 156 msec. In this plot, it is seen that absolute displacement during the middle portion of the movement (at 156 msec following onset) shows a strong relationship with wave amplitude at a lag of about 50 msec from the sampled movement point (or $156 + 50 = 200$ msec into the wave following movement onset), since there is a negative peak in the correlation function at lag $= 50$ msec. It is seen that this extreme value was replicable. One sees also a positive peak in the function at lag $= 140$ msec where $r = +.367$ on one replication and $+.671$ on another. These data indicate that there was a less reliable tendency for absolute displacement at 156 msec to correlate with MEP amplitude at 246 msec ($= 156 + 140$). The curve of Fig. 1B oscillated about $r = 0$ after lag $= 320$ msec (not shown in the figure). In summary, absolute displacement early in movement (8 msec) was maximally and discretely correlated with an MEP component at 48 msec but not with later components, whereas the same movement variable later in the course of the reaching behavior correlated with amplitude of later components in the MEP. The latencies between early movement point and early wave point and between later movement point and the first correlated later wave point were comparable, 40 and 50 msec, respectively. Evidence of independent representation of early and later movement segments in early and late wave components as described was seen in four of five of the animals. The early MEP component of the fifth cat was small and early movement

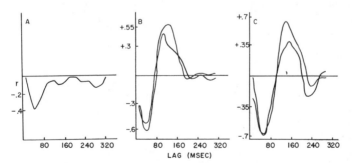

FIG. 1. *A:* correlation coefficient (r) between early (8 msec) displacement value with various 28-msec MEP segments lagged from the movement locus and plotted as a function of lag for cat 7. *B:* same as *A* except a point midway through the movement (at 108 msec after movement onset) is the reference point from which MEP segments are lagged. Data from cat *9;* two replications taken 1 week apart are plotted. *C:* same as *B* except the movement point is at 156 msec, which was midway through the movement for this cat (No. 7). Each curve is based on at least 200 trials.

showed no distinct correlation with any subsequent wave points.

Figure 1C is the same kind of correlation function as is Fig. 1B, except that 1) the data in the two sections of the figure were derived from two different animals, and 2) the time locus of movement sampled in Fig. 1C is 108 msec rather than the 156-msec locus of Fig. 1B. The lag from movement point to the first wave locus of maximal (absolute) correlation is 40 msec. It is worth emphasizing the remarkable similarity in topography between Fig. 1B and C, despite the fact that the data come from two different cats and that two different time loci in movement were represented in each section of the figure.

It was a typical finding in all animals that a time locus in the displacement variables midway through the movement, from 100 to 200 msec, depending on individual response speed characteristics of the subjects, correlated maximally with amplitude of a late positive MEP component with a latency of 140–240 msec, the precise time locus again depending on the overall response speed of each cat. Thus, the larger the mean overall response duration of a cat, the later in the wave would be the point of maximum correlation with displacement at a point midway through the movement. There was also a tendency for lag between maximally related movement and wave points to increase with overall response duration. Thus for the two animals whose performance is recorded in Fig. 1B and C, the lags are in the 40- to 50-msec range. These two animals had typical response duration means of 140–200 msec over repeated days. For two other cats whose typical mean response durations were 220–280 msec, the latencies between middle movement displacement values and maximally related MEP amplitude loci were 90–250 msec. As we have previously reported (6) the topography of MEPs of all cats was similar, each MEP having a surface negativity at 30–50 msec, followed by a later positive-negative complex whose latency was a function of response duration within and between cats (see Fig. 4). It is noted that the biphasic late component explains the oppositely directed pair of extrema in Fig. 1B and C since the two MEP phases were negatively correlated; for the cats represented in Fig. 1B and C, the correlation between these phases varied across sessions from $-.6$ to $-.74$.

Although midmovement displacement correlated with both phases of a later, biphasic MEP component, it is emphasized that early and later movements were independently related to early and later MEP components. Figure 2A and B emphasizes the apparently independent neural representation of early and late portions of the voluntary movement in the present research. (This figure is from two of the five cats; similar data obtained from two of the remaining subjects.) In this figure is shown 1) the correlations (circled) between early movement points with early and later wave points, 2) the correlations between later movement with early and later wave points, 3) the correlations between early and late movement points, and 4) the correlations between early and later wave points. For each section, early movement (displacement) is more strongly correlated with early wave than with later wave, and later movement is more strongly correlated with later wave than with early wave. It is noted in Fig. 2B, however, that early movement and later wave and early wave and later movement are significantly correlated. We believe that these values are inflated, not due to causal relations between the correlated variables but due to the extremely high correlation (uniquely in this cat) between early and late movement. Partial correlation analysis could have confirmed this hypothesis; this, unfortunately, was not done. However, it is noted in Fig. 2A (and in similar data from the other animals not shown) that when the correlation between early and late movement values was low, there was no tendency for high correlation between early movement and late wave or between early wave and late movement, although early movement and early wave and late movement remained highly correlated. The high correlation between early and late movement in the one cat is consistent with our observation that this animal was the fastest in the study (its mean response durations on various days were in the range 120–160 msec). Obviously, the faster or more ballistic the movement the more difficult it is to modify separate portions of the trajectory as independent segments, and so early and late portions tend to be highly correlated.

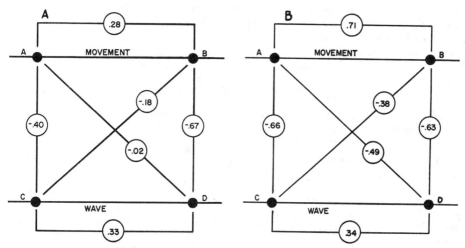

FIG. 2. Correlations between members of various pairs of wave and movement loci. The movement variable is absolute displacement. Circled numbers are Pearson r values. A: cat 7; A = 8 msec, B = 156 msec, C = 42 msec, D = 200 msec (all times referenced from movement onset). B: cat 9; A = 8 msec, B = 108 msec, C = 42 msec, D = 150 msec.

For all animals, absolute displacement was not the only movement variable whose lagged correlation function with MEP loci showed distinct maxima or minima. Vertical and horizontal displacements yielded, within each cat, correlation functions with topography quite similar to the correlation functions with absolute displacement. That is, if absolute displacement at M msec correlated optimally at lag L, so would the correlation functions with vertical and horizontal displacement show peaks at lag = L msec. Absolute displacement was most often the function yielding the highest absolute r values in all cats and over repeated days.

The correlation functions relating time loci in the three speed variables to MEP loci were similar to the correlation functions on the displacement variables, although the size of extreme values (maximum and minimum r values in the correlation functions) in speed data was always considerably lower than in corresponding displacement data. The loci of peaks in correlation functions of speed and wave typically fell at the peak loci of the correlation functions on displacement. Thus, although it was, in principle, possible for speed and displacement to be independent over trials and thus independently represented in brain wave variables, such did not obtain. Typically, speed and displacement were found to be correlated ($|r| \geq .5$) in all cats at all points during the movements.

Nonspecific effects on movement of operant conditioning of movement-evoked potentials

In order to experimentally validate the observed correlations between movement parameters and the amplitude of specific components of movement-evoked potentials, direct manipulation of movement was attempted by instrumental conditioning at specific components of the correlated movement-evoked potential. The amplitude of chosen components of the MEP were conditioned in two polarities, both in the direction of the observed correlation between MEP and movement and in the opposite direction to the observed correlation. In this way the manipulation of movement by wave conditioning would be complete with bidirectional predictions. Further, irrelevant or classically conditioned effects which were not related to contingent reinforcement or learning could be ruled out since such nonassociational effects would be expected to operate in one direction only.

It is recalled that for each cat, a 28-msec segment of the MEP was chosen for conditioning, the chosen segment being that segment which had the highest correlation with an 8-msec segment of a movement variable studied in the previous phase of the

Animal Studies

TABLE 1. *The 28-msec segments of movement-evoked potentials chosen for operant conditioning on basis of best relations with an 8-msec segment in a movement variable*

Cat	Movement Variable and Latency From Movement Onset of Movement Segment*	Correlated Wave Segment, msec	Correl Coef
5	Vertical displacement, 172 msec	288–312	−.60
6	Absolute displacement, 148 msec	360–384	−.61
7	Absolute displacement, 156 msec	188–212	−.67
9	Horizontal displacement, 108 msec	128–152	−.63
10	Absolute displacement	164–188	+.40

* Maximally correlated with evoked potential segment.

research. Table 1 lists the chosen segments for each cat.

Of the five animals with which we started, one (no. 7) was eliminated from the conditioning studies due to a cranial infection. Another (no. 10) failed to respond to the first reinforcement contingency (decreasing the amplitude of the chosen MEP segment). He was therefore not subjected to the second neural reinforcement contingency (increasing amplitude of the chosen MEP segment), since the latter task was designed to serve as a control procedure for nonassociative factors operating in the former. Therefore, the movement records studied in conjunction with brain wave conditioning are for the three remaining cats: 5, 6, and 9. The successful conditioning of chosen MEP segments in these animals was indicated by the change in mean percent criterion responses from our base-line phase to our training phases: for the −.85 SD criterion, the change was 19.96–47.44% correct; for the +.85 SD criterion, the change was 21.49–62.59% criterion-reaching responses. (These percentages and subsequent data are based on the last 400 trials of training in each contingency.)

Table 1 indicates that the correlations between chosen wave segments and correlated movement-variable segments were negative (in cats 5, 6, 9). Thus, the conditioned decreases in wave-segment amplitude (under the −.85 SD contingency) should have generated an increase in movement-segment value and the conditioned increases (with the +.85 SD contingency) in wave-segment amplitude should have yielded a decrease in movement values. Figure 3 represents the typical finding showing distributions of movement-segment values in the base line and two training conditions for one cat. It is seen that the distribution of movement-segment values is centered on the abscissa during base line, moved to the right under the −.85 SD wave training contingency, and moved to the left under the +.85 SD contingency. An analysis of variance performed on the medians of such distributions for all successfully trained cats

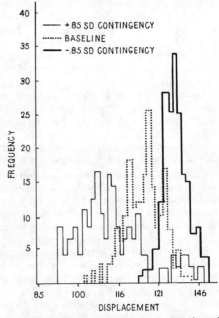

FIG. 3. Distributions of the movement point values (on abscissa) maximally correlated with the criterion brain wave segment during the three training conditions for cat 6.

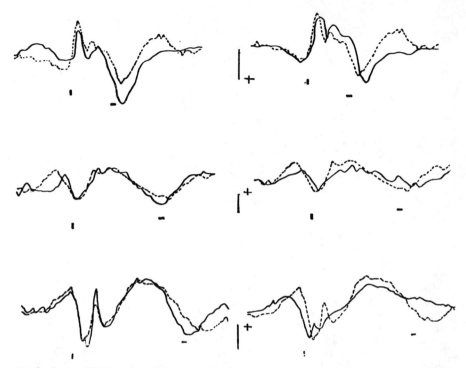

FIG. 4. Averaged MEPs collected from base line and the training conditions in each cat. Each superimposition is of base line and training, with −.85 SD MEP contingency waves in the left column and +.85 SD contingency waves in the right. Dashed curves are base-line data. Each row represents a different cat; top = cat 9, middle = cat 5, bottom = cat 6. Height of surface polarity marker = 60 μv. Vertical bars are movement onset; horizontal bars indicate locations of conditioned segments.

yielded $F = 25.566$ ($P = .0053$). A critical value of the smallest mean difference between any pair of means (of medians) chosen for comparison was calculated for two-tailed statistical significance at the $P = .05$ level of confidence. The smallest difference between any obtained pair of means was greater than the calculated criterion. Thus, the brain wave segment-conditioning training had the predicted effects on the correlated movement segments.

The operant reinforcement contingencies, in general, had predictable effects on chosen MEP segments and their correlated movement segments, however a comparison of base line with training data on movement as well as on brain potentials indicates that the changes in these variables were extensive. Figure 4 shows various superimposed base line and training average MEPs. It is seen that although the changes in MEP topography between conditions show time localization in two of the superimposed pairs of MEPs, each of the four remaining pairs of traces are different all along the epoch. Figure 5A, B, and C are plots of average movement variables as a function of time following movement onset in each successively trained cat, in each training condition. For each cat, the movement variable plotted is that variable containing the segment maximally correlated with the conditioned MEP segment (the center of which is indicated). While there is some tendency for the largest differences among these curves to fall in the region of the MEP-correlated segments, it is clear that the movements differ at many time loci.

Specific movement changes related to conditioning of movement-evoked potentials

The large changes in movement topography between base line and training conditions could have been due to a variety of factors, such as increased efficiency of move-

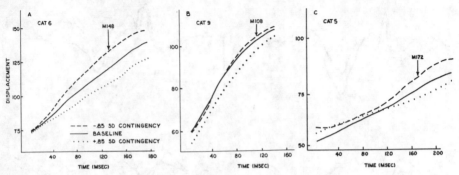

FIG. 5. Displacement as a function of time following onset of movement during base line and two MEP training contingency conditions. Origin of 8-msec segment expected (from correlation studies) to show change following conditioning is indicated with arrow.

ment with practice, the development of superstitious behavior, and so forth. Such extensive changes (in neural and behavioral variables) for whatever reasons, obscured any relations between specific changes in the MEP and specific changes in movement. We therefore employed a simple sorting analysis on data derived from the final day of each training condition: *1*) MEPs were averaged into two buffers yielding *a*) the average MEP of criterion responses (in which amplitude during the chosen segment met the .85 SD neural reinforcement contingency), and *b*) the average MEP of noncriterion responses; *2*) movement variables (as in Fig. 5 *A, B, C*) were sorted into two buffers yielding *a*) the average displacement function evoking a criterion MEP, and *b*) the average displacement function evoking noncriterion MEPs.

Typical results of these procedures are Figs. 6 and 7. In Fig. 6 it is seen that averaged MEPs, sorted as just described, can differ at exquisitely localized times corresponding to the conditioned segments. The upper two, superimposed sorted pairs in the right-hand column of the figure, in particular, demonstrate that early and late segments of the wave are independent. Figure 7*A, B, C* are representative superimposed averaged movement-function pairs sorted as just described. The movement variables shown are those containing segments which, for each cat, were maximally related to conditioned MEP segments. Corresponding to the discretely localized MEP differences in Fig. 6, the differences between the sorted movement function are localized to the region of the indicated correlated movement segments (compare Figs. 7 and 5). The localization is not perfect in Fig. 7*C* and the extent of localization is greater than 8 msec in Fig. 7*A* and *B*. The nature and significance of these imperfect outcomes will be considered later. It should be emphasized however that the sorted movement functions in Fig. 7*A* and *B* are statistically superimposable for the first two-thirds of their epochs. The sorted MEP pairs corresponding to Fig. 7*A* and *B* are in the upper two panels in the left-hand column of Fig. 6. These traces are likewise superimposable up to the region of the conditioned segments. More generally, gross changes in MEPs from base line to training were paralleled by gross changes in evoking movements, and discrete changes in MEPs sorted on reinforcement criteria were paralleled by more discrete differences in corresponding sorted movements. In only one case did the sorting procedure not eliminate gross MEP differences, which are evident in the bottom left-hand panel in Fig. 6. Figure 7*E* shows the sorted movements corresponding to this exceptional MEP pair and it is clear that gross differences in sorted movements, particularly evident in comparison with Fig. 7*A, B, C*, again parallel the gross MEP differences.

Conditioning effects on other movement variables

Inasmuch as the speed variables were observed in the cross-correlational data to be correlated with the displacement variables, it would be expected that speed variables, like the displacement variables, would be affected by MEP conditioning procedures.

Operant Conditioning of Wave Amplitude 347

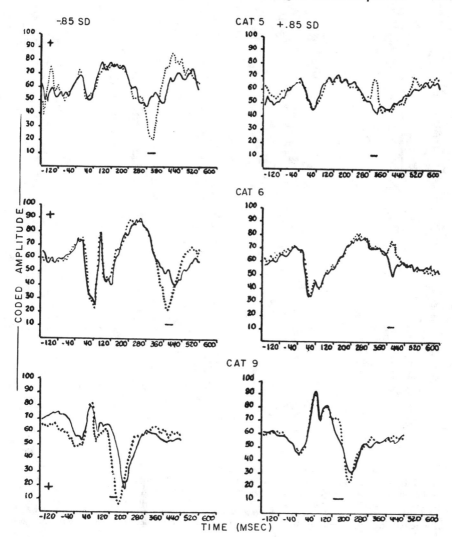

FIG. 6. MEPs collected during training days of near-asymptotic performance and sorted into separate superimposed averages (shown) on the basis of whether or not a wave to be sorted met the MEP criterion. −.85 SD contingency data is in the left column, +.85 SD contingency data is in the right column. Each row represents a different cat; cat 5 = top, cat 6 = middle, cat 9 = bottom. Horizontal line indicates trained segment; movement begins at time = 0. Ten coded amplitude units = 15 μv. Plus signs indicate surface positivity. Dotted curves are criterion responses; solid lines are failures.

In general, this obtained. All movement variables were grossly different between base line and training conditions. Additionally, no movement variables sorted as described above (on MEP criteria) were superimposable. Figure 7D is a plot of sorted, sequential, instantaneous horizontal speeds as a f(time) for the same cat whose absolute displacements are shown in Fig. 7C. (The two sections are based on simultaneously collected data sets.) Segments of difference are seen in both sections of the figure at 80–100 msec, although the speed function contains other difference segments.

DISCUSSION

Our laboratory has previously reported (2) that sequential components in photic evoked potentials show statistical independence. Independent conditioning of such

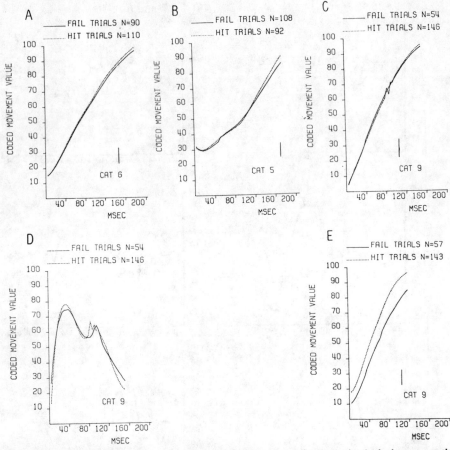

FIG. 7. Sorted average movement functions sorted during training on the basis of whether or not the associated MEPs of each average met or failed to meet the MEP criterion. Ordinate values are arbitrary. Dashed lines are associated with successful MEPs; straight lines are associated with failure. Vertical marker indicates center of 8-msec segment expected to show change in panels *A, B, C,* and *E,* which are displacements. *D* is the horizontal speed record collected simultaneously with data of *C*. Panels *E, B,* and *A* were collected during −.85 SD MEP contingency; *C* and *D* were collected during +.85 SD contingency.

components suggests that sequential segments in slow waves are capable of independent representation of behavior-relevant information. The present report anchors the earlier findings in a behavioral situation with evidence that different temporal portions of a voluntary reaching trajectory are related correlationally to different portions of the associated sensorimotor cortical macropotential. More impressive, operant conditioning of discrete components of the cortical macropotential results in predictable, discrete, and finely detailed alterations of the behavioral movements. Finally, the present experiments demonstrate, by bidirectional operant conditioning of the amplitude of cortical potential components, bidirectional and predicted changes in movement. That is, we have shown that opposite polarity of the conditioned components actually represent opposite directions along parameters of behavioral movement. Thus, not only do sequential moments in time along the cortical waves represent sequential behaviors, but the opposing polarities represent opposing directions along the movement parameters. Supplementary to these findings, the present experiments support our previous demonstrations that small or rare events of EEG do not constitute noise but that in contrast, under suitable conditions such events may become dominant. These findings reemphasize *1*) the state-dependence of wave components, and *2*) the fact that conditioning procedures may optimize the conditions for the occurrence of these highly variable events which would, under other conditions, be averaged out of the record.

The present findings were most forceful in the case of animals whose movements were slower. Although the data of even the fastest of our animals showed higher correlations between early movement with early wave and late movement with late wave than between early movement with late wave, the latter correlation was still quite high. While this correlation is evidence against independent representation, it is reasonably interpreted as spurious. The faster the movement, the more likely it is that early and late portions of it will correlate. Thus, early movement and late wave are correlated to the extent that early movement correlates with late movement. Late movement was observed to be uniquely correlated with late wave in animals generating less ballistic behaviors.

Sorting movements associated with criterion and noncriterion MEPs into separate categories yielded sorted pairs of movement records showing differences near or at the region of the movement segment where differences were predicted from the cross-correlation data. The localization was not perfect in two ways:

1) In Fig. 7A and B, the sorted pairs differed during time segments longer than the 8-msec segment used in correlation analysis. This lack of localization is actually quite consistent with the correlational data. Figure 1 shows that peaks in the lagged correlation functions were 40–80 msec wide. Although wave and movement segments sampled in all phases of the study were, respectively, 24 and 8 msec wide, the data of Fig. 1 indicate that such neighboring time segments in either variable autocorrelate. Thus, sorting movements on wave criteria should have resulted in segments of differences greater than 8 msec. Again autocorrelation in both variables effectively determines the width of independent sequential segments observable.

2) In Fig. 7C, the localization of difference in the sorted movement pairs is earlier than the segment locus predicted to show a difference from the cross-correlation analysis; the movements are close to superimposed at this predicted segment at 104–112 msec. However, the obtained difference locus at 83–92 msec is very close to the predicted segment. It is reasonable to assume that the 83- to 92-msec segment would have correlated with the MEP criterion segment nearly as well as did the 104- to 112-msec segment since such closely neighboring segments typically showed autocorrelations of .7 and higher. We chose segment widths for wave and movement in this first study of wave-movement relations on a relatively arbitrary basis. Otherwise stated, the widths we employed were probably not the widths of physiologically determined segments capable of completely independent representation. In extensions of this kind of work, it would probably be better to use segment widths for movement equivalent to autocorrelation bands (e.g., widths of peaks in autocorrelograms on 1-msec movement points). With very rapid movements—the data of Fig. 7C are based on our fastest moving cat—such bands are at least 80 msec wide; the obtained locus of difference in Fig. 7C is well within 80 msec of the predicted locus. Also important in accounting for the imperfect locus of difference in Fig. 7C is the fact that the figure is based on a 200-trial sample, during which 73% of the cat's MEPs were at criterion level or better. (For Fig. 7A and B, the corresponding values are 45 and 55%, respectively.) Since this animal (of Fig. 7C) responded so well to the neural reinforcement contingency, it is reasonable to assume that its noncriterion responses were probably very close to criterion. Since correlations between wave and movement segments in this study never exceeded |.7|, movement value accounts for no more than about 49% of wave variance. Thus, if small differences between criterion and noncriterion MEPs in cat 9 were accompanied by small differences in associated movements, the later differences could easily have been obscured by noise in the system. It might have been better to have performed the sorting analysis on data collected before the cat mastered the operant.

Taken together in perspective, the findings reported here strongly suggest that sequential components in reaching trajectory variables are represented sequentially and in detail in the associated EEG potentials. It remains to find methods of eliminating factors which, in the present preliminary paradigm, were a source of spurious correlations. Moreover, because the correlations between wave and movement existed for all movement variables, it is not possible to positively state which movement variable or combination of variables is represented in wave amplitude. The relationships found strongly suggest that some function of our movement variables could be so specified, perhaps with the aid of multivariate curvilinear and partial regression techniques. Such finer-grained analysis could yield higher correlations. Our first paradigm, however, had certain built-in aspects which, in retrospect, encouraged autocorrelations within and between movement variables. The animals were trained to reach from the floor to an elevated hole in the wall before them. (This seemed to us a rather natural movement to shape in the cat.) Thus, both horizontal and vertical displacements were likely to increase together since the time the cats had to complete the behavior was limited. Such time limits are likely to cause increased response speeds with concommitant autocorrelations within movement variables as well as correlations among speeds and displacements over trials. While it is possible to remove some of the restrictions on the behavior in the hope that increased degrees of freedom in the situation may reveal independence among movement variables, our feeling is that the more powerful manipulation in this research is the manipulation of reinforcement contingency. Thus, for example, if both vertical displacement and vertical speed appear to correlate with the amplitude of an MEP segment, the best demonstration of their independent representation would involve requiring the subject to change the MEP segment without changing one or the other of the movement variables. In this connection, we point out that whatever the neural event under study may be, the manipulation of conjoint reinforcement contingencies on wave and movement has, uniquely, the capability of decoding the neural representation of voluntary movement in the unrestrained, chronic animal.

SUMMARY

Five cats were trained to make stereotyped reaching movements. The movements were tracked and recorded simultaneously with the recording of contralateral, sensorimotor cortical macropotentials apparently evoked by the movements. Cross-correlational analysis between the neural and behavioral analogs provided evidence of the representation at particular times in the cortical potentials of values of localized time segments in reaching trajectories. Subsequent instrumental training of cortical potential values were accompanied by predictable changes in reaching behaviors. The results were discussed in terms of *1*) the capacity of sequential segments of evoked potentials for representation of independent segments of analog movement, and *2*) the limitations of the present paradigm in revealing such representation.

ACKNOWLEDGMENTS

These studies are supported by grants from the Graduate College, University of Iowa to S. S. Fox and Public Health Service Interdisciplinary Mental Health Predoctoral Fellowship (Grant 5T0MH10641) to J. P. Rosenfeld.

Present address of J. P. Rosenfeld: Dept. of Psychology, Cresap Laboratory, Northwestern University, Evanston, Ill. 60201.

REFERENCES

1. Fox, S. S. AND RUDELL, A. P. Operant controlled neural event: formal and systematic approach to electrical coding of behavior in brain. *Science* 162: 1299–1302, 1968.
2. Fox, S. S. AND RUDELL, A. P. Operant controlled neural event: functional independence in behavioral coding by early and late components of visual cortical evoked response in cats. *J. Neurophysiol.* 32: 458–561, 1970.
3. MOUNTCASTLE, V. B. Modality and topographic properties of single neurons of cat's somatic sensory cortex. *J. Neurophysiol.* 20: 408–434, 1957.
4. MOUNTCASTLE, V. B., DAVIES, P. W., AND BERMAN, A. L. Response patterns of cortical neurons. *J. Neurophysiol.* 20: 374–407, 1957.
5. ROSENFELD, J. P., RUDELL, A. P., AND FOX, S. S. Operant control of neural events in humans. *Science* 165: 821–823, 1969.
6. ROSENFELD, J. P. AND FOX, S. S. Movement-related macropotentials in cat cortex. *Electroencephalog. Clin. Neurophysiol.* 7: 489–494, 1972.

IX

MEDITATION AND HYPNOSIS

The Physiology of Meditation 28
Robert K. Wallace and Herbert Benson

How capable is the human organism of adjusting to psychologically disturbing changes in the environment? Our technological age is probably testing this capacity more severely than it was ever tested in the past. The impact of the rapid changes—unprecedented in scale, complexity and novelty—that technology is bringing about in our world seems to be having a deleterious effect on the mental and physical health of modern man. Some of the common disorders of our age, notably "nervous stomach" and high blood pressure, may well be attributable in part to the uncertainties that are burgeoning in our environment and daily lives. Since the environment is not likely to grow less complex or more predictable, it seems only prudent to devote some investigative attention to the human body's resources for coping with the vicissitudes of the environment.

There are in fact several ways in which an individual can control his physiological reactions to psychological events. Among the claims for such control the most notable have come from practitioners of meditation systems of the East: yoga and Zen Buddhism. This article will review and discuss recent studies of the effects of meditation that have been made by ourselves and by other investigators.

Yogis in India have long been reputed to perform phenomenal feats such as voluntarily stopping the heartbeat or surviving for extended periods in an "airtight" pit or in extreme cold without food or in a distorted physical posture. One of the first investigators to look into these claims in an objective way was a French cardiologist, Thérèse Brosse, who went to India in 1935 equipped with a portable electrocardiograph so that she could monitor the activity of the heart. Brosse concluded from her tests that one of her subjects actually was able to stop his heart. In 1957 two American physiologists, M. A. Wenger of the University of California at Los Angeles and B. K. Bagchi of the University of Michigan Medical School, conducted a more extensive investigation in collaboration with B. K. Anand of the All-India Institute of Medical Sciences in New Delhi. None of the yogis they studied, with more elaborate equipment than Brosse had used, showed a capability for stopping the heart. Wenger and Bagchi concluded that the disappearance of the signal of heart activity in Brosse's electrocardiogram was probably an artifact, since the heart impulse is sometimes obscured by electrical signals from contracting muscles of the thorax. (In attempting to stop the heart the yogis usually performed what is called the Valsalva maneuver,

Reprinted with permission from *Scientific American*, 1972, Vol. 226, No. 2. Copyright © 1972 by Scientific American, Inc. Copies of this article can be obtained from W. H. Freeman and Company, 660 Market Street, San Francisco, Cal. 94104 (Offprint No. 1242).

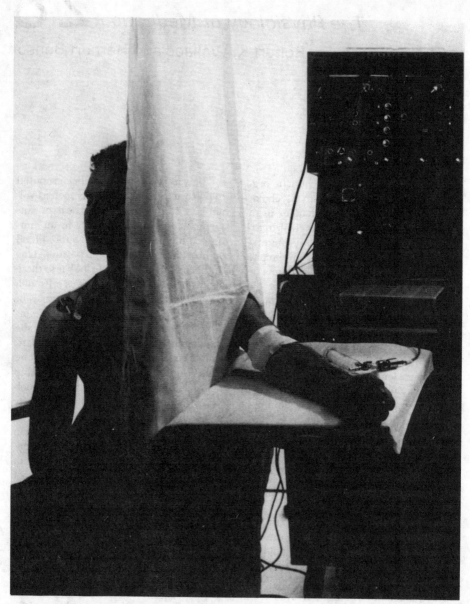

ISOLATED SUBJECT is connected with an instrument array that continuously records such physiological variables as heart rate and blood pressure. A catheter in the subject's left arm draws samples of arterial blood at 10-minute intervals; these samples are analyzed for oxygen and carbon dioxide content and for blood acidity and blood-lactate level. The subject's arm is screened from his view to minimize the psychological effects of blood withdrawal. Each subject first sat quietly for an interval and then was invited to meditate for a 30-minute period. At the end of the period the subject was asked to stop meditating but to continue sitting quietly during a further recording interval. Thirty-six qualified "transcendental" meditators from 17 to 41 years old volunteered as subjects for the study, which was conducted both at the Harvard Medical Unit of the Boston City Hospital and at the University of California at Irvine.

which increases the pressure within the chest; it can be done by holding one's breath and straining downward.) Wenger, Bagchi and Anand did find, however, that some of the yogis could slow both heartbeat and respiration rate.

Reports of a number of other investigations by researchers in the 1950's and 1960's indicated that meditation as practiced by yoga or Zen meditators could produce a variety of physiological effects. One of the demonstrated effects was reduction of the rate of metabolism. Examining Zen monks in Japan who had had many years of experience in the practice of deep meditation, Y. Sugi and K. Akutsu found that during meditation the subjects decreased their consumption of oxygen by about 20 percent and reduced their output of carbon dioxide. These signs of course constitute evidence of a slowing of metabolism. In New Delhi, Anand and two collaborators, G. S. Chhina and Baldeu Singh, made a similar finding in examination of a yoga practitioner; confined in a sealed metal box, the meditating yogi markedly reduced his oxygen consumption and carbon dioxide elimination.

These tests strongly indicated that meditation produced the effects through control of an "involuntary" mechanism in the body, presumably the autonomic nervous system. The reduction of carbon dioxide elimination might have been accounted for by a recognizably voluntary action of the subject—slowing the breathing—but such action should not markedly affect the uptake of oxygen by the body tissues. Consequently it was a reasonable supposition that the drop in oxygen consumption, reflecting a decrease in the need for inhaled oxygen, must be due to modification of a process not subject to manipulation in the usual sense.

Explorations with the electroencephalograph showed further that meditation produced changes in the electrical activity of the brain. In studies of Zen monks A. Kasamatsu and T. Hirai of the University of Tokyo found that during meditation with their eyes half-open the monks developed a predominance of alpha waves—the waves that ordinarily become prominent when a person is thoroughly relaxed with his eyes closed. In the meditating monks the alpha waves increased in amplitude and regularity, particularly in the frontal and central regions of the brain. Subjects with a great deal of experience in meditation showed other changes: the alpha waves slowed from the usual frequency of nine to 12 cycles per second to seven or eight cycles per second, and rhythmical theta waves at six to seven cycles per second appeared. Anand and other investigators in India found that yogis, like the Zen monks, also showed a heightening of alpha activity during meditation. N. N. Das and H. Gastaut, in an electroencephalographic examination of seven yogis, observed that as the meditation progressed the alpha waves gave way to fast-wave activity at the rate of 40 to 45 cycles per second and these waves in turn subsided with a return of the slow alpha and theta waves.

Another physiological response tested by the early investigators was the resistance of the skin to an electric current. This measure is thought by some to reflect the level of "anxiety": a decrease in skin resistance representing greater anxiety; a rise in resistance, greater relaxation. It turns out that meditation increases the skin resistance in yogis and somewhat stabilizes the resistance in Zen meditators.

We decided to undertake a systematic study of the physiological "effects," or, as we prefer to say, the physiological correlates, of meditation. In our review of the literature we had found a bewildering range of variation in the cases and the results of the different studies. The

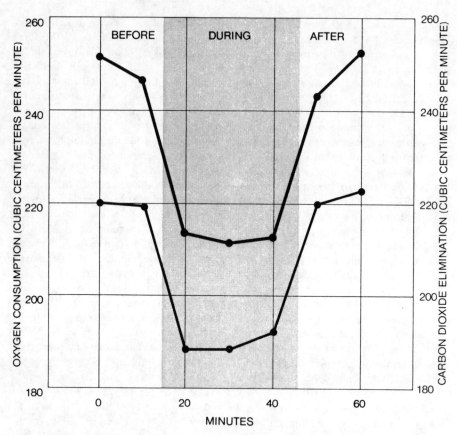

EFFECT OF MEDITATION on the subjects' oxygen consumption *(upper)* and carbon dioxide elimination *(lower)* was recorded in 20 and 15 cases respectively. After the subjects were invited to meditate both rates decreased markedly *(colored area)*. Consumption and elimination returned to the premeditation level soon after the subjects stopped meditating.

subjects varied greatly in their meditation techniques, their expertise and their performance. This was not so true of the Zen practitioners, all of whom employ the same technique, but it was quite characteristic of the practice of yoga, which has many more adherents. The state called yoga (meaning "union") has a generally agreed definition: a "higher" consciousness achieved through a fully rested and relaxed body and a fully awake and relaxed mind. In the endeavor to arrive at this state, however, the practitioners in India use a variety of approaches. Some seek the goal through strenuous physical exercise; others concentrate on controlling a particular overt function, such as the respiratory rate; others focus on purely mental processes, based on some device for concentration or contemplation. The difference in technique may produce a dichotomy of physiological effects; for instance, whereas those who use contemplation show a decrease in oxygen consumption, those who use physical exercise to achieve yoga show an oxygen-consumption increase. Moreover, since most of the techniques require rigorous discipline and long training, the range in abilities is wide, and it is difficult to know who is an "expert" or how expert he may be. Obviously all these complications made the problem of selecting suitable subjects for our systematic study a formidable one.

Fortunately one widely practiced yoga technique is so well standardized that it enabled us to carry out large-scale studies under reasonably uniform conditions. This technique, called "transcendental meditation," was developed by Maharishi Mahesh Yogi and is taught by an organization of instructors whom he personally qualifies. The technique does not require intense concentration or any form of rigorous mental or physical control, and it is easily learned, so that all subjects who have been through a relatively short period of training are "experts." The training does not involve devotion to any specific beliefs or life-style. It consists simply in two daily sessions of practice, each for 15 to 20 minutes.

The practitioner sits in a comfortable position with eyes closed. By a systematic method that he has been taught, he perceives a "suitable" sound or thought. Without attempting to concentrate specifically on this cue, he allows his mind to experience it freely, and his thinking, as the practitioners themselves report, rises to a "finer and more creative level in an easy and natural manner." More than 90,000 men and women in the U.S. are said to have received instruction in transcendental meditation by the organization teaching it. Hence large numbers of uniformly trained subjects were available for our studies.

What follows is a report of the detailed measurements made on a group of 36 subjects. Some were observed at the Thorndike Memorial Laboratory, a part of the Harvard Medical Unit at the Boston City Hospital. The others were observed at the University of California at Irvine. Twenty-eight were males and eight were females; they ranged in age from 17 to 41. Their experience in meditation ranged from less than a month to nine years, with the majority having had two to three years of experience.

During each test the subject served as his own control, spending part of the session in meditation and part in a normal, nonmeditative state. Devices for continuous measurement of blood pressure, heart rate, rectal temperature, skin resistance and electroencephalographic events were attached to the subject, and during the period of measurement samples were taken at 10-minute intervals for analysis of oxygen consumption, carbon dioxide elimination and other parameters. The subject sat in a chair. After a 30-minute period of habituation, measurements were started and continued for three periods: 20 to 30 minutes of a quiet, premeditative state, then 20 to 30 minutes of meditation, and finally 20 to 30 minutes after the subject was asked to stop meditating.

The measurements of oxygen consumption and carbon dioxide elimination confirmed in precise detail what had been reported earlier. Oxygen consumption fell sharply from 251 cubic centimeters per minute in the premeditation period to 211 cubic centimeters during meditation, and in the postmeditation period it rose gradually to 242 cubic centimeters. Similarly, carbon dioxide elimination decreased, from 219 centimeters per minute beforehand to 187 cubic centimeters during meditation, and then returned to about the premeditation level afterward. The ratio of carbon dioxide elimination to oxygen consumption (in volume) remained essentially unchanged throughout the three periods, which indicates that the controlling factor for both was the rate of metabolism. The reduction in metabolic rate (and hence in the need for oxygen) during meditation was reflected in a decrease, essentially involuntary, in the rate of respiration (off two breaths per minute) and in the volume of air breathed (one liter less per minute).

For the measurement of arterial blood

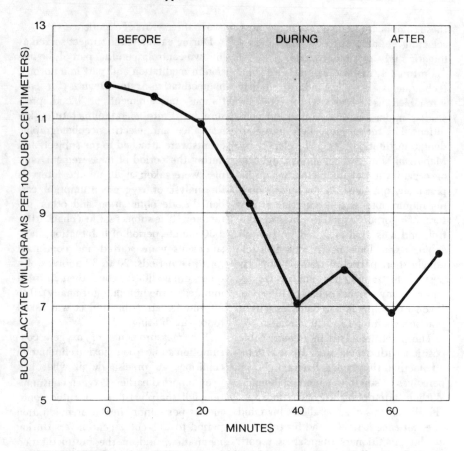

RAPID DECLINE in the concentration of blood lactate is apparent following the invitation to start meditating (colored area). Lactate is produced by anaerobic metabolism, mainly in muscle tissue. Its concentration normally falls in a subject at rest, but the rate of decline during meditation proved to be more than three times faster than the normal rate.

pressure and the taking of blood samples we used a catheter, which was inserted in the brachial artery and hidden with a curtain so that the subject would not be exposed to possible psychological trauma from witnessing the drawing of blood. Since local anesthesia was used at the site of the catheter insertion in the forearm, the subject felt no sensation when blood samples were taken. The blood pressure was measured continuously by means of a measuring device connected to the catheter.

We found that the subjects' arterial blood pressure remained at a rather low level throughout the examination; it fell to this level during the quiet premeditation period and did not change significantly during meditation or afterward. On the average the systolic pressure was equal to 106 millimeters of mercury, the diastolic pressure to 57 and the mean pressure to 75. The partial pressures of carbon dioxide and oxygen in the arterial blood also remained essentially unchanged during meditation. There was a slight increase in the acidity of the blood, indicating a slight metabolic acidosis, during meditation, but the acidity was within the normal range of variation.

Measurements of the lactate concentration in the blood (an indication of

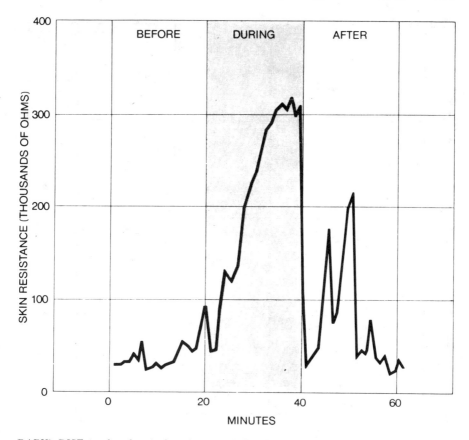

RAPID RISE in the electrical resistance of the skin accompanied meditation (colored area) in a representative subject. The 15 subjects tested showed a rise of about 140,000 ohms in 20 minutes. In sleep skin resistance normally rises but not so much or at such a rate.

anaerobic metabolism, or metabolism in the absence of free oxygen) showed that during meditation the subjects' lactate level declined precipitously. During the first 10 minutes of meditation the lactate level in the subjects' arterial blood decreased at the rate of 10.26 milligrams per 100 cubic centimeters per hour, nearly four times faster than the rate of decrease in people normally resting in a supine position or in the subjects themselves during their premeditation period. After the subjects ceased meditating the lactate level continued to fall for a few minutes and then began to rise, but at the end of the postmeditation period it was still considerably below the premeditation level. The mean level during the premeditation period was 11.4 milligrams per 100 cubic centimeters, during meditation 8.0 milligrams and during postmeditation 7.3 milligrams.

How could one account for the fact that lactate production, which reflects anaerobic metabolism, was reduced so much during meditation? New experiments furnished a possible answer. These had to do with the rate of blood flow in meditating subjects; the explanation they suggest appears significant with respect to the psychological benefits that can be obtained from meditation.

In studies H. Rieckert conducted at the University of Tübingen, he reported

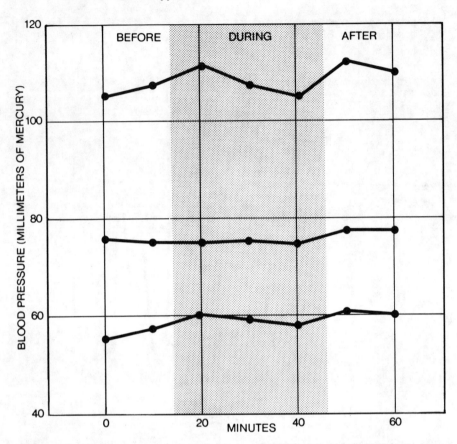

NO SIGNIFICANT CHANGE was observed in nine subjects whose arterial blood pressure was recorded before, during and after meditation. Systolic pressure (*top*), mean pressure (*middle*) and diastolic pressure (*bottom*), however, stayed relatively low throughout.

that during transcendental meditation his subjects showed a 300 percent increase in the flow of blood in the forearm. In similar measurements on our subjects we found the increase in forearm blood flow to be much less: 32 percent. Still, this increase was interesting, and it offered an explanation of the relatively large decrease in blood-lactate concentration. The main site of lactate production in the body is the skeletal muscle tissue. Presumably the observed acceleration of blood flow to the forearm muscles during meditation speeds up the delivery of oxygen to the muscles. The resulting gain in oxidative metabolism may substitute for anaerobic metabolism, and this would explain the sharp drop in the production of lactate that accompanies meditation.

The intriguing consequence of this view is that it brings the autonomic nervous system further into the picture. In a situation of constant blood pressure (which is the case during meditation) the rate of blood flow is controlled basically by dilation or constriction of the blood vessels. The autonomic nervous system, in turn, controls this blood-vessel behavior. One element in this system, a part of the sympathetic nerve network, sometimes gives rise to the secretion of acetylcholine through special fibers and thereby stimulates the blood vessels to dilate. Conversely, the major part of the sympathetic nerve network stimulates the secretion of norepinephrine and thus causes constriction of the blood vessels. Rieckert's finding of a large increase in blood flow during meditation suggested that meditation increased the activity of the sympathetic nerve network that secretes the dilating substance. Our own finding of a much more modest enhancement of blood flow indicated a different view: that meditation reduces the activity of the major part of the sympathetic nerve network, so that its constriction of the blood vessels is absent. This interpretation also helps to account for the great decrease in the production of lactate during meditation; norepinephrine is known to stimulate lactate production, and a reduction in the secretion of norepinephrine, through inhibition of the major sympathetic network, should be expected to diminish the output of lactate.

Whatever the explanation of the fall in the blood-lactate level, it is clear that this could have a beneficial psychological effect. Patients with anxiety neurosis show a large rise in blood lactate when they are placed under stress [see "The Biochemistry of Anxiety," by Ferris N. Pitts, Jr.; SCIENTIFIC AMERICAN, February, 1969]. Indeed, Pitts and J. N. McClure, Jr., a co-worker of Pitts's at the Washington University School of Medicine, showed experimentally that an infusion of lactate could bring on attacks of anxiety in such patients and could even produce anxiety symptoms in normal subjects. Furthermore, it is significant that patients with hypertension (essential and renal) show higher blood-lactate levels in a resting state than patients without hypertension, whereas in contrast the low lactate level in transcendental meditators is associated with low blood pressure. All in all, it is reasonable to hypothesize that the low level of lactate found in subjects during and after transcendental meditation may be responsible in part for the meditators' thoroughly relaxed state.

Other measurements on the meditators confirmed the picture of a highly relaxed, although wakeful, condition. During meditation their skin resistance to an electric current increased markedly, in some cases more than fourfold. Their heart rate slowed by about three beats per minute on the average. Electroencephalographic recordings disclosed a marked intensification of alpha waves in all the subjects. We recorded the waves from seven main areas of the brain on magnetic tape and then analyzed the patterns with a computer. Typically there was an increase in intensity of slow alpha waves at eight or nine cycles per second in the frontal and central regions of the brain during meditation. In several subjects this change was also accompanied by prominent theta waves in the frontal area.

To sum up, our subjects during the practice of transcendental meditation manifested the physiological signs of what we describe as a "wakeful, hypometabolic" state: reductions in oxygen consumption, carbon dioxide elimination and the rate and volume of respiration; a slight increase in the acidity of the arterial blood; a marked decrease in the blood-lactate level; a slowing of the heartbeat; a considerable increase in skin resistance, and an electroencephalogram pattern of intensification of slow alpha waves with occasional theta-wave activity. These physiological modifications, in people who were practicing the easily learned technique of transcendental meditation, were very similar to those that have been observed in highly trained experts in yoga and in Zen monks who have had 15 to 20 years of experience in meditation.

How do the physiological changes during meditation compare with those in other relaxed states, such as sleep and

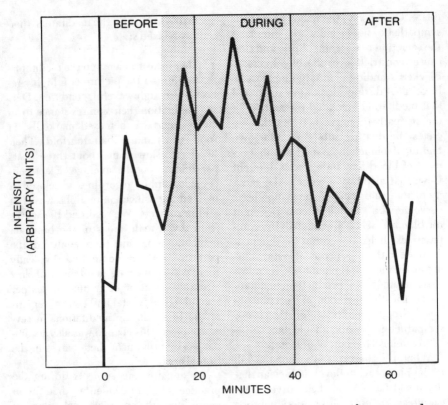

INCREASE IN INTENSITY of "slow" alpha waves, at eight to nine cycles per second, was evident during meditation (*colored area*) in electroencephalograph readings of the subjects' frontal and central brain regions. This is a representative subject's frontal reading. Before meditation most subjects' frontal readings showed alpha waves of lower intensity.

hypnosis? There is little resemblance. Whereas oxygen consumption drops rapidly within the first five or 10 minutes of transcendental meditation, hypnosis produces no noticeable change in this metabolic index, and during sleep the consumption of oxygen decreases appreciably only after several hours. During sleep the concentration of carbon dioxide in the blood increases significantly, indicating a reduction in respiration. There is a slight increase in the acidity of the blood; this is clearly due to the decrease in ventilation and not to a change in metabolism such as occurs during meditation. Skin resistance commonly increases during sleep, but the rate and amount of this increase are on a much smaller scale than they are in transcendental meditation. The electroencephalogram patterns characteristic of sleep are different; they consist predominantly of high-voltage (strong) activity of slow waves at 12 to 14 cycles per second and a mixture of weaker waves at various frequencies—a pattern that does not occur during transcendental meditation. The patterns during hypnosis have no relation to those of the meditative state; in a hypnotized subject the brain-wave activity takes the form characteristic of the mental state that has been suggested to the subject. The same is true of changes in heart rate, blood pressure, skin resistance and respiration; all these visceral adjustments in a hypnotized person

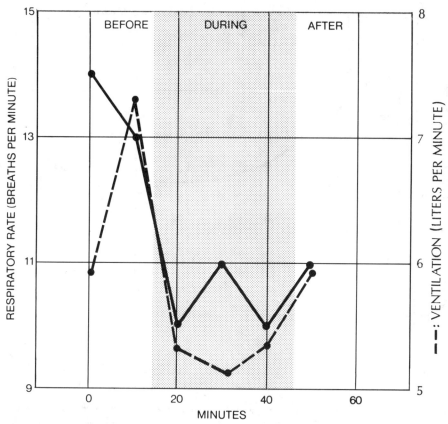

DECREASES OCCURRED in respiratory rate (*black*) and in volume of air breathed (*broken*) during meditation. The ratio between carbon dioxide expired and oxygen consumed, however, continued unchanged and in the normal range during the entire test period.

merely reflect the suggested state.

It is interesting to compare the effects obtained through meditation with those that can be established by means of operant conditioning. By such conditioning animals and people have been trained to increase or decrease their heart rate, blood pressure, urine formation and certain other autonomic functions [see "Learning in the Autonomic Nervous System," by Leo V. DiCara; SCIENTIFIC AMERICAN, January, 1970]. Through the use of rewards that act as reinforcers a subject is taught to make a specific visceral response to a given stimulus. This procedure and the result are quite different, however, from what occurs in transcendental meditation. Whereas operant conditioning is limited to producing specific responses and depends on a stimulus and feedback of a reinforcer, meditation is independent of such assistance and produces not a single specific response but a complex of responses that marks a highly relaxed state.

The pattern of changes suggests that meditation generates an integrated response, or reflex, that is mediated by the central nervous system. A well-known reflex of such a nature was described many years ago by the noted Harvard physiologist Walter B. Cannon; it is called the "fight or flight" or "defense alarm" reaction. The aroused sympathetic nervous

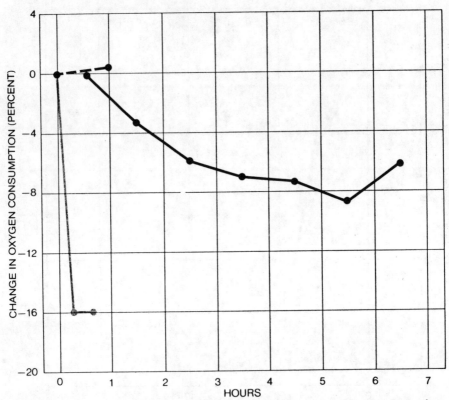

CONSUMPTION OF OXYGEN is compared in three different circumstances: during hypnosis *(broken)*, sleep *(black)* and meditation *(gray)*. No significant change occurs under hypnosis. One study shows that oxygen consumption is reduced by about 8 percent after five hours' sleep. Meditation brings twice the reduction in a fraction of the time.

system mobilizes a set of physiological responses marked by increases in the blood pressure, heart rate, blood flow to the muscles and oxygen consumption. The hypometabolic state produced by meditation is of course opposite to this in almost all respects. It looks very much like a counterpart of the fight-or-flight reaction.

During man's early history the defense-alarm reaction may well have had high survival value and thus have become strongly established in his genetic makeup. It continues to be aroused in all its visceral aspects when the individual feels threatened. Yet in the environment of our time the reaction is often an anachronism. Although the defense-alarm reaction is generally no longer appropriate, the visceral response is evoked with considerable frequency by the rapid and unsettling changes that are buffeting modern society. There is good reason to believe the changing environment's incessant stimulations of the sympathetic nervous system are largely responsible for the high incidence of hypertension and similar serious diseases that are prevalent in our society.

In these circumstances the hypometabolic state, representing quiescence rather than hyperactivation of the sympathetic nervous system, may indicate a guidepost to better health. It should be well worthwhile to investigate the possibilities for clinical application of this state of wakeful rest and relaxation.

A Theory of Hypnotic Induction Procedures

T. X. Barber and W. D. DeMoor

The first part of the paper delineates nine variables in hypnotic induction procedures that give rise to heightened responsiveness to test-suggestions: (a) defining the situation as hypnosis; (b) removing fears and misconceptions; (c) securing cooperation; (d) asking the subject to keep his eyes closed; (e) suggesting relaxation, sleep, and hypnosis; (f) maximizing the phrasing and vocal characteristics of suggestions; (g) coupling suggestions with naturally-occurring events; (h) stimulating goal-directed imagining; and (i) preventing or reinterpreting the failure of suggestions. Data are presented to support the theory that the nine variables augment responsiveness to test-suggestions by giving rise to positive attitudes, motivations, and expectancies which, in turn, tend to produce a willingness to think with and vividly imagine those things that are suggested. The second part of the paper specifies situational variables and variables involved in induction procedures that produce a trance-like appearance, changes in body feelings, and reports of having been hypnotized.

Hypnotic induction procedures tend to give rise to four distinguishable phenomena: heightened responsiveness to test-suggestions, a trance-like appearance, changes in body feelings, and reports of having been hypnotized. To formulate a general theory of hypnotism, it is necessary to specify the variables in hypnotic induction procedures (the antecedent variables) that give rise to each of the four phenomena (consequent variables). Although many types of induction procedures are described

---

[1] This research was supported, in part, by a grant (MH-19152) from the National Institute of Mental Health, U.S. Public Health Service. Some of the material for this paper was derived from chapters by T. X. Barber in a book (Barber, Spanos, & Chaves, in press) which is to be published by Aldine-Atherton, Inc. We are grateful to Aldine-Atherton for permission to use material from the text.

[2] Reprint requests to T. X. Barber, Ph.D., Medfield Foundation at Medfield State Hospital, Harding, Massachusetts 02042.

[3] Also at Medfield Foundation.

in the vast literature on hypnotism (Arons, 1961; Erickson, Hershman, & Secter, 1961; Hartland, 1966; Teitelbaum, 1965; Weitzenhoffer, 1957), the critical variables in induction procedures have not, as yet, been clearly specified. Weitzenhoffer (1957) notes that the failure to pinpoint the critical variables is probably due to the fact that there are still too many imponderables in the domain of hypnotism. Moss (1965) finds it difficult to delineate the antecedent variables because "there are literally hundreds of methods of induction, often standing in considerable contrast to each other, such that it is difficult to abstract and identify any common elements [p. 18]." Although much more work is needed to specify all of the crucial variables, we believe that sufficient data are available to formulate a theory of hypnotic induction procedures that can organize the data into a coherent framework and can stimulate further research.

Originally published in *The American Journal of Clinical Hypnosis,* 1972, Vol. 15, 112-135. Copyright 1972 by the American Society of Clinical Hypnosis.

Induction procedures tend to produce heightened responsiveness to test-suggestions of the type traditionally associated with hypnotism—e.g., test-suggestions for arm levitation, arm heaviness, limb or body rigidity, anesthesia, analgesia, age-regression, hallucination, and amnesia (Barber, 1969; Barber & Calverley, 1968; Hilgard & Tart, 1966). In Part 1 of this paper we shall delineate nine variables involved in induction procedures that play a role in enhancing responsiveness to such test-suggestions. These nine variables are listed in Table 1, column 1. We shall place these variables in a theoretical framework by focusing on the following pivotal question: Why are the nine variables effective in heightening responsiveness to test-suggestions? We shall provide data to support our theory that the nine variables augment responsiveness to test-suggestions by converging on two sets of mediating variables: (a) attitudes, motivations, and expectancies toward the test-situation and (b) willingness to think with and vividly imagine the suggested effects (see Table 1, column 2).

Induction procedures also tend to give rise to three additional phenomena: a trance-like appearance, changes in body feelings, and reports of having been hypnotized (Gill & Brenman, 1959, pp. 13–19; Weitzenhoffer, 1957, pp. 210-211). These three phenomena have been traditionally viewed as indices of a hypnotic trance state, and the hypnotic trance state has been viewed as a critical factor in producing heightened response to test-suggestions. We disagree with the traditional viewpoint. We do not view these three phenomena as indices of a hypnotic trance state, and we do not see them as playing a crucial role in producing heightened response to test-suggestions. From our viewpoint, the trance-like appearance, the changes in body feelings, and the reports of having been hypnotized are conceptualized as consequent (dependent) variables that are primarily due to the antecedent variables (the suggestions and instructions) that are included in induction procedures. We shall focus on these three phenomena in Part 2 of this paper.

Part 1.

VARIABLES THAT AUGMENT RESPONSIVENESS TO TEST-SUGGESTIONS[4]

We shall first discuss nine variables in induction procedures that appear to produce heightened responsiveness to test-suggestions (see Table 1).

Variable 1. Defining the Situation as Hypnosis

Induction procedures explicitly or implicitly define the situation to the subject as hypnosis. Simply labeling the situation as hypnosis is sufficient, by itself, to raise responsiveness to test-suggestions in most subjects who participate in present-day experiments.

The effects on suggestibility of simply defining the situation as hypnosis were demonstrated in two experiments by Barber and Calverley (1964d, 1965a). In both of these experiments, the subjects were randomly assigned to one of two groups. Subjects allocated to one group were told that they were participating in a hypnosis experiment and subjects allocated to the other group were told that they were control subjects. After the situation had been defined in these different ways, subjects in both groups were treated *identically*, that is, they were tested individually on response to the eight standardized test-suggestions that comprise the Barber Suggestibility Scale (Barber & Calverley, 1963; Barber, 1969).[5] In each of the two experiments, subjects who were told they were participating in a hypnosis experiment showed a significantly higher level of suggestibility as compared to those told that they were control subjects.

---

[4] We shall discuss the variables involved in hypnotic induction procedures when they are administered in experimental or clinical settings. Although stage hypnosis involves some of the same variables, the stage situation also includes other variables that are discussed elsewhere (Meeker & Barber, 1971).

[5] The Barber Suggestibility Scale measures overt and subjective responses to eight test-suggestions: arm lowering; arm levitation; hand lock (inability to unclasp hands); thirst hallucination; verbal inhibition (inability to say one's name); body immobility; posthypnotic-like response; and selective amnesia.

TABLE 1
INDUCTION VARIABLES, MEDIATING VARIABLES, AND RESPONSES TO TEST-SUGGESTIONS

(1) Variables in Induction Procedures	(2) Mediating Variables	(3) Consequent Variables
1. Defining the situation as hypnosis 2. Removing fears and misconceptions 3. Securing cooperation 4. Asking the subject to keep his eyes closed 5. Suggesting relaxation, sleep, and hypnosis 6. Maximizing the phrasing and vocal characteristics of suggestions 7. Coupling suggestions with naturally-occurring events 8. Stimulating goal-directed imagining 9. Preventing or reinterpreting the failure of suggestions	1. Positive attitudes, motivations and expectancies 2. Thinking with and vividly imagining the suggested effects	Responses to test-suggestions (for arm levitation, limb or body rigidity, age-regression, analgesia, hallucination, amnesia, etc.)

The experiments described above raise an important question: When other variables are held constant, why are subjects on the average somewhat more responsive to test-suggestions when the situation is defined to them as hypnosis?

One factor that seems to play a role is the subject's assumptions about hypnosis. For more than a century practically everyone has assumed that subjects in a hypnosis experiment manifest a high level of response to test-suggestions. Subjects participating in present-day experiments, typically college students, also accept this assumption (London, 1961). Consequently, when present-day subjects are told that they are in a hypnosis experiment, they typically construe this to mean that (a) they are in a special kind of situation in which a high level of response to test-suggestions for limb rigidity, analgesia, age-regression, etc., is desired and expected, and (b) if they do not try to experience those things that the hypnotist suggests, they will be considered as poor or uncooperative subjects and the hypnotist will be disappointed. On the other hand, when subjects are told that they have been assigned to a control condition, they do not necessarily expect to manifest a high level of response to test-suggestions of the type traditionally associated with the word *hypnosis*.

In brief, simply telling subjects that they are in a hypnosis situation is generally sufficient to produce a small but statistically significant enhancement in their responsiveness to test-suggestions. Most subjects enter a situation that is defined to them as hypnosis with positive attitudes and expectations. They seem to associate hypnosis with fascinating experiences and they seem to expect that the hypnotist's suggestions will lead to interesting changes in their private experiences and overt behavior.

Variable 2. Removing Fears and Misconceptions

Although simply defining the situation as hypnosis is sufficient to produce a significant enhancement of responsiveness to test-suggestions in most subjects, the hypnotist typically attempts to enhance further the subject's readiness to respond. Before beginning his more formal procedures, the well-trained hypnotist administers a "pre-induction education" that aims, in part, to remove fears and misconceptions.

Some subjects have inhibitory fears about hypnosis. Authors such as Hartland (1966), Weitzenhoffer (1957), and Wolberg (1948) discuss in detail what should be done in such cases. The hypnotist should try to find the reasons for the fears and misconceptions and should try to surmount them. Since positive attitudes and motivations are necessary for the subject to participate actively in the production of the desired responses, these authors advocate that, before beginning the more formal procedures, the hypnotist should give a reassuring talk concerning the usual misgivings, false ideas, and prejudices about hypnosis. Hartland (1966) pointed out that most failures "are due to lack of adequate preparation of the subject, and lack of adequate discussion before induction is attempted [p. 20]." This preparation or "pre-induction education" should include removing misconceptions about being dominated, about loss of will-power, and about what occurs during induction including fears of entering a zombie-like state or becoming unaware or unconscious. The importance of this kind of pre-induction education has been demonstrated in recent studies by Macvaugh (1969), Cronin, Spanos, and Barber (1971), and Diamond (1971). Each of these studies showed that subjects are more responsive to test-suggestions after the hypnotist has given favorable information about hypnosis with the intention of removing inhibitory fears and misconceptions.

Variable 3. Securing Cooperation

Even when the subject's fears and misconceptions have been removed, additional measures are usually needed to secure a high level of motivation and complete cooperation. Wolberg (1948) notes that subjects' responsiveness in a hypnotic situation depends on their motivation to be hypnotized: "Before hypnosis is attempted, the physician must build up motivations for hypnosis [p. 111]." Hartland (1966) also emphasizes that the subject "must either want to comply with the suggestions of the hypnotist or must feel that, regardless of his own will, he cannot resist [p. 18]."

How do hypnotists go about enhancing the subject's motivation and securing his cooperation? The hypnotist may explicitly ask the subject for his cooperation. He may inform the subject that his willingness to imagine vividly or to think with those things that are suggested is crucial. He may also inform the subject that, if he cooperates and tries to experience those things that are suggested, he will find that it is not especially difficult to have very interesting experiences. A series of experiments by Barber and associates (summarized by Barber, 1965a, 1965b, 1969) indicate that these elements of the "pre-induction education" are very important in heightening responsiveness to test-suggestions. In these experiments, subjects were given a set of task-motivational instructions that asked the subject to cooperate, to try to experience those things that would be suggested, and to try to imagine vividly. Furthermore, the subjects were told that, if they cooperated and tried to imagine vividly, they would find that it is not especially difficult to have interesting experiences. In a series of experiments, these task-motivational instructions were effective in producing a significant enhancement of suggestibility. In fact, the task-motivational instructions were generally as effective in raising responsiveness to test-suggestions as repeated suggestions of relaxation, drowsiness, sleep, and hypnosis.

Variable 4. Asking the Subject to Keep His Eyes Closed

At the beginning of the formal procedures, the hypnotist either asks the subject to close his eyes or suggests that the eyes are becoming heavy and are closing. Typically, the subject then keeps his eyes closed during the remainder of the session. Keeping the eyes closed appears to play a role in removing visual distractions, in enhancing the subject's ability to imagine vividly and to engage in fantasy, and in augmenting responsiveness to some types of test-suggestions.

It appears likely that test-suggestions which involve vivid imagining and fantasying, such as suggestions for age-regression or age-progression, may be more effectively carried out when the subject's eyes are closed. It also appears that keeping the eyes closed may be necessary to carry out other types of test-suggestions, such as suggestions to have a dream on a specified topic. However, experimental evidence indicates that whether or not the subject's eyes are closed does not significantly affect responsiveness to more "simple" or "motoric" test-suggestions such as arm heaviness, arm levitation, or body immobility (Barber & Calverley, 1965a). Further research is clearly needed to determine more precisely what types of test-suggestions are more easily carried out and what types are not affected when the subject keeps his eyes closed.

Variable 5. Suggesting Relaxation, Sleep, and Hypnosis

Most present-day induction procedures include repeated suggestions of relaxation, drowsiness, sleep, and hypnosis (henceforth labeled as *relaxation-sleep-hypnosis suggestions*). The relevant question here is whether such suggestions heighten responsiveness to test-suggestions for arm levitation, limb rigidity, analgesia, hallucination, amnesia, and so on and, if so, why?

In a previous section (Variable 1), we mentioned that, when other variables are held constant, subjects are more responsive to test-suggestions if they are told they are participating in a hypnosis experiment rather than in a control experiment. In the present section, we shall discuss a supplementary experimental finding: when the situation is defined to all subjects as hypnosis, the subjects are more responsive to test-suggestions if they are also given relaxation-sleep-hypnosis suggestions.

In three experiments (Barber & Calverley, 1965a, 1965b—Experiments 1 and 2), subjects who were randomly assigned to one experimental treatment were tested individually on the Barber Suggestibility Scale after they had been told that they were participating in a hypnosis experiment and, in addition, had been exposed to repeated relaxation-sleep-hypnosis suggestions (*e.g.*, "Relax completely ... completely relaxed ... comfortable and relaxed ... breathing regularly and deeply ... drowsy and sleepy ... listening only to my voice .. Soon you will be deeply asleep, but you will have no trouble hearing me ... deep, sound sleep ... deepest trance ... "). Subjects who were randomly assigned to another experimental treatment were tested individually on the Barber Suggestibility Scale after they had been told simply that they were participating in a hypnosis experiment. Each of the three experiments indicated that, when other factors are held constant, repeated relaxation-sleep-hypnosis suggestions produce a statistically significant gain in suggestibility which is above that found when subjects are simply told that they are participating in a hypnosis experiment.

Why do subjects generally show a higher level of responsiveness to test-suggestions when they are given repeated suggestions of relaxation, drowsiness, sleep, and hypnosis? At least part of the gain in suggestibility associated with relaxation-sleep-hypnosis suggestions appears to be due to (a) the special effectiveness of these kinds of suggestions in defining the situation to the subject as "truly hypnosis" which (b) tends to produce more positive attitudes, motivations, and expectancies toward the situation and a concomitant willingness to think with the suggestions. There are four interrelated aspects to this interpretation:

1. Most subjects who participate in present-day experiments believe that relaxation-sleep-hypnosis suggestions are part of a hypnosis experiment.

2. If these subjects are told that they are participating in a hypnosis experiment, but the hypnotist does *not* go on to say the kinds of things that hypnotists are supposed to say (*e.g.*, does not repeatedly suggest that the subject is becoming relaxed and is entering a deep sleep), the situation is not very effectively defined as hypnosis. In other words, some subjects are not convinced that the situation is "truly hypnosis" and they typically state afterwards that the experimenter was not a good hypnotist.

3. Subjects' doubts as to whether they are actually participating in a hypnosis experiment are effectively removed when the experimenter goes on to suggest repeatedly that they are becoming relaxed and drowsy and are entering a state of sleep or hypnosis.

4. When the situation is *effectively* defined as hypnosis, that is, when the subjects perceive the situation as unusual, different, and special (as "truly hypnosis"), they

concomitantly tend to perceive it as one in which high responsiveness to test-suggestions is desired and expected. Also, when the situation is effectively defined as hypnosis, subjects are aware that if they resist or do not think with the suggestions, they will negate the purpose of the experiment, the experimenter's time and effort will be wasted, the experimenter will be disappointed, and they will be considered as poor or uncooperative subjects. In other words, it appears that relaxation-sleep-hypnosis suggestions tend to raise responsiveness to test-suggestions primarily by producing more positive attitudes, motivations, and expectancies toward the test-situation and a concomitant willingness to think with and vividly imagine those things that are suggested.

In addition to augmenting responsiveness to test-suggestions, relaxation-sleep-hypnosis suggestions also play an important role in producing three additional phenomena—a trance-like appearance, changes in body feelings, and reports of having been hypnotized. The second part of this paper discusses the latter three phenomena and shows how they relate to suggestibility.

Variable 6. *Maximizing the Phrasing and Vocal Characteristics of Suggestions*

The precise wording of suggestions appears to play an important role in determining the subject's response. Hypnotists have learned through practice that a simple suggestion that something is occurring—e.g., "You cannot bend your arm", "You are insensitive to pain", "You are six years of age"—is rarely effective. To produce arm rigidity, analgesia, or age-regression, more elaborate suggestions are needed. For instance, to produce arm rigidity, the trained hypnotist will intone a series of suggestions, such as "As I stroke your arm ... you will feel that it is becoming much stiffer and straighter. The stiffness is increasing.... You can feel all the muscles tightening up ... stiffer and straighter ... it is beginning to feel just as stiff and rigid as a steel poker.... Picture a steel poker in your mind ... you will feel that your arm has become just as stiff and rigid as that steel poker ... [Hartland, 1966, pp. 81–82]." Similarly, the hypnotist will not simply suggest analgesia; instead, he will give more extended and elaborate suggestions such as, "Novocain has been injected into the hand.... You feel the numbness spreading.... The hand is dull, numb, and insensitive.... It is a piece of rubber, a lump of matter, without feeling, without sensation...." Along similar lines, the trained hypnotist will not simply suggest to the subject that he is six years old, but will give more elaborate suggestions intended to induce the subject to focus his thoughts on the past, to imagine vividly events that occurred at the age of six, and to put aside thoughts about the present. In fact, learning to be an effective hypnotist consists primarily of learning the art of giving suggestions which include elaborate descriptive imagery. Some of this elaboration of suggestions involves guiding the subject to imagine vividly in a goal-directed way. We shall discuss this in some detail below (Variable 8).

Whether the suggestions are worded permissively or authoritatively also appears to be important. Recent studies indicate that permissively worded suggestions tend to be more effective than those that are worded authoritatively. For instance, more subjects testified that they had forgotten specified material when they were given permissive suggestions ("Try to forget.") rather than authoritative suggestions ("You will forget.") (Barber & Calverley, 1966a). Similarly, when awakened at night during rapid eye movement (REM) periods, more subjects testified that they were dreaming about a suggested topic if they had been given a permissive suggestion ("Try to think about and to dream about [the selected topic] all through the night....") rather than if they had been given an authoritatively-worded suggestion ("You will think about and dream about [the selected topic] all through the night.") (Barber, 1969, pp. 66–67).

Of course, the suggestions should be phrased in such a way that they are clear to the subject. The subject cannot respond to words he does not comprehend. Gindes (1951, p. 104) illustrated this in an amusing way. A friend of his iterated and reiterated the suggestion that his subject was becoming more and more lethargic. After an hour of futile effort, the subject opened his eyes and asked, "What is 'lethargic' anyway?"

In addition to the wording of the test-suggestions, hypnotists also stress the importance of vocal characteristics such as intonations, inflections, and volume. Hartland (1966) has placed heavy emphasis on these aspects of the suggestions and has given many examples of the following variations in the proper use of vocal expressions: (a) alterations in the volume of the voice, (b) changes in the inflection and modulation of the voice, (c) changes in the rate of delivery, (d) stressing of particular words, and (e) insertion of suitable pauses between suggested ideas.

Although we agree with Hartland that the method of presenting test-suggestions is important in determining the subject's response, this variable has not as yet been evaluated experimentally. In fact, we know of only one experiment which assessed the effect on suggestibility of variations in the hypnotist's tone of voice. In this experiment (Barber & Calverley, 1964a) all of the subjects were given the eight test-suggestions from the Barber Suggestibility Scale, but half of the subjects received the suggestions in a lackadaisical tone of voice and the others received the same suggestions in a firm, forceful tone. A higher level of response to the test-suggestions was elicited when the suggestions were presented forcefully rather than lackadaisically. These results were presumably due to the following: a forceful tone reflects an attitude of confidence on the part of the experimenter which conveys to the subject that what the experimenter suggests ought to happen. Contrariwise, a lackadaisical tone reflects a non-confident or an uninvolved attitude on the part of the experimenter and this unfavorable attitude presumably transfers to the subject.

In brief, it appears that variations in the wording and tone of test-suggestions play an important role in determining the subject's responsiveness. However, further research is needed to specify more precisely the specific variations in the phrasing and vocal characteristics of suggestions that exert effects on specific kinds of responses.

Variable 7. Coupling Suggestions with Naturally-Occurring Events

Common to most induction procedures is the coupling of suggestions with events that are naturally occurring in the situation. The general principle involved here has been formulated by Hartland (1966) as follows: "You should always couple an effect that you want to produce with one that the subject is actually experiencing at the moment [p. 35]." Spiegel (1959) refers to this variable as "... the art of induction ... tacitly implying that the phenomena are due to the signal of the hypnotist."

As Weitzenhoffer (1957, p. 273) has noted, Erickson's "specialized techniques" (Haley, 1967) seem to pivot around this variable in three interrelated ways: (a) whatever technique is employed is adapted to the ongoing behavioral activities of the subject, (b) whatever behavior the subject manifests is interpreted as a successful response (leading the subject to believe he is responding appropriately whether this is actually true or not), and (c) suggestions are given which anticipate behaviors or experiences which will occur naturally. The major purpose of these techniques appear to be to lead the subject to believe that he is responding to suggestions and thus may expect that he will continue to respond.

Gindes (1957) referred to this variable as an "innocent artifice" and commented as follows: "The therapist's words must closely follow the subject's actions without his growing aware of it. The interaction of these two factors leads the subject to believe that he is following the operator's suggestions, although the reverse is true [p. 90]."

Let us now briefly glance at some of the techniques that use naturally-occurring processes to enhance suggestibility.

In the Eye-Fixation technique, the subject is asked to look at an object held a little above his eyes. While the subject stares at the fixation object, a strain upon his eyes tends to occur. Suggestions for eye fatigue are timed to coincide with the natural fatigue. When using Wolberg's or Erickson's Hand Levitation technique, the hypnotist gives suggestions which coincide with sensations that tend to occur naturally as a person focuses on his hand—*e.g.*, slight motions, separations, or jerkings of the fingers. In one variant of the Postural Sway technique, the subject is asked to stand with his eyes closed and, since no one stands perfectly still when the eyes are closed, the trained hypnotist detects the rhythm of the swayings. The suggestions of sway are then timed to coincide with the naturally-occurring sway (Watkins, 1949). When using the Metronome technique, the hypnotist gives repeated suggestions of eye-heaviness, drowsiness, and eye-closure while the subject focuses his gaze at a blinking light and listens to the monotonous sound of the metronome which tends to produce a feeling of eye-heaviness. Similarly, in drug hypnosis, the drugs used tend to produce drowsiness, narcosis, or confusion and the hypnotist times his suggestions of drowsiness to coincide with the chemically induced effects (Weitzenhoffer, 1957, pp. 255–263).

Wilson (1967) provided experimental evidence indicating that responsiveness to test-suggestions is increased when the suggestions are linked with naturally-occurring events. In this experiment, subjects who were randomly assigned to an experimental group were asked to imagine various suggested effects while ingenious methods were used to help them experience the effects without their knowing that they were receiving such aid. For instance, each subject was asked to imagine that the room was red while a tiny bulb was lit secretly which provided a very faint red tinge to the room. Following these procedures, each subject was assessed individually on his response to the 8-item Barber Suggestibility Scale. Other subjects, who were randomly assigned to a control group, were tested individually on the same suggestibility scale with no prior attempt made to pair suggestions with naturally-occurring events. The experimental treatment—coupling suggestions with naturally-occurring events—was effective in raising responsiveness to the Barber Suggestibility Scale which was given subsequently. Subjects in the experimental group obtained an average score of 5, whereas those in the control group obtained a significantly lower average score of 3 on the 8-point suggestibility scale.

Variable 8. Stimulating Goal-Directed Imagining

The kinds of test-suggestions used in induction procedures usually do *not* directly ask the subject to carry out an overt behavior. Instead, they commonly ask the subject to imagine a situation which, if it were to occur in reality, would produce the desired behavior. For example, the criterion for passing the arm heaviness suggestion of the Stanford Hypnotic Susceptibility Scale (Form C) consists of the subject lowering his outstretched arm (Weitzenhoffer & Hilgard, 1962). However, this suggestion does not directly ask the subject to lower his arm. Instead, the subject is asked to imagine a situation which, if it were to occur in reality, would produce a lowering of the arm. The suggestion for arm heaviness is worded as follows: "... Imagine you are holding something heavy in your hand.... Now the hand and arm feel heavy as if the [imagined] weight were pressing down ... and as it feels heavier and heavier the hand and arm begin to move down...." If the situation which the subject is asked to imagine actually occurred (if the subject actually held a heavy weight in his hand), then his arm would become heavy and would move downward.

Many other test-suggestions used in induction procedures also instruct the subject to imagine situations which, if they actually occurred, would produce the desired response: "I want you to imagine a force acting on your hands to push them apart ... [Weitzenhoffer & Hilgard, 1962]." "Imagine your hands are two pieces of steel that are welded together so that it is impossible to get them apart [Barber, 1965b]." "Think of your arm becoming stiffer and stiffer ... as if it were in a splint so the elbow cannot bend [Weitzenhoffer & Hilgard, 1962]." Each of these test-suggestions clearly instructs the subject to engage in *goal-directed imagining* or *goal-directed fantasy*—that is, *the subject is asked to imagine a situation which, if it actually transpired, would give rise to the behavior that is suggested* (Spanos, 1971).

An experiment by Spanos (1971) studied the relationship between goal-directed imagining and responding to test-suggestions. After the subjects (female student nurses) were exposed to repeated suggestions of relaxation, drowsiness, sleep, and hypnosis, they were given a series of standardized test-suggestions—*e.g.*, arm levitation, arm heaviness, hand lock, and inability to say one's name. Immediately after completing each test-suggestion, the subject was asked to report what was passing through her mind during the time she was receiving the suggestion. In most instances, the subjects who passed the test-suggestions reported that they were imagining a situation which, if it actually transpired, would give rise to the behavior that was suggested. Typical examples of this type of imagining, which Spanos labeled as *goal-directed fantasy*, were as follows: One subject who passed a test-suggestion that her arm was light and was rising reported that "I imagined that my arm was hollow, there was nothing in it, and somebody was putting air into it." Another subject who passed a test-suggestion for arm heaviness reported that "I imagined that there were all kinds of rocks tied to my arm. It felt heavy and I could feel it going down."

An important finding in the Spanos (1971) experiment was that subjects who passed the test-suggestions at times disregarded the specific things they were told to imagine. Instead, they imagined something different that was, nonetheless, goal-directed in that it implied the performance of the appropriate overt response. For instance, a subject who was asked to imagine that her throat and jaw were in a vise that prevented her from saying her name, reported imagining instead that "Somebody had their hands around my neck ... it was a man whose hands were big and hairy and dirty." Reports of this type indicated that the subjects did not necessarily regard the specific contents of the suggestions as directives that they had to follow in detail. Instead, they tended to view the contents of the suggestions as guides around which they might construct goal-directed fantasies, that is, construct imaginary situations which, if they actually transpired, would result in the suggested behavior. In another experiment (Spanos & Ham, 1971), a suggestion for selective amnesia (to forget the number 4) was given to a group that had been exposed to a hypnotic induction procedure and to a no induction control group. With very few exceptions, subjects in both groups who appeared to manifest selective amnesia (who did not verbalize the number 4) reported a goal-directed fantasy; for example, they imagined the numbers 1 through 5 on a blackboard and then imagined that the number 4 had been erased.

Subjects in the Spanos (1971) experiment who did not pass a test-suggestion either reported an unwillingness to cooperate or to respond to the suggestion (*e.g.*, "I was kind of hoping that I would be able to pull my hands apart") or indicated that they were motivated to cooperate but did not imagine the suggested effects or did not carry out goal-directed imagining. An illustrative report that falls into the latter category (a motivated subject who did not imagine) was as follows: "When you said that my hands would be stuck together, I tried to make them so. I held them real tight and pushed hard ... but it just didn't work."

In another experiment by Spanos and Barber (in press), subjects were exposed to repeated relaxation-sleep-hypnosis suggestions and then were given suggestions for arm levitation (worded in one of four ways). To ascertain whether the subjects had engaged in goal-directed imagining during the arm levitation suggestion, each subject was asked to report what was passing through his mind when he was given the suggestion. Subsequently, each subject who passed the test-suggestion completed a rating scale which asked to what extent the arm levitation was experienced as involuntary ("I experienced the arm rising completely by itself"), partially involuntary, or totally voluntary ("I only had the experience of causing the arm to rise"). Considering only those subjects who passed the test-suggestion for arm levitation, goal-directed imagining and the experience of involuntariness were related as follows: (a) All of the subjects who carried out goal-directed imagining—who imagined a situation which, if it actually transpired, would result in the arm moving up—felt that the arm levitation occurred involuntarily. (b) Of those who did not carry out goal-directed imagining, only 40% felt that the arm levitation occurred involuntarily.

In brief, suggestions used in induction procedures commonly attempt to stimulate goal-directed imagining by asking the subject to imagine events which, if they actually occurred, would give rise to the response that is being suggested. The experiments by Spanos (1971), Spanos and Ham (1971), and Spanos and Barber (in press) indicate that stimulating goal-directed imagining is an important variable in enhancing overt responsiveness to test-suggestions and in producing the experience that the overt responses are involuntary occurrences.

Variable 9. Preventing or Reinterpreting the Failure of Suggestions

Hypnotists, at times, challenge the subject to overcome a suggested effect. For instance, the subject may be challenged to open his eyes while suggestions are given that he cannot do so. If the subject overcomes the challenge and opens his eyes, his expectancy that he can pass test-suggestions will probably be reduced and he will tend to be less responsive to subsequent test-suggestions. Consequently, skillful hypnotists at times phrase the challenge in such a way that the subject will not have the opportunity to discover that he has failed a test-suggestion. Instead of directly suggesting eye-catalepsy, for example, the hypnotist might give suggestions as follows: "Your eyes are stuck tight ... so tight that *if* you tried to open them, *you could not*. But you will *not* try to open them. *You have no desire* to open your eyes, but only want to sleep deeper [Weitzenhoffer, 1957, p. 214]." London (1967) has stated the underlying principle thus: "The thrust of the patter most commonly used in inductions therefore deliberately soft-pedals the occurrence of failure on the subject's part by variously justifying it, approving it, denying it, or retrospectively confusing the issue so that the subject cannot be sure he has failed at all [p. 70]."

Although preventing or reinterpreting the failure of suggestions appears to be important, this factor has not yet been clearly demonstrated experimentally. Studies are needed to determine the precise effects on suggestibility of preventing the subject from failing test-suggestions and of reinterpreting failures as non-failures.

Mediating Variables

We have listed nine variables associated with induction procedures that tend to raise responsiveness to test-suggestions. This list is not exhaustive. Many other variables might also be operative in certain induction procedures. For instance, suggestibility might be enhanced in some cases if the hypnotist reinforces the subject from time to time for desired responses by stating "Good," "Fine," or "Excellent." Additional variables which are not an integral part of induction procedures but may affect the subject's responsiveness, such as the hypnotist's prestige and personality characteristics, have been discussed elsewhere (Barber, 1969, 1970a, 1970b). Further research is warranted to determine (a) to what degree each of the nine variables, and also other variables, facilitate suggestibility, (b) which variables, when applied in combination, have an additive effect, and (c) what combination of variables are sufficient to maximize responsiveness.

From our viewpoint, the nine variables give rise to heightened suggestibility by converging on a set of mediating variables that we have labeled as positive attitudes, motivations, and expectancies—the subject views his responding to suggestions or his being hypnotized as interesting or worthwhile (positive attitude), desires and tries to experience the suggested effects (positive motivation), and believes that he can and will experience those things that are suggested or believes that he can be hypnotized (positive expectancy). Also, as outlined in Table 1, when positive attitudes, motivations, and expectancies have been produced, the subject is apt to think with and vividly imagine those things that are suggested. Furthermore, when the subject thinks with the suggestions and vividly imagines, he tends to perform the suggested overt behaviors and to experience the suggested effects.

Let us now look more closely at the first set of mediating variables (positive attitudes, motivations, and expectancies) and then at the second set (thinking with and vividly imagining the suggested effects).

Attitudes, motivations, and expectancies. The important role played by the subject's attitudes, motivations, and expectancies has been stressed previously by White (1941), Sarbin (1950), and Pattie (1956). White (1941) noted that some subjects "who at first are insusceptible become excellent subjects when changes are made in the pattern of motives." Sarbin (1950) and Pattie (1956) also observed that some subjects were unresponsive to test-suggestions in one hypnotic session and very responsive in another session and offered presumptive evidence that such changes were due to alterations in the subjects' attitudes and motivations toward the test-situation.

Recent studies have confirmed the important role played by attitudes, motivations, and expectancies. In these studies, the subjects rated their attitudes, motivations, or expectancies with regard to hypnosis prior to the induction procedure and their ratings were then correlated with their responsiveness to test-suggestions during the hypnotic session. In general, the subjects' level of responsiveness to test-suggestions was positively correlated with (a) their attitudes toward hypnosis (Andersen, 1963; Barber & Calverley, 1966b; Melei & Hilgard, 1964), (b) their motivation to be hypnotized (J. R. Hilgard, 1970), and (c) their expectancies (self-predictions) of their own hypnotizability (Barber & Calverley, 1966b, 1969; Dermen & London, 1965; Melei & Hilgard, 1964; Unestahl, 1969). In general, other investigators (London, Cooper, & Johnson, 1962; Rosenhan & Tomkins, 1964; Shor, 1971; Shor, Orne, & O'Connell, 1966) similarly found that responsiveness to test-suggestions in a hypnotic situation is positively correlated with subjects' pre-experimental attitudes toward hypnosis or their pre-experimental expectations of their own hypnotizability.

An additional series of studies, which tried to produce positive (or negative) attitudes, motivations, or expectancies toward the test-situation, further indicated that these factors play an important role in determining response to test-suggestions. In a number of experiments, Barber (1969) showed that responsiveness to test-suggestions is enhanced by instructions which aim to produce both positive motivations (*e.g.*, "What I ask is your cooperation in helping this experiment by trying to imagine vividly what I describe to you....") and positive expectancies (*e.g.*, "Everyone passed these tests when they tried ... if you try to the best of your ability, you can easily ... do the interesting things I tell you...."). Other studies (Cronin, Spanos, & Barber, 1971; Diamond, 1971; Macvaugh, 1969) indicated that suggestibility is enhanced when an attempt is made to obviate negative attitudes toward hypnosis by removing misconceptions. Experiments by Barber and Calverley (1964b, 1964c) demonstrated that subjects show very little response to test-suggestions when an attempt is made to produce *negative* attitudes and motivations toward the test-situation, for example, by defining the situation as a test of gullibility. Also, studies by Barber and Calverley (1964d), Klinger (1970), and Wilson (1967) appear to indicate that suggestibility is enhanced when an attempt is made to produce a positive expectancy that the suggested effects can be experienced.

Thinking with and vividly imagining. When the suggestibility-enhancing factors have given rise to positive attitudes, motivations, and expectancies, the subject is ready to think with and vividly imagine those things that are suggested. The more a subject thinks with and vividly imagines the suggested effects, the less he tells himself that the suggested events cannot or will not occur, that is, the less he covertly contradicts the suggestions. We see responsiveness to test-suggestions as directly related to thinking with and vividly imagining those things that are suggested and inversely related to thinking against (covertly contradicting) the suggestions.

Focused thinking and imagining have wide ramifications. Gindes (1951), who also views *focused thoughts* as of cardinal significance in producing the phenomena associated with induction procedures, describes their potential effects as follows:

> They can cause changes in temperature, make us perspire or break out into goose-flesh. They can alter the regularity of a heart-beat, or the rate of blood-pressure and respiration. A fear thought compels the blood to leave the brain; if severe enough, the victim faints. Psychosomatics, a branch of medical science that recognizes a mental basis for physical ailments, makes the bold assertion that eighty-five percent of illnesses, hitherto regarded as organic, are actually functional in origin; i.e., they result directly from the impact upon the body of a thought charged with emotion. This claim would seem fantastic, were it not backed up by incontrovertible clinical data [p. 4].

What is the experimental evidence that thinking with and imaginatively focusing on suggestions leads to the overt behaviors and private experiences that have been historically associated with hypnotism? An important but neglected paper by Arnold (1946) provided some preliminary answers to this question.

First, Arnold pointed out that "words may be considered as symbols which stand for the situation or activity they refer to". If we hear words that describe a previous experience, the earlier experience tends to be reinstated in a sketchy, fragmentary way; we tend to visualize or feel ourselves in the previous situation, and we tend to reexperience our attitudes and reactions in the previous situation. When the experimenter repeatedly suggests to the subject that his arm is feeling heavy or he is falling forward, he is inviting the subject to imagine vividly an earlier situation in which he had these experiences and, as a result of the imaginative focusing, to reinstate the attitudes, feelings, and reactions that he had previously experienced.

Secondly, Arnold summarized a series of studies which indicated that thinking about and imaginatively focusing on a suggested movement or activity tends to bring about that movement or activity. Arnold referred to the well-known experiments of Jacobson (1930, 1932) which showed that an imagined movement, for example, bending an arm, is associated with small but measurable contractions of the flexor muscles of the arm. If the subject is relaxed when he is imagining the movement, the minimal muscular contractions occur in the limb which is imagined as moving and do not occur in the other limbs or the trunk. Schultz (1932), Hull (1933), Arnold (1946), and Mordey (1960) asked subjects to think or to imagine that they were falling forward or backward, that an arm was moving up or down, or that a pendulum (which they held by two fingers of one hand) was moving to the right or left. The subjects tended to show the movement either slightly or markedly in the imagined direction. In general, subjects who showed the most marked movements in the imagined direction reported afterwards that they had imagined vividly, whereas subjects who showed the least movement reported afterwards that they either did not imagine the movement or they thought of other things.

Thirdly, Arnold (1946) summarized studies by Schultz (1932) which showed that thinking about and vividly imagining a sensation can produce physiological changes which are associated with the sensation. Working with subjects who had been trained to remain relaxed while they imagined various situations (autogenic training), Schultz found that instructions to imagine vividly that the forehead is cool gave rise to a drop in the temperature of the forehead in about one-third of the subjects. Conversely, Schultz found that relaxed subjects who were asked to imagine vividly that a hand was exposed to heat tended to experience a sensation of warmth which was associated, in about 80% of the subjects, with a rise in skin temperature up to about 2°C.

Other investigators have presented similar results. For instance, Hadfield (1920) and Menzies (1941) found that some subjects who had learned to relax showed vasoconstriction and a drop in the temperature of the skin when they were instructed to think about and to imagine vividly that a limb was cold and showed vasodilation and a rise in skin temperature when asked to think about and vividly imagine that a limb was warm. Similarly, Harano, Ogawa, and Naruse (1965) asked subjects who had been trained in relaxation to repeat to themselves and to focus on the idea, "My arms are warm." When imaginatively focusing on these words, the subjects generally exhibited several interrelated changes in the arms—a change in experienced warmth, an increase in blood volume, and an increase in the surface temperature. It is important to note that Harano et al., did *not* find these changes in the arms when the subjects tried deliberately to raise the temperature of the arms without thinking or imagining that the arms were warm.

To recapitulate, experiments summarized by Arnold (1946), and a series of more recent experiments, indicate that instructions to think about and vividly imagine movements or sensations tend to bring about the actual movements or sensations. Also, Arnold presented data indicating that, when a subject experiences a suggested effect, he tends to think with and imaginatively focus on the suggestion. Conversely, when a subject fails to experience suggested effects, "he reports afterwards either that he could not concentrate on the experimenter's words, perhaps because he could not forget the absurdity of the situation, perhaps because he kept thinking of something else; or he admits that he deliberately resisted either by criticizing the experimenter's procedure to himself, or by starting to think of something else [Arnold, 1946]."

Arnold also presented data suggesting that when a subject vividly imagines that his arm is rising, and then experiences arm levitation, he feels that he did not make the arm move deliberately or voluntarily; instead, he feels that it moved involuntarily by itself. However, subsequent studies by Spanos (1971) and Spanos and Barber (in press) indicated that the experience of involuntariness is related not simply to imagining *per se*, but to a certain kind of imagining which can be labeled as *goal-directed imagining* or *goal-directed fantasy*. As pointed out previously in this paper (Variable 8), a series of studies which pertained to goal-directed imagining also provided evidence indicating that vivid imagining plays a role in responding to test-suggestions.

Part 2.

VARIABLES THAT PRODUCE A TRANCE-LIKE APPEARANCE, CHANGES IN BODY FEELINGS, AND REPORTS OF HAVING BEEN HYPNOTIZED

In Part 1 of this paper, we discussed the antecedent and mediating variables that augment responsiveness to test-suggestions for limb rigidity, analgesia, age-regression, amnesia, etc. In addition to raising overt and subjective response to test-suggestions, hypnotic induction procedures also tend to produce three additional phenomena—a trance-like appearance, changes in body feelings, and reports of having been hypnotized. From the traditional viewpoint, the latter three phenomena have been considered to be indices of a hypnotic trance state, and the hypnotic trance state has been viewed as a critical factor in producing heightened response to test-suggestions. From our viewpoint, the trance-like appearance, the changes in body feelings, and the reports of having been hypnotized (a) are due, primarily, to suggestions and instructions included in induction procedures and (b) they are *not* necessary, although they may be helpful, in producing heightened response to test-suggestions. Let us again look at induction procedures to ascertain which antecedent variables give rise to a trance-like appearance, changes in body feelings, and reports of having been hypnotized.

Trance-Like Appearance

As Weitzenhoffer (1957, pp. 210–211) and other writers have pointed out, at the completion of the induction procedure, subjects typically manifest a trance-like appearance. For instance, they appear limp and relaxed, lack facial expression, and manifest a disinclination to talk, psychomotor retardation, and immobility or passivity. Also, when asked to open their eyes, they at times show a fixed stare (a "trance stare"). Which of the many variables included in induction procedures give rise to the trance-like appearance?

An experiment by Barber and Calverley (1969) indicated that the trance-like appearance—as indicated by limpness-relaxation, lack of spontaneity, psychomotor retardation, and a "fixed stare"—is due, in part, to Variable 4—asking the subject to keep his eyes closed for a period of time. The same experiment also indicated that an even greater proportion of subjects manifest a trance-like appearance if they not only keep their eyes closed for a period of time, but, in addition, are given repeated suggestions of relaxation, drowsiness, and sleep or are told to put themselves in hypnosis.

Since the trance-like appearance is primarily due to suggestions for relaxation, drowsiness, sleep, and/or hypnosis that are given to the subject while his eyes are closed, one wonders whether such a trance-like appearance or the variables leading to it are essential for the subject to manifest a high level of response to test-suggestions for limb rigidity, body immobility, analgesia, age-regression, amnesia, and the like. This does not seem to be the case.

The trance-like appearance can be removed by suggestions or instructions and the subject can continue to show a high level of response to test-suggestions. For instance, several years ago, we carried out the following informal study with eight responsive subjects. The subjects were first exposed to a hypnotic induction procedure which focused on repeated relaxation-sleep-hypnosis suggestions. The subjects manifested a trance-like appearance and also responded to test-suggestions for arm heaviness, arm levitation, inability to take their clasped hands apart, and thirst hallucination. Next, the subjects were told to be awake and alert, to stop appearing as if they were in a trance, but to continue to remain responsive to test-suggestions. The subjects remained highly responsive to test-suggestions for inability to say their name, body immobility, and selective amnesia, but they no longer showed a trance-like appearance; in fact, they appeared as alert and awake as any other normally-awake individual.

Instructing subjects to be alert is not the sole means of removing the trance-like appearance. It may also be removed by (a) administering test-suggestions that require activity, effort, or alertness, for instance, test-suggestions for heightened strength and endurance or (b) instructing the subject to respond to the test-suggestions that he will receive in the post-experimental period and then stating, "Wake up—the experiment is over [Barber, 1958, 1962]." After the trance-like appearance has been removed by these suggestions or instructions, the subject may continue to manifest a high level of response to test-suggestions. In other words, the hypnotist can first give suggestions—for example, relaxation-sleep-hypnosis suggestions—that tend to produce a trance-like appearance, and then he can give suggestions to remove the trance-like appearance, and neither the suggestions which give rise to a trance-like appearance nor those which remove it necessarily affect the subject's responsiveness to test-suggestions.

Of course, a high level of responsiveness to test-suggestions can be elicited without first inducing a trance-like appearance. If subjects are not asked to close their eyes and are not given relaxation-drowsiness-sleep suggestions, many subjects will show a high level of response to test-suggestions, especially if the experimenter implements some of the other variables that are commonly included in induction procedures—e.g., securing cooperation, maximizing the phrasing and vocal characteristics of suggestions, coupling suggestions with naturally-occurring events, stimulating goal-directed imagining, and preventing or reinterpreting the failure of suggestions (Arons, 1961; Barber, 1969, 1970a, 1970b; Klopp, 1961; Meeker & Barber, 1971; Wells, 1924).

Changes in Body Feelings

Subjects who have been exposed to a hypnotic induction procedure typically report several kinds of changes in body feelings. Some of these changes are direct responses to the suggestions that the subjects have received. For instance, subjects may report that a hand was felt to be rising involuntarily, a limb was experienced as being rigid, and a hand was experienced as being numb and insensitive. These kinds of changes in body feelings are, of course, direct responses to suggestions for arm levitation, limb rigidity, and hand anesthesia.

However, subjects also report changes in body feelings that have not been directly suggested by the hypnotist. For instance, Gill and Brenman (1959, pp. 13–19) note that a substantial proportion of subjects who have been exposed to an induction procedure report changes in the size of the body or body parts, "disappearance" of the body or parts of the body, changes in equilibrium such as giddiness or dizziness, changes in feelings of reality, changes in experienced temperature (feeling either very hot or very cold), and changes in the way the experimenter's voice is experienced (as either very near or very far).

A study by Barber and Calverley (1969) indicated that these kinds of changes in body feelings are due, in part, to Variable 4 (keeping the eyes closed for a period of time) and, in part, to Variable 5 (receiving suggestions of relaxation, drowsiness, sleep and/or hypnosis). There were three groups of subjects in this experiment. Subjects in a control group (Close-Your-Eyes group) were individually asked to close their eyes and to await further instructions.[6] After five minutes had elapsed, they were asked to open their eyes and told that the experiment was over. Subjects in a second group (Hypnotic Induction group) were individually exposed to repeated suggestions of relaxation, drowsiness, sleep, and hypnosis and to a series of test-suggestions. Subjects in a third group (Place-Yourself-in-Hypnosis group) were asked individually to put themselves in hypnosis and, after five minutes had elapsed, were assessed on response to test-suggestions. At the end of each individual session, each subject completed a rating scale that described changes in body feelings that might have occurred during the experiment. The results are presented in Table 2. As this table shows, subjects in the Close-Your-Eyes group did not differ significantly from those in the Hypnotic Induction group or the Place-Yourself-in-Hypnosis group on 3 of the 6 items that assessed changes in body feelings (changes in the size of the body or body parts, changes in equilibrium, and changes in experienced temperature). The Close-Your-Eyes group obtained lower scores than the other two groups on the remaining three items that assessed changes in body feelings ("disappearance" of the body or body parts, changes in feelings of reality, and changes in the way the experimenter's voice is experienced). It should be noted, however, that 26% of the subjects in the Close-Your-Eyes group reported changes in feelings of reality and 34% testified that the experimenter's voice seemed either very near or very far away. Post-experimental interviews indicated that the changes in body feelings reported by subjects in the Close-Your-Eyes group were due to sitting quietly with eyes closed for a period of time while expecting *something* to occur but *not* due to their defining the situation as hypnosis or expecting hypnosis.

In brief, the study summarized above indicated that some of the changes in body feelings of the type described by Gill and Brenman (1959) are due to the subject keeping his eyes closed for a period of time while expecting something to happen. However, the study also indicated that relaxation-sleep-hypnosis suggestions or suggestions to enter hypnosis also play an important role in producing these changes in body feelings. Thus, it would appear that these kinds of changes—*e.g.*, changes in the size of the body or body parts, changes in equilibrium, changes in feelings of reality—are a complex function of several interacting variables including the following: keeping the eyes closed for a period of time; expect-

---

[6] These control subjects were recruited from a school in which no previous experiments had been conducted. They were told that they were participating in a psychological experiment, and they accepted this definition of the situation—there was no indication that they defined the situation to themselves as hypnosis.

TABLE 2

PERCENTAGE OF SUBJECTS IN CLOSE-YOUR-EYES, HYPNOTIC INDUCTION, AND PLACE-YOURSELF-IN-HYPNOSIS GROUPS REPORTING CHANGES IN BODY FEELINGS

	Close-Your-Eyes Group (N = 50)	Hypnotic Induction Group (N = 55)	Place Yourself-in-Hypnosis Group (N = 55)
Changes in size of body or body parts	36	51	47
"Disappearance" of body or body parts	4*	20	16
Changes in equilibrium (giddiness, dizziness, etc.)	66	85	76
Changes in feelings of reality	26*	69	69
Changes in experienced temperature (feeling either very hot or very cold)	34	36	42
Changes in the way the experimenter's voice is experienced (as very near or very far)	34*	71	64

Note.—The asterisk indicates that the percentage is significantly smaller than the other two percentages in the same *row* at the .05 level of confidence.

ing something to happen or expecting changes in body feelings; receiving or giving oneself suggestions of relaxation, drowsiness, sleep, and hypnosis; and expectantly focusing on and magnifying the body feelings and sensations that are produced when the eyes are closed and when suggestions of sleep or hypnosis are administered.

Do these changes in body feelings *directly* affect the subject's response to test-suggestions? Although the changes in body feelings do not appear to have a *direct* effect, they may *indirectly* affect the subject's response to test-suggestions by further heightening his expectancies. When the subject finds that he is experiencing changes in body feelings as he receives suggestions, his expectancy that he can be affected by suggestions may increase, and his heightened expectancy may enhance his response to subsequent test-suggestions.

Reports of Having Been Hypnotized

At the end of the session, subjects are often asked if they felt they were hypnotized. Since the situation is defined as hypnosis, and since the subjects are given only two major categories for classifying their experiences—as not hypnotized or as hypnotized to some degree—, their reports will not fall outside of the two categories. For instance, they will not state simply that they felt relaxed, or they felt sleepy, or they were ready and willing to respond to suggestions. Instead, they will report either that they were not hypnotized or that they were hypnotized to a light, medium, or deep level.

Whether or not and to what extent subjects report that they were hypnotized is dependent on (a) the degree of congruence between what they experienced and what they think hypnosis involves, (b) whether the hypnotist states they were hypnotized, and (c) the wording of the questions that are submitted to them to elicit their reports. We shall discuss each of these variables in turn.

Congruence between experiences and conceptions of hypnosis. Barber and Calverley (1969) presented data indicating that subjects' reports that they were or were not hypnotized are dependent on whether the subjects' experiences dovetailed with their conceptions of what hypnosis involves. Many highly responsive subjects who participated in this investigation gave reasons such as the following for believing that they were not truly hypnotized: they were aware of what they were doing, were aware of their surroundings, could think of extraneous things, could hear extraneous sounds, or did not have complete amnesia for the session. Although these subjects were highly responsive to test-suggestions (and they also showed a trance-like appearance and experienced changes in body feelings), they did not believe they were truly hypnotized because their experiences did not dovetail with their conceptions of what true hypnosis is supposed to involve. They apparently believed that a person is truly hypnotized only if he is unaware of himself and his surroundings.

Although subjects' conceptions of hypnosis vary widely, a study by Barber, Dalal, and Calverley (1968) indicated that (a) most subjects believe that hypnosis involved relaxation, and/or changes in body feelings, and/or responsiveness to test-suggestions, and (b) they judge the degree to which they were hypnotized by noting the degree to which they experienced relaxation, changes in body feelings, or responsiveness to test-suggestions. In this study, subjects who stated that they were hypnotized to a light, medium, or deep level were asked, "What basis did you use to judge your level of hypnotic depth?" Most of the replies could be classified into the following three categories:

1. Some of the subjects stated that they judged their level of hypnotic depth from the degree to which they felt relaxed or sleepy. Of course, feeling relaxed or sleepy was a direct response to the suggestions for relaxation, drowsiness, and sleep that were used in the induction procedure.

2. Other subjects stated that they judged their degree of hypnosis on the basis of changes in body feelings. These changes in body feelings were of two types. One type (e.g., "I felt my hand rising by itself.") was a direct response to test-suggestions (e.g., suggestions for hand levitation). A second type was similar to the kinds of changes in body feelings that have been specified by Gill and Brenman, for example, feelings of dizziness or giddiness, "disappearance" of the body or body parts, changes in experienced temperature, changes in feelings of reality, etc. As stated earlier, these kinds of changes in body feelings are also closely associated with suggestions, namely, with suggestions for relaxation, drowsiness, sleep, and/or hypnosis given to an expectant subject when his eyes are closed.

3. Most commonly, the subjects judged the degree to which they were hypnotized by observing the degree to which they experienced the effects that were suggested. Also, failure to experience some or all of the test-suggestions was commonly given as a reason by the subjects for believing that they were not hypnotized.

In line with the preceding paragraph, Gill and Brenman (1959) have noted that, when subjects state that they are hypnotized, they typically seem to mean that they are ready and willing to respond to suggestions. These authors documented this observation as follows:

First, we would induce hypnosis in someone previously established as a "good" subject; then we would ask him how he knew he was in hypnosis. He might reply that he felt relaxed. Now we would suggest that the relaxation would disappear *but he would remain in hypnosis*. Then we would ask again how he knew he was in hypnosis. He might say because his arm "feels numb"—so again, we would suggest the disappearance of this sensation. We continued in this way until finally we obtained the reply, "I know I am in hypnosis because I *know* I will do what you tell me." This was repeated with several subjects, with the same results [p. 36].

Effect of cues from the hypnotist on the subject's report. An experiment presented by Barber et al. (1968) showed that whether or not and to what degree subjects believe they were hypnotized is affected by cues from the hypnotist. In this experiment, 63 subjects were individually exposed to an induction procedure and to the Barber Suggestibility Scale. Next, the hypnotist spoke to each subject individually as follows: one-third of the subjects, selected at random, were told, "I observed you closely during the experiment and, from what I observed, you were hypnotized," and another one-third of the subjects, also selected at random, were told, "I observed you closely during the experiment and, from what I observed, you were not hypnotized." The hypnotist did not say anything to the remaining subjects. Since the subjects were exposed to the hypnotist's statements at random, the hypnotist's statements were not dependent on the subject's actual performance. Despite the fact that the three sets of subjects had performed very similarly during the experiment—for instance, they obtained practically identical scores on the Barber Suggestibility Scale—, their post-experimental self-ratings of their hypnotic depth differed markedly and were in line with the hypnotist's statements about their performance.

Although, in the above experiment, the cues fom the hypnotist were explicit, there is reason to believe that the hypnotist may also transmit cues to his subjects through more subtle means such as the manner in which he talks to them, the tone and inflections of his voice, and the type of questions that he asks (Rosenthal, 1968; Barber, 1969; Barber, in press; Barber & Calverley, 1964a; Barber & Silver 1968a, 1968b). In other words, subjects' reports pertaining to having been hypnotized are influenced by explicit cues from the hypnotist and may also be influenced by more subtle cues emanating from the hypnotist.

Wording of the questions. Whether subjects state that they were or were not hypnotized is also dependent on the wording of the questions they are asked. The importance of the wording of the questions was illustrated in another experiment presented by Barber et al. (1968). In this experiment, 53 subjects were exposed to the induction procedure and to the test-suggestions that comprise the Stanford Hypnotic Susceptibility Scale. Upon completion of the session, the subjects were randomly assigned to three groups and were individually asked the following questions:

Group A: "Did you experience the hypnotic state as basically *similar* to the waking state?"

Group B: "Did you experience the hypnotic state as basically *different* from the waking state?"

Group C: "Did you or did you not experience the hypnotic state as basically *different* from the waking state?"

Even though the subjects in Groups A, B, and C had performed in the same way during the session—for instance, their scores on the Stanford Hypnotic Susceptibility Scale were practically identical—, the differently worded questions elicited markedly different replies. Only 17% of the subjects in Group A (who received the question that included the word *similar*) reported that the hypnotic state was experienced as basically *different* from the waking state as compared to 72% in Group B and 64% in Group C. In brief, this experiment showed that subjects' reports pertaining to their experience of the hypnotic state are markedly influenced by the wording of the questions that are submitted to them.

How would we interpret the foregoing findings? Our post-experimental interviews with the subjects suggested the following interpretation: Many subjects were willing to categorize their experiences in contradictory ways because their experiences were ambiguous or multifaceted and they did not know how to classify them. For instance, a subject could answer "Yes" to the question, "Did you experience the hypnotic state as basically different from the waking state?" because he experienced various suggested effects during the session. However, the same subject could also answer "Yes" to the antithetical question, "Did you experience the hypnotic state as basically *similar* to the waking state?" because he was aware of himself, of his role in the situation, and of extraneous events and hence did not feel he was in a "hypnotic state" or a "trance" as it is popularly conceived.

Comment. Let us now summarize and draw out some implications from this discussion pertaining to subjects' reports that they were hypnotized. These reports are dependent, in part, on whether the subjects' direct or indirect responses to suggestions dovetailed with their conceptions of what hypnosis involves. Since subjects typically perceive hypnosis as involving relaxation, and/or changes in body feelings, and/or responsiveness to test-suggestions, they typically report that they were hypnotized if they responded to the relaxation-sleep-hypnosis suggestions (which tend to give rise to relaxation and to changes in body feelings) and/or to the test-suggestions. Also, the subjects' reports that they were hypnotized are dependent, in part, on whether or not the hypnotist states they were hypnotized and on the implicit suggestions that are present in the questions they are asked.

Since the subjects' reports of having been hypnotized are closely related to their conceptions of hypnosis and to their responsiveness to explicit or implicit suggestions, the factors that determine whether they respond to suggestions also indirectly determine whether they state that they were hypnotized. What factors determine whether subjects respond to suggestions? Although it has been commonly assumed that the effective factor is a "hypnotic trance state," we see the effective factors as follows: The degree to which subjects respond to suggestions—including the relaxation-sleep-hypnosis suggestions and the test-suggestions—is dependent on (a) the subjects' pre-existing attitudes, motivations, and expectancies toward the test-situation and (b) the effectiveness of the nine factors involved in induction procedures in producing more positive attitudes, motivations, and expectancies and a concomitant willingness to think with and imagine vividly those things that are suggested.

Does the subject's belief that he is or is not hypnotized affect his responsiveness to test-suggestions? Although it is questionable whether this belief exerts a *direct* effect, it may affect his responsiveness *indirectly*. If the subject responds to suggestions and judges from his response that he is hypnotized, this might indirectly heighten his expectancy that he will respond to further suggestions, and the heightened expectancy may influence his subsequent response.

Summary

This paper has attempted to relate the variables involved in induction procedures (antecedent variables) to four consequent variables: responses to test-suggestions, trance-like appearance, changes in body feelings, and reports of having been hypnotized. Four questions were at the forefront of discussion.

1. *Which of the many variables involved in induction procedures are effective in enhancing responsiveness to test-suggestions?* Nine variables were delineated that appear to have this suggestibility-enhancing effect: (a) defining the situation as hypnosis; (b) removing fears and misconceptions; (c) securing cooperation; (d) asking the subject to keep his eyes closed; (e) suggesting relaxation, sleep, and hypnosis; (f) maximizing the phrasing and vocal characteristics of suggestions; (g) coupling suggestions with naturally-occurring events; (h) stimulating goal-directed imagining; and (i) preventing or reinterpreting the failure of suggestions.

2. *Why do these nine variables raise responsiveness to test-suggestions?* Data were presented supporting our theory that the nine variables augment suggestibility because they give rise to positive attitudes, motivations, and expectancies toward the test-situation which, in turn, give rise to a willingness to think with and vividly imagine those things that are suggested.

3. *What variables in the induction procedure or in the situation give rise to a trance-like appearance, changes in body feelings, and reports of having been hypnotized?* The trance-like appearance was shown to be due, in part, to variable 4 (keeping the eyes closed for a period of time) and, in part, to variable 5 (suggestions for relaxation, sleep, and hypnosis). Some of the changes in body feelings (*e.g.*, arm levitation or limb heaviness) were shown to be direct responses to the test-suggestions and others (*e.g.*, feelings of dizziness or "disappearance" of parts of the body) were shown to be due to the subjects expectantly focusing on and magnifying the sensations and feelings that are produced when one is told repeatedly, while his eyes are closed, that he is becoming relaxed, drowsy, sleepy, and is entering a hypnotic state. Subjects' reports that they were hypnotized were shown to be dependent upon (a) the degree of congruence between what they experienced and what they think hypnosis involves, (b) whether or not the hypnotist stated that they were hypnotized, and (c) the wording of the questions that were submitted to them to elicit their reports.

4. *How do the latter three phenomena—trance-like appearance, changes in body feelings, and reports of having been hypnotized—relate to responsiveness to test-suggestions?* These three phenomena have been traditionally viewed as indices of a hypnotic trance state, and the hypnotic trance state, in turn, has been viewed as a critical factor in producing heightened response to test-suggestions. From our viewpoint, the three phenomena are not necessary, although two of the three—changes in body feelings and the subject believing that he is hypnotized—tend to be helpful, in augmenting response to test-suggestions. If a subject responds to the relaxation-sleep-

hypnosis suggestions or to some of the other variables involved in induction procedures and, consequently, experiences changes in body feelings and judges from his responses that he is hypnotized, his expectancy that he can be affected by suggestions is enhanced. His enhanced expectancy, in turn, tends to heighten his responsiveness to subsequent test-suggestions.

REFERENCES

ANDERSEN, M. L. Correlates of hypnotic performance: An historical and role-theoretical analysis. Unpublished doctoral dissertation, University of California, Berkeley, 1963.

ARNOLD, M. B. On the mechanism of suggestion and hypnosis. *Journal of Abnormal and Social Psychology*, 1946, 41, 107–128.

ARONS, H. *The new master course in hypnotism.* Irvington, N.J.: Power Publishers, 1961.

BARBER, T. X. Hypnosis as perceptual-cognitive restructuring: II. "Post"-hypnotic behavior. *Journal of Clinical and Experimental Hypnosis*, 1958, 6, 10–20.

BARBER, T. X. Toward a theory of hypnosis: Post-hypnotic behavior. *Archives of General Psychiatry*, 1962, 7, 321–342.

BARBER, T. X. Experimental analysis of "hypnotic" behavior: A review of recent empirical findings. *Journal of Abnormal Psychology*, 1965, 70, 132–154. (a)

BARBER, T. X. Measuring "hypnotic-like" suggestibility with and without "hypnotic induction;" psychometric properties, norms, and variables influencing response to the Barber Suggestibility Scale (BSS). *Psychological Reports*, 1965, 16, 809–844. (b)

BARBER, T. X. *Hypnosis: a scientific approach.* New York: Van Nostrand Reinhold, 1969.

BARBER, T. X. *LSD, marihuana, yoga, and hypnosis.* Chicago: Aldine, 1970. (a)

BARBER, T. X. *Suggested ('hypnotic') behavior: The trance paradigm versus an alternative paradigm.* Harding, Mass.: Medfield Foundation, 1970. (b)

BARBER, T. X. Pitfalls in research: Nine investigator and experimenter effects. In R. M. W. Travers (Ed.) *Handbook of research on teaching.* Chicago: Rand McNally, in press.

BARBER, T. X., & CALVERLEY, D. S. "Hypnotic-like" suggestibility in children and adults. *Journal of Abnormal and Social Psychology*, 1963, 66, 589–597.

BARBER, T. X., & CALVERLEY, D. S. Effect of *E*'s tone of voice on "hypnotic-like" suggestibility. *Psychological Reports*, 1964, 15, 139–144. (a)

BARBER, T. X., & CALVERLEY, D. S. Empirical evidence for a theory of "hypnotic" behavior: Effects of pretest instructions on response to primary suggestions. *Psychological Record*, 1964, 14, 457–467. (b)

BARBER, T. X., & CALVERLEY, D. S. The definition of the situation as a variable affecting "hypnotic-like" suggestibility. *Journal of Clinical Psychology*, 1964, 20, 438–440. (c)

BARBER, T. X., & CALVERLEY, D. S. Toward a theory of hypnotic behavior: Effects on suggestibility of defining the situation as hypnosis and defining response to suggestions as easy. *Journal of Abnormal and Social Psychology*, 1964, 68, 585–592. (d)

BARBER, T. X., & CALVERLEY, D. S. Empirical evidence for a theory of "hypnotic" behavior: Effects on suggestibility of five variables typically included in hypnotic induction procedures. *Journal of Consulting Psychology*, 1965, 29, 98–107. (a)

BARBER, T. X., & CALVERLEY, D. S. Empirical evidence for a theory of "hypnotic" behavior: The suggestibility-enhancing effects of motivational suggestions, relaxation-sleep suggestions, and suggestions that the subject will be effectively "hypnotized." *Journal of Personality*, 1965, 33, 256–270. (b)

BARBER, T. X., & CALVERLEY, D. S. Toward a theory of "hypnotic" behavior: Experimental analyses of suggested amnesia. *Journal of Abnormal Psychology*, 1966, 71, 95–107. (a)

BARBER, T. X., & CALVERLEY, D. S. Toward a theory of hypnotic behavior: Experimental evaluation of Hull's postulate that hypnotic susceptibility is a habit phenomenon. *Journal of Personality*, 1966, 34, 416–433. (b)

BARBER, T. X., & CALVERLEY, D. S. Toward a theory of "hypnotic" behavior: Replication and extension of experiments by Barber and co-workers (1962–65) and Hilgard and Tart (1966). *International Journal of Clinical and Experimental Hypnosis*, 1968, 16, 179–195.

BARBER, T. X., & CALVERLEY, D. S. Multidimensional analysis of "hypnotic" behavior. *Journal of Abnormal Psychology*, 1969, 74, 209–220.

BARBER, T. X., DALAL, A. S., & CALVERLEY, D. S. The subjective reports of hypnotic subjects. *American Journal of Clinical Hypnosis*, 1968, 11, 74–88.

BARBER, T. X., & SILVER, M. J. Fact, fiction, and the experimenter bias effect. *Psychological Bulletin* (Monograph Supplement), 1968, 70 (No. 6, Pt. 2), 1–29. (a)

BARBER, T. X., & SILVER, M. J. Pitfalls in data analysis and interpretation: A reply to Rosenthal. *Psychological Bulletin* (Monograph Supplement), 1968, 70 (No. 6, Pt. 2), 48–62. (b)

BARBER, T. X., SPANOS, N. P., & CHAVES, J. F. *Hypnotism and human potentialities.* Chicago: Aldine-Atherton, in press.

CRONIN, D. M., SPANOS, N. P., & BARBER, T. X. Augmenting hypnotic suggestibility by providing favorable information about hypnosis. *American Journal of Clinical Hypnosis*, 1971, 13, 259–264.

DERMEN, D., & LONDON, P. Correlates of hypnotic susceptibility, *Journal of Consulting Psychology*, 1965, 29, 537–545.

DIAMOND, M. J. The use of observationally-presented information to modify hypnotic susceptibility. Paper presented at Eastern Psychological Association, New York, April 17, 1971.

ERICKSON, M. H., HERSHMAN, S., & SECTER, I. I. *The practical application of medical and dental hypnosis.* New York: Julian Press, 1961.

GILL, M. M., & BRENMAN, M. *Hypnosis and related states.* New York: International Universities Press, 1959.

GINDES, B. C. *New concepts of hypnosis.* New York: Julian Press, 1951.

HADFIELD, J. A. The influence of suggestion on

body temperature. *Lancet*, 1920, 2, 68–69.
HALEY, J. (Ed.) *Advanced techniques of hypnosis and therapy: selected papers of Milton H. Erickson.* New York: Grune & Stratton, 1967.
HARANO, K., OGAWA, K., & NARUSE, G. A study of plethysmography and skin temperature during active concentration and autogenic exercise. In W. Luthe (Ed.) *Autogenic training.* New York: Grune & Stratton, 1965.
HARTLAND, J. *Medical and dental hypnosis and its clinical applications.* Baltimore: Williams & Wilkins, 1966.
HILGARD, E. R., & TART, C. T. Responsiveness to suggestions following waking and imagination instructions and following induction of hypnosis. *Journal of Abnormal Psychology*, 1966, 71, 196–208.
HILGARD, J. R. *Personality and hypnosis: A study of imaginative involvement.* Chicago: University of Chicago Press, 1970.
HULL, C. L. *Hypnosis and suggestibility: An experimental approach.* New York: Appleton-Century, 1933.
JACOBSON, E. Electrical measurements of neuromuscular states during mental activities. *American Journal of Physiology*, 1930, 91, 567.
JACOBSON, E. Electrophysiology of mental activities. *American Journal of Physiology*, 1932, 44, 677–694.
KLINGER, B. I. Effect of peer model responsiveness and length of induction procedure on hypnotic responsiveness. *Journal of Abnormal Psychology*, 1970, 75, 15–18.
KLOPP, K. K. Production of local anesthesia using waking suggestion with the child patient. *International Journal of Clinical and Experimental Hypnosis*, 1961, 9, 59–62.
LONDON, P. Subject characteristics in hypnosis research: I. A survey of experience, interest, and opinion. *International Journal of Clinical and Experimental Hypnosis*, 1961, 9, 151–161.
LONDON, P. The induction of hypnosis. In J. E. Gordon (Ed.) *Handbook of clinical and experimental hypnosis.* New York: Macmillan, 1967. Pp. 44–79.
LONDON, P., COOPER, L. M., & JOHNSON, H. J. Subject characteristics in hypnosis research: II. Attitudes toward hypnosis, volunteer status, and personality measures. III. Some correlates of hypnotic susceptibility. *International Journal of Clinical and Experimental Hypnosis*, 1962, 10, 13–21.
MACVAUGH, G. S. *Hypnosis readiness inventory.* Chevy Chase, Md.: G. S. Macvaugh (4402 Stanford Street), 1969.
MEEKER, W. B., & BARBER, T. X. Toward an explanation of stage hypnosis. *Journal of Abnormal Psychology*, 1971, 77, 61–70.
MELEI, J. P., & HILGARD, E. R. Attitudes toward hypnosis, self-predictions, and hypnotic susceptibility. *International Journal of Clinical and Experimental Hypnosis*, 1964, 12, 99–108.
MENZIES, R. Further studies of conditioned vasomotor responses in human subjects. *Journal of Experimental Psychology*, 1941, 29, 457–482.
Moss, C. S. *Hypnosis in perspective.* New York: Macmillan, 1965.
MORDEY, T. R. The relationship between certain motives and suggestibility. Unpublished Master's thesis, Roosevelt University, 1960.
PATTIE, F. A. Methods of induction, susceptibility of subjects, and criteria of hypnosis. In R. M. Dorcus (Ed.) *Hypnosis and its therapeutic applications.* New York: McGraw-Hill, 1956. Chap. 2.
ROSENHAN, D. L., & TOMKINS, S. S. On preference for hypnosis and hypnotizability. *International Journal of Clinical and Experimental Hypnosis*, 1964, 12, 109–114.
ROSENTHAL, R. Experimenter expectancy and the reassuring nature of the null hypothesis decision procedure. *Psychological Bulletin* (Monograph Supplement), 1968, 70 (No. 6, Pt. 2), 30–47.
SARBIN, T. R. Contributions to role-taking theory: I. Hypnotic behavior. *Psychological Review*, 1950, 57, 255–270.
SCHULTZ, J. H. *Das Autogene Training.* Leipzig: G. Thieme Verlag, 1932.
SHOR, R. E. Expectancies of being influenced and hypnotic performance. *International Journal of Clinical and Experimental Hypnosis*, 1971, 19, 154–166.
SHOR, R. E., ORNE, M. T., & O'CONNELL, D. N. Psychological correlates of plateau hypnotizability in a special volunteer sample. *Journal of Personality and Social Psychology*, 1966, 3, 80–95.
SPANOS, N. P. Goal-directed fantasy and the performance of hypnotic test suggestions. *Psychiatry*, 1971, 34, 86–96.
SPANOS, N. P., & HAM, M. L. Cognitive activity in response to hypnotic suggestion: Goal-directed fantasy and selective amnesia. Department of Sociology, Boston University, 1971.
SPANOS, N. P., & BARBER, T. X. Cognitive activity during "hypnotic" suggestibility: Goal-directed fantasy and the experience of non-volition. *Journal of Personality*, in press.
SPIEGEL, H. Hypnosis and transference. *Archives of General Psychiatry*, 1959, 1, 634–639.
TEITELBAUM, M. *Hypnosis induction technics.* Springfield, Ill.: C C Thomas, 1965.
UNESTAHL, L.-E. Hypnosis and hypnotic susceptibility. *Scandinavian Journal of Clinical and Experimental Hypnosis*, November, 1969, No. 2.
WATKINS, J. G. *Hypnotherapy of war neuroses.* New York: Ronald Press, 1949.
WEITZENHOFFER, A. M. *General techniques of hypnotism.* New York: Grune & Stratton, 1957.
WEITZENHOFFER, A. M., & HILGARD, E. R. *Stanford hypnotic susceptibility scale: Form C.* Palo Alto, Calif.: Consulting Psychologists Press, 1962.
WELLS, W. R. Experiments in waking hypnosis for instructional purposes. *Journal of Abnormal and Social Psychology*, 1924, 18, 389–404.
WHITE, R. W. A preface to the theory of hypnotism. *Journal of Abnormal and Social Psychology*, 1941, 36, 477–505.
WILSON, D. L. The role of confirmation of expectancies in hypnotic induction. Unpublished doctoral dissertation, University of North Carolina, 1967.
WOLBERG, L. R. *Medical hypnosis. Vol. 1. The principles of hypnotherapy.* New York: Grune & Stratton, 1948.

Suggested ("Hypnotic") 30
Behavior: Trance Paradigm versus an Alternative Paradigm

T. X. Barber

It has been traditionally assumed that certain types of procedures, labeled as "trance inductions," give rise to a special state of consciousness (hypnotic trance) in some individuals. It has also been assumed that as the hypnotic trance becomes deeper or more profound, the subject becomes more responsive to suggestions for age regression, analgesia, hallucinations, deafness, amnesia, and so on. In this chapter, I will critically analyze these and other assumptions that underly the traditional (trance) paradigm and I will present an alternative paradigm for conceptualizing the experiences and behaviors that have been historically subsumed under the term "hypnotism" or "hypnosis."

Paradigms in Science

In a cogent analysis of the history of science, Kuhn (1962) has shown that scientists working in an area of inquiry usually share common basic assumptions pertaining to the nature of the phenomena they are investigating. The shared assumptions, which are often more implicit than explicit, together with a related set of criteria for asking meaningful questions and for selecting research topics, are termed a "paradigm."

Work on this chapter (Medfield Foundation Report #103) was supported by research grants (MH-11521 and MH-19152) from the National Institute of Mental Health, U.S. Public Health Service.

Reprinted from Erika Fromm and Ronald E. Shor, editors, *Hypnosis: Research Developments and Perspectives,* Aldine•Atherton, Inc. Chicago. Copyright © 1972 by Erika Fromm and Ronald E. Shor.

Each paradigm may give rise to more than one theory that aims to explain the phenomena. Although theories deriving from any one paradigm differ in various aspects, they share the same basic assumptions, the same methodological criteria, and the same framework for asking meaningful questions. As Braginsky, Braginsky, and Ring (1969) have pointed out, "In academic psychology, for example, competing behavioristic theories of learning were for a long time able to flourish despite widespread agreement concerning how the phenomenon [of learning] should be approached—a consensus that was particularly likely to be evident when such theories were challenged by the radically different assumptions of cognitively oriented theories (p. 30)."

In the history of science, there are many important instances when the consensually-shared paradigm could not easily explain new research data. In these instances, a few scientists began to question the basic assumptions underlying the traditional paradigm, a new way of viewing the phenomena (an alternative paradigm) was slowly developed, and after a period of debate, misunderstandings, and acrimony, the alternative paradigm was accepted by new generations of investigators and slowly became dominant. In astronomy, for example, the geocentric view of the planetary system was replaced by the heliocentric view; in chemistry, the phlogiston conception of combustion was replaced by the oxygen conception; in physics, theories pertaining to the ether were replaced by conceptions that did not postulate an ether; and, in psychology, introspective analysis gave way to behaviorism.

As Chaves (1968) has pointed out, it appears that a paradigm shift may be occurring at the present time in the area of inquiry historically subsumed under the term "hypnosis." In the next section, I will briefly describe the underlying assumptions of the traditional (trance) paradigm that has dominated this area of inquiry for more than a hundred years. Following this, I will formulate some of the postulates of an alternative paradigm.

The Special State (Trance) Paradigm

During the past century, terms such as "hypnotic trance state" (or "trance," "hypnosis," "hypnotic state," and "hypnotized") have been widely used by both scientists and laymen and have become part of the everyday vocabulary of children and adults. Although the implications of these terms have been slowly changing over the years, they seem to refer to some kind of fundamental change in the state of the organism.

During the nineteenth century, terms such as "hypnotic trance" or "hypnosis" typically implied that the subject resembled the sleepwalker or somnambule, that is, resembled the person who arises from his bed at night, walks around while "half asleep," and responds in a dissociated, rather automatic way to a narrow range of stimuli. Some present-day investigators also think of the "hypnotic trance" subject as resembling a sleepwalker. As Hilgard (1969a) has pointed out, "Hypnosis is commonly considered to be a 'state' perhaps resembling the state in which the sleepwalker finds himself, hence the term 'somnambulist' as applied to the deeply hypnotized person (p. 71)." Other present-day investigators who utilize the terms "hypnotic trance" or "hypnosis" do not seem to mean that the subject resembles the sleepwalker. Although, as Bowers (1966) has noted, "Most [present-day] investigators interested in hypnosis believe that there is an hypnotic state which fundamentally differs from the waking state" (p. 42), they differ among themselves as to the exact meaning to be assigned to the terms.

Bowers (1966) views hypnosis as "an altered state within which suggestions have a peculiarly potent effect" (p. 50). However, Gill and Brenman (1959) use the term "hypnotic state" to refer to an "induced psychological regression, issuing, in the setting of a particular regressed relationship between two people, in a relatively stable state which includes a subsystem of the ego with various degrees of control of the ego apparatuses" (p. xxiii). Other investigators attach different connotations to the term. For instance, among the essential characteristics of the hypnotic state, Orne (1959) includes a tolerance for logical inconsistencies ("trance logic") and alterations in subjective experiences induced by suggestions. Evans (1968) views hypnosis as an altered subjective state of awareness in which dissociative mechanisms are operating, Meares (1963) sees the basic element in hypnosis as an atavistic regression to a primitive mode of mental functioning, and Shor (1962) views the hypnotic state as having three dimensions—hypnotic role-taking, trance, and archaic involvement.

Although the above and other theoretical formulations attribute somewhat different properties to the hypnotic state, they derive from a common set of basic assumptions (an underlying paradigm). Some of the underlying assumptions of the hypnosis or trance paradigm appear to include the following:

1. There exists a state of consciousness, a state of awareness, or a state of the organism that is fundamentally (qualitatively) different from other states of consciousness such as the waking state, the deep sleep state, and the state of unconsciousness. This distinct state is labeled "hypnosis," "hypnotic state," "hypnotic trance," or simply "trance."

2. The state of hypnotic trance may occasionally occur spontaneously, but it is usually induced by special types of procedures that are labeled "hypnotic inductions" or "trance inductions." Although trance induction procedures vary in content—for example, they usually include, but they need not include, fixation of the eyes, suggestions of relaxation, and suggestions of drowsiness and sleep—they all appear to have two essential features

in common: they suggest to the subject that he is entering a special state (hypnotic trance) and investigators who adhere to the traditional paradigm agree that the procedures are capable of producing hypnotic trance.

3. The hypnotic trance state is not a momentary condition that the subject enters for only a few seconds. On the contrary, when a person has been placed in a hypnotic trance, he remains in it for a period of time and he is typically brought out of it by a command from the hypnotist, such as "Wake up!"

4. Subjects who are in a hypnotic state are responsive, both overtly and subjectively, to test suggestions for rigidity of the muscles or limbs, age regression, analgesia and anesthesia, visual and auditory hallucination, deafness, blindness, color blindness, negative hallucination, dreaming on a specified topic, heightened performance (on physical or cognitive tasks), amnesia, and posthypnotic behavior.[1]

5. As Sutcliffe (1960) pointed out, some investigators who adhere to the trance paradigm believe the suggested phenomena are "genuine" or "real," whereas others are far more skeptical. For example, some investigators who accept the trance paradigm view hypnotic deafness as indistinguishable from actual deafness, and the hypnotic dream as indistinguishable from the nocturnal dream. However, other investigators who accept the trance paradigm view the hypnotic deaf subject as a person who is able to hear but thinks that he cannot, and they perceive the hypnotic dream as differing in essential respects from the night dream. Although investigators who adhere to the trance paradigm disagree on the "reality" of the suggested phenomena, the important point to emphasize is that they all view the phenomena as associated with hypnotic trance, and they consequently label the phenomena as "hypnotic phenomena," not simply as "suggested phenomena."

6. There are levels or depths of hypnotic trance; that is, hypnotic trance can vary from light, to medium, to deep, to very deep (somnambulism).

7. As the depth of hypnotic trance increases, the subject's ability to experience suggested phenomena vividly and intensely also increases. For example, as the subject becomes more deeply hypnotized, he is more able to have a vivid and intense experience of age regression, analgesia, hallucination, or amnesia.

In brief, the dominant (trance) paradigm sees the person who responds to test suggestions as being in a fundamentally different state from the person who is unresponsive to test suggestions. The construct "hypnotic state," "trance," or "hypnosis" is used to refer to this state, which is conceived to differ, not simply quantitatively, but in some basic, qualitative way, from normal waking states and from states of sleep.

1. Henceforth, in this chapter, the term "response" or "responsiveness to test suggestions" will be used as a shorthand term to refer to both overt and subjective responses to each of the types of suggestions mentioned in this paragraph.

An Alternative Paradigm: The Member of the Audience Analogy

There is another way of viewing responsiveness to test suggestions[2] that does not involve special state constructs such as "hypnosis," "hypnotized," "hypnotic state," or "trance." This alternative paradigm does not see a qualitative difference in the "state" of the person who is and the one who is not responsive to test suggestions. Although the alternative paradigm has many historical roots (discussed by Sarbin, 1962), it derives primarily from my more recent theoretical endeavors and those of Sarbin (Barber, 1964a, 1967, 1969b, 1970a; Sarbin, 1950; Sarbin & Andersen, 1967; Sarbin & Coe, in press). An analogy to members of an audience watching a motion picture or a stage play may clarify the paradigm.

One member of an audience may be attending a performance with the purpose of having new experiences. His attitude is that it is interesting and worthwhile to feel sad, to feel happy, to empathize, and to have the other thoughts, feelings, and emotions the actors are attempting to communicate. He both desires and expects the actors to arouse in him new or interesting thoughts and emotions. Although he is aware that he is watching a contrived performance and that he is in an audience, he does not actively think about these matters. Since this member of the audience has "positive" attitudes, motivations, and expectancies toward the communications emanating from the stage, he lets himself imagine and think with the statements and actions of the actors; he laughs, weeps, empathizes and, more generally, thinks, feels, emotes, and experiences in line with the intentions of the actors.

Another member of the audience had an anxious and tiring day at the office, wanted to go to bed early in the evening, and came to the performance unwillingly, in order to avoid an argument with his wife. He is not interested in having the emotions and experiences the actors are attempting to communicate. He does not especially desire and does not expect to feel empathic, happy, sad, excited, or shocked. He is continually aware that he is in an audience and that he is observing a deliberately contrived performance. Given this set of attitudes, motivations, and expectancies, this member of the audience does not let himself imagine and think with the statements and actions of the actors; he does not laugh, weep, empathize or, more generally, think, feel, emote, and experience in line with the communications from the actors.

The implications of this analogy are:

1. The experimental subject who is highly responsive to test suggestions resembles the member of the audience who experiences the thoughts, feelings,

---

2. As stated in footnote 1, in the remainder of this chapter the term "responsiveness" or "response to test suggestions" will refer to both overt and subjective responses to suggestions for limb rigidity, age regression, analgesia, hallucination, amnesia, postexperimental ("posthypnotic") behavior, and so on.

and emotions that the actors are attempting to arouse. The very suggestible subject views his responding to test suggestions as interesting and worthwhile; he desires and expects to experience those things that are suggested. Given these underlying "positive" attitudes, motivations, and expectancies, he lets himself imagine and think with the things suggested and he experiences the suggested effects.

2. The experimental subject who is very unresponsive to test suggestions resembles the member of the audience who does *not* experience the thoughts, feelings, and emotions that the actors are attempting to arouse. The very nonsuggestible subject views his responding to test suggestions as not desirable; he neither wants nor expects to experience those things that are suggested. Given these underlying "negative" attitudes, motivations, and expectancies, he does not let himself imagine and think with the things suggested and he does not experience the suggested effects.

3. It is misleading and unparsimonious to label the member of an audience, who is thinking, feeling, and emoting in line with the communications of the actors, as being in a special state (hypnotic trance) that is fundamentally different from the waking state. In other words, it is misleading and unparsimonious to restrict our conceptions of normal conditions or waking conditions to such an extent that they exclude the member of the audience who is having various experiences as he listens to the communications from the stage. Furthermore, since the member of the audience, who is responding to the words of the actors, and the experimental subject, who is responding to the words (test suggestions) of the experimenter, do not differ in any important way in their attitudes, motivations, and expectancies toward the communications or in the way they think along with the communications, it is also misleading and unparsimonious to label the subject who is responding to test suggestions as being in a special state (hypnotic trance).

4. Although the member of the audience who is responding to the words of the actors and the experimental subject who is responding to the test suggestions of the experimenter have similar attitudes, motivations, and expectancies toward the communications and are similarly "thinking with" the communications, *they are being exposed to different types of communications*. The messages or communications from the actors are intended to elicit certain types of thoughts, feelings, and emotions—to empathize, to feel happy or sad, to laugh or to weep, to feel excited or shocked—whereas the messages or communications (test suggestions) from the experimenter are intended to elicit somewhat different types of thoughts, feelings, or emotions—to experience an arm as heavy, to experience oneself as a child, to forget preceding events, and so forth. From this viewpoint, the member of the audience and the subject who is responding to test suggestions are having different experiences, *not because they are in different "states" but because they are receiving different communications*.

The above analogy exemplifies some of the basic assumptions underlying the alternative paradigm. These assumptions include the following:

1. It is unnecessary to postulate a fundamental difference in the "state" of the person who is and the one who is not responsive to test suggestions.

2. Both the person who is and the one who is not responsive to test suggestions have attitudes, motivations, and expectancies toward the communications they are receiving.

3. The person who is very responsive to test suggestions has "positive" attitudes, motivations, and expectancies toward the communications he is receiving. That is, he views his responding to test suggestions as interesting or worthwhile and he wants to, tries to, and expects to experience the suggested effects. Given these "positive" attitudes, motivations, and expectancies, he lets himself think with and imagine those things that are suggested.

4. The person who is very unresponsive to test suggestions has "negative" attitudes, motivations, and expectancies toward the communications he is receiving. That is, he views his responding to test suggestions as not interesting or worthwhile and he neither tries to nor expects to experience the suggested effects. Given these "negative" attitudes, motivations, and expectancies, he does not let himself imagine or think with the suggestions; instead, he verbalizes to himself such statements as "This is silly" or "The suggestion won't work."

5. The three factors—attitudes, motivations, and expectancies—vary on a continuum (from negative, to neutral, to positive) and they converge and interact in complex ways to determine to what extent a subject will let himself think with and imagine those things that are suggested. The extent to which the subject thinks with and vividly imagines the suggested effects, in turn, determines his overt and subjective responses to test suggestions.

6. Concepts derived from abnormal psychology—such as "trance," "somnambulism," and "dissociation"—are misleading and do not explain the overt and subjective responses. Responsiveness to test suggestions is a normal psychological phenomenon that can be conceptualized in terms of constructs that are an integral part of normal psychology, especially of social psychology. Social psychology conceptualizes other social influence processes, such as persuasion and conformity, in terms of such mediating variables as attitudes, motivations, expectancies, and cognitive processes. In the same way, the mediating variables that are relevant to explaining responsiveness to test suggestions include attitudes, motivations, expectancies, and cognitive-imaginative processes.

7. The phenomena associated with test suggestions are considered to be within the range of normal human capabilities. However, whether or not the suggested phenomena are similar to or different from phenomena occurring in real-life situations that bear the same name, is viewed as an open question that needs to be answered empirically. For example, such questions as the

following are open to empirical investigation: To what extent is suggested analgesia similar to the analgesia produced by nerve section or by anesthetic drugs? What are the similarities and differences between suggested and naturally-occurring (nonsuggested) blindness, color blindness, deafness, hallucination, dreaming, and amnesia? The empirical evidence at present indicates that, although there are some similarities between the suggested and nonsuggested phenomena, they also differ in very important respects. For example, suggested color blindness has only superficial resemblances to actual color blindness, and suggested amnesia is much more labile or transient than actual amnesia. (Since these issues have been discussed in detail elsewhere—Barber, 1959b, 1961b, 1962a, 1962b, 1963, 1964b, 1964c, 1965b, 1969b, 1970a—they will be discussed only peripherally here.)

Which paradigm is more successful in explaining responsiveness to test suggestions for limb rigidity, age-regression, analgesia, hallucination, amnesia, etc.—the traditional one that postulates a special state that is fundamentally different from the waking state, or the alternative one that focuses on attitudes, motivations, expectancies, and thinking with and imagining those things that are suggested? I will next summarize experimental data, pertaining to responsiveness to test suggestions under control ("waking") conditions, which indicate that the alternative paradigm provides a more successful and more parsimonious explanation.

Response to Test Suggestions Without "Hypnosis"

A substantial number of subjects are highly responsive to test suggestions when no attempt is made to place them in a "hypnotic trance state." Let us look at a few examples.

HUMAN-PLANK FEAT

The stage hypnotist suggests to a selected subject that his body is becoming stiff and rigid. When the subject appears rigid, the stage hypnotist and an assistant place him between two chairs, one chair beneath the subject's head and the other beneath his ankles. The subject typically remains suspended between the two chairs for several minutes, as if he were a human plank. The traditional paradigm assumes that the subject is able to perform the human-plank feat because he is in a state—a hypnotic trance state—that is qualitatively different from ordinary states of consciousness. This notion is not supported by the empirical data.

Collins (1961) demonstrated conclusively that, when male and female control subjects are told directly (without any special preliminaries) to keep their bodies rigid, practically all perform the human-plank feat, that is, they remain suspended between two chairs for several minutes, one chair beneath the head and the other beneath the ankles. In fact, Collins demonstrated that control subjects are able to perform the feat just as easily as subjects who have been exposed to a trance induction procedure and who are ostensibly in "hypnotic trance." The control ("awake") subjects and also the experimental ("hypnotized") subjects stated, at the conclusion of Collins' experiment, that they were surprised at their own performance because they did not believe initially that they could so easily perform the human-plank feat.

At times, stage hypnotists ask a person to stand on the chest of the subject who is rigidly suspended between two chairs, one chair beneath his shoulders and the other beneath his calves. The traditional paradigm assumes that the suspended subject is able to support the weight of a man on his chest because he is in a special state of consciousness—a hypnotic trance state. The empirical evidence does not support this assumption. In my laboratory, six unselected male subjects were told under control conditions (without any special preliminaries) to make their body rigid and to keep it rigid. They were then suspended between two chairs, one chair beneath the shoulders and the other beneath the calves. Each subject was able to support the weight of a man on his chest. All subjects were surprised that they could so easily support the weight of a man and all disagreed vehemently with the statement that they were in a trance.

RESPONSE TO OTHER TEST SUGGESTIONS

Experimental studies that I have summarized elsewhere (Barber, 1965a) have demonstrated that a substantial proportion of individuals are responsive to various kinds of test suggestions when no attempt is made to place them in a "hypnotic trance." In these experiments, 62 unselected college students were assigned at random to a control condition (they were simply told that they were to receive a test of imagination). They were then assessed individually on objective and subjective responses to the eight standardized test suggestions of the Barber Suggestibility Scale: Arm Lowering (the subject's right arm is heavy and is moving down); Arm Levitation (the left arm is weightless and is moving up); Hand Lock (the clasped hands are welded together and cannot be taken apart); Thirst "Hallucination" (he is becoming extremely thirsty); Verbal Inhibition (his throat and jaw muscles are rigid and he cannot speak his name); Body Immobility (his body is heavy and he cannot stand up); "Posthypnotic-Like" Response (when he hears a click postexperimentally, he will cough automatically); and Selective Amnesia (when the experiment is over, he will not remember one specific test-suggestion).[3]

---

3. The subject receives a maximum Objective score of 8 points on the Barber Suggestibility Scale (one point for each of the eight test suggestions) if: the right arm moves down 4 or more inches; the left arm rises 4 or more inches; the sub-

Suggested ("Hypnotic") Behavior

As Table 5.1, column 1, shows, about one-fourth of these control subjects, who were given the eight test suggestions immediately after they were simply told that they were to receive a test of imagination, passed the Arm Lowering, Arm Levitation, Verbal Inhibition, and Body Immobility items both objectively (manifesting the suggested overt behavior) and subjectively (testifying postexperimentally that they actually experienced the suggested effect). In addition, nearly half of these control subjects passed the Thirst "Hallucination" item and 40 per cent passed the Hand Lock item (that is, they tried to unclasp their hands but had not succeeded after 15 seconds, and they testified that they actually felt that their hands were stuck). Furthermore, about 13 per cent of these control subjects passed the "Posthypnotic-Like" Response and the Selective Amnesia items.

Although a surprisingly high proportion of subjects were responsive to the test suggestions under the control condition, even more dramatic results were obtained when another group of 62 subjects, randomly selected from the same college population, were tested individually on the same test suggestions after receiving task-motivational instructions for 45 seconds. These task-motivational instructions, which aimed to produce favorable motivations, attitudes, and expectancies toward the test situation and to heighten the subject's willingness to imagine and think about those things that would be suggested, were worded as follows:

> In this experiment I'm going to test your ability to imagine and to visualize. How well you do on the tests which I will give you depends entirely upon your willingness to try to imagine and to visualize the things I will ask you to imagine. Everyone passed these tests when they tried. For example, we asked people to close their eyes and to imagine that they were at a movie theater and were watching a show. Most people were able to do this very well; they were able to imagine very vividly that they were at a movie and they felt as if they were actually looking at the picture. However, a few people thought that this was an awkward or silly thing to do and did not try to imagine and failed the test. Yet when these people later realized that it

---

> ject tries to but fails to unclasp his hands; he shows swallowing, moistening of lips, or marked mouth movements and states postexperimentally that he became thirsty during this test; he tries but does not succeed in saying his name; he tries but does not succeed in standing fully erect; he coughs or clears his throat when the cue is presented postexperimentally; and he does not refer to the critical item during the postexperimental interview but recalls at least four other items and then recalls the critical item when told "Now you can remember."
>
> In addition to the Objective scores, assigned as described above, the subject also receives a maximum Subjective score of 8 points on the Barber Suggestibility Scale (one point for each of the eight test suggestions) if he states, during the standardized post experimental interview, that he actually experienced each of the suggested effects and that he did not respond overtly to the test suggestion simply to follow instructions or to please the experimenter.

wasn't hard to imagine, they were able to visualize the movie picture and they felt as if the imagined movie was as vivid and as real as an actual movie. What I ask is your cooperation in helping this experiment by trying to imagine vividly what I describe to you. I want you to score as high as you can because we're trying to measure the maximum ability of people to imagine. If you don't try to the best of your ability, this experiment will be worthless and I'll tend to feel silly. On the other hand, if you try to imagine to the best of your ability, you can easily imagine and do the interesting things I tell you and you will be helping this experiment and not wasting any time (Barber & Calverley, 1962, p. 366).

The subjects who received these task-motivational instructions showed a dramatically high level of objective and subjective responsiveness to the test suggestions (manifesting the suggested overt behaviors and testifying that they subjectively experienced the suggested effects). As Table 5.1, column 2, shows, from 56 per cent to 69 per cent of the subjects who received task-motivational instructions passed the Arm Lowering, Arm Levitation, Verbal Inhibition, and Body Immobility items, 76 per cent passed the Thirst "Hallucination" item, and 81 per cent passed the Hand Lock item. In addition, around 40 per cent of the subjects who received task-motivational instructions passed the Selective Amnesia and "Posthypnotic-Like" Response items.

As stated above, 62 subjects were tested individually under the control condition and 62 were tested under the task-motivational condition. In addition, 62 subjects, randomly chosen from the same population of college students, were assessed individually on response to the same test suggestions after they were exposed to a standardized 15-minute procedure of the type traditionally labeled as a "trance induction." This trance induction proce-

TABLE 5.1 Percentage of subjects passing each test suggestion both objectively and subjectively

Test Suggestion	Per Cent of Subjects Passing		
	control	task-motivational instructions	trance induction procedure
1. Arm lowering	26$_b$	61$_a$	72$_a$
2. Arm levitation	24$_b$	56$_a$	56$_a$
3. Hand lock	40$_b$	81$_a$	69$_a$
4. Thirst hallucination	48$_b$	76$_a$	74$_a$
5. Verbal inhibition	27$_b$	69$_a$	64$_a$
6. Body immobility	27$_b$	66$_a$	63$_a$
7. Posthypnoticlike response	14$_b$	42$_a$	29$_{ab}$
8. Selective amnesia	13$_b$	39$_a$	35$_a$

SOURCE: Barber, 1965a.
NOTE: Percentages in the same row containing the same subscript letter do not differ from each other at the .05 level of confidence.

dure, which is presented verbatim elsewhere (Barber, 1969b), included the following salient features: (a) Instructions were administered to produce favorable attitudes, motivations, and expectancies (for example, "Hypnosis is nothing fearful or mysterious. . . . Your cooperation, your interest, is what I ask for. . . . Nothing will be done that will in any way cause you the least embarrassment . . . you will be able to experience many interesting things."). (b) The subject was asked to fixate on a light blinking in synchrony with the sound of a metronome and was given suggestions of eye heaviness and eye closure (for example, "The strain in your eyes is getting greater and greater. . . . You would like to close your eyes and relax completely"). (c) Suggestions of relaxation, drowsiness, and sleep were administered repeatedly ("comfortable, relaxed, thinking of nothing, nothing but what I say . . . drowsy . . . deep sound comfortable sleep . . . deeper and deeper. . ."). (d) It was suggested to the subject that he was entering a unique state, a deep trance, in which he would be able to have interesting and unusual experiences

As table 5.1, column 3, shows, subjects exposed to the trance induction procedure were generally as responsive to the test suggestions as those subjects who had received the brief task-motivational instructions under waking conditions. Also, Table 5.1 shows that the subjects who received the trance induction procedure as well as those who received the task-motivational instructions were significantly more responsive to the test suggestions than the control group.

Table 5.2 shows the number of test suggestions that were passed by subjects in each of the three experimental groups. The reader will note that 13 per cent and 10 per cent of the subjects under the task-motivational instructions and trance induction conditions, respectively, and none of the controls,

TABLE 5.2 Number of test suggestions passed (both objectively and subjectively) by subjects in control, task-motivational instructions, and trance induction groups

Number of Test Suggestions Passed	Per Cent of Subjects Passing		
	control group	task-motivational group	trance induction group
8	0 ⎫	13 ⎫	10 ⎫
7	3 ⎬ 16	16 ⎬ 60	16 ⎬ 53
6	11 ⎥	15 ⎥	16 ⎥
5	2 ⎭	16 ⎭	11 ⎭
4	11 ⎫	16 ⎫	15 ⎫
3	6 ⎬ 38	8 ⎬ 27	13 ⎬ 36
2	21 ⎭	3 ⎭	8 ⎭
1	16 ⎫ 45	10 ⎫ 13	6 ⎫ 11
0	29 ⎭	3 ⎭	5 ⎭

SOURCE: Barber, 1965a.

passed all eight of the test suggestions. Also, 16 per cent, 60 per cent, and 53 per cent of the subjects under the control, task-motivational instructions, and trance induction condition, respectively, were relatively highly responsive to test suggestions, passing at least five of the eight items.

The data presented above indicate the following:

1. When subjects are tested on response to test suggestions under a control condition (immediately after they are simply told that they are to be given a test of imagination), the majority respond to some test suggestions and a small proportion manifest a rather high level of response. Under the control condition, subjects typically passed two of the eight test suggestions and 16 per cent passed at least five of the eight.

2. Although most subjects respond to some test suggestions under a control condition, a markedly higher level of response is found when subjects are given task-motivational instructions, that is, instructions designed to produce positive motivations, attitudes, and expectancies toward the suggestive situation and a consequent willingness to think with and imagine those things that are suggested.

3. A trance induction procedure, which focuses on repeated suggestions of eye heaviness, relaxation, drowsiness, and sleep, also raises response to test suggestions above the control or base level.

4. Comparable high levels of response to test suggestions of arm heaviness, body immobility, inability to say one's name, selective amnesia, and so forth, are produced when task-motivational instructions are given alone (without a trance induction procedure) and when a trance induction procedure is given alone (without explicit task-motivational instructions).[4]

Why is enhanced responsiveness to test suggestions produced both by task-motivational instructions and also by a trance induction procedure? There are at least three possible interpretations:

1. From the traditional (special state) paradigm, one might hypothesize that both task-motivational instructions and a trance induction procedure

---

4. Additional experiments, summarized elsewhere (Barber, 1969b, pp. 60–70), also found comparable high levels of response to test suggestions (suggestions for analgesia, gustatory "hallucination," enhanced cognitive proficiency, dreaming on a specified topic, time distortion, color blindness, visual-auditory "hallucination," and amnesia) in subjects exposed to task-motivational instructions alone and in those exposed to a trance induction procedure. However, several considerations noted by Hilgard and Tart (1966) and by Edmonston and Robertson (1967) led to an additional experiment (Barber & Calverley, 1968), which indicated that task-motivational instructions given alone are slightly less effective in facilitating suggestibility than task-motivational instructions given together with a trance induction procedure. Barber and Calverley (1968) hypothesized that the slightly higher level of suggestibility that was found when the task-motivational instructions were combined with the trance induction pro-

give rise to a hypnotic trance state. However, with few exceptions, subjects who have received task-motivational instructions appear awake, claim they are awake, and do not show a limp posture, passivity, a blank stare, or any other sign of trance. Are task-motivated subjects, who show no signs of being in a trance, actually in a trance? This question, which derives from the traditional paradigm, cannot be answered by any empirical method available at the present time.

2. Also, from the special state paradigm, one might hypothesize that (a) subjects who are highly responsive to test suggestions after they have received a trance induction procedure are in a hypnotic trance state whereas (b) those who are highly responsive after receiving task-motivational instructions are not in a hypnotic trance but are responsive for other reasons, for example, because they are highly motivated to respond. This interpretation also leads to an anomaly for the special state paradigm because it is now being said that the kind of high response to test suggestions that has been traditionally associated with hypnotic trance can also be produced as easily without hypnotic trance.

3. From the alternative paradigm, which does not postulate a special state, one could hypothesize that both task-motivational instructions and a trance induction procedure raise response to test suggestions above the level found under a control condition because they produce more positive attitudes, motivations, and expectancies toward the suggestive situation and a greater willingness to think with and to imagine those things that are suggested. This hypothesis can be empirically confirmed or disconfirmed by (a) assessing attitudes, motivations, expectancies, and willingness to think with the suggestions prior to and also after the administration of a control, a task-motivational, and a trance induction treatment to three random groups of subjects and (b) testing responsiveness to suggestions after the second assessment (of attitudes, motivations, etc.).

A PERSONAL REPORT ON RESPONDING TO TEST SUGGESTIONS

As stated above, some individuals manifest a high level of response to test suggestions when no attempt is made to hypnotize them. I also manifest a high level of response. Let me now give a personal, phenomenological re-

---

cedure was due to the fact that under this condition the situation was defined to the subjects as "hypnosis" and, vice versa, the slightly lower level of suggestibility found with task-motivational instructions alone was due to the fact that under this condition the situation was defined as a "test of imagination." This hypothesis clearly merits testing, especially since earlier experiments (Barber & Calverley, 1964e, 1965) indicated that, with everything else constant, a higher level of responsiveness to test suggestions is produced when subjects are told they are participating in a "hypnosis" experiment rather than in an "imagination" experiment.

port of the factors underlying my own responsiveness to test suggestions (Barber, 1970b).

An experimenter states that he would like to assess my responsiveness to suggestions and I agree to be tested. Since I believe that it is an interesting and worthwhile learning experience to respond to the kinds of test suggestions that I expect he will give me, I have a positive attitude toward the test situation and am motivated to experience those things that will be suggested. Furthermore, I expect that suggested effects, such as arm levitation, age regression, and amnesia can be experienced. Since I have positive attitudes, motivations, and expectancies, I will not evaluate, analyze, or think contrary to those things that are suggested; for example, I will not say to myself, "This suggestion is not worded correctly," "The suggestion will not be effective," or "This is just an experiment." On the contrary, I will let myself think with, imagine, and visualize those things that the experimenter will describe.

The experimenter asks me to extend my right arm and then suggests repeatedly that it is solid, rigid, like a piece of steel. If I had a reason not to respond to the suggestion, I could prevent myself from thinking of the arm as rigid. However, since there is no reason to resist, on the contrary, since I am motivated to experience the suggested effects, I let myself think with the suggestion—I verbalize to myself that the arm is rigid and I imagine it as a piece of steel. When the experimenter then states, "Try to bend the arm, you can't," I do not say to myself, "Of course I can bend it." Instead, I continue to think of the arm as rigid, I continue to picture it as a piece of steel, and when I make an attempt to bend it, I find that I cannot.

(Thinking back to the suggestion, after the experiment is over, I realize the following: (a) when I was imagining my arm as rigid, I involuntarily contracted the muscles in the arm, (b) the involuntary muscular contractions made the arm feel rigid, (c) the actual rigidity in the arm reinforced the thought that the arm was rigid and immovable, and (d) when told to try to bend the arm, I continued to think and imagine that the arm was a piece of steel and I continued to maintain the involuntary muscular contraction. Although these considerations are clear to me retrospectively, during the experiment I was picturing the arm as a piece of steel and I was not actively thinking about these underlying mechanisms.)

The experimenter next suggests repeatedly that my left hand is dull, numb, a piece of rubber, a lump of matter without feelings or sensations. I think with the suggestions and I picture the hand as a rubbery lump of matter that is separated from the rest of my body. The experimenter then places the hand in a pain-producing apparatus that brings a heavy weight to bear upon a finger. Although this heavy weight normally produces an aching pain in the finger, I do not think of the stimulation as pain. Instead I continue to think of the hand and finger as a rubbery lump of matter "out there" and I think of the sensations produced by the heavy weight *as sensations* that have their own unique and interesting properties. Specifically, I think of the sensations as a series of separate sensations—as a sensation of pressure, a cutting sensation, a numbness, a feeling of heat, a pulsating sensation. Although under other circumstances I would label these sensations as pain, I do not let myself think of the sensations in this way; instead, I think of them as a complex of varying sensations in a dull, rubbery hand and I state honestly that although I experience a variety of unique sensations I do not experience anxiety, distress, or pain.

The experimenter then suggests that I see a cat in the corner of the room. Since I have a positive attitude toward the suggestive situation and am motivated to experience the suggested effects, I inhibit the thought that there is no cat in the room. Instead, I let myself vividly visualize a black cat that I have often seen before and I think of it as being in the corner of the room. Since I continue to think of the cat as being "out there," and since I inhibit the thought that I am visualizing it in my mind's eye, I state that I see the cat in the corner of the room.

The experimenter next instructs me to close my eyes (presumably to remove visual distractions that might interfere with the forthcoming tasks), and then suggests that I am in Boston Symphony Hall and I hear the orchestra playing. I let myself vividly imagine that I am at the symphony and that the orchestra is playing Beethoven's Fifth Symphony. I inhibit the thought that I am really in an experimental situation, I focus on the idea that I am in Symphony Hall, and I "hear" the music, which becomes continually more vivid.

(Afterwards, thinking back to the suggestion for auditory hallucination, I realize that I was "making the music in my head." However, at the time I received the suggestion I was vividly imagining and thinking about Beethoven's symphony and I was not thinking about such matters as where the music was coming from. Although I could have stopped thinking with the suggestion, for instance, I could have said to myself that I was actually in an experimental situation, I had no reason to verbalize such contrary thoughts to myself and I continued to imagine vividly and to think of myself as being in Symphony Hall.)

The experimenter next suggests that time is going back, my body is becoming small, and I am a child of six years of age. I do not say to myself that I am an adult, that I cannot become a child, or that this suggestion won't work. On the contrary, since I have positive attitudes, motivations, and expectancies toward the suggestive situation, I think with the suggestion. I let myself imagine vividly that my body is small and tiny (and I begin to feel that I am actually tiny), I think of myself as a child, and I vividly imagine myself in the first grade classroom. I then let the imaginative situation "move" by itself; the first-grade teacher talks to the students, two boys in the back of the room throw spitballs when the teacher turns her back to the class, and later the bell rings for recess. Since I focused on the idea that I was a child, since I felt myself as small and tiny, since I could "see" the events occurring in the classroom, since I did not say to myself "I am really an adult," I testify afterwards that I actually felt that I was six years old and that I found this part of the experiment vivid and very interesting.

Later, the experimenter suggests that when the session is over I will not remember anything that occurred. Soon afterwards he states that the experiment is over and asks me what I remember. Since I have no reason to resist the suggestion for amnesia, I say to myself that I do not remember what occurred, I keep my thoughts on the present, I do not think back to the preceding events, and I state that I do not remember. The experimenter subsequently states, "Now you can remember." I now let myself think back to the preceding events and I verbalize them.

In summary, speaking personally and phenomenologically, I can experience arm rigidity, hand levitation, analgesia, visual and auditory hallucination, age regression, amnesia, and other suggested effects that have been traditionally thought to be associated with hypnotism. I do not need a "trance induction procedure" in order to experience these effects. Since I am ready at any time to adopt a positive attitude, motivation, and expectancy toward the suggestive situation, I am ready at any time to think with and to imagine or visualize those things that are suggested. On the other hand, if I had a reason not to respond to the test suggestions, I could adopt a quite different set of attitudes, motivations, and expectancies toward the situation, I could tell myself that I shall not respond, and I could easily prevent myself from thinking with and vividly imagining those things that are suggested.

Three additional points should be emphasized:

1. If the experimenter first suggests to me, as has happened on several occasions, that I am becoming relaxed, drowsy, sleepy, and am entering a hypnotic trance state, I can think with these suggestions and can feel relaxed, dowsy, sleepy, and passive. However, when the experimenter subsequently gives suggestions that involve effort, for example, suggestions of arm rigidity or analgesia, I no longer feel relaxed, sleepy, or passive and I may, in fact, feel very alert and aroused. The traditional "trance induction procedure" comprised of repetitive suggestions of relaxation, drowsiness, sleep, and hypnosis, appears to me to be just another set of suggestions that I can accept and it is not necessary or especially important in determining my responsiveness to test suggestions.

2. When I am experiencing suggested analgesia, age regression, amnesia, and so on, I do not feel that I am in a special state—a hypnotic state or a trance—that is discontinuous with or qualitatively different from my ordinary state of consciousness. In fact, when I am responding to test sugges-

tions I do not feel that my "state" differs in any important way from the state I am in when I watch a motion picture or stage play. When I am in an audience, I let myself imagine and think with the communications from the stage and I empathize, laugh, feel sad, cry, and have the other emotions, feelings, and vicarious experiences that the actors are attempting to communicate. In essentially the same way, when I am in an experimental situation and am being assessed for response to test suggestions, I let myself think with and vividly imagine those things that are suggested and I have the experiences that the experimenter is attempting to communicate.

3. If I wish, I can give myself the same suggestions that are given by the experimenter. For instance, I can suggest to myself that time is going backwards, my body is becoming small and tiny, and I am a child of a certain age. I can then think about and vividly imagine a situation that occurred when I was a child and I can inhibit contrary thoughts. Similarly, in a dental situation I can give myself suggestions that the sensations are interesting and not uncomfortable and by thinking of each of the varying sensations (drilling, pressure, pricking, heat, and so forth) as sensations per se, I can inhibit anxiety, distress, and pain. I have also found that the same technique—focusing on the sensations as sensations—is sufficient to block the pain and distress associated with various methods used in the laboratory to produce pain, for example, pain produced by immersing a limb in ice water and pain produced by using a tourniquet to cut off the blood supply to an arm. Although these experiences have been traditionally subsumed under the term "autohypnosis," I do not feel that I am in a special state (a hypnotic trance state) fundamentally different from my ordinary state of consciousness; on the contrary, when I am having these experiences I feel as normal and as awake as when I am watching a movie, a stage play, or a television show.

Data Ostensibly Supporting the Traditional (Special State) Paradigm

At first glance, the traditional notion that a special state underlies high responsiveness to test suggestions appears to be supported by data such as the following:

1. Stage hypnotists appear to elicit unique or special behaviors from subjects who seem to be in a special state (hypnotic trance).

2. Experimenters have reported that a variety of amazing or special effects can be elicited from subjects who are ostensibly in a hypnotic trance.

3. High response to test suggestions is associated with observable trance-like characteristics.

4. Some highly responsive subjects testify that they experienced a special state of consciousness.

5. Some highly responsive subjects do not "come out of it" immediately

—they seem to remain in a trance after the experiment is over.

6. Some highly responsive subjects spontaneously forget the events and spontaneous amnesia is a critical indicant of a special state.

7. Highly responsive subjects show a special type of logic—"trance logic"—which indicates that they are in a special state.

Let us look at each of these sets of data in turn.

A complete discussion of the data is found in the original article.

Résumé and Prospects

As Kuhn (1962) pointed out, a change in scientific paradigm is preceded by a period in which research yields data that do not fit into the prevailing paradigm. Recent research has produced data incongruous with the prevalent trance paradigm. Some of the anomalous data include:

1. Some individuals are very responsive, both overtly and subjectively, to test suggestions when they are tested under a base level (control) condition (without any special instructions). Also, when unselected subjects are simply exposed to brief instructions intended to produce positive attitudes, motivations, and expectancies toward the test situation ("task-motivational instructions"), they are about as responsive to test suggestions for body immobility, hallucination, age regression, analgesia, amnesia, etc. as unselected subjects who have been exposed to a procedure of the type traditionally termed a "trance induction" and who are, presumably, in a hypnotic trance. The trance paradigm could not have predicted these results and it requires ad hoc assumptions in order to explain them. It has to assume, after the fact, that highly responsive control subjects or task-motivated subjects who have not been exposed to a trance-induction procedure, who do not appear to be in a trance, and who do not think they are in a trance are actually in a hypnotic trance.

2. The anomaly mentioned above appeared earlier in the work of hypnotic state theorists such as Erickson. To maintain the logic of the trance paradigm, Erickson was compelled to contend that some subjects are in a deep hypnotic trance even when they do not think they are in a trance, are judged by psychologists and psychiatrists as being in a normal waking state, and are even judged by Erickson himself as being "seemingly awake and functioning . . . in a manner similar to that of a nonhypnotized person operating at the waking level" (Erickson, 1967a, p. 13). Why was Erickson compelled to categorize subjects as being in a deep hypnotic trance even though the subjects appeared, to objective observers and to themselves, to

be normally awake? Because Erickson was certain that the subjects would respond to his suggestions. This logic led to another serious anomaly for the trance paradigm: subjects were judged to be in a hypnotic trance because they would show high response to test suggestions and, turning around circularly, the high response to test-suggestions was explained as due to the presence of hypnotic trance (Barber, 1964a).

3. Since all investigators conceive of the human organism as a psychophysiological unity, special states of the organism are expected to have some physiological concomitants. For more than 50 years, investigators have been trying to find a physiological concomitant or index of the presumed special state labeled "hypnosis" or "trance." Not only have they failed to unearth any special physiological change associated with the presumed special state, but they have also consistently found that physiological functioning during the postulated special state varies in the same way as in nonspecial or ordinary states. Of course, many failures over many years to find a physiological index does not prove that such an index does not exist or will never be found. Nevertheless, consistent findings that physiological functions during the presumed special state vary in the same way as in nonspecial states are becoming more and more anomalous for the trance paradigm as time goes on.

A traditional paradigm is not overthrown simply because some of the relevant data are incongruous with it (Kuhn, 1962). In addition to pointing out data that are anomalous for the traditional formulation, an alternative paradigm must also be able to explain all of the relevant phenomena at least as well if not better than the traditional one. The underlying contention of this chapter has been that the alternative paradigm, which has had a rather brief period of development (Barber, 1961b, 1964a, 1967, 1969b, 1970a; Barber & Calverley, 1962; Sarbin, 1950, 1956, 1962, 1964; Sarbin & Anderson, 1967; Sarbin & Coe, in press), can explain the relevant phenomena more satisfactorily than the trance paradigm which has had more than a hundred years of development. Specifically, this chapter has shown how the alternative paradigm explains not only the "amazing" phenomena of stage hypnosis, such as, the human-plank feat, stopping the pulse in the arm, and performance by the subject of weird antics such as dancing with an invisible partner, but also explains the following phenomena that can be elicited by suggestions in experimental or clinical situations: production of blisters, removal of warts, analgesia, age regression, age-progression, visual and auditory hallucinations, deafness, trancelike characteristics, difficulty or delay in coming out of trance, suggested and also spontaneous amnesia, and trance logic. These phenomena, which were explained in this chapter from the alternative paradigm, were not selected at random. On the contrary, they were selected as representing the strongholds of the trance paradigm—as representing the phenomena that had been universally accepted as explainable only by positing a special state of the organism.

Looking to the future, I will venture four predictions:

1. As more investigators adopt the alternative paradigm, the kinds of questions that are asked and the focus of research will change. Instead of asking what the most effective methods for inducing a deep hypnotic trance are or how the hypnotic state differs from the waking state, researchers will ask questions such as the following:

a. What kinds of instructions are most effective in eliciting positive attitudes, motivations, and expectancies toward the test situation?

b. Are all three factors—positive attitudes, positive motivations, positive expectancies—equally necessary for high response to test suggestions? How does a subject respond to test suggestions when he believes it is interesting and worthwhile to be responsive (positive attitude), when he tries to experience those things that are suggested (positive motivation), but when he does not believe that he can experience either a specific suggested effect, such as visual hallucination or amnesia, or all of those things that are suggested (negative expectancies)? A rather large number of additional questions can be formulated along similar lines. For instance, how does a subject respond to test suggestions when he believes he can experience the suggested effects (positive expectancy) but he has negative attitudes and motivations toward the test situation?

c. How can subjects be helped to imagine vividly those things that are suggested?

d. How can subjects be given practice in thinking with suggestions? Stated somewhat differently, how can they be given practice in covertly verbalizing the suggestions to themselves while, at the same time, inhibiting contrary thoughts such as "This suggestion won't work," or "It's impossible to experience this [suggested effect]"?

2. As the alternative paradigm becomes accepted by more researchers, the kind of response to test suggestions that has been traditionally subsumed under the term "hypnosis" will no longer be viewed as closely related to abnormal phenomena such as sleepwalking and fugue states (Gill & Brenman, 1959). Instead, the processes involved in responding to test suggestions will be analyzed in a similar manner as, and will be found partially to overlap with, such social psychological influence processes as conformity, attitude change, and persuasion (Barnlund, 1968; Bettinghaus, 1968; Hartley & Hartley, 1958, pp. 15–158; McGuire, 1969; Secord & Backman, 1964, pp. 93–231). The recent formulation by Sarbin and Coe (in press), which subsumes "hypnotic" behavior under the psychology of influence communication, will be viewed as a major turning point in this area. I will also venture to predict that, in the more distant future, a unified theory of social influ-

ence processes will be used to explain not only conformity, attitude change, and persuasion but also responses to test suggestions and other types of responses that were previously thought to be associated with a qualitatively distinct state (trance).

3. Conceptions of *normal human abilities* or "human potentialities" (Otto, 1966) will be markedly broadened when the subject who is responsive to communications (test suggestions) from an experimenter is seen to be as normal and as awake as the member of the audience who is responsive to communications from the actors. Investigators will no longer think in terms of rare individuals ("somnambulists") who possess unusual capacities, who differ in some basic way from other human beings, who are able to enter a special state ("deep somnambulistic trance"), and who are able to have experiences that other human beings find it very difficult if not impossible to have. On the contrary, investigators will think in terms of a wide range of normal human abilities that can be manifested when individuals adopt positive attitudes, motivations, and expectancies toward the test situation. These abilities, which will be viewed as within the normal human repertoire, will include: the ability to perform feats such as the human-plank feat; the ability to control or block pain (analgesia) by thinking of other things or by thinking of the sensations as sensations; the ability to imagine and to visualize vividly (hallucination and suggested dream); the ability to imagine or fantasy events that occurred at an earlier time (age regression) or that may occur in the future (age progression); and the ability to block or stop thinking about earlier events (amnesia).

4. As Chaves (1968) has pointed out, attempts have already been made to subsume the alternative paradigm under the traditional (trance) paradigm. I expect that further attempts will be made along these lines (Kuhn, 1962). That is, adherents of the traditional paradigm may contend that hypnotic trance (or hypnosis or hypnotic state) refers to (or is) positive attitudes, motivations, and expectancies and thinking with and imagining the suggested effects. To be consistent, adherents of the traditional paradigm may also contend that all procedures, instructions, and experimental manipulations that aim to produce positive attitudes, motivations, etc—for example, task-motivational instructions (Barber & Calverley, 1962), or having the subject first observe another person who is highly responsive to test suggestions (Klinger, 1970)—are actually trance induction procedures. Also, to be consistent, they may contend that the member of the audience who is laughing, crying, empathizing, and so on, as he receives communications from the actors, is also in a hypnotic trance. These contentions will change the meaning of the term "hypnotic trance"; the term will no longer refer to a special state basically different from ordinary states of consciousness. Of course, such attempts to change the meaning of the central construct (trance) will be self-defeating for the traditional paradigm. Since the

construct "hypnotic trance" (or "hypnosis" or "hypnotic state") has always referred to some kind of basic, qualitative change in the organism and has accreted many associated connotations (including connotations of somnambulism or sleepwalking), attempts to give it a new meaning and new connotations will lead to confusion rather than clarity and, sooner or later, the construct and its many associated assumptions will be viewed as a historical curiosity by students of human behavior.

References

Barber, T.X. Toward a theory of pain: Relief of chronic pain by prefrontal leucotomy, opiates, placebos, and hypnosis. Psychological Bulletin, 1959b, 56, 430-460.

Barber, T.X. Physiological effects of "hypnosis". Psychological Bulletin, 1961b, 58, 390-419.

Barber, T.X. Hypnotic age regression: A critical review. Psychosomatic Medicine, 1962a, 24, 286-299.

Barber, T.X. Toward a theory of hypnosis: Posthypnotic behavior. Archives of General Psychiatry, 1962b, 7, 321-342.

Barber, T.X. The effects of "hypnosis" on pain: A critical review of experimental and clinical findings. Psychosomatic Medicine, 1963, 25, 303-333.

Barber, T.X. "Hypnosis" as a causal variable in present-day psychology: A critical analysis. Psychological Reports, 1964a, 14, 839-842.

Barber, T.X. Hypnotic "colorblindness", "blindness", and "deafness": (A review of research findings). Diseases of the Nervous System, 1964b, 25, 529-538.

Barber, T.X. Toward a theory of "hypnotic" behavior: Positive visual and auditory hallucinations. Psychological Record, 1964c, 14, 197-210

Barber, T.X. Measuring "hypnotic-like" suggestibility with and without "hypnotic induction"; psychometric properties, norms, and variables influencing response to the Barber Suggestibility Scale (BSS). Psychological Reports, 1965a, 16, 809-844.

Barber, T.X. Physiological effects of "hypnotic suggestions": A critical review of recent research (1960-64). Psychological Bulletin, 1965b, 63, 201-222.

Barber, T.X. "Hypnotic" phenomena: A critique of experimental methods. In J. E. Gordon (Ed.), 1967, pp. 444-480.

Barber, T.X. Hypnosis: A Scientific Approach. New York: Van Nostrand Reinhold Co., 1969b.

Barber, T.X. *LSD, Marihuana, Yoga, and Hypnosis.* Chicago: Aldine Publishing Co., 1970a.

Barber, T.X. The phenomenology of ('hypnotic') suggestibility. Harding, Mass.: The Medfield Foundation, 1970b.

Barber, T.X., & Calverley, D.S. "Hypnotic behavior" as a function of task motivation. *Journal of Psychology*, 1962, 54, 363-389.

Barber, T.X., & Calverley, D.S. Toward a theory of hypnotic behavior: Effects on suggestibility of defining the situation as hypnosis and defining response to suggestions as easy. *Journal of Abnormal and Social Psychology*, 1964e, 68, 585-592.

Barber, T.X., & Calverley, D.S. Empirical evidence for a theory of hypnotic behavior: Effects on suggestibility of five variables typically included in hypnotic induction procedures. *Journal of Consulting Psychology*, 1965, 29, 98-107.

Barber, T.X., & Calverley, D.S. Toward a theory of "hypnotic" behavior: Replication and extension of experiments by Barber and co-workers (1962-65) and Hilgard and Tart (1966). *International Journal of Clinical and Experimental Hypnosis*, 1968, 16, 179-195.

Barnlund, D.C. *Interpersonal Communication: Survey and Studies.* Boston: Houghton Mifflin, 1968.

Bettinghaus, E.P. *Persuasive Communication.* New York: Holt, Rinehart & Winston, 1968.

Bowers, K.S. Hypnotic behavior: The differentiation of trance and demand characteristic variables. *Journal of Abnormal Psychology*, 1966, 71, 42-51.

Braginsky, B.M., Braginsky, D.D., & Ring, K. *Methods of Madness: The Mental Hospital as a Last Resort.* New York: Holt, Rinehart & Winston, 1969.

Chaves, J.F. Hypnosis reconceptualized: An overview of Barber's theoretical and empirical work. *Psychological Reports*, 1968, 22, 587-608.

Collins, J.K. Muscular endurance in normal and hypnotic states: A study of suggested catalepsy. Honors thesis, Department of Psychology, University of Sydney, Australia, 1961.

Edmonston, W.E., Jr., & Robertson, T.G., Jr. A comparison of the effects of task motivational and hypnotic induction instructions on responsiveness to hypnotic suggestibility scales. *American Journal of Clinical Hypnosis*, 1967, 9, 184-187.

Erickson, M.H. *Advanced Techniques of Hypnosis and Therapy: Selected Papers of Milton H. Erickson.* (Edited by J. Haley) New York: Grune & Stratton, 1967a.

Evans, F.J. Recent trends in experimental hypnosis. *Behavioral Science*, 1968, 13, 477-487.

Gill, M.M., & Brenman, Margaret. *Hypnosis and Related States: Psychoanalytic Studies in Regression.* New York: International Universities Press, 1959.

Gordon, J.E. (Ed.) *Handbook of Clinical and Experimental Hypnosis.* New York: Macmillan, 1967.

Hartley, E.L., & Hartley, Ruth E. *Fundamentals of Social Psychology.* New York: Knopf, 1958.

Hilgard, E.R. Altered states of awareness. *Journal of Nervous and Mental Disease*, 1969a, 149, 68-79.

Hilgard, E.R., & Tart, C.T. Responsiveness to suggestions following waking and imagination instructions and following induction of hypnosis. *Journal of Abnormal Psychology*, 1966, 71, 196-208.

Klinger, B.E. Effect of peer model responsiveness and length of induction procedure on hypnotic responsiveness. *Journal of Abnormal Psychology*, 1970, 75, 15-18.

Kuhn, T.S. *The Structure of Scientific Revolutions.* Chicago: University of Chicago Press, 1962.

McGuire, W.J. The nature of attitudes and attitude change. In G. Lindzey & E. Aronson (Eds.), *The Handbook of Social Psychology, III: The Individual in a Social Context.* (2nd edition). Reading, Mass.: Addison-Wesley Publishing Co., 1969, pp. 136-314.

Meares, A. *Theories of hypnosis.* In J. M. Schneck (Ed.), 1963a, pp. 390-405.

Orne, M.T. The nature of hypnosis: Artifact and essence. *Journal of Abnormal and Social Psychology*, 1959, 58, 277-299.

Otto, H.A. *Explorations in Human Potentialities.* Springfield, Ill.: Charles C. Thomas, 1966.

Sarbin, T.R. Contributions to role-taking theory: I. Hypnotic behavior. *Psychological Review*, 1950, 57, 255-270.

Sarbin, T.R. Physiological effects of hypnotic stimulation. In R. M. Dorcus, (Ed.), *Hypnosis and its Therapeutic Applications.* New York: McGraw-Hill, 1956, pp. 1-57.

Sarbin, T.R. Attempts to understand hypnotic phenomena. In L. Postman (Ed.), *Psychology in the Making: Histories of selected research problems.* New York: Knopf, 1962, pp. 745-785.

Sarbin, T.R. Role theoretical interpretation of psychological change. In P. Worchel & D. Byrne (Eds.), *Personality Change.* New York: Wiley, 1964, pp. 176-219.

Sarbin, T.R., & Andersen, M.L. Role-theoretical analysis of hypnotic behavior. In J. Gordon (Ed.), 1967, pp. 319-344.

Sarbin, T.R., & Coe, W.C. Hypnotic Behavior: The Psychology of Influence Communication. New York: Holt, Rinehart & Winston, (in press).

Schneck, J.M. (1953). Hypnosis in Modern Medicine. Springfield, Ill.: C. C. Thomas, third edition, 1963a.

Secord, P.F., & Backman, C.W. Social Psychology. New York: McGraw-Hill, 1964.

Shor, R.E. Three dimensions of hypnotic depth. International Journal of Clinical and Experimental Hypnosis, 1962, 10, 23-38.

Sutcliffe, J.P. "Credulous" and "skeptical" views of hypnotic phenomena: A review of certain evidence and methodology. International Journal of Clinical and Experimental Hypnosis, 1960, 8, 73-101.

The Effect of Suggestion on Visual Acuity

31

Charles Graham and Herschel W. Liebowitz

Abstract: The effect of positive suggestion on myopic, or nearsighted, visual acuity was assessed under optimum physiological conditions. The acuity of hypnotically susceptible myopes significantly improved after presentation of direct hypnotic and posthypnotic suggestion, with improvement being greatest initially and the hypnotic procedure most effective for those Ss with the poorest acuity. Transfer to the normal waking state outside the experimental situation was demonstrated in this instance. Similar suggestions given without hypnosis to different but equally susceptible myopes resulted in an equal magnitude of visual improvement in the laboratory but little transfer outside the experimental situation. No significant effects were demonstrated when testing hypnotically insusceptible myopes or Ss with normal or hyperopic (farsighted) vision. Optometric examinations, conducted before, during, and after the experimental treatment, indicated that alterations in the refractive power of the eye could not account for the dramatic changes in visual acuity observed.

Manuscript submitted March 1, 1971.

[1] The research was conducted at The Pennsylvania State University and was supported by the Pennsylvania Transportation and Traffic Safety Center and by grant MH 08061 from the National Institute of Mental Health. The final version of this manuscript was prepared at the Unit for Experimental Psychiatry, Institute of the Pennsylvania Hospital, and supported in part by Contract Nonr-4731 (00) from the Office of Naval Research. Reprint requests should be sent to Dr. Graham, Unit for Experimental Psychiatry, 111 North 49th Street, Philadelphia, Pa. 19139.

[2] We wish to acknowledge the assistance of Dr. G. B. Stein, optometrist, for conducting the visual examinations outside the laboratory; and also of Richard A. Olsen, Nina A. Lasserson, and Robert T. Hennessy during various phases of the study. We also wish to thank Frederick J. Evans, A. Gordon Hammer, Emily C. Orne, and Martin T. Orne for their helpful comments during the preparation of this report.

Reprinted from the July 1972 *International Journal of Clinical and Experimental Hypnosis* (Vol. 20, 169-186). Copyrighted by The Society for Clinical and Experimental Hypnosis, July 1972.

Myopia, or nearsightedness, is a common visual abnormality resulting from excessive refractive power of the lens and cornea in relation to the length of the eyeball, i.e., the image falls in front of the retina, the plane of best focus. The simple and effective correction of the condition is to wear spectacles to reduce the refractive power, or error, of the eye. A number of studies, however, have indicated that the visual performance of myopes may be improved without corrective lenses, through optometric training techniques (Hildreth, Meinberg, Milder, Post, & Sanders, 1947; Woods, 1946). In their review of the literature, Sells and Fixott (1957) concluded that these techniques do not alter the refractive power of the eye and that the mechanism probably involves more adequate utilization of perceptual cues available in the test situation.

Not so easily accounted for, however, is the transcendence of visual functioning which is reported to occur while under hypnosis. Improvement, in this instance, is not based on long and involved training procedures but occurs almost immediately as a result of direct or indirect suggestion and in some cases, spontaneously, when testing in different experimental contexts. Erickson (1943) and LeCron (1952), for example, report that hypnotic age regression to a period prior to the wearing of spectacles improved both near and far vision. A transient improvement of visual acuity was also observed by Browning & Crasilneck (1957) while working with nine cases of suppression amblyopia. Spontaneous improvement of acuity under hypnosis has been reported by Weitzenhoffer (1951) and also by Kline (1953) when testing Ss on unrelated tasks.

The mechanism of such improvement is not known. It has been postulated that the increased ability of the good hypnotic S to focus his attention is the relevant variable (Crasilneck & Hall, 1959). However, in a series of carefully designed optometric studies, Kelley (1958) has shown that significant improvement through suggestion occurs just as easily and quickly in waking Ss as in hypnotized Ss. He presents evidence that suggestion acts to reduce the refractive power of the eye. A recent case study by Davison and Singleton (1967), however, questioned traditional views concerning visual acuity when it was demonstrated that visual acuity can be improved under hypnosis in the absence of alterations in refractive power.

There are a number of questions raised by a careful reading of the literature. The acuity tests involved did not always provide control of a number of relevant parameters which, while not critical clinically, must be taken into account in experimental studies. Secondly, the evidence regarding refractive power is obscured by the previous technical inability to assess refractive power adequately *during* suggested improvement of visual acuity. In view of this, the objectives of the present study were: (a) to determine whether myopic visual acuity can be significantly improved through the use of hypnosis and suggestion under rigorous laboratory conditions and (b) to determine whether there are simultaneous changes in the refractive power of the eye using the recently developed laser scintillation technique (Hennessy & Leibowitz, 1970).

In line with these objectives, the present study examined the effectiveness of suggestions given to highly hypnotizable individuals while in deep hypnosis, both within the laboratory and outside the experimental situation. The generalizability of the phenomenon to a population differing in susceptibility to hypnosis was tested as well as whether suggestions would be effective when given without a prior induction of hypnosis. The refractive power of the eye was assessed simultaneously with suggested improvement of myopic visual acuity using the laser technique. These parameters were sequentially investigated in the experiments to be described below.

Experiment I

Subjects

The nine volunteers for the experimental group were selected from the student population according to the following criteria: (a) attainment of the maximum objective score on the Barber Suggestibility Scale (BSS; Barber, 1969), (b) a subjective refractive error which did not vary more than 0.5 diopter from the static refractive error, i.e., true myopia,[3] (c) less than 1.5 diopter of astigmatism. The four male and five female Ss had no prior experience with hypnosis and constituted approximately 20% of those who had volunteered for screening. Three Ss were highly myopic (−4.00, −2.25, and −1.75 diopter), three were slightly myopic (−0.75, −0.50, −0.25 diopter), while the remaining three exhibited no myopia (0.00, + 0.25, + 0.50 diopter). For cases in which the two eyes differed in refractive error (one eye requiring more correction than the other), the above values represent the eye with the lesser error.

The Ss were first screened for hypnotic susceptibility, then given a visual examination in the office of a local optometrist. This examination included an initial determination of the present spectacle prescription. Acuity and astigmatism were assessed with and without corrective

---

[3] Subjective refractive error is the best discrimination made by S while reading a standard clinical acuity chart. Static, or objective refractive error is the plane within the relaxed eye at which the light rays from a distant target are actually focused, as determined by retinoscopic examination. True myopia is indicated when close agreement exists between the subjective and objective measurements.

lenses, both objectively and subjectively. A final test evaluated phoria (degree of eye muscle imbalance) and amplitude of accommodation at the far and near point. The experimental procedure involved three weekly sessions in the psychology laboratory, followed by an identical examination in the optometrist's office. All optometric examinations were conducted in the normal waking state.

Two control groups, composed of volunteer students who habitually wore spectacles, were selected on the basis of self-report of myopia. The first group ($N = 5$) allowed an evaluation of possible visual improvement in the laboratory due solely to memorization of the acuity chart, and the second group ($N = 4$) provided information on the contribution of motivational and performance feedback during suggested visual improvement. These Ss received neither hypnotic screening nor optometric examination.

Apparatus

Because the standard clinical Snellen letters are not equally discriminable (Duke-Elder, 1942), the Landolt "C" was used as the test object. A special chart consisting of black, single break Landolt Cs viewed against a white background was produced photographically and arranged in 19 rows of 10 Cs each. Each C had eight possible positions, vertical and at 45 degree intervals. These were reported by reference to a cue card consisting of a sample C illustrating the possible break positions with corresponding numbers. The break in the Cs ranged from 0.5 to 9.8 minutes of arc (20/10 to 20/200) at the viewing distance of 20 ft.

The greatest variation in acuity results from changes in illuminance (Shlear, 1937). Similarly, contrast between the target and background (Cobb & Moss, 1927) is also an important variable. In order to maintain these parameters at their optimal values, the luminance level of the light portion of the acuity test chart was maintained at 100 ft. L, which is on the flat portion of the acuity-luminance function. The contrast between the test object and background was 7 to 1. No specular reflection was visible from S's position.

Procedure

The S, without his glasses and in the normal waking condition, was seated in a comfortable chair and viewed the acuity chart binocularly from a distance of 20 ft. An observer was present to counter any tendency toward squinting or head movement. In order to familiarize S with the reporting procedure, he was instructed to read the 20/200 line first, and then to read the chart from left to right, beginning with the line above the lowest line he reported seeing clearly and distinctly.

Upon indication that he could no longer make the necessary discriminations, he was told to close his eyes and relax as this would enable him to see more clearly. The S was then told to open his eyes and continue reporting, or, if necessary, to guess the break positions in the next lower line. If the percentage correct was at least 50% he was asked to continue. Every effort was made to obtain an accurate measure of the maximum acuity. At no time was S rushed to respond more quickly than he wished. Upon failure to meet the 50% criterion, S was asked to replace his glasses, close his eyes and relax, after which acuity was again determined in the same manner as above.

The chart lights were then turned off and S, now without his glasses, faced away from the chart and was hypnotized. After the induction, specific suggestions were given to feel completely relaxed, refreshed, and alert, to concentrate on the muscles around and behind the eyes and notice them becoming as relaxed and weightless as the rest of the muscles in the body. It was explained that the degree of relaxation attained would affect the muscles which controlled the lens, thus changing the focus of the eye and permitting clearer vision.

The chart lights were turned on and the acuity testing procedure without glasses repeated until performance fell below the 50% criterion. The final acuity test was then made in the hypnotic condition while wearing glasses.

Prior to awakening S, a posthypnotic suggestion was given to the effect that he now knew how well he really could see and that this was dependent on his ability to relax his eye muscles. The "control" over these muscles was given to him by means of a signal (1 ... 5, relax) which he would want to practice during the following week. It was explained that the improvement would transfer to the waking state. The S was then awakened and given performance feedback in terms of percentage improvement over the four experimental conditions. It was stressed that the improvement was contingent upon his ability to relax his eye muscles and that hypnosis was simply a tool which enabled him to do this more effectively. Further improvement could be expected if he practiced.

This procedure was repeated twice again at weekly intervals, after which S reported to the optometrist for an examination identical to that given prior to the experimentation. The optometrist had no information regarding the results of the experimental treatment.

Since S viewed the same acuity test chart during all three sessions, the possibility exists that some memorization could have taken place. As a more stringest test of improvement due to memorization, the acuity of Control Group I was tested for three *consecutive* days in the same manner as the experimental group. A relaxation period was substituted for the hypnotic induction. No information was provided as to the correctness of responses nor was the group given the rationale for visual improvement. Control Group II was tested under the same duration conditions, but received limited information in that they were told previous Ss tested in the same situation had improved with practice and, in addition, this group was shown their results at the conclusion of each daily session.

Results

Acuity in terms of threshold visual angle for the three sessions is plotted for the experimental group in Figure 1 and for the control groups in Figure 2. These data reflect the test condition without glasses and are plotted on a logarithmic scale to emphasize relative changes. For both the highly and slightly myopic groups, the hypnosis procedure produces a lowering of the threshold visual angle (i.e., improvement in vision). There is also a marked improvement over sessions for the highly myopic Ss. Moderately myopic Ss showed only a slight improve-

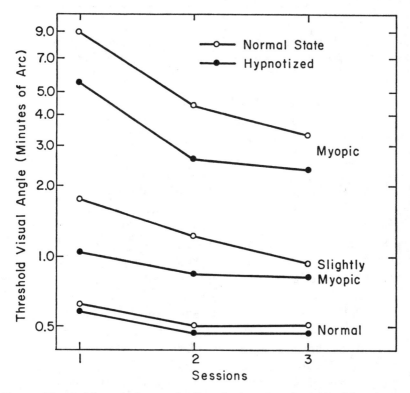

FIG. 1. Threshold resolution angle (in minutes of arc) attained by the experimental groups in the waking and hypnotic conditions in Experiment I.

ment over sessions. For the normal Ss, neither hypnosis nor repetition of sessions made much difference in acuity.

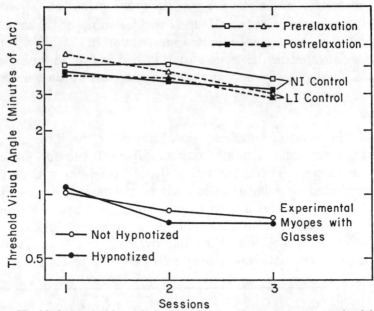

FIG. 2. Unaided threshold resolution angle (in minutes of arc) attained by the no information (NI) and limited information (LI) control groups during pre- and postrelaxation conditions and by all experimental myopes while wearing glasses in Experiment I.

It should be noted that the threshold visual angle for the normal Ss in the present study is 0.5 minute of arc. This is the physiological limit of visual acuity as determined theoretically by the separation of the cones in the central fovea (Shlear, 1937). The fact that the theoretically best resolution angle was obtained for these Ss confirms that acuity was tested under optimum physiological conditions.

Figure 2 presents the data for the control Ss. A slight improvement in acuity can be observed over sessions but it is not comparable to the changes observed with the experimental myopic group.

The results of an analysis of variance reflect the efficacy of the hypnotic procedure in improving myopic visual acuity, both within a given session ($F = 14.33, p < .01$) and as a function of sessions ($F = 6.05, p < .01$). Since both the myopic experimental and control groups improved, the groups were compared directly. Improvement under hypnosis was significantly greater ($t = 2.89, p < .01$) than improvement after rest for both control groups.

Acuity as measured in the optometrist's office also showed an improvement, as indicated in Table 1. This result was due mainly to the highly myopic group who appear to have transferred the experience gained in the laboratory. Changes in the optometrist's office, however, were not as great as in the laboratory and these were not reflected by the refractive error data. No changes were observed in any of the other visual measures taken.

Thus, Experiment I indicated that myopic visual acuity was significantly improved through the use of hypnosis and positive suggestion. Neither memorization nor feedback could account for the observed effect. The improvement in acuity was not accompanied by concomitant changes in the refractive power of the eye when measured outside the laboratory.

TABLE 1

MEASUREMENT OF UNAIDED VISUAL ACUITY OBTAINED DURING OPTOMETRIC EXAMINATION BEFORE AND AFTER THE LABORATORY PROCEDURE IN EXPERIMENT I

Degree of Myopia	Subjects	Unaided Visual Acuity			
		Right eye		Left eye	
		Before	After	Before	After
Emmetropic (normal)	1	20/15	20/15−	20/25	20/25−
	2	20/15	20/15	20/15	20/15
	3	20/15	20/15	20/20+	20/15
Slightly myopic	4	20/200	20/200	20/40	20/30−
	5	20/200	20/80+	20/300	20/100−
	6	20/30−	20/30−	20/25−	20/30−
Highly myopic	7	Fingers 5'	20/300	Fingers 9'	20/200
	8	20/400	20/300	20/400	20/400−
	9	Fingers 7'	20/400−	Fingers 7'	20/400

Note.—20/20 acuity is based on a resolution angle of one minute of arc. The threshold angle can be computed from the above Snellen Notations by dividing the numerator into the denominator. The notation "fingers" indicates vision impaired to the extent of only being able to discriminate the number of fingers held up at the indicated distance.

EXPERIMENT II

It has been demonstrated that suggestions given to highly susceptible Ss while in deep hypnosis resulted in improvement of visual acuity. Further investigation was needed to determine whether the phenomenon is confined to highly susceptible Ss only or is dependent on a prior hypnotic induction. Kelley (1958) has suggested that neither susceptibility nor hypnosis is a necessary prerequisite. However, of the 14 Ss tested in his study, only one was considered deeply hypnotized and this

S performed the acuity tests in an atypical fashion. Experiment II explores in more detail the generalizability of the phenomenon to individuals who differ in their susceptibility to hypnosis and the extent to which suggestions are effective when administered in the absence of hypnosis.

Subjects

Eleven college-age volunteers, who had not participated in Experiment I, were divided into hypnotically susceptible (maximum objective score of eight on the BSS) and insusceptible (objective score of zero on the BSS) groups. The visual criteria were the same as those of the myopic group in Experiment I, and all underwent the same optometric examination. The six Ss in the susceptible group and the five in the insusceptible group had no previous experience with hypnosis.

Procedure

The apparatus and procedure were the same as in Experiment I with the following exceptions:

1. After the initial test of hypnotic susceptibility, administered by E outside of the experimental laboratory, no S in either group received a hypnotic induction for the remainder of the experiment. All testing in the laboratory and the optometrist's office occurred in the normal waking state.

2. In Experiment I Ss viewed the acuity chart four times in each session, twice with their glasses on and twice with them off. As a more stringent test, Ss in the present study saw the chart only with their glasses off. They viewed it initially without glasses and, after eye closure and relaxation for a period of time corresponding to the hypnotic induction, again viewed the chart without glasses. No S ever saw the chart while wearing glasses or from a distance less than 20 ft.

3. The suggestions given to each group differed in the following respects. It was explained to both groups that relaxation of the lens muscles could lead to visual improvement through a change in the focal point of the lens. This rationale was identical to that given to the myopes in the first experiment. However, for the susceptible Ss, it was further stated that various studies had demonstrated that being hypnotized was not a prerequisite for obtaining improvement. The important factor was the increased capacity for relaxation possessed by good hypnotic Ss, and for these individuals it would be an easy matter to improve their vision in a waking relaxed state. The insusceptible Ss were told that acuity improved under hypnosis, but like many other phenomena associated with hypnosis, improvement in vision was also

well within the capabilities of the nonhypnotizable S if he merely learned to relax his eyes. Great importance was attached to relaxation, and emphasis was placed on the equivalence of susceptible and insusceptible Ss in improving their vision. Both groups were asked to practice eye-muscle relaxation between weekly experimental sessions.

Results

The data are plotted in Figure 3. Approximately the same magnitude of improvement occurred for the susceptible Ss who were not hypnotized as for the hypnotized Ss in Experiment I. As in the previous study, positive changes as a function of sessions were observed for both the susceptible and insusceptible Ss. Comparison of the experimental performances in the first prerelaxation session and in the third postrelaxation session indicated significant improvement ($F = 11.53$, $p < .01$) in acuity for the susceptible group. For the insusceptible Ss, the improvement is not significant ($F = 1.91$, $p > .20$). Results of the optometric examinations given before and after the experimental treatment were similar to those obtained in Experiment I, i.e., no significant change or trend was evident to account for the improvement of the susceptible Ss. This group, however, did not demonstrate the degree of transfer evident in Experiment I.

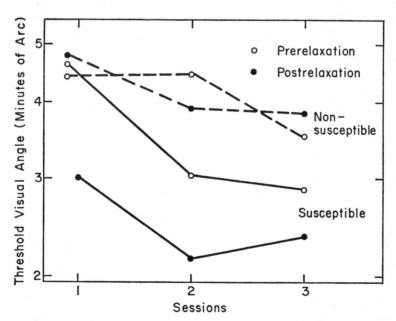

FIG. 3. Threshold resolution angle (in minutes of arc) attained by the hypnotically susceptible and insusceptible groups in Experiment II when tested without hypnosis in the laboratory.

Thus, Experiment II indicated that myopic visual acuity was significantly improved in the absence of a formal hypnotic induction. This improvement occurred for hypnotically susceptible Ss only and did not transfer to outside the experimental situation. Again, no change in the refractive power of the eye was observed in the optometrist's office.

Experiment III

The present data confirm previous observations that visual acuity can be improved, at least temporarily, by suggestion. The question which remained to be answered concerned the mechanism of this improvement. A logical possibility is a change in the magnitude of accommodation of the eye, which would easily account for the effects observed in this and previous experiments. The range over which a myopic S can, with effort, keep an object in focus does not extend as far into space as does the range of an individual with normal vision; thus, distant objects are seen indistinctly. A shift in this range, which is essentially the rationale given to Ss in the previous phases, could conceivably account for the observed improvement.

The method of choice is to evaluate accommodation *simultaneously* with the acuity measurements. Unfortunately previous techniques were not completely satisfactory in this respect. The common use of the retinoscope introduces E into the visual field and the flashing of a light into S's eye. The procedure, in other words, introduces distracting influences into the test situation. Thus, an observed change in accommodation may simply be the result of S's shifting his focus from the distant vision, which is required in the test procedure, to a focus for near vision, in order to take note of the actions of the examiner. Recently, a method has been developed—the laser scintillation technique—which obviates these difficulties.

The diverged beam from a laser has a peppery, granular appearance caused by the constructive and destructive interferences of light from the coherent source. When S moves his head, or the surface on which the image is reflected is moved, the granular pattern may also seem to move. If S is overaccommodated (myopic), he will perceive "against" motion, i.e., the apparent motion of the granular pattern is in opposition to the motion of the head. If underaccommodated, S will perceive the motion to be in the same direction of the head movement. When S is accommodated for the ideal focal plane, no motion or an indistinct swirling motion will be apparent. Through the use of suitable lenses the motion of the granular pattern can be neutralized and the lens power required taken as a measure of the refractive state of the eye.

As used in this study, the laser beam was flashed in the visual field after being reflected from a rotating drum and S was required to indicate the direction of movement of the granularity. The details of the procedure and apparatus employed to determine refractive error are described elsewhere (Hennessy & Leibowitz, 1970). In a control experiment, in which accommodation was monitored by means of a continuously recording infrared optometer, the accommodative response was found to be relatively slow. Thus, the measurement indicated by the laser technique represents the accommodation in force at the time of the brief presentation of the laser. If the laser technique is employed during a period of improved vision, it is therefore possible to determine if concomitant changes occur in accommodation.

Subjects

Five college-age volunteer Ss, who had previously attained the maximum score on both the Harvard Group Scale of Hypnotic Susceptibility, Form A (Shor & Orne, 1962), and the BSS, were tested. All were optically corrected myopes with low levels of astigmatism and no previous experience with the experiment.

Procedure

The procedure involved three daily sessions. The first session consisted of refraction determination in the waking state using the laser; the second and third, of simultaneous refraction and acuity measurements taken in both the waking and hypnotic conditions. The apparatus was the same as in the first two phases of the experiment with the addition of the laser.

Simultaneous measurement was accomplished by having S read the acuity chart and, while doing so, report the direction of movement within the randomly presented laser pattern which appeared to be superimposed briefly on selected portions of the acuity chart. The Ss were first tested without their glasses in the normal waking state, then hypnotized and given the rationale for visual improvement used in the first two phases of the experiment. Measurements were than taken in the hypnotic condition. No posthypnotic suggestions were given.

Results

Table 2 presents the acuity measurements taken while awake and during hypnosis for all Ss. Values larger than one in the table indicate progressively greater degrees of myopia. In line with the findings in the two previous studies, visual acuity was seen to improve, being significant in the second session ($p = .03$) but not over all sessions, due to the

TABLE 2
VISUAL ACUITY IN TERMS OF THE THRESHOLD RESOLUTION ANGLE (MINUTES OF ARC) FOR MYOPIC Ss UNDER NORMAL AND HYPNOTIC VIEWING CONDITIONS IN EXPERIMENT III

Subjects	2nd Session		3rd Session	
	Awake	Hypnotic	Awake	Hypnotic
1	6.18	4.08	4.89	3.61
2	4.88	3.61	4.08	2.99
3	2.03	1.30	1.50	1.00
4	1.69	0.90	1.30	0.90
5	1.69	1.50	1.69	1.96
$\bar{X}$	3.29	2.28	2.69	2.09

TABLE 3
SPHERICAL REFRACTIVE ERROR (IN DIOPTERS) FOR MYOPIC Ss TESTED WHILE AWAKE AND DURING HYPNOSIS IN EXPERIMENT III

Subjects	1st Session Awake	2nd Session Hypnotic	3rd Session Hypnotic
1	−1.99	−2.12	−2.06
2	−1.62	−1.62	−1.56
3	−0.69	−0.62	−0.76
4	−0.69	−0.74	−0.68
5	−0.69	−0.31	−0.62
$\bar{X}$	−1.13	−1.08	−1.14

reversal of one S. The rank-order correlation between initial and final acuity levels was .975 ($p < .001$), indicating the effect of suggestion was selective, the hypnotic procedure being more effective for the more myopic.

Table 3 presents the data on refractive error which is relevant to the physiological purpose of this study. Negative values in the table represent progressively greater degrees of myopia. As can be seen, there were no more changes in refractive error for the myopes who obtained a large improvement in vision than for those who did not. The individual changes in refraction which did occur were neither large enough nor consistent enough to account for the improved acuity. Although Ss reported feelings of deep visual relaxation in responding to the hypnotic suggestion, this was not translated into any appreciable change from their waking refractive state. Thus, over all the experiments reported here, refractive error was seen not to change before, during, or after improvement in visual acuity.

Discussion

The findings of the present study confirm earlier reports concerning improvement of myopic visual acuity under hypnosis. Direct hypnotic suggestion, in conjunction with posthypnotic suggestion, led to a significant improvement in myopic vision. The magnitude of improvement was greatest during the initial stages of treatment, and the procedure proved most effective for those individuals with the poorest initial acuity. When the hypnotic induction was part of the experimental treatment, acuity improved in the laboratory and, when tested in the optometrist's office, was found to have transferred to the normal waking state outside the experimental situation. Without the formal hypnotic induction, acuity improved in the laboratory but did not transfer. It cannot be inferred, however, that hypnosis is the significant variable in achieving transfer of visual improvement. Differences undoubtedly exist in motivation for participation in the study, the diligence with which the relaxation exercises were practiced, and the relevance improved vision held for individual Ss. Since these factors were not assessed in the present study, the question remains open. The experimental treatment had no significant effect on those individuals with normal or farsighted vision.

In the hypnotic condition, Ss seemed to respond in a more confident manner, answering more quickly and with fewer hesitations. Their eyes did not appear different; no rigidity or dullness of gaze was evident. However, Ss reported that for brief periods of time the chart would appear "crystal clear," then rapidly become blurred again. These "flashes of clear vision" were not confined solely to the hypnotic condition. They occurred for all myopic Ss in both the waking and hypnotized conditions and have been reported previously in the literature (Gregg, 1946; Kelley, 1958; Marg, 1952). No adequate explanation of this phenomenon has as yet been advanced. The Ss tested under hypnosis attributed the "flashes" to hypnosis; those tested in the waking state attributed them to the degree of eye muscle relaxation. The phenomenon was used under both conditions to effectively reinforce the rationale.

The necessity of a formal hypnotic induction was questioned when motivated Ss, who were susceptible to hypnosis, also obtained a significant improvement when tested in the normal waking state in the laboratory. However, the relationship between suggested improvement and hypnotic susceptibility needs further study since E was aware, and, more important, S was aware, that the individual conducting the experiment knew who was and who was not susceptible to hypnosis. The problems of E-bias and S-motivation in the experimental situation require clarification.

An important point underlying the present study is that vision involves a series of complex processes, only part of which are anatomical. It is not the refractive state of the eye alone, but the interaction of the dioptric and neural mechanisms which enables us to see as well as we do. This interaction was apparent when it was demonstrated that although the eye remained essentially constant before, during, and after the experimental treatment in terms of where it focused the image of the acuity targets, i.e., no change occurred in refractive power, vision improved significantly.

The rationale given the myopic Ss required that they relax the muscles controlling the shape of the lens. Subjectively, Ss reported experiencing a sensation of relaxation in their eyes. Objectively, however, the laser technique indicated no significant change in the refractive power of the eye. Thus, the mechanism involved in improvement was not the one suggested by the rationale. The validity of the rationale, however, was not the criterion for its selection. The induction procedure employed muscle relaxation suggestions, and the rationale was viewed by E merely as a convenient and logical extension of these suggestions.

Various other explanatory mechanisms have also not been supported by the present study or in previous research. Vision may be improved through a reduction in pupillary diameter since this reduction decreases the effect of optical aberrations. In the present study, pupillary size was maintained at the minimum diameter at the luminance levels used (Leibowitz, 1952). Visual improvement has also occurred under controlled laboratory conditions when the pupil has been expanded and held constant through the use of mydriatic agents (Kelley, 1958) or, conversely, when Ss were required to view the acuity targets through a pinhole (Copeland, 1967). These studies also exclude explanations involving squinting or sighting over the bridge of the nose, both of which will improve vision. Davison and Singleton (1967) report that acuity improves under hypnosis when the muscles which control the accommodation response are paralyzed by cycloplegic drugs. The majority of studies in the area report no change in refractive power which is consistent or large enough to account for the obtained improvement. These findings render explanations involving a change in the axial length of the eyeball, the curvature of the cornea, or the shape of the lens all the more dubious.

Neither memorization nor performance feedback could account for the magnitude of improvement obtained by the myopic Ss in the present study. Further evidence contrary to an explanation involving memorization is provided by Kelley (1958) and Davison and Singleton (1967) who demonstrated an increase in acuity using different acuity targets and multiple acuity charts, respectively. The issue has been raised that susceptible Ss may unconsciously perform below capacity in the waking state, knowing that their performance is to be tested under hypnosis at a later time (Zamansky, Scharf, & Brightbill, 1964). This effect is not likely to be relevant to the results obtained in the present study since susceptible Ss, irrespective of exposure to hypnosis, obtained a significant improvement in vision. In addition, the visual capacity of the myopes who participated in the present study was assessed during the optometric examination conducted independently and prior to the experimental treatment. The population thus consisted of true myopes who had been recently and correctly prescribed for.

In general, the findings of the present study parallel the results obtained through longer and more complicated optometric training procedures, suggesting perhaps, that the myopic individual sets his internal standard of daily visual performance lower than necessary. Through failure to see distant objects as clearly as desired he may become progressively more dependent upon the corrective lenses and no longer try to exceed the internal standard, except under unusual circumstances. In this regard, Helson (1964) has demonstrated that internal criteria play a significant role in determining performance levels in a variety of situations. This possibility does not, of course, answer the basic question of how the suggested improvement of myopic visual acuity is mediated physiologically. Whatever the mechanism, it can not be fitted neatly into the traditional schema of a causal relationship among factors in the peripheral visual anatomy. These explorations in the use of hypnosis and suggestion have demonstrated the existence of as yet unexplained aspects of the visual system. Ultimately, investigation of the more central components of the system may lead to therapeutic techniques of greater effectiveness.

References

BARBER, T. X. *Hypnosis: A scientific approach.* New York: Van Nostrand, 1969.

BROWNING, C. W., & CRASILNECK, H. B. The experimental use of hypnosis in suppression amblyopia: A preliminary report. *Amer. J. Ophthal.*, 1957, *44*, 468–476.

COBB, P. W., & MOSS, F. K. The relation between extent and contrast in the liminal stimulus for vision. *J. exp. Psychol.*, 1927, *10*, 350–364.

COPELAND, V. L. Increased visual acuity of myopes while in hypnosis. *J. Amer. optomet. Ass.*, 1967, *38*, 663–664.

CRASILNECK, H. B., & HALL, J. A. Physiological changes associated with hypnosis: A review of the literature since 1948. *Int. J. clin. exp. Hypnosis*, 1959, *7*, 9–50.

DAVISON, G. C., & SINGLETON, L. A preliminary report of improved vision under hypnosis. *Int. J. clin. exp. Hypnosis*, 1967, *15*, 57–62.

DUKE-ELDER, W. S. *Text-book of ophthalmology.* Vol. 1. London: Henry Klimpton, 1932.

ERICKSON, M. H. Hypnotic investigation of psychosomatic phenomena: Psychosomatic interrelationships studied by experimental hypnosis. *Psychosom. Med.*, 1943, *5*, 51–58.

GREGG, J. R. Variable acuity. Paper presented at the meeting of the West. Visual Train. Conf., Los Angeles, August 1946.

HELSON, H. *Adaptation-level theory: An experimental and systematic approach to behavior.* New York: Harper & Row, 1964.

HENNESSY, R. T., & LEIBOWITZ, H. W. Subjective measurement of accommodation with laser light. *J. opt. Soc. Amer.*, 1970, *60*, 1700–1701.

HILDRETH, H. R., MEINBERG, W. H., MILDER, B., POST, L. T., & SANDERS, T. E. The effect of visual training on existing myopia. *Amer. J. Ophthal.*, 1947, *30*, 1563–1576.

KELLEY, C. R. Psychological factors in myopia. Unpublished doctoral dissertation, New School for Social Research, 1958.

KLINE, M. V. The transcendence of waking visual discrimination capacity with hypnosis: A preliminary case report. *Brit. J. med. Hypnotism*, 1952–53, *4*, 32–33.

LECRON, L. M. A study of age regression under hypnosis. In L. M. LeCron (Ed.), *Experimental hypnosis.* New York: Macmillan, 1952. Pp. 155–174.

LEIBOWITZ, H. W. The effect of pupil size on visual acuity for photometrically equated test fields at various levels of luminance. *J. opt. Soc. Amer.*, 1952, *42*, 416–422.

MARG, E. "Flashes" of clear vision and negative accommodation with reference to the Bates method of visual training. *Amer. J. Optom.*, 1952, No. 128.

SELLS, S. B., & FIXOTT, R. S. Evaluation of research on effects of visual training on visual functions. *Amer. J. Ophthal.*, 1957, *44*, 230–236.

SHLEAR, S. The relation between visual acuity and illumination. *J. gen. Physiol.*, 1937, *21*, 165–188.

SHOR, R. E., & ORNE, E. C. *Harvard Group Scale of Hypnotic Susceptibility, Form A.* Palo Alto, Calif.: Consulting Psychologists Press, 1962.

WEITZENHOFFER, A. The discriminatory recognition of visual patterns under hypnosis. *J. abnorm. soc. Psychol.*, 1951, *46*, 388–397.

WOODS, A. C. Report from the Wilmer Institute on the results obtained in the treatment of myopia by visual training. *Amer. J. Ophthal.*, 1946, *29*, 28–57.

ZAMANSKY, H. S., SCHARF, B., & BRIGHTBILL, R. The effect of expectancy for hypnosis on prehypnotic performance. *J. Pers.*, 1964, *32*, 236–248.

Hypnotic Control of Peripheral Skin Temperature

Christina Maslach, Gary Marshall, and Philip Zimbardo

ABSTRACT

In an exploratory study on the specificity of autonomic control, subjects attempted to simultaneously change the skin temperature of their two hands in opposite directions. Subjects who were trained in hypnosis were successful in achieving this bilateral difference, while waking control subjects were not. These findings demonstrate the powerful influence that cognitive processes can exert on the autonomic nervous system and also suggest the possibility of more effective therapeutic control of psychosomatic problems.

DESCRIPTORS: Hypnosis, Skin temperature, Autonomic control, Psychosomatic medicine.

Maintenance of a relatively constant level of body temperature is a vital physiological function. It is so efficient and automatic that we become aware of the process only when pathological internal conditions cause us to react with fever or chills, and when extremes of environmental conditions markedly alter the skin temperature of our limbs. To what extent can such a basic regulatory function be brought under volitional control?

Luria (1969) performed an experiment dealing with this question, in which he studied the mental feats of a man with eidetic imagery. Apparently, his subject could induce such vivid visual images that they exerted a profound influence on his behavior. When he was instructed to modify the skin temperature in his hands, he was able to make one hand hotter than it had been by two degrees, while the other became colder by one and a half degrees. These bilateral changes were attributed by the subject to the "reality" of his visual images, which consisted of putting one hand on a hot stove while holding a piece of ice in the other hand. Is such a phenomenon replicable with "normal" individuals not born with the remarkably developed eidetic ability of this man? We were led to

---

This study was financially supported by an Office of Naval Research grant NOOO 14-67-A-0112-0041 to Philip G. Zimbardo, supplemented by funds from an NIMH grant 03859-09 to Ernest R. Hilgard.

Address requests for reprints to: Christina Maslach, Department of Psychology, 3210 Tolman Hall, University of California, Berkeley, California 94720.

Reprinted with permission of the publisher and author from *Psychophysiology*, Vol. 9, 600-605. Copyright 1972, The Society for Psychophysiological Research.

believe so on the basis of converging research findings in the areas of visceral learning, cognitive control of motivation, and hypnosis.

Neal Miller and his associates at Rockefeller University (1969a, 1969b) have recently demonstrated that the control over skeletal muscle responses through operant conditioning procedures can be extended to responses of the glands and viscera. Their work has generated the powerful conclusion that any discriminable response which is emitted by any part of the body can be learned if its occurrence is followed by reinforcement. These results are extended in the work of Zimbardo and his colleagues (1969) which experimentally demonstrates that biological drives, as well as social motives, may be brought under the control of cognitive variables such as choice and justification, even in the absence of external reinforcers.

It appeared to us that hypnosis: a) is a state in which the effects of cognitive processes on bodily functioning are amplified; b) enables the subject to perceive the locus of causality for mind and body control as more internally centered and volitional; c) is often accompanied by a heightened sense of visual imagery; and d) can lead to intensive concentration and elimination of distractions. For these reasons, it should be possible for well-trained hypnotic subjects to gain control over regulation of their own skin temperature without either external reinforcement or even external feedback. While there have been some attempts to control temperature through hypnosis or other methods (Barber, 1970; Green, Green, & Walters, 1970), they have often lacked adequate controls and tend to focus only on unilateral changes.

Our present study was exploratory in nature and attempted to demonstrate that hypnotic subjects would be able to achieve simultaneous alteration of skin temperature in opposite directions in their two hands, while waking control subjects would not. The bilateral difference of one hand becoming hotter than normal, while the other gets colder, was chosen in order to rule out any simple notion of general activation or prior learning and to control for any naturally occurring changes in skin temperature. We also attempted to rule out other alternative explanations of changes in skin temperature by keeping environmental conditions constant and by minimizing overt skeletal responses on the part of the subjects.

Method

Subjects

All of the subjects (with the exception of the junior author—PGZ) were undergraduate paid volunteers from the introductory psychology course at Stanford University. Three of the Ss received hypnotic training prior to the experiment, while the remaining 6 Ss did not. The training averaged about 10 hrs per person and was usually conducted in small groups. It was permissive in orientation, stressing the S's ability to achieve self-hypnosis, and involved several criterion tests.

Procedure

The Ss were individually tested in the Laboratory of Dermatology Research at the Stanford Medical Center. The ambient temperature in this room was automatically regulated to maintain a constant level. Ten thermocouples of copper constantin were taped to identical sites on the ventral surface of the two hands and forearms of the S. Both room and skin temperatures were continuously monitored by a Honeywell recording system. The Ss lay on a bed with their arms resting comfortably at their sides and with open palms extended upward in exactly the same position. This posture was maintained throughout the session, and there was no overt body movement.

For the hypnotic Ss, the experiment began with approximately 10 min of hypnotic induction. The remainder of the session was identical for both hypnotic and waking control Ss. They were first asked to focus attention on their hands, and were then told to make an arbitrarily selected hand hotter, and the other colder, than normal. Accompanying this last, brief instruction were suggestions of several images which could be useful in producing this effect, as well as encouragement to generate personal imagery and commands which might be necessary to achieve the desired result. The S lay in silence for the duration of the testing session (which averaged about 10 min). The final instruction was to normalize the temperature in both hands by returning it to the initial baseline level. Each of the Ss participated in 2 such sessions. In addition, 1 of the Ss completed 2 sessions utilizing auto-hypnosis, a procedure in which the S provides the instructions to himself.

Results

All of the hypnotic Ss demonstrated the ability to produce bilateral changes in skin temperature. Large differences (as much as 4° C) between identical skin sites on opposite hands appeared within 2 min of the verbal suggestion, were maintained for the entire testing period, and then were rapidly eliminated upon the suggestion to normalize skin temperature. Temperature decreases in the "cold" hand were generally much larger than the increases in the "hot" hand, the largest decrease being 7° C, while the largest increase was 2° C. In contrast, none of the waking control Ss were able to achieve such significant bilateral changes in the temperature of their hands. Any temperature change that they did exhibit was usually in the same direction for both hands (rather than in opposite directions), thus yielding close to a zero score for bilateral change (see Fig. 1). The difference between these control scores and the consistently large bilateral changes of the hypnotic Ss is highly significant ($t = 14.27$, $df = 7$, $p < .001$). All of the hand thermocouples reflected these successful bilateral changes, while the forearm thermocouples showed no temperature changes at all, thus indicating the specificity of this hypnotic control process. Also, the performance of the hypnotic Ss showed an improvement from the first to the second session; this was not true of the control Ss.

When the individual patterns of reaction in the hypnotized Ss are examined,

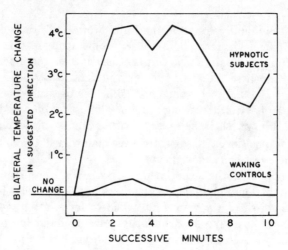

Fig. 1. Mean algebraic sum of bilateral skin temperature differences ("successful" directional changes in each hand were weighted positively, while changes which were opposite to the suggested direction were weighted negatively).

the degree of control that they were able to exert becomes even more apparent. The S's data shown in Fig. 2 reveals how, following the suggestion to make her left hand colder and right hand hotter (opposite to their relative baseline position), she rapidly "drove" them in the appropriate directions. After maintaining the separation for more than 10 min, she re-established the initial baseline difference as soon as she was given the instruction to normalize her skin temperature. Since there was no overlap in the temperature distributions of the two hands, the obtained differences from min 4 to min 16 were extremely significant (within-subject $t = 20.18, df = 12, p < .001$).

Both the hypnotic and waking control Ss reported trying hard to meet the experimental demand. Several of the control Ss even believed that they had successfully completed the task, although as noted earlier, their largest bilateral difference was very slight. All Ss also reported that they had generated assorted imagery to help them produce changes in their skin temperature. Some of the imagery involved realistic experiences, such as having one hand in a bucket of ice water and the other under a heat lamp, while other imagery had a more symbolic or fantasy quality. In addition, Ss also used image-less "commands" given independently to each hand (i.e. "you become hot, you become cold").

In the initial pre-test, verbal feedback was given to the Ss when they had succeeded in producing the bilateral difference in temperature. Such feedback had an unexpected negative effect, resulting in the "loss" of the attained difference, and was subsequently eliminated in the experimental sessions. It may be that the intensive concentration required to achieve the unusual performance demanded in this study was disturbed by having to attend to and process the informational input from the experimenter. In a sense, the feedback, although supportive, operated as a distractor to attenuate the obtained differences in skin

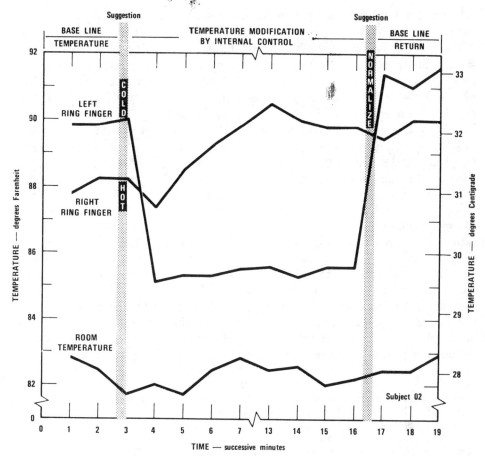

Fig. 2. Simultaneous modification of skin temperature in opposite directions in the right and left hands (omitted min 8–12 are no different from the rest of the modification period).

temperature. The ability of hypnotic Ss to successfully perform this task without feedback is particularly evident in the data of the S using auto-hypnosis, who was able to produce bilateral differences in skin temperature without the aid of any external demands, feedback, or extrinsic sources of reinforcement.

Discussion

Although we are not in a position to characterize the underlying physiological mechanisms responsible for the bilateral control of skin temperature which we have shown, we believe that the role of hypnosis in the process is quite understandable. The research by Miller on visceral learning has stressed the important function served by curare in paralyzing the skeletal musculature of the animals. At first, this methodological control was thought to be necessary only to rule out possible influences of skeletal musculature on glandular and visceral responding. However, it now appears that curarizing the animals "may help to maintain a constant stimulus situation and/or to shift the animal's attention from distracting skeletal activities to the relevant visceral ones [Miller, 1969b, p. 19]."

We would argue that the effects of hypnosis are analogous to those of curare, since hypnosis provides a set of training conditions which permit a greater than normal degree of generalized relaxation, removal of distracting stimuli, and enhanced concentration upon a given, relevant dimension. Hypnotic training may also aid in the control of experiential, behavioral, and physiological processes by increasing the subject's confidence in his ability to exert such control, and by altering consciousness to the point that words and images can be more readily translated into a code language to which he is physiologically responsive.

To us, the significance of research in this area is less in understanding how hypnosis per se operates, but rather how human beings "naturally" learn to induce ulcers, tachycardia, excessive and uncontrolled sweating, and other forms of psychosomatic illness. Miller's work suggests that the intervention and modification of such reactions follow principles of operant conditioning. Our work adds the possibility that the sources of reinforcement in both producing and changing psychosomatic symptomatology may be cognitive in nature. In a recent clinical application of these ideas,[1] patients are trying to reduce their migraine headaches by learning how to voluntarily control their blood flow and skin temperature via biofeedback techniques. Therapeutic control may thus be best achieved by combining the precision of reinforcement contingencies with the power of a more pervasive cognitive approach to dealing with such mind-body interactions.

REFERENCES

Barber, T. X. *LSD, marihuana, yoga and hypnosis.* Chicago: Aldine Publishing Co., 1970.
Green, E. E., Green, A. M., & Walters, E. D. Self-regulation of internal states. In J. Rose (Ed.), *Progress of cybernetics: Proceedings of the International Congress of Cybernetics, London, 1969.* London: Gordon and Breach, 1970.
Luria, A. R. *The mind of a mnemonist.* New York: Discus Books, 1969.
Miller, N. E. Learning of visceral and glandular responses. *Science,* 1969, *163*, 434–445. (a)
Miller, N. E. Autonomic learning: Clinical and physiological implications. Invited lecture at the XIX International Congress of Psychology, London, 1969. (b)
Zimbardo, P. G. *The cognitive control of motivation.* Glenview, Ill.: Scott, Foresman and Co., 1969.

---

[1] Sargent, J. D., Green, E. E., & Walters, E. D. Unpublished research report entitled, "Preliminary Report on the Use of Autogenic Feedback Techniques in the Treatment of Migraine and Tension Headaches," 1971.

X

CLINICAL APPLICATIONS

33 Biofeedback Techniques in Behavior Therapy

Thomas H. Budzynski and Johann Stoyva

One of the most encouraging innovations in contemporary psychotherapy has been Wolpe's (1958) development of systematic desensitization. In essence, this technique involves substituting a relaxation response for an anxiety response. When desensitization is complete, the patient is able to think calmly about things which formerly made him extremely anxious.

Contained within this novel and ingenious technique of Wolpe's is the germ of a far-reaching idea - - the view that man may be able to acquire voluntary control over a variety of physiological functions and, in so doing, to alter his psychological states for the better. This idea is by no means of recent vintage - - witness the disciplines of Yoga and Zen. And, even in the Western world, techniques such as autogenic training and Jacobson's (1938) progressive relaxation were already well-launched by the second quarter of this century. At the present time, however, two developments promise to extend vastly the entire range and power of such an approach: (a) A major factor is the accelerated growth of electronic instrumentation; researchers are now able to measure physiological events in the intact human which formerly were difficult or impossible to measure. (b) A related factor is the introduction of the psychophysiological feedback loop. By precisely detecting a physiological event and then converting the resulting electronic signal into either auditory or visual feedback, a subject can be made immediately and continuously aware of the level of a physiological event. This is the core of the biofeedback technique. It is an approach which promises to extend greatly the ability of man to acquire voluntary control over a variety of physiological functions.

This research was supported by National Institute of Mental Health Grant MH-15596.
Johann Stoyva is supported by National Institute of Mental Health Research Scientist Development Award, KO1-M4-43361.

This paper is an English translation of a chapter from *Die Bewaltigung von Angst. Beitrage der Neuropsychologie zur Angstforschung. Reihe Fortschritte der Klinischen Psychologie*, Bd. 4, 1973, edited by N. Birnbaumer. It is printed here with permission of the publisher, Urban & Schwarzenberg, the authors and the editor.

The thesis advanced in this paper is that biofeedback techniques have considerable clinical potential. This potential seems likely to be realized by integrating biofeedback techniques with certain procedures in behavior therapy. An expected additional source of clinically useful procedures are the recent experiments on the modification of autonomic responses (see Miller, 1969).

In the behavior therapy technique known as systematic desensitization, the basic procedure is to replace an anxiety response with a relaxation response. As Jacobson (1938) and many others have observed, the anxious patient typically shows elevated levels of muscle tension, particularly during an anxiety episode. Wolpe (1958), drawing on Jacobson's extensive work with progressive relaxation, postulated that a condition of muscle relaxation is physiologically incompatible with an anxiety response. On the strength of this postulate, muscle relaxation has come to play a central role in systematic desensitization.

The patient undergoing desensitization is first taught muscle relaxation by means of an abbreviated Jacobson progressive relaxation technique (usually two to six training sessions). He is next presented with a graded series of anxiety scenes, known as a "hierarchy." A given hierarchy is focussed around a single theme, such as public-speaking anxiety, and the items are ranked from the one which is the least anxiety-producing to that which is most anxiety-producing. When the actual desensitization is begun, the least anxiety-evoking scene is used first. The task of the patient is to maintain a state of deep relaxation at the same time that he imagines the anxiety item. After he masters a given scene, the patient progresses to a more difficult one until finally he is able to maintain calmness even while visualizing the scene which formerly made him the most anxious. Wolpe (1958) reports that generalization from the clinic to life situations occurs readily, and in the cases of specific anxieties such as phobias, Wolpe and Lazarus (1966) report an impressive successful outcome rate of 80 to 90 per cent.

Desensitization with Alpha Feedback

Our first attempt at using biofeedback techniques in behavior therapy, stimulated by Kamiya's early work, involved the feedback control of alpha. We surmised that alpha might be useful as an indicator of relaxation, and as a means of quantifying certain aspects of the desensitization process. In particular, we felt that training in the feedback control of alpha might prove valuable in helping a patient to regain a state of calm after an anxiety episode. However, as will be detailed shortly, there were a number of difficulties with alpha. These difficulties caused us later to shift our attention to EMG feedback as a means of inducing muscle relaxation.

Our initial results with alpha feedback were quite encouraging. In agreement with Kamiya (1968) and Stoyva & Kamiya (1968), we found that when subjects were provided with immediate tone feedback as to the presence or absence of alpha they were able to increase their alpha levels (per cent of time EEG record shows alpha rhythms). As Kamiya had indicated, subjects said the alpha condition was associated

with feelings of tranquility and relaxation. They also reported that
the production of alpha was facilitated by the suppression of visual
imagery, but that any feelings of anxiety would immediately block the
alpha rhythm.

Since it seemed to be associated with relaxation, we decided
to use the alpha state as a counterconditioner in desensitizing a
patient. This man was an architect who suffered unusually strong
anxiety concerning a variety of death-related themes. His anxiety
about funeral houses, graveyards, coffins and corpses had generalized
to a fear of the dark so that he felt himself unable to enter the
darkened basement of his own house. He also found himself filled
with a dread when driving home after dark, imagining there might be
a body in the back seat of the car.

The patient (J.L.) was first given feedback training in producing
alpha, and in two sessions he learned to increase his alpha level
from 20 per cent to 80 per cent. He then made up a hierarchy of four
death-related scenes (see Figure 1 for hierarchy items).

J.L.'s task was to visualize the successive hierarchy items,
and at the same time to remain relaxed. The criterion for successful
performance was set at 20 seconds of visualization without anxiety.
Unfortunately, as soon as the patient began visualizing any scene,
whether anxious or pleasant in nature, the alpha rhythm disappeared.
Consequently, it was not possible to use the absence of alpha as an
indicator of anxiety _during_ visualization of the scenes, since any
kind of visualization, whether anxious or neutral, would cause alpha
to disappear. (Subsequently, presence of anxiety was indicated by
a finger signal.) However, alpha feedback was found to be useful
between visualizations. The patient reported that such feedback helped
him to banish the anxiety-evoking scenes very quickly and to recapture the state of deep relaxation.

Certain features of the patient's responses to the death-hierarchy scenes are worthy of note. J.L. reported that his first visualization of scene 1 (day 1), "Sleeping in the attic," was not very
clear - - although as Figure 1 shows, he had, in fact, reached the
20 second criterion. During his second imagining of the same scene
(also on day 1), his visualization was clearer and there was an associated increase in anxiety and a failure to reach criterion. (Note
that the ordinate in Figure 1 does not represent the amount of alpha,
but the _time_ elapsed before the patient gives the anxiety signal.)

The patient's reaction to scene 2, "Calumet Cemetery," was also
interesting. Several hours after his first desensitization session,
he remembered that his grandmother was buried in this cemetery. This
recollection seemed to increase the anxiety associated with the scene,
and J.L. felt that it ought to be changed from second to third place
in the hierarchy.

As Figure 1 indicates, J.L. was successfully desensitized to
each of the scenes by day 4. Six months later, a follow-up showed
that the phobia had not recurred. The patient was now able to go

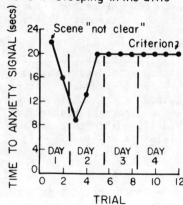
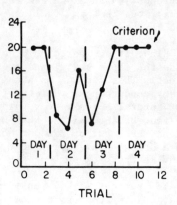
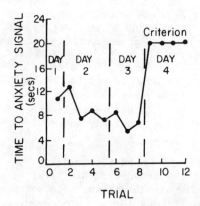
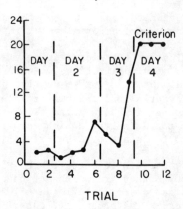

Figure 1 Systematic desensitization using alpha EEG feedback. Criterion for successful desensitization was patient's ability to remain calm for a 20 second visualization period.

into places such as basements, attics and cemeteries where previously he had been afraid to venture.

The most important result of this case study was that it revealed certain difficulties in applying alpha feedback in desensitization. One of these difficulties has already been described. J.L., it will be recalled, showed ample resting alpha; but during both unpleasant _and_ neutral visualizations, suppressed alpha completely. With such individuals, the absence of alpha cannot be used as an indicator of anxiety during visualization. Another problem is that some subjects show a high amount of alpha most of the time - - there seems to be little change in per cent alpha as the subject switches from pleasant imagery to anxiety-evoking imagery, or to a condition of no imagery. Finally, there are some subjects who show little or no alpha, even when they are relaxed, so that shaping is either extremely difficult or impossible.

In the hopes of resolving some of the shortcomings present with alpha feedback, we turned our attention to muscle activity.

EMG Feedback in Behavior Therapy

A major reason for choosing to work with EMG activity was the evidence from therapies such as progressive relaxation (Jacobson, 1938), autogenic training (Schultz & Luthe, 1959), and behavior therapy (Wolpe, 1958) attesting to the therapeutic usefulness of muscle relaxation in a variety of anxiety and stress-related disorders. Wolpe (1958), for example, wrote that:

In selecting responses to oppose the anxiety responses I was guided by the presumption that responses that largely implicate the parasympathetic division of the autonomic nervous system would be especially likely to be incompatible with the predominately sympathetic responses of anxiety (p. 72).

Explicit in Wolpe's formulation is the concept that muscle relaxation has extensive effects on the autonomic nervous system - - in general, muscle relaxation is associated with parasympathetic activity; whereas muscle tension is associated with sympathetic activation. There exists considerable evidence in support of this position; in particular, see Germana's (1969) fine review article on autonomic-somatic integrations. Evidence indicating that the autonomic and somatic systems act in an integrated way has been provided by the experiments of W.R. Hess (1954) who conceived of the "ergotropic" and "trophotropic" systems. Hess discovered that when he electrically stimulated certain diencephalic regions he observed simultaneous behavioral arousal and sympathetic activation, an ergotropic response. Stimulation of other diencephalic regions produced parasympathetic activity and a behaviorally calm animal - - a trophotropic response.

The essential validity of the "ergotropic-trophotropic" distinction has been demonstrated by Gellhorn (1964, 1967). According to Gellhorn, the "ergotropic syndrome" consists of sympathetic - adrenal events - - increased heart rate and blood pressure, adrenomedullary secretion, sweat secretion, pupil dilatation, EEG desynchronization, and increased somato-motor activity. Generally, the trophotropic responses are opposite in nature.

Gellhorn also states that the organism's response is graded, depending on the type and intensity of stimulation. For example, when the animal is exposed to a moderate stimulus, only the sympathetic or "neurogenic" responses may occur. The adrenomedullary component is absent. But with more intense stimulation, both the sympathetic and adrenomedullary responses typically appear.

Difficulties with Muscle Relaxation

Granted that considerable evidence supports the view that the somatic musculature has effects on the autonomic nervous system, there are several problems with muscle relaxation as it is customarily used in behavior therapy: (a) The tense, anxious patient is likely to find it difficult to relax; in striving to relax he might, in fact, become even more tense. (b) The abbreviated progressive relaxation training (2 to 6 training sessions) generally used by behavior therapists may frequently not result in thorough muscle relaxation; both Jacobson (1938) and Schultz and Luthe (1959) emphasize that many months of training may be required. (c) The demand characteristics of the desensitization procedure are such that the patient is inclined to say he is relaxed even when he is not. In the light of these difficulties, we felt that a technique for physiological monitoring of relaxation would be extremely useful in behavior therapy.

In our laboratory, we conjectured that these difficulties could be largely overcome with an EMG feedback system. The basic technique would be to detect accurately the level of tension in a muscle and then to feed back to the patient an auditory signal with a frequency proportional to the amount of tension in the muscle. A key assumption was that a feedback approach would enable patients to attain quicker and more thorough muscle relaxation than has been possible with the traditional relaxation techniques. If electronically feasible, such a system would be useful not only for teaching relaxation, but would enable continuous monitoring of EMG levels as well as their quantification on a trial-by-trial basis. Continuous monitoring of EMG levels would be a useful aid to the therapist conducting desensitization; e.g., as a guide to pacing the rate at which scenes from the anxiety hierarchy are presented. Quantification of EMG levels, say, on a trial-by-trial basis, would be a valuable research tool. Speed and thoroughness of muscle relaxation could be studied, as well as the relation of EMG levels to other physiological and behavioral variables. Such a development would be in keeping with Lang's (1969) suggestion that further exploration of the mechanism of muscle relaxation in systematic desensitization is a much-needed line of inquiry.

Instrumentation and Validation Studies

Our first objective was to develop a high performance EMG feedback unit. This task was much more demanding electronically than constructing a unit for the feedback of EEG alpha activity, which has been the bioelectric signal most commonly used in feedback research with human subjects (see Kamiya, 1968). A main reason for the greater difficulty in using EMG activity is that the integrated EMG signal produced by the relaxed muscle is exceedingly small, generally only a fraction of the EEG alpha rhythm (1-4 μv for EMG versus 20-80 μv for alpha). Therefore, the signal-to-noise ratio of the preamplifier must be a great deal higher in EMG work than in alpha training. Also, the vastly greater range of frequencies produced by muscle tissue, 10-10,000 Hz for EMG as opposed to 8-12 for alpha, complicates the task of accurately detecting and processing the signal.

These two major difficulties were eventually overcome in the course of developing a unit which provided the subject with a crisp, sensitive feedback of his own muscle activity as detected by surface electrodes.

The basic function of the EMG feedback unit (see Figure 2) is to provide the subject with a tone, the frequency of which is proportional to the EMG activity in a particular muscle group. If EMG activity is high, then the tone frequency is high. As EMG activity decreases, the tone simultaneously decreases in frequency. In effect, the feedback tone tracks the fluctuating level of EMG activity in the muscle. The task of the subject, who has EMG electrodes applied to the surface of the skin over a particular muscle (e.g., frontalis or forearm extensor) and who hears the tone through his headphones, is to keep the tone at a low frequency by relaxing that muscle.

As the subject becomes better at relaxing, the task is gradually made more difficult for him through shaping. The shaping procedure is accomplished by increasing the gain of the feedback loop so that the subject must relax the muscle even more deeply in order to produce the desired low-frequency tone. Thus, the difficulty of the task can be adjusted to form a series of finely graded steps. Each step is designed to match the trial-by-trial performance of the subject.

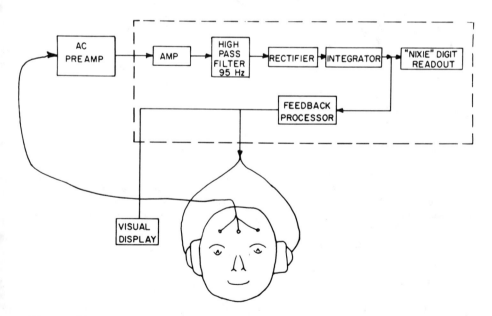

Figure 2 Functional diagram of BIFS EMG feedback unit.*
*Bio-Feedback Systems, Inc., 2736 47th Street, Boulder, Colorado 80301

An extremely useful research feature of the EMG feedback unit (see Figure 2) is its quantification of EMG activity, achieved by means of a constant reset level integration technique (see Shaw, 1967). The resulting figure is available to the experimenter as a four digit NIXIE tube readout at the end of each trial. The integration process is such that the final figure which has accumulated on the readout panel (and is held for 20 seconds) represents the average level of EMG activity in microvolts for that particular one-minute trial. Thus, over a series of one-minute trials, EMG levels might be 5.12, 4.32, 3.17, 2.68, 2.72 and 1.87 μv for six consecutive trials. The resulting data, minute-by-minute EMG levels, can readily be represented in graphic form and are extremely useful in studying the dynamics of the human in a feedback loop.

Once a suitable unit had been developed, the next step was to determine whether subjects would relax better with the help of EMG feedback than without it (Budzynski & Stoyva, 1969). Subjects in the experimental group received EMG feedback and the shaping procedure as well. Subjects in the first control group heard a steady low tone; those in the second control group received no feedback at all (silent condition). Each volunteer was told to relax as deeply as possible, and that the better his performance the more he would be paid. The results spoke clearly in favor of the experimental subjects, whose post-training EMG levels (on the frontalis) were about 50 per cent lower than their pre-training levels. In the steady low tone group, post-training EMG levels increased by about 28 per cent; in the silent group they decreased by about 24 per cent.

Another study (Budzynski, 1969) demonstrated that the effects of feedback-assisted muscle relaxation are not limited to the particular muscle generating the feedback signal, but are manifested in CNS changes as well. This experiment drew on the work of Venables and Wing (1962) which had shown that the fusion threshold for *paired flashes of light* may be used to determine cortical activation level. Generally, individuals who are anxious or hyperalert are able to discriminate smaller differences in the time interval between the successive light flashes than are individuals with more normal arousal levels.

In Budzynski's (1969) experiment, 12 normal subjects were given EMG feedback training on the forearm and frontalis muscles. Each subject showed an increased two-flash threshold (the discrimination interval was longer) in the deep relaxation condition.

Systematic Desensitization with EMG Feedback

We have employed EMG feedback for both the relaxation training and the desensitization phases of behavior therapy. A valuable feature of the EMG feedback unit described above is that by means of its quantification circuitry it provides both a continuous monitoring of arousal levels during visualization of anxiety scenes, and yields quantitative data on the desensitization process.

The patient customarily first receives training in the relaxation of the frontalis. A valuable property of this muscle is that when it is deeply relaxed, there seems to be good generalization to the upper body - - in most subjects. But should the patient have difficulty in the rather subtle task of learning to relax his frontalis muscle, feedback training can be shifted to either the masseter or the forearm muscle. Even the tensest patients are able to make some progress with the masseter or forearm.

An unanticipated advantage of the feedback technique in behavior therapy, is that it can be used to break down the relaxation training into such small steps that patients are virtually certain of at least some success. And a modest success experience at the outset of therapy - - as demonstrated by the patients' decreasing EMG scores - - can serve as a useful motivator.

On the day desensitization is to begin, the patient first receives about 10 minutes of feedback in order to help him reach EMG levels comparable to those achieved in his training sessions. After the patient reaches a suitable EMG level, his therapist tells him to imagine a pleasant scene. The frontalis EMG levels for this pleasant scene generally run 10 to 15 per cent higher than during straight relaxation trials. This EMG level is then used as a criterion of successful relaxation during visualization of anxious scenes. A light display on the console of the EMG unit is adjusted so that whenever the patient's EMG level remains below criterion, the therapist sees a green light. If the patient becomes slightly tense, the green light goes out and an amber one comes on. Should the patient become even more tense, the amber light also goes out and a red light comes on. At this point the therapist terminates the scene. Generally, the termination of the scene by the therapist occurs before the patient becomes aware of any anxiety.

Having desensitized approximately 20 patients with the foregoing technique, we believe that EMG feedback-assisted desensitization provides several advantages - - although controlled studies are not yet available to document this point. First, the initial relaxation training assures a high probability of thorough relaxation during the desensitization proper. Second, since the EMG indicator can be set at a high sensitivity, most visualizations can be terminated before the patient is aware of rising anxiety. Third, the EMG feedback quantification capability should prove useful in exploring some important aspects of the desensitization procedure. Fourth, the use of the feedback between visualizations helps the patient to return quickly to a relaxed condition. This fourth feature may prove to be of great value with anxious patients, since these are individuals who, once aroused, tend to stay aroused. Malmo (1966), for example, has written that anxious patients suffer from a deficiency in _internal_ feedback mechanisms which prevents their bodily systems from readily returning to an equilibrium or baseline condition. Perhaps training procedures involving _external_ feedback techniques could be useful in alleviating such difficulties.

Some case descriptions are of interest at this point. Each of the two following patients had failed to make progress in conventional behavior therapy because of a marked inability to relax; but each eventually proved able to learn relaxation with the help of feedback training.

Case 1: A 45-year old woman was referred to us because her marked inability to learn muscle relaxation had prevented her from making progress in the desensitization phase of behavior therapy. She suffered from extreme anxiety at social gatherings. At such times she also showed a tremor (functional in nature) of the right hand so marked as to prevent her from shaking hands with anyone or from even holding a glass. Feedback training was started with the frontalis, but since this task proved frustrating, her training was shifted to the easier-to-control forearm muscle, where she made good progress. Later the training was shifted to the masseter and, finally, to the frontalis. After she had learned to relax well, her regular therapist desensitized her to her social anxiety (without using any feedback devices). She now participated in social occasions, and was no longer afflicted by the tremor.

Case 2: This individual, a 42 year-old management consultant, had developed a marked anxiety for public speaking. His anxiety had become increasingly severe, and had forced him to refuse several lucrative speaking engagements. At this point he sought help. For the first several weeks he made little progress with his feedback-assisted relaxation training. EMG levels from his right forearm, for example, remained at high levels. We therefore decided to vary our approach. Reasoning that since he was right-handed, his right arm would probably be more strongly involved in anxiety and stress-related reactions than his non-dominant arm, we shifted the training to his left forearm. He proved able to relax this muscle and was later able to progress to the relaxation of other muscles.

Another useful observation provided by this patient is that his learning of relaxation was hindered by his persisting in an incorrect strategy. He thought that extremely deep breathing would be the way to produce deep relaxation, and he persisted in this approach, ignoring what the feedback tone was telling him, until the experimenter suggested that shallower breathing might be better. He was also told to pay careful attention to the feedback tone as a means of guiding his response. Subsequently, the patient was desensitized and, up to the point of this writing, has overcome his public speaking phobia - - although he still experiences occasional brief episodes of anxiety when addressing large or disputatious audiences. At these times, by using his relaxation response, he is generally able to recover before a cycle of anxiety-leading-to-more-anxiety can begin.

Sensitivity of the EMG Indicator

Two additional observations show that EMG activity, particularly of the facial musculature, can be a highly sensitive indicator of anxiety during desensitization.

The first observation indicates an anticipatory rise in muscle tension in the patient who is about to begin desensitization trials. With each of six phobic patients who had his frontalis or masseter EMG levels monitored during desensitization trials, it was strikingly apparent that frontalis EMG levels began to rise during the preliminary period when the subject was supposed to be relaxing. This observation - - which would go undetected with desensitization performed in the usual way - - strongly suggests that desensitization has probably often been attempted despite lack of muscle relaxation on the part of the patient.

A second observation pertains to EMG activity <u>during</u> the visualization of anxious scenes. In each of six patients who was desensitized with the aid of EMG feedback (and for whom polygraphic write-out of EMG activity was also available), muscle tension built up rapidly when the patient began to feel anxious during the visualization of a scene. Surprisingly, the build-up of EMG activity actually preceded the verbal report of anxiety by about 5-15 seconds for these individuals. An example:

The patient in question suffered from pervasive anxiety sufficiently disabling to make him quit work. After 10 to 12 sessions of feedback training he learned to achieve deep muscle relaxation. He was then instructed to visualize a scene ("being watched") and to use a finger switch to indicate both when he felt relaxed and when he later felt the slightest bit anxious. Thus, the procedure for the patient was: (<u>a</u>) to press the switch as a signal of relaxation, (<u>b</u>) to start visualizing the scene, (<u>c</u>) to press the switch again as soon as he felt the slightest twinge of anxiety. Immediately following his anxiety signal, the patient would be told to stop visualizing the scene. No EMG feedback was available to the subject during this procedure.

The fascinating, and perhaps controversial, observation which emerged was that the EMG indicator of anxiety was considerably more sensitive than the individual's verbal report. For instance, in Figure 3, the patient's finger signal lagged 8-10 seconds behind the EMG indicator of anxiety.

So far we have asked six patients, trained in the deep relaxation through information feedback, to indicate as quickly as possible when they felt anxious in visualizing a certain scene. For each patient, his EMG response was more sensitive than his verbal report. Even the best subject was able to signal only approximately five seconds after the first detectable increase in EMG levels.

In addition to being of interest in their own right, the two foregoing observations raise certain doubts with regard to the desensitization technique as it is customarily practiced in behavior therapy. One unsettling possibility is that in the past, desensitization may frequently have been attempted without the patient's being truly relaxed. If Wolpe's basic tenets about anxiety and deep muscle relaxation are correct, then the presence of muscle tension would interfere with the process of desensitization.

448 *Clinical Applications*

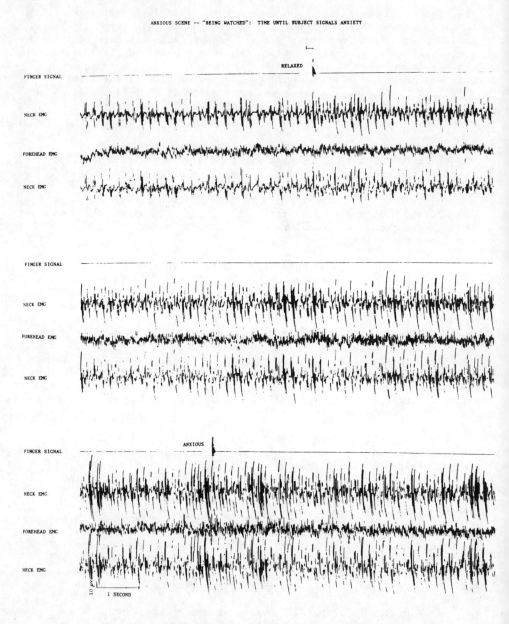

Figure 3 Forehead and neck EMG during visualization of anxious scene, "being watched." A continuous recording from top to bottom. Both neck records are from same electrode site, but bandpass widths are different. The first finger switch signal indicates that the patient is relaxed and is beginning to visualize the scene. The second signal represents his report of slight anxiety.

These observations also suggest that <u>one reason</u> desensitization procedures have frequently been unsuccessful with pervasive anxiety cases is that these patients have not properly learned thorough relaxation. Moreover, these individuals may be people in whom the anxiety response is quite resistant to ordinary relaxation procedures. Once the anxiety response is triggered, arousal levels (including muscle tension) may remain high. Feedback techniques should be useful in ensuring that desensitization is neither attempted nor continued without proper relaxation.

Note on Assumptions

Throughout this paper we have maintained a number of assumptions regarding muscle relaxation and its effect on the individual. Before concluding, we wish to make these assumptions explicit, to sketch out some of the evidence for them, and to point out where further investigation is necessary.

1. <u>Does muscle relaxation inhibit anxiety?</u> As mentioned previously, a central assumption in Wolpe's (1958) desensitization technique is that thorough muscle relaxation is a condition incompatible with the presence of anxiety. We have found this assumption ordinarily to be true. As indicated in the preceding section, patients desensitized with EMG monitoring often report having felt no anxiety at all.

A crucial aspect of the above assumption is that muscle relaxation affects the autonomic nervous system, and brings about a condition in which parasympathetic responses are dominant. Implicit in this last statement is the assumption that anxiety has both muscular and autonomic components. If either of these components remains at a high or activated level, then the high arousal condition necessary for the experience of anxiety - - and probably for other powerful emotions as well - - is still present.

For most individuals, a relaxed musculature appears sufficient to dampen the activity of the autonomic nervous system; that is, to produce a shift towards parasympathetic responding. When muscle activity is reduced, autonomic **activity** is reduced and vice versa. In other words, the two systems act in tandem. As stated earlier, there is considerable evidence in support of this proposition - - see Germana (1969), Hess (1954), Gellhorn (1964), Obrist (1970).

With some individuals, however, the muscle and autonomic systems are more loosely coupled; one system has little effect on the other. Cameron's (1944) "autonomic responders" may behave in this fashion. In his study of the physiological manifestations of anxiety, Cameron found that anxious patients fell into three groups: Those in whom the skeletal musculature was mainly involved, those in whom muscular and autonomic responses appeared equally involved, and those in whom autonomic responses were most prominent (usually either cardiovascular or gastrointestinal symptoms). The latter individuals - - the "autonomic responders" - - may be the people for whom muscle relaxation does not diminish anxiety. We have occasionally encountered patients like this; even though they are muscularly relaxed, they still feel anxious.

The matter is certainly open to empirical test. Specifically, are the patients who still feel anxious, despite being muscularly relaxed, those who are autonomic responders in stress situations? With these patients, for whom muscle relaxation is not very useful, perhaps there are other ways of dampening visceral activity.

2. Is muscle relaxation needed in systematic desensitization? An additional assumption embodied in this paper is that muscle relaxation is necessary in systematic desensitization. Actually, we prefer not to put the matter so dogmatically. We would sooner say that muscle relaxation is a useful means of accomplishing systematic desensitization, and is backed up by a more convincing body of studies than any other method (see Paul, 1969). To be sure, there may be other ways of conducting desensitization. Anger, for example, is another response incompatible with the feeling of anxiety, as Wolpe and Lazarus have indicated (1966).

Rachman (1968), in particular, has expressed doubts as to the need for muscle relaxation in desensitization. He states that if a patient is able to acquire a feeling of "calm," regardless of the state of his musculature, this condition is adequate for desensitization purposes.

Our position, based on experience with patients in our laboratory, is that the hypothesis of Lader and Matthews (1968) will help to clarify matters. They view desensitization of anxiety as an habituation process. And they postulate that habituation occurs optimally in a low arousal condition - - such as can be produced by muscle relaxation.

On the basis of this conception, it seems quite plausible that some patients would not require relaxation training in order for desensitization to occur. Take, for example, healthy young normal subjects "suffering" from circumscribed phobias such as fear of snakes or spiders - - long the favorite target population of controlled studies on systematic desensitization. These individuals may be close enough to optimal arousal levels for desensitization to occur readily without any special relaxation training. Supporting evidence for this assertion was found in a recent study in this laboratory (unpublished observations). We found that in a young adult population, most of the subjects relaxed very well even prior to any EMG feedback training.

However, we predict that, in the case of patients suffering from severe anxiety, arousal levels (including muscle tension) will be well above those at which desensitization readily occurs. With such patients, training in muscle relaxation would be important for successful desensitization, since muscle relaxation would produce the lowered arousal levels (lowered muscle activity and its attendant dampening of autonomic responses) required for habituation of the anxiety response.

The foregoing prediction could fairly easily be subjected to empirical test. And, if the outcome were positive, this would be of value in defining the patient population for whom muscle relaxation is useful. A negative result would document the assertion that muscle relaxation is not required for systematic desensitization. It would be important to do this study with sophisticated physiological monitoring. During desensitization, measures from the three major bodily systems should be continuously sampled; from the musculature, the autonomic nervous system, and the CNS.

Evidence consistent with the two foregoing predictions was obtained in a recent study by Farmer & Wright (1971). Patients whose muscular reactivity was high - - as estimated by scores on the Fisher & Cleveland (1958) body-image Barrier test, a modification of the Rorshach - - benefited significantly more from the use of muscle relaxation in desensitization than they did from a muscle activity (see Lazarus, 1965) condition. On the other hand, subjects with low Barrier scores, and presumably low muscle reactivity under stress, did just as well with the muscle activity condition as with the muscle relaxation condition.

Farmer & Wright put forward an intriguing point of view - - and one highly congenial to us. They interpret their observations as supporting the idea that the most effective way to extinguish conditioned anxiety is to inhibit activity in the physiological response mode which, for that patient, is maximally reactive under stress. For some patients this would be the skeletal musculature, for others the cardiovascular system, for yet others the gastrointestinal system. Farmer & Wright (1971) state that: ". . . if a patient's skeletal musculature were relatively more reactive to stress than his other physiological response systems, muscle relaxation would be expected to be an effective anxiety-inhibitor (p. 2)."

Perhaps in systematic desensitization of the future, at least in difficult cases, the first step would be to determine polygraphically the individual's pattern of physiological responding under psychological stress. Next he would be given feedback training to moderate his physiological response pattern under stress. Special emphasis would be placed on teaching him to moderate his maximally active response. Finally, desensitization would take place.

<u>3. Is there bodily generalization of feedback-induced muscle relaxation?</u> This question concerns whether feedback training in relaxing one muscle group generalizes to other muscle groups and acts to dampen activity in other bodily systems as well.

Our observations are that such generalization occurs, though not invariably. Further, our experience over the past several years has been that if a single muscle is to be used for purposes of relaxation training, then the frontalis is the muscle of choice. Once subjects are able to master this difficult-to-relax muscle, they can usually apply their newly-acquired skill to other muscle groups which have not received specific feedback training. Thus, the subject himself can deliberately produce generalization to other muscles.

There is also evidence that muscle relaxation influences other bodily systems. For example, Budzynski's (1969) experiment showed that muscle relaxation has CNS effects - - in the relaxed subject there was a decrease in the ability to discriminate temporally paired flashes of light. In the relaxation condition, the time interval between the paired light flashes had to be longer in order for the subject to see the flashes as two rather than one, an indication of lowered cortical arousal.

Especially in young normal subjects, generalization to other muscle groups appears to occur easily - - at least in the sense that tension levels decline in other bodily muscle groups as well as in the one which is receiving feedback training. However, young normal subjects are often skilled at relaxing even without EMG feedback. A recent study in our laboratory showed that subjects whose frontalis EMG levels were low prior to training (lower half of sample) did not get much lower as a result of feedback training (decline of 13%). But subjects with comparatively high initial EMG levels (upper half of sample) showed substantial decrements both during and after training (decline of 46%) - - a result clearly in keeping with the law of initial values. See Figure 4 for graph of these data. The same observation has been independently made at the University of Düsseldorf by Mr. Rolf Engel and Miss Pola Sittenfeld (personal communication).

These two independent studies underscore what will probably emerge as a major principle in the clinical application of EMG feedback training: Feedback-assisted muscle relaxation training is valuable for high EMG subjects; i.e., those who are muscularly tense. Those who are relaxed already have no need for it. Presumably, this muscularly tense population would include most anxiety neurotics (see Jacobson, 1938) and patients with stress-related disorders such as tension headache, insomnia, arthritis, ulcers, essential hypertension. Note that this conception would be in keeping with Farmer & Wright's (1971) idea of modifying the patient's most reactive stress-related response.

However, to return to the issue of generalization. What if a subject relaxes well on the muscle which is receiving feedback, but remains tense elsewhere? This is especially likely to happen in the case of pervasive anxiety patients, in whom anxiety is intense and nearly always present.

In such instances, the subject can be given feedback training on several muscle groups; e.g. forearm first, then masseter, then frontalis and, finally, all three in combination. To ensure a low arousal condition favorable for desensitization (in accordance with Lader & Mathews' theory, 1968), the patient can subsequently be given training on the feedback control of theta. This 4-7 Hz rhythm, associated with sensations of drowsiness and warmth, occurs just prior to sleep. Theta rhythms are the immediate precursor of sleep (see Rechtschaffen & Kales, 1968), and represent the furthest one can go in the direction of low arousal and still yet remain awake.

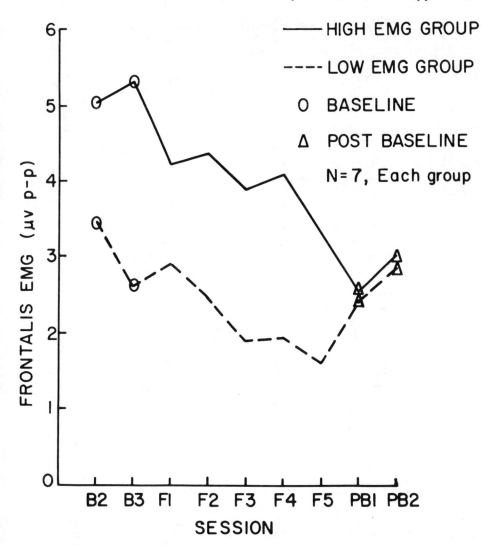

Figure 4 Frontalis EMG levels across sessions for high and low EMG subjects (all received frontalis EMG feedback).

In our laboratory, we are currently training normal subjects to produce theta voluntarily. The training, which aims at the shaping of a low arousal condition, begins with feedback-induced muscle relaxation. After the subject has become proficient at attaining muscle relaxation, a condition which is frequently associated with spontaneous increase in theta, he is shifted to training on feedback control of his theta rhythm.

Teaching patients to become skilled at readily attaining a low-arousal condition - - which is essentially a parasympathetic response - - could have many applications in anxiety and stress-related disorders, since many of these afflictions seem to involve a sustained or excessive sympathetic over-activation. If a "cultivated" parasympathetic response can be used for countering such sympathetic over-activation, then the clinical applications are considerable.

Future Directions

An exciting aspect of the biofeedback techniques is that they seem likely to become sources of innovation in therapy - - particularly when integrated with behavior therapy or autogenic training, themselves vigorous areas and fertile in ideas. One example of an innovation may be seen in a recent development in our own laboratory - - the "pervasive anxiety technique." Basically, this approach capitalizes on the semi-automatic nature of the EMG feedback procedure in order to devise a form of self-desensitization in pervasive anxiety - - a disorder which has proven quite refractory to the usual behavior therapy techniques, partly because the patient is often unable to specify the things which made him anxious. Therefore, it becomes almost impossible to construct hierarchies for use in desensitization.

Rather than attempting to construct hierarchies, our procedure involves first training the patient in thorough relaxation with feedback. This is followed by a number of sessions of self-desensitization aided by EMG feedback. In these sessions the patient initially spends 10 to 15 minutes becoming deeply relaxed - - at which time the pitch of the feedback tone indicates a low EMG level. The patient then begins to think about whatever seems to be bothering him that day. Whenever the rising pitch of the feedback tone indicates a build-up of tension, the patient stops thinking about the anxiety-evoking situation and makes use of the feedback tone in order to guide himself back to a relaxed condition. Once back to a low level, he again resumes thinking about the bothersome situation. This cycle is repeated until he can remain relaxed while thinking of this situation. The patient then shifts to some other anxiety-evoking thought. After the session the patient informs the therapist of the situations he found to be associated with increased tension. Once identified, these anxiety sources also can be broken down into hierarchies for more specific desensitization.

Another possibility in behavior therapy, as already mentioned in the previous section, is that several physiological parameters could be simultaneously monitored during desensitization. The pace of desensitization could be regulated by the patient's performance on a <u>cluster</u> of physiological measures rather than only by EMG levels. In some cases, a computer might be useful. For a patient undergoing desensitization, his responses on a cluster of physiological responses - - and perhaps even his verbal responses as well - - could be processed by an on-line computer program. This same computer would supply various types of visual and auditory feedback to the patient. Some ingenious work along these lines, focussing particularly on the use of computers in exploring systematic desensitization, has recently been described by Lang (1969).

An attractive practical aspect of the feedback technique is that it would allow a division of labor in the therapy process. Feedback-assisted relaxation training, for example, could readily be carried out by a technician rather than by an already-overburdened professional. Moreover, a number of patients could be learning the relaxation technique simultaneously. Such a development has already taken place in autogenic training (Stokvis and Wiesenhütter, 1963).

Biofeedback techniques could readily be applied to autogenic training. Rather than relying primarily on the patient's verbal report of progress, feedback methods could be used both to assist in training and to quantify the patient's developing ability at controlling particular responses.

The applicability of feedback methods is specifically relevant in several of the exercises. For example, in the first standard exercise, in which the patient aims at relaxing his muscles sufficiently to produce a feeling of heaviness in the limbs, EMG feedback could be advantageous. In the second exercise, designed to produce sensations of bodily warmth, surface temperature monitoring and temperature feedback would probably be useful. Green, Green, and Walters (1970) have noted that, with feedback, subjects are able to raise finger temperature by about 3° F. after three or four training sessions. Heart rate feedback may prove useful in the third standard exercise, which is designed to produce slowing of the heart beat. Feedback of respiratory activity should be very useful in the fourth standard exercise, in which the patient aims at producing calm and regular breathing. Research in this area has been actively pursued by Green et al (1970) at the Menninger Foundation.

Feedback training of certain brain wave rhythms might also be profitably integrated with either behavior therapy or autogenic training. Wolpe and Lazarus (1966), for example, have used a thought-stopping technique in cases of compulsive ideation. The patient is commanded to "stop thinking!" Training on the feedback control of alpha, a rhythm which is associated with a relaxed attitude, a feeling of "letting go," and a dampening of visual imagery (Nowlis and Kamiya, 1970), might be useful for such patients. To stop his obsessive ruminations, the patient would "go into alpha." Similarly, feedback training on theta (4-7 Hertz), a rhythm associated with drowsiness, might be useful in alleviating sleep-onset insomnia, a disorder to which the relaxation and warmth exercises of autogenic training have frequently been applied (Luthe, 1969). Training in theta control has been initiated in several laboratories; Green, Green and Walters (1970), Brown (1970), and in our Colorado Laboratory.

Another promising source of clinical innovation is the work on autonomic conditioning, a vigorous research area. Miller (1969), DiCara (1970) and their associates have shown that the range of autonomic responses which can be instrumentally conditioned is truly astonishing. Using intracranial electrical stimulation as reinforcement, these investigators in an extensive series of experiments have instrumentally conditioned heart rate increases and decreases, blood pressure increases and decreases, stomach contractions, intestinal motility, rate of urine formation in the kidney, and regional bloodflow. In one experiment, rats were even able to learn the remarkably specific response of increasing or decreasing the bloodflow to one ear.

A significant feature of these experiments is that the animals were heavily curarized, a condition which unexpectedly resulted in a far stronger degree of autonomic conditioning than occurred in non-curarized preparations. Consequently, a major challenge in this area is whether the dramatic autonomic conditioning achieved in curarized animal preparations can be attained in the intact, non-curarized human. Several laboratories are active in this endeavor; e.g., Shapiro, Tursky, Gershorn and Stern (1969), Engel and Hansen (1966).

It seems likely that the next few years will see the evolution of a combination of techniques for dealing with psychosomatic or stress-related disorders. Such a combination of techniques might be attempted, for example, with essential hypertension, particularly if the disease is in its early stages prior to any permanent renal or vascular changes. First, systematic training in general muscle relaxation could be used in training patients to lower their arousal levels. Sources in the autogenic training literature (see Luthe, 1969, Vol. II, p. 70) report that relaxation training often produces a reduction in blood pressure levels (see also Jacobson, 1938, p. 423). Second, patients could be trained with specific feedback of blood pressure to help them to produce lower pressures, as in the work of Shapiro et al (1969). Finally, if these patients are afflicted by various anxieties, which aggravate their disorder, then systematic desensitization could be used to help them moderate their anxieties.

If the biofeedback techniques are able to prove their mettle in anxiety and stress-related disorders (see Budzynski, Stoyva, & Adler, 1970, for a biofeedback application to tension headache), then the practical consequences are considerable. These techniques may well evolve into a means of modifying man's "defense-alarm" reaction. This powerful, sympathetico-adrenal response appears to be an integral part of the various stress disorders. Though adaptive under conditions of primitive living where strenuous physical exertions are necessary for survival, under conditions of civilized living, the sustained evocation of the defense-alarm reaction seems likely to lead to stress-related disorders. Evidence supporting this idea has been advanced by Charvat, Dell, Folkow (1964), Simeon (1962), Wolff (1968). Methods of modifying the reaction to stress, such as those suggested by biofeedback techniques, could have widespread applications in the area of psychosomatic disorders.

References

Brown, B.B. Recognition of aspects of consciousness through association with EEG alpha activity represented by a light signal. Psychophysiology, 1970, 6, 442-452.

Budzynski, T.H. Feedback-induced muscle relaxation and activation level. Unpublished doctoral dissertation, University of Colorado, 1969.

Budzynski, T.H., & Stoyva, J.M. An instrument for producing deep muscle relaxation by means of analog information feedback. Journal of Applied Behavior Analysis, 1969, 2, 231-237.

Budzynski, T.H., Stoyva, J.M., & Adler, C.S. Feedback-induced muscle relaxation: Application to tension headache. Behavior Therapy and Experimental Psychiatry, 1970, 1, 205-211.

Cameron, D.E. Observations on the patterns of anxiety. American Journal of Psychiatry, 1944, 101, 36.

Charvat, J., Dell, P., & Folkow, B. Mental factors and cardiovascular disorders. Cardiologia, 1964, 44, 124-141.

DiCara, L. Learning in the autonomic nervous system. Scientific American, 1970, 222, 30-39.

Engel, B.T., & Hansen, S.P. Operant conditioning of heart rate slowing. Psychophysiology, 1966, 3, 176-187.

Farmer, R.G., & Wright, J.M.C. Muscular reactivity and systematic desensitization. Behavior Therapy, 1971, 2, 1-10.

Fisher, S., & Cleveland, S.E. Body image and personality. Princeton: VanNostrand, 1958.

Gellhorn, E. Motion and emotion. Psychological Review, 1964, 71, 457-472.

Gellhorn, E. Autonomic-somatic integrations. Minneapolis: University of Minnesota Press, 1967.

Germana, J. Central efferent processes and autonomic-behavioral integration. Psychophysiology, 1969, 6, 78-90.

Green, E., Green, A., & Walters, D. Voluntary control of internal states: Psychological and physiological. Journal of Transpersonal Psychology, 1970, 1, 1-26.

Hess, W.R. Diencephalon: Autonomic and extrapyramidal functions. New York: Grune & Stratton, 1954.

Jacobson, E. Progressive Relaxation. (2nd ed.) Chicago: University of Chicago Press, 1938.

Jacobson, E. Modern treatment of tense patients. Springfield, Ill.: Charles C. Thomas, 1970.

Kamiya, J. Conscious control of brain waves. Psychology Today, 1968, 1, 57-60.

Lader, M.H., & Mathews, A.M. A physiological model of phobic anxiety and desensitization. Behaviour Research and Therapy, 1968, 6, 411-421.

Lang, P.J. The on-line computer in behavior therapy research. American Psychologist, 1969, 24, 236-239.

Lazarus, A.A. A preliminary report on the use of directed muscular activity in counterconditioning. Behaviour Research and Therapy, 1965, 2, 301-303.

Luthe, W. (Ed.) Autogenic therapy. New York: Grune & Stratton, 1969, (Vols. I-V).

Malmo, R.B. Studies of anxiety: Some clinical origins of the activation concept. In C.D. Spielberger (Ed.), Anxiety and behavior. New York: Academic Press, 1966, 157-177.

Miller, N. Learning of visceral and glandular responses. Science, 1969, 163, 434-445.

Nowlis, D.P., & Kamiya, J. The control of electroencephalographic alpha rhythms through auditory feedback and the associated mental activity. Psychophysiology, 1970, 6, 476-484.

Obrist, P.A., Webb, R.A., Sutterer, J.R., & Howard, J.L. The cardiac-somatic relationship: Some reformulations. Psychophysiology, 1970, 6, 569-587.

Paul, G.L. Outcome of systematic desensitization. II: Controlled investigations of individual treatment, technique variations, and current status. In C.M. Franks (Ed.), Behavior therapy: Appraisal and status. New York: McGraw-Hill, 1969, 105-159.

Rachman, S. The role of muscular relaxation in desensitization therapy. Behaviour Research and Therapy, 1968, 6, 159-166.

Rechtschaffen, A.,& Kales, A. A manual of standardized terminology, techniques and scoring system for sleep stages of human subjects. Washington, D.C.: United States Government Printing Office, 1968.

Schultz, J.H., & Luthe, W. Autogenic training: A psychophysiological approach in psychotherapy. New York: Grune & Stratton, 1959.

Shapiro, D., Tursky, B., Gershon, E., & Stern, M. Effects of feedback and reinforcement on the control of human systolic blood pressure. Science, 1969, 163, 588-590.

Shaw, J.C. Integration technique. In P.H. Venables and I. Martin (Eds.), Manual of psychophysiological methods. New York: John Wiley & Sons, 1967, 403-465.

Simeons, A.T.W. Man's presumptuous brain: An evolutionary intrepretation of psychosomatic disease. New York: E.P. Dutton & Co., 1962.

Stokvis, E., & Wiesenhütter, E. Der Mensche in der Entspannung, (2 Auflage), Stuttgart: Hippokrates Verlag, 1963.

Stoyva, J.M., & Kamiya, J. Electrophysiological studies of dreaming as the prototype of a new strategy in the study of consciousness. Psychological Review, 1968, 75, 192-205.

Venables, P.H., & Wing, J.K. Level of arousal and the subclassification of schizophrenia. Archives of General Psychiatry, 1962, 7, 114-119.

Wolff, H.G. In S. Wolf (Ed.), Harold G. Wolff's 'Stress and disease.' (2nd Ed.) Springfield, Ill.: Charles C. Thomas, 1968.

Wolpe, J. Psychotherapy by reciprocal inhibition. Stanford: Stanford University Press, 1958.

Wolpe, J., & Lazarus, A.A. Behavior therapy techniques. New York: Pergamon Press, 1966.

Interactions Between Learned and Physical Factors in Mental Illness

Neal E. Miller

SINCE THIS PAPER BRIEFLY COVERS some 30 years of work, my discussion will be limited to a few unrelated high points with an effort to relate them to our interdisciplinary goals.

The basic hypothesis that John Dollard and I[4] formulated many years ago was that, to the extent that functional neuroses are acquired during one's lifetime, they must be learned either by known laws of learning or by new laws yet to be discovered. In this review, primary emphasis will be placed on the neuroses rather than on the psychoses. Since it is obvious that there is no completely clear-cut distinction between the two, in that learning does enter into both, it remains to be seen whether its role is primary or secondary. The effect of learning on psychoses is demonstrated quite clearly by the contents of delusions. People no longer are likely to be deluded into thinking that they are Napoleon Bonaparte or Franklin Delano Roosevelt, but their delusions have other, more modern contents.

FEAR AS A LEARNED DRIVE MOTIVATING FURTHER LEARNING

Figure 1 shows the apparatus used in an experiment illustrating some of the properties of fear. The left side is white and has a grid; the right side is black and has a smooth floor. If animals are placed within this apparatus, with the door separating the two compartments open, they will wander casually about. But their behavior will be radically changed if for a number of trials, they are put into the left side, where a strong shock is sent through the grid and where they are allowed to escape through the open door to the other compartment. On subsequent trials where the shock is absent they will run rapidly from the left to the right. But is this running the automatic persistence of a habit or is a learned drive involved? From the urination and defecation of the animals and their generally agitated behavior, one might think that they had

---

Supported by USPHS Grants MH 13189, MH 19183, and MH 19991.

*Although the original experiment described here is subject to the criticism that the electric shocks might potentiate an innate preference for a specific compartment or that the motivation might come from the blocking of a learned habit of running, instead of from fear, similar results have been secured in a subsequent experiment on cats that was designed to eliminate both of these alternative explanations.[17]

Reprinted by permission of Grune & Stratton, Inc. and the author from *Seminars in Psychiatry*, Vol. IV, No. 3, 1972.

learned to be afraid. The way to test this is to prevent the door from dropping open. With his escape blocked, the animal will indeed get very excited. He will scramble around and eventually turn the wheel which is connected so that it causes the door to drop open. Figure 2 shows that during a series of such trials the rats will learn the new habit of promptly rotating the wheel.*

LEARNING OF SYMPTOMS

If you observe the animal, you will see that his behavior is very peculiar—he has a compulsion to rotate the wheel and run through into the other compartment. But, if you know the history, this bizarre behavior is perfectly understandable. It is compatible with the idea that the rat is motivated by a learned drive to fear the white compartment, which acts just the same as any other drive such as hunger, and that the escape from this fear functions as a reward for rotating the wheel just as food does for a hungry animal. As the animal becomes skillful at rotating the wheel, his behavior becomes more and more casual, so that you would not notice any overt symptoms of fear at all and might indeed be puzzled by the way the animal calmly and persistently performs this act. One might note an analogy with the "belle indifference" of some hysterical patients about their symptoms. But, if you block the habit so that animal can no longer get to the wheel or if the wheel no longer works to drop the door open, the fear immediately returns, as is shown by urination, defecation, and agitated behavior. Again one can note an analogy to the fact that the blocking of at least certain neurotic symptoms produces an increase in anxiety which is concealed, so to speak, behind the symptom.

Figure 3 shows that if you make the wheel nonfunctional, the habit of turning the wheel extinguishes, demonstrating that the reduction in fear by escaping to the other compartment was essential for maintaining this habit. But, as Fig. 3 also shows, if you

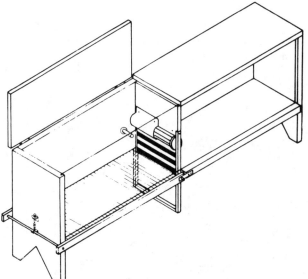

Fig. 1. Apparatus for demonstrating that fear functions as a learned drive and a reduction in fear as a reward. (By permission.[12])

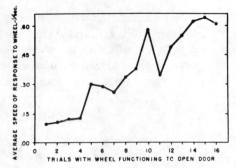

Fig. 2. Learning of a new response, that of rotating a wheel, which allows the rat to escape from a situation that elicits fear. (By permission.[12])

make the bar functional so that it will cause the door to drop, the animal will learn a new habit of pressing the bar. This seems like the symptom substitution which sometimes occurs if a symptom is merely blocked without solving the patient's emotional problem.

If you merely continue to give the animal additional trials without electric shocks, it takes an enormous length of time for this habit to extinguish. With the rat, the habit may last for as many as 500 trials. In a similar habit with a higher organism, such as a dog, the habit may persist for thousands of trials without extinguishing.[26] But if eventually you do extinguish or counter-condition the fear, not only the performance of the original habit disappears but also the tendency to learn any new substitute disappears. I think that this removal of the underlying drive of fear may be analogous to the goal of various types of psychotherapy, be they Freudian or behavioral. It is interesting that as the behavior therapists are becoming more experienced, they are finding that it is necessary to locate and deal with the patient's real phobia rather than some subsidiary symptom.[20]

The animal model just described is quite applicable to phobias and also to certain compulsions whose interruption induces fear.[4] It also seems to be readily applicable to combat neuroses, where the source of the fear often is quite clear and the fear-reducing value of the symptom that allows the soldier to escape from combat provides a clear explanation for the reinforcement of that symptom. The hysterical paralysis of the trigger finger or of the legs of an infantryman or an interference with the depth perception of a pilot are examples. Certain types of combat amnesia that are not

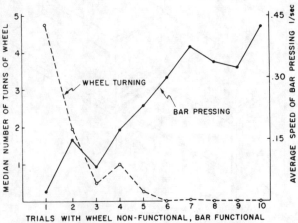

Fig. 3. Extinction of the response of rotating the wheel when it is no longer rewarded by escape from fear; learning of a second response, pressing a bar, after it begins to be rewarded by escape from fear. (By permission.[12])

readily explainable by head injury are another example. Our analysis is that the intense fear aroused by the terrifying memories motivates one to stop thinking about them and the consequent relief reinforces the inhibition of thought. If those of you who have a mild fear of heights had to jump across a 5-foot gap from the roof of a skyscraper to a ledge, you might well run up to the edge and suddenly find yourself unable to jump. Our hypothesis is that a train of thought is a series of responses just like running and that stopping a specific train of thought is a response just like the stopping of running.

Now, if you were given a drug that reduced your fear enough, it is conceivable that you would be able to jump. But, if you landed on a precarious ledge, you might find yourself still more frightened. Grinker and Spiegel[6] found that intravenous injections of a barbiturate often could relax a patient with combat amnesia so that he could recover his memories which were indeed terrifying. To quote their vivid description:

> The terror exhibited in the moments of supreme danger, such as at the imminent explosion of shells, the death of a friend before the patient's eyes, the absence of cover under a heavy dive-bombing attack, is electrifying to watch. The body becomes increasingly tense and rigid; the eyes widen and the pupils dilate, while the skin becomes covered with fine perspiration. The hands move about convulsively, seeking a weapon, or a friend to share the danger. The breathing becomes incredibly rapid and shallow. The intensity of the emotion sometimes becomes unbearable; and frequently, at the height of the reaction, there is a collapse and the patient falls back in bed and remains quiet a few minutes, usually to resume the story at a more neutral point.*

According to the theory that Dollard and I have put forward, other forms of repression are motivated in very much the same way as is combat amnesia. When the subject loses the ability to remember and to think about certain topics, he reduces his ability to make fine discriminations and to solve problems adaptively. In our analysis, we have emphasized fear (or, anxiety as it is called when its source is vague) as the motivation and a reduction in fear as the reinforcement, or, as psychiatrists frequently call it, the secondary gain. But other drives and rewards may be involved, such as anger, the need for love, the need for social approval, which may even be based on an innate human gregariousness, guilt, and the needs for achievement and self-assertiveness. Most of these motivations still are not well understood; studying them rigorously in the laboratory presents a highly significant challenge. (See Miller,[13] pp. 262-272.)

As Grey Walter contends, one way of reducing an overwhelming and incapacitating chronic fear is by minute lesions at appropriate points in the brain. We would expect such a reduction to eliminate a considerable variety of symptoms. But it is interesting that he found that the points that were most effective for eliminating extreme obsessive-compulsive behavior were somewhat different from those that were most effective for anxiety. This finding seems to rule out mere placebo effects. It suggests either that there may be different places in the brain for fears of different origins or, more probably, that some other kind of motivation besides pure fear is involved in obsessive-compulsive behavior. Additional investigation should be fruitful.

---

*For a detailed discussion of the mechanism for such an increase in fear, commonly called the negative therapeutic effect, and for the rationale of analyzing resistances first, see Miller.[13,15]

DRUG EFFECTS AND ADDICTION

The foregoing analysis has assumed that barbiturates reduce fear, and, indeed, the clinical observations by Grinker and Spiegel were the basis of a series of experiments in my laboratory, which showed that sodium amytal can reduce the fear in rats.[14] But, if this barbiturate specifically reduces fear, then we might expect a quick painless dose of it, via a chronic catheter into a vein, to act as a reward for a frightened animal. Thus, such an animal should learn to press a bar to administer the drug to itself. And, as Fig. 4 shows, this is the case.[3] While rats without shocks did not learn to press the bar for either of the two barbiturates, in two separate experiments rats, frightened by occasional inescapable shocks, did learn to press the bar for amobarbital. But, after

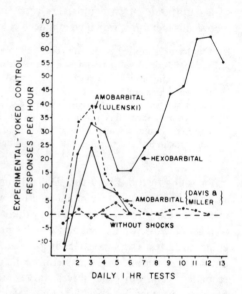

Fig. 4. Learning to press a bar reinforced by an injection of a fear-reducing drug. When an experimental rat pressed a bar both he and a yoked control (whose bar functioned only to record his presses) received an immediate injection via a chronic catheter into a vein. The experimental rats learned when periodic inescapable brief electric shock created a chronic fear situation but did not learn if they received no shocks. The control rats did not learn. In two experiments on sodium amytal the rewarding effect of the drug wears off after several days while the effect of the faster-acting hexobarbital appears to be more permanent.[3]

several days, the rewarding effect weakened and disappeared. Unfortunately, in clinical practice the initial excellent fear-reducing properties of a drug often weaken with repeated administration.

Since immediate rewards are more effective than delayed ones, we would expect a quicker-acting barbiturate to be more effective than a slower-acting one. And the quicker-acting of the two drugs, hexobarbital, was more effective as a persistent reward.

I believe that the foregoing experiment illustrates one of the mechanisms that may be involved in addiction. Escape from the aversive withdrawal symptoms is another well-known mechanism. It may be that some drugs have also a direct rewarding effect.[22] The foregoing mechanisms do not need to be mutually exclusive; in some cases they may potentiate each other.

The social and behavioral effects of alcohol have been a puzzle. It is supposed to be a central nervous system depressant, but the increase in the decibel level at a cocktail party does not seem to be an obvious symptom of such depression. It is supposed to depress higher functions first, but it is not clear that standing mutely and shyly in a corner involves higher cortical functions than becoming the life of the party. Conger[1] solved some of these paradoxes by showing that in an approach-avoidance conflict, alcohol reduces the fear-motivating avoidance more than the appetitive drive of the hunger-motivating approach. Thus, hungry animals that have been prevented from approaching food by receiving an electric shock at the goal resume running completely up to the goal if they are given a mild dose of alcohol. Figure 5 shows the results of an experiment in which separate groups of animals were trained either to approach or to avoid, and the strength of each tendency was measured by the force which the animal exerted when temporarily restrained. It is clear that the alcohol reduced the avoidance motivated by fear more than the approach motivated by hunger. Many other experiments on conflict behavior are relevant to mental illness, but it would require a separate paper to do them justice.[13]

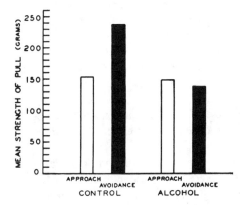

Fig. 5. The effect of 1.5 ml of 10% alcohol per 100 g of body weight given intraperitoneally on approach and avoidance responses measured separately by strength of pull exerted during temporary restraint halfway down a runway. (By permission.[1])

VISCERAL LEARNING AND CHOICE OF PSYCHOSOMATIC SYMPTOM

I would like you to recall the first experiment in which rats learned to rotate the wheel to escape from the white compartment into the black one. This experiment involved two stages of learning: (1) By a classical conditioning procedure, the fear originally elicited by an electric shock was transferred to the white compartment; and (2) by an instrumental training procedure, the response of rotating the wheel was learned in order to escape from the fear-inducing white compartment. This second step is what Thorndike[27] called trial-and-error learning, and what Skinner[24] has renamed and popularized as operant conditioning.

It is widely believed that classical conditioning is a simpler type of learning and that the autonomic nervous system that mediates glandular and visceral responses is less intelligent so that it can be modified only by classical conditioning. Conversely, it is believed that only the skeletal responses mediated by the presumably brighter somatic nervous system are subject to the more sophisticated type of instrumental learning. Similarly, many, but not all, psychiatrists believe that the psychosomatic symptoms mediated by the autonomic nervous system are a more primitive type of symptom that cannot be influenced by symbolic factors or the secondary gains that I would call rewards. The problem is important because a classically conditioned response must be reinforced by an unconditioned stimulus that elicits exactly the same response that is to be learned. But in instrumental learning, the reward does not elicit the response to be learned; it reinforces any immediately preceding response. Thus the same reward may be used to reinforce any one of a number of different responses, as when one uses food to train a dog to sit up or lie down. Similarly, a given response may be reinforced by a number of different rewards—food for a hungry dog or water for a thirsty one. If the instrumental learning of glandular and visceral responses is possible, there are many more chances for psychosomatic symptoms to be reinforced and learned.

Determining whether or not visceral responses are subject to instrumental learning runs into the problem of separating out the effects of the skeletal muscles since the two systems are intimately and adaptively interrelated in maintaining homeostasis during normal behavior. If I offered you $100 to speed up your heart rate, you would probably think for a moment and then run up a flight of stairs and come back to collect your money. But from my point of view that would be cheating because you would be using the control of your skeletal muscles to produce an indirect effect on your heart rather than directly controlling your heart rate. A subtler method of affecting your heart rate is to change the depth or rate of your respiration. In order to rule out such indirect effects, we paralyzed the skeletal muscles of our rats with curare, which left the autonomic responses relatively unaffected, and maintained them on artificial respiration. It was shown later that curarization also improved visceral learning, presumably by removing a lot of "noise" from the situation and reducing the possibility that rewarding skeletal responses produced indirect effects that masked the direct ones.

Our procedure was to record accurately the visceral response, and immediately reward those small spontaneous fluctuations that were in the desired direction. After the animal had learned small changes, we shaped him by gradually requiring larger and larger ones to reach the criterion level that automatically and immediately delivered the reward.

But how does one reward an animal that is paralyzed by curare? We have used two different types of reward: the direct electrical stimulation of rewarding areas in the brain, and the escape from mild electric shocks to the tail.

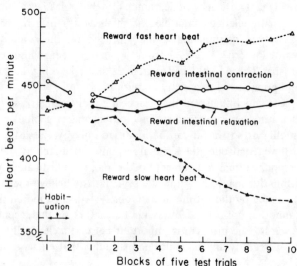

Fig. 6. Curarized rats rewarded for increase in heart rate will increase it but those rewarded for decrease will decrease it. (By permission.[18])

Figure 6 shows an experiment in which half of the rats were rewarded for changes in heart rate and the other half were rewarded for changes in intestinal contraction. Half of each group were rewarded for increases and the other half for decreases. The animals that were rewarded for a fast heart rate learned to increase their heart rate while those that were rewarded for a slow heart rate learned to decrease it. This technique of training groups in opposite directions controls for classical conditioning and for other effects of the procedure. You will notice that those rats rewarded for increased or decreased intestinal contraction did not change their heart rates at all.

Figure 7 shows that if the animals were rewarded for increasing their intestinal contractions, they learned to increase them; if they were rewarded for decreases, they learned to decrease. But the rats rewarded for changes in heart rate did not change their intestinal contractions. This specificity of the learning controls for a number of other factors, e.g., that we were not just increasing the general arousal of the rats.

To summarize a large amount of other work in this area,[16] we have successfully

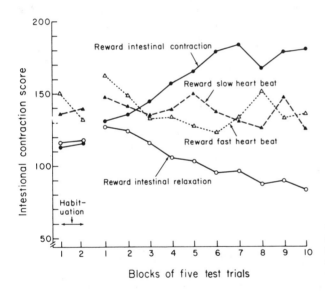

Fig. 7. Curarized rats rewarded for increased intestinal contraction learn to increase while those rewarded for decreases learn to decrease. (By permission.[18])

trained both increases and decreases in blood pressure without any changes in heart rate. Similarly, we have trained rats to increase and decrease urine formation without any changes in heart rate or blood pressure. But the blood flow through the kidneys was increased when we rewarded increases and decreased when we rewarded decreases in urine formation. Other responses have been trained, such as contraction of the uterus and vasomotor responses. Perhaps the most interesting study was an experiment where we specifically rewarded and trained differential vasomotor responses in two ears. I think you will agree that it is hard to imagine any skeletal response or even any thoughts which would cause the rat to blush in one ear but not in the other. With the specificity that it is possible to achieve in visceral training, I hope that such training will be useful in investigating the mechanisms of the control of these responses and in investigating the effects of drugs on them.

So far, the results of visceral learning are good. But now we come to an extraordinarily perplexing and vexing phenomenon, which is shown in Fig. 8. In this summary of the results of a number of studies on heart rate, most of them involving relatively similar procedures in our own laboratory, it can be seen that as time has marched on the learned differences have become progressively smaller! In fact, right now the differences that we are obtaining are still smaller, as might be expected from an extrapolation of the curve.

In addition to the decline in learned changes of heart rate, the baseline heart rates before training are from 50-100 beats per min higher than those in the earliest experiments, and at this higher rate they are much less variable, which may be a part of the difficulty. The high initial rate can be reduced by respirating the rats on a mixture of 95% air, and 5% carbon dioxide. This procedure seems to restore the general reactivity of the system and at first also seemed definitely to improve the learning, although that now seems to be another false lead.

Gorski[5] has found that the amount of testosterone necessary to sterilize a neonatal female rat has changed over the past 10 yr from an initial level of approximately $9\mu g$ to a present one of approximately $90\mu g$. I would be very interested to learn of any

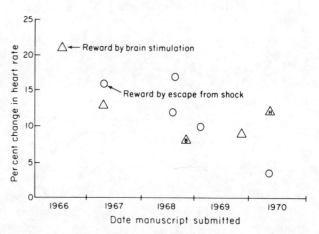

Fig. 8. Progressive decline in amount of learned change in heart rate during a 5-yr period. Experiments are performed under approximately similar conditions in the author's laboratory, except for point B, which is from data by Hothersall and Brener[7] and point H, which is from data by Slaughter et al.[25] For each experiment, the increases and decreases are converted to percentage changes and then averaged.

similar changes that others have observed in the behavioral, pharmacologic, or physiologic reactions of rats during the last 10 yr. Are there strain differences caused by the procedure of mass breeding for rats that do not bite experimenters? Can the widespread use of antibiotics in animal husbandry be a culprit? Is there some subtle form of contamination in the environment that is polluting our results?

One aspect of Fig. 8 is reassuring. The two points distinguished by letters represent results from other laboratories of replications that are statistically highly reliable.

Now let us turn to the really interesting problem, namely, evidence on human visceral learning. A number of ingenious and courageous investigators have shown that people can learn to change their heart rate, blood pressure, or vasomotor responses. In general, the changes have been smaller than those achieved in our earlier experiments on curarized animals. It has been harder to rule out mediation via changes in breathing and other skeletal responses, so the interpretation has been more controversial.[2,8,9] On the other hand, recent experiments have shown that certain selected subjects can produce small changes in blood pressure without changes in heart rate, or changes in heart rate without changes in blood pressure.[23] This specificity suggests that the changes are not indirect ones unless there is some way of using skeletal responses to produce independently these two changes.

Some of the best evidence for human visceral learning, available for some time, has been neglected because of our disciplinary compartmentalization. The urethral sphincters are innervated exclusively by the autonomic nervous system but man has learned to control his urination. The conditions are favorable for such learning; there is strong motivation and immediate knowledge of results. But there has been some controversy about whether skeletal musculature may be involved indirectly in such control. This point has been resolved by a heroic experiment in which Lapides et al.[10] completely paralyzed 16 human subjects—some by curare and some by succinylcholine—so that they had to be maintained on artificial respiration. As has been known by clinicians who have used curare for treating tetanus, these subjects did not become incontinent. The new finding was that these subjects could initiate urination on command about as fast as ever, and could terminate it in about twice the time required when they were not paralyzed, stopping in spite of the fact that a large amount of fluid still remained in the bladder. Thus, it is quite clear that these people could control urination when all effects of contractions of the skeletal musculature were ruled out.

Other evidence, which has not been verified in such elegant detail, comes from the observation that many small boys learn to control their tears in the presence of their peers but may release them in the presence of a more sympathetic audience, such as their mothers. Furthermore, some actors and actresses can cry at will. Some of these claim they do it by recreating the details of an extremely sad situation, but others claim that they do not need to feel sad in order to cry, any more than they need to feel mad to clench their fists convincingly. It would be interesting to study the pattern of autonomic responses of these two schools of thought to see if the responses of the latter would be more specific than those of the former.

I believe that further study of animal and human visceral learning, and its possible therapeutic application, is a highly significant area of research. But I deplore the exaggerated publicity in the newspapers about this and other kinds of work commonly called biofeedback. Such exaggerated articles are raising impossible hopes which will result in a premature disillusionment, thus preventing the hard work necessary to prove what therapeutic usefulness or etiologic understanding can be derived from this type of research.

It is quite natural that the first investigations have concentrated on producing an effect. The experienced investigator has learned that it is a waste of time to devise elaborate, time-consuming controls before he has evidence that there may be a phenomenon to control. But once we reach the stage of being able to get clinically significant effects, it will be extremely important to put in careful controls for placebo effects, i.e., the effects of the hope and the suggestion produced by the impressive apparatus and of the personal attention that the therapeutic coach gives to the patient.

EFFECTS OF LEARNING ON STRESS; PSYCHOSOMATIC CONSEQUENCES

In conclusion, I would like to talk about another way in which learning can affect psychosomatic symptoms. The right side of Fig. 9 illustrates the kinds of learned effect that we have just been discussing; e.g., a rat may learn to escape fear by either increasing or decreasing his heart rate, depending on which type of change is rewarded

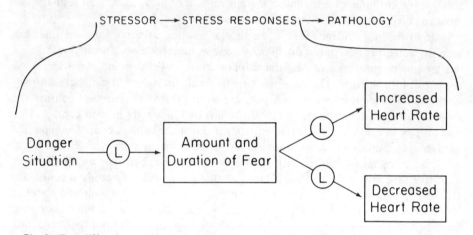

Fig. 9. Two different ways in which learning may affect a psychosomatic symptom: (1) it may affect the amount and duration of an emotional response such as fear, which in turn may have a direct innate tendency to produce psychosomatic change; and (2) it may affect the type and amount of change that is elicited in response to a given emotional stimulus.

by a reduction in the fear. The left side of Fig. 9 shows that it is possible also for learning to affect the amount and duration of the fear elicited in a given danger situation. Fear can innately produce psychosomatic effects, such as changes in heart rate or even lesions in the stomach. It is the type of learning illustrated in the left part of the diagram that will concern us here.

Observations of combat indicate that fear in situations of intermittent danger can be reduced by learning exactly what to expect and what to do (Miller,[13] p. 268). The first part of this proposition has been verified experimentally by Myers.[19] He used a conditional emotional response type of test, in which the animal's fear inhibited drinking. Animals that had a signal that enabled them to learn when to expect an electric shock (and hence when they were safe) showed less chronic fear than those for which the shocks were completely unpredictable.

Figure 10 shows the apparatus used by Weiss[28] to verify and extend this finding. The rats are semirestrained and have electrodes on their tails. The electrodes of the first two rats are wired in series, so that they must receive physically exactly the same amount of electric current. Such rats are called yoked partners. The lucky third rat is the control and receives no shock. Myers[19] used a generally similar procedure.

The rats are placed in separate, soundproof compartments. In one type of experiment,[29] the shock for one rat is predictable because a signal precedes it, but for the other rat the shock is not predictable since the signal is given at random. Figure 11, from the experiment by Weiss,[28] shows the effects of this procedure as measured by total length of lesions in the stomach. You can see that the rats with the predictable, signaled shock scarcely have more lesions than the control rats, which are confined for an equal time without food or shock. The large lesions are in the stomachs of the rats

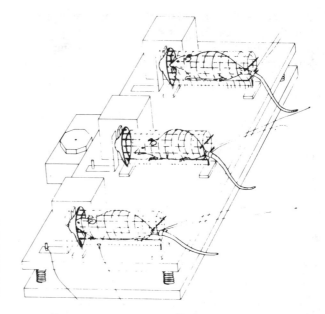

Fig. 10. Apparatus for studying psychologic factors affecting psychosomatic effects of shocks that are equally strong because the electrodes on the tails of the two yoked animals are wired in series. (By permission.[32])

with the unsignaled and hence unpredictable shock. When both rats receive physically the same electric shock, the psychologic variable of knowing when the shock is coming produces a large difference.

Returning again to Fig. 10, in a different use of the apparatus, the first rat can actuate an electronic relay by poking his nose through a hole in the tip of the cone to touch the L-shaped metal plate in front of the hole. If he performs this coping

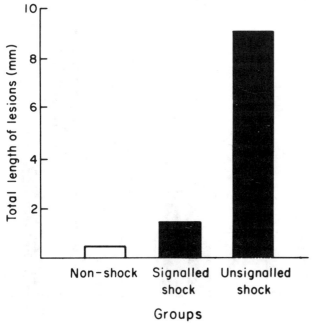

Fig. 11. The amount of stomach lesions that are produced by signaled (predictable) as compared with unsignaled (unpredictable) electric shocks of equal physical intensity.[29]

response soon enough after the danger signal comes on, he turns off that signal and avoids the shock. If he waits too long, he gets the shock but can turn it off by touching the plate. The second, or yoked, animal can perform the same response but, since the plate is not connected to any relay, the response does not have any effect. He

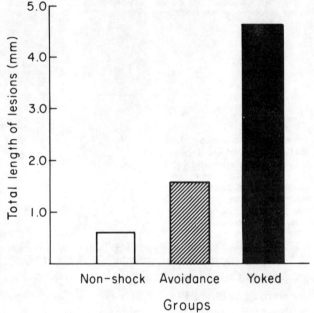

Fig. 12. Effect of being able to perform an avoidance-coping response on the amount of stomach lesions. Each yoked rat received exactly the same electric shocks as his avoidance partner, because the electrodes on their tails were wired in series.[29]

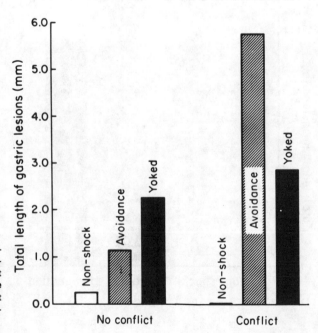

Fig. 13. Being the "executive" rat that learns the avoidance task reduces the amount of stomach lesions when the task is simple and clear-cut but increases it when the task involves conflict.[30]

is at the mercy of his partner. Figure 12 shows that this helpless, yoked rat gets by far the most stomach lesions. Although both rats receive exactly the same shocks, being able to perform the avoidance-coping response produces a great reduction in the stomach lesions.[28]

If the avoidance-coping rats are given a feedback signal which clearly informs them when they have performed a correct response, the number of stomach lesions is reduced still more.[31] The opposite is true if a conflict is introduced by giving a brief electric shock to rats who correctly perform the response that avoids the long train of shocks. This procedure reduces the safety-signal value of performing the correct response. Under this condition, the avoidance-coping rats have many more stomach lesions than their yoked controls. This result is shown in Fig. 13, where the left-hand bars are a replication of the preceding experiment and the right-hand bars show the effects of conflict.[30] In short, having to perform the response that controls the situation is greatly beneficial if the task is simple and straightforward but can be greatly detrimental if the task is difficult and produces conflict. Again we see the extreme importance of psychologic factors even though for the two groups the physical strength of the shocks, including those inducing the conflict, are held equal.

EFFECTS OF LEARNED COPING VS HELPLESSNESS ON NOREPINEPHRINE

In these experiments the psychosomatic effects of stress were measured by stomach lesions, but the effects are similar for temperature changes and also for levels of plasma corticosterone. One of the most interesting of the additional measures is the level of norepinephrine in the brain. Figure 14 shows that, compared with the normal control animals, the rats that can perform a successful escape response in a simple, nonconflictual situation have a higher level of norepinephrine in the brain. Conversely, those who are helpless because none of their responses can affect the shock have norepinephrine levels in the brain that are lower than that of the nonshocked control rats.[32] With the help of a pharmacologist, Dr. Larissa Pohorecky, Dr. Weiss is beginning to replicate these results and to extend them by determining the effects on synthesis and utilization. We also plan to see if there are any effects on the other biogenic amines.

In conclusion, I would like to point out the relationship of these last results to those presented by Dr. Axelrod in this volume and to clinical observations of the type summarized by Schildkraut.[21] As you know from such work, the drugs that are useful in treating some, but unfortunately not all, cases of depression are those that increase the effectiveness of norepinephrine (and possibly other biogenic amines) at the synapse. The drugs that have the opposite effect of reducing the effectiveness of norepinephrine (and possibly other amines) at the synapse also have the opposite effect of causing or intensifying depressions if they are given to the wrong patient. Schildkraut[21] has hypothesized that situationally produced depressions may involve the same reduction in the effectiveness of norepinephrine at the synapse. (For a study on the more complex effects of chlorpromazine, see Leibowitz and Miller.[11])

In Fig. 14 we have experimental evidence that a hopeless situation does indeed produce a reduction in the norepinephrine level in the brain, which presumably will decrease its effectiveness at the synapse. Perhaps this is a normal mechanism for producing a mildly depressed mood, which often is adaptive in preventing the animal

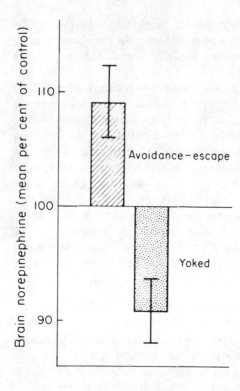

Fig. 14. Compared with nonshocked control rats, those that are able to perform an avoidance-escape response have an increased level of norepinephrine in their brain while their helpless yoked partners, who have no coping response available, have a decreased level of norepinephrine.[32]

from wasting too much energy by struggling with a hopeless situation. But perhaps when this mechanism is intensified by either a biochemical error or an extraordinarily hopeless environmental situation, it may lead to a maladaptive level of depression which in turn may create further failure, thus producing a vicious circle of reduced norepinephrine levels and continued depression. We hope to verify experimentally the other links in this hypothetical chain.

Finally, the increased level in norepinephrine in the animals that can perform a coping response may be a normal mechanism for producing an elevation of mood that helps to keep successful responses going at a high level. To continue with a possibly ridiculous speculation, many of our college students today seem to have vague symptoms of depression. They ask "Is there anything worthwhile?" But, when I grew up during the Great Depression after the stock market crash of 1929, I don't remember any of my young friends having that feeling. It seemed to me that we were so busy scrambling that we didn't get depressed. Perhaps, in an affluent section of society, a lack of need for and successful performance of coping responses results in a failure to raise the level of norepinephrine to adequate levels, and hence leads to feelings of vague depression. Perhaps we have evolved so that our mental health depends on an environment that forces us to perform coping responses.

Whether or not this last speculation is true, I trust that it is clear that learned and innate factors can interact to produce effects that are important for both mental and physical health and that the problem of understanding the details of such interactions requires interdisciplinary research of the type exemplified by the contributors to this volume.

REFERENCES

1. Conger, J. J.: The effects of alcohol on conflict behavior in the albino rat. Quart. J. Stud. Alcohol 12:1, 1951.
2. Crider, A., Schwartz, G. E., and Shnidman, S.: On the criteria for instrumental autonomic conditioning: A reply to Katkin and Murray. Psychol. Bull. 71:455, 1969.
3. Davis, J. D., Lulenski, G. C., and Miller, N. E.: Comparative studies of barbiturate self-administration. Int. J. Addictions 3:207, 1968.
4. Dollard, J., and Miller, N. E. Personality and Psychotherapy. New York, McGraw-Hill, 1950.
5. Gorski, R. A. Personal communication.
6. Grinker, R. R., and Spiegel, J. P.: War Neurosis. New York, Blakiston, 1945, p. 80.
7. Hothersall, D., and Brener, J. Operant conditioning of changes in heart rate in curarized rats. J. Comp. Physiol. Psychol. 68:338, 1969.
8. Katkin, E. S., and Murray, E. N. Instrumental conditioning of autonomically mediated behavior: Theoretical and methodological issues. Psychol. Bull. 70:52, 1968.
9. —, —, and Lachman, R.: Concerning instrumental autonomic conditioning: A rejoinder. Psychol. Bull. 71:462, 1969.
10. Lapides, J., Sweet, R. B., and Lewis, L. W.: Role of striated muscle in urination. J. Urol. 77:247, 1957.
11. Leibowitz, S. F., and Miller, N. E.: Unexpected adrenergic effect of chlorpromazine: Eating elicited by injection into rat hypothalamus. Science 165:609, 1969.
12. Miller, N. E.: Studies of fear as an acquirable drive: I. Fear as motivation and fear reduction as reinforcement in the learning of new responses. J. Exp. Psychol. 38:89, 1948.
13. —: Liberalization of basic S-R concepts: Extensions to conflict behavior, motivation and social learning. In Koch, S. (Ed.): Psychology: A Study of a Science, Study 1, Vol. 2. New York, McGraw-Hill, 1959, p. 196.
14. —: The analysis of motivational effects illustrated by experiments on amylobarbitone sodium. In Steinberg, H., de Rueck, A. C. S., and Knight, J. (Eds). Animal Behaviour and Drug Action. London, Churchill, 1964, p. 1.
15. —: Some implications of modern behavior therapy for personality change and psychotherapy. In Byrne, D., and Worchel, P. (Eds.): Personality Change. New York, Wiley, 1964, p. 149.
16. —: Learning of visceral and glandular responses. Science 163:434, 1969.
17. —: Neal E. Miller: Selected Papers. Chicago, Aldine-Atherton, 1971, p. 576.
18. —, and Banuazizi, A.: Instrumental learning by curarized rats of a specific visceral response, intestinal or cardiac. J. Comp. Physiol. Psychol. 65:1, 1968.
19. Myers, A. K.: The effects of predictable vs. unpredictable punishment in the albino rat. Ph.D. thesis. New Haven, Conn., Yale University, 1956.
20. Porter, R. (Ed.): The Role of Learning in Psychotherapy. London, Churchill, 1968.
21. Schildkraut, J. J.: Neuropsychopharmacology and the Affective Disorders. Boston, Little, Brown, 1969.
22. Schuster, C. R., Jr., and Thompson, T.: Self administration of and behavioral dependence on drugs. Ann. Rev. Pharmacol. 10:483, 1969.
23. Shapiro, D., Tursky, B., and Schwartz, G. E.: Differentiation of heart rate and systolic

blood pressure in man by operant conditioning. Psychosom. Med. 32:417, 1970.

24. Skinner, B. F.: The Behavior of Organisms. New York, Appleton-Century, 1938.

25. Slaughter, J., Hahn, W., and Rinaldi, P.: Instrumental conditioning of heart rate in the curarized rat with varied amounts of pretraining. J. Comp. Physiol. Psychol. 72:356, 1970.

26. Solomon, R. L., Kamin, L. J., and Wynne, L. C.: Traumatic avoidance learning: The outcomes of several extinction procedures with dogs. J. Abnorm. Psychol. 48:291, 1953.

27. Thorndike, E. L.: Animal intelligence: An experimental study of the associative processes in animals. Psychol. Rev. Monogr. *No. 2.*, 1898.

28. Weiss, J. M.: Effects of coping responses on stress. J. Comp. Physiol. Psychol. 65:251, 1968.

29. —: Somatic effects of predictable and unpredictable shock. Psychosom. Med. 32:397, 1970.

30. —: Effects of punishing the coping response (conflict) on stress pathology in rats. J. Comp. Physiol. Psychol. 77:14, 1971.

31. —: Effects of coping behavior with and without a feedback signal on stress pathology in rats. J. Comp. Physiol. Psychol. 77:22, 1971.

32. —, Stone, E. A., and Harrell, N.: Coping behavior and brain norepinephrine level in rats. J. Comp. Physiol. Psychol. 72:153, 1970.

Biofeedback and Visceral Learning: Clinical Applications

David Shapiro and Gary E. Schwartz

IN THE PAST DECADE, we have witnessed a major breakthrough in research on the behavioral regulation of visceral and neural processes. Contrary to previous assumptions, the accumulated evidence strongly suggests that these processes may be subject to essentially the same kind of voluntary regulation as activity of the skeletal muscles. The essence of the experimental technique in human studies is to give the person information about changes in some particular function, such as heart rate, electrodermal activity, muscle potentials, or blood pressure, and to reward him for responses of a given amplitude or direction. In studies of heart-rate control, for example, the physiologic information may be a visual display using a meter of the cardiotachometer output that indicates beat-to-beat variations in heart rate, with a flash of light or a tone for each beat-to-beat interval faster or slower than a given rate, a click for each heart beat, or the auditory output of a stethoscope. The rewards used may be interesting pictures, money, avoidance of shock, or simply success in the task of controlling an internal response. After a period of training with feedback, and reinforcement for desired responses, observations are made of changes in response rate as compared with certain base-line or control conditions.

Although there has been considerable debate over whether this process of visceral learning should be thought of as simple conditioning or some kind of cognitive control, this is a theoretical issue that needs to be clarified by more extensive studies.[8,15,30] We take the position that conditioning terminology adds precision to the design and analysis of experiments in the field. Now that the research has advanced beyond the stage of demonstrational studies, we need to investigate experiential cognitive, affective, and physiologic mechanisms of learned control of internal responses. The fact that the rate of internal physiologic processes can be increased or decreased by various individual and environmental conditions is, of course, well known.[51,53] We know very little, however, about the processes by which experience and training bring about relatively permanent modifications of visceral and other physiologic functioning. Operant feedback methods open up many new opportunities for this kind of investigation.

Supported by NIMH Grants K3-MH-20,476, MH-08853, and Office of Naval Research Contract N00014-67-A-0298-0024.

Reprinted by permission of Grune & Stratton, Inc. and the authors from Seminars in Psychiatry, *Vol. 4, 1972, 171-184.*

In the past, systematic research on learning in the autonomic nervous system has been limited largely to techniques of Pavlovian conditioning. Although it is a significant means of studying visceral learning, Pavlovian conditioning cannot account for the diverse variations in direction and pattern of visceral responses that occur with development and experience. A major failing of this model is that it does not deal adequately with individual variations in spontaneous reactivity or in sustained levels of visceral function, such as persistent muscle tension, high blood pressure, and vasomotor constriction.

In the feedback-operant model of visceral learning, the reinforcer (reward) serves to strengthen the response that it follows. What is learned need not depend on the specific response elicited by the reinforcer itself. With this model, variations in visceral learning are possible inasmuch as any given reward can strengthen any number of different visceral responses, and any given visceral response can be strengthened by any number of different rewards. The selection of physiologic response characteristics for modification, whether amplitude, waveform, latency, or direction, depends on the imagination and ingenuity of the investigator, the capability of instrumentation to identify the response of interest and provide a feedback display, and the relevance of the response—theoretical, clinical, or otherwise. It is well known that operant conditioning techniques have been used to produce variations in fine control and regulation of ongoing skeletal motor behaviors.[58] Used in conjunction with feedback methods, similar fine control would seem feasible in the case of visceral and neural processes. Naturally, built-in constraints of organ anatomy and physiology must be taken into account in considering any particular manipulation or clinical application; e.g., natural physiologic limits and homeostatic mechanisms.[47]

BASIC FINDINGS

The initial demonstrations of the phenomenon of learned operant control of internal physiologic responses were made in human subjects and, quite early, it was noted that control was facilitated if the subject was provided feedback information of continuous changes in the particular functions.[43] Also, critical to the growth of this research was the development of sensitive recording devices, electronic filters, special-purpose computers, and logic circuitry that made it possible to detect small changes in the particular function studied and to provide immediate feedback to subjects about the responses. Because of the rapidity and complexity of certain physiologic responses, such as blood pressure changes at each beat of the heart or millisecond variations in latency of the evoked cortical response, automatic instrumentation is absolutely essential in the production of required feedback displays and presentation of immediate rewards for preselected responses.

As the field developed, a number of critical issues emerged. One concerned the possible mediation of the observed visceral learning by more readily controllable somatic activities such as breathing, gross bodily movements, and isometric muscular contractions. If these changes, which we know to be readily influenced by operant conditioning, in turn elicit the appropriate physiologic changes, a simple explanation can be given for the observed learning. This problem was effectively resolved in numerous animal experiments by paralyzing skeletal muscle responses with curare, which blocks impulses at the neuromuscular function. With this procedure, and a variety of powerful conditioning techniques not possible in humans, operant learning of visceral responses in animal preparations was shown not to depend on feedback from overt motor activity. This animal research has been reviewed by DiCara and by Miller.[16,39,40]

Reviews of the human studies are found in Katkin and Murray and in Kimmel.[30,31] Selected papers have been reprinted in Barber et al., Kamiya et al., and Stoyva et al.,[3,29,60] and these collections also include related papers on hypnosis, meditation, yoga, drugs and altered states of consciousness, and cognitive control of internal and emotional states. The list of variables so far shown to be modified by operant-feedback methods in man include electrodermal activity; heart-rate speeding, slowing, and stabilization; systolic and diastolic blood pressure; gross muscle potentials; single motor unit activity; alpha, beta, and theta rhythms of the EEG; evoked cortical responses; skin temperature; peripheral vasomotor activity; and salivation.

VISCERAL LEARNING RESEARCH

The typical study in the field of visceral learning research employs a bidirectional conditioning design consisting of two groups, one reinforced for increasing the particular visceral activity and the other for decreasing the activity. The extent that the same reward can serve to increase as well as to decrease the same function rules out the simple interpretation that the obtained results depend on classical conditioning—the mere presentation of the reinforcer.

Operant control of autonomic responses is not necessarily mediated by peripheral changes in activity of the skeletal muscles. In the normal (nonparalyzed) human, it would seem reasonable to expect certain muscular changes to accompany increases or decreases in cardiovascular or other functions. However, where data have been monitored on bodily movements, muscle tension, or respiration in operant autonomic studies, most of the studies show little relationship between motor activity and the learned autonomic changes.

No clear-cut simple relationships have been demonstrated between operant autonomic control and cognitive or thought processes. There is some conjecture that the kind of thinking involved in controlling internal responses is relatively undeliberate or unconscious, a kind of passive volition or indirect process of cognitive control.[25] In the production of increased EEG alpha activity through feedback, the associated mental activity seems to be related to a state of relaxation, a feeling of letting go, and pleasant effects.[42] Although it seems reasonable to assume that mental imagery and thoughts can trigger differences in visceral activity, there have been few systematic experimental demonstrations of relationships between mental processes and specific visceral responses.[46,48] Cognitive–physiologic correlations may be difficult to tease out these human studies of autonomic control since usually only one session of training is used and the effects tend to be relatively small. In long-term studies, should larger differences be obtained, these subjective correlations may become more evident.

Conditioned changes in visceral responses are not necessarily correlated with overall changes in autonomic arousal. In fact, apparently correlated physiologic functions can be dissociated from one another by means of feedback and operant conditioning. A striking example derives from a series of studies on blood pressure and heart rate. In the first experiment normotensive subjects were given feedback and reward for increasing or decreasing systolic blood pressure while heart rate was simultaneously monitored.[52] Significant pressure changes were obtained between increase and decrease subjects without corresponding differences in heart rate. After replicating this result in a second sample of subjects, a similar experiment was performed, except that this time feedback and reward were given for increasing or decreasing heart rate while systolic blood pressure was simultaneously monitored.[54,55] This time heart rate was conditioned without corresponding changes in blood pressure. Clearly, conditioned changes in blood pressure are not dependent on heart rate, and vice versa. However, the role of other determinants of blood pressure (stroke volume, peripheral resistance) as possible mechanisms of learned blood pressure control has not been assessed.

Schwartz et al. reasoned that the degree of dissociation obtained in the foregoing experiments depends on the natural association or correlation between systolic blood pressure and heart rate.[47,49] They found that on each beat of the heart both functions are relatively independent. When either function is reinforced, the other is randomly related to it and therefore does not change. They went on to show that patterns of systolic blood pressure and heart rate could be brought under experimental control. Subjects rewarded for simultaneous increases in heart rate and systolic blood pressure showed comparable increases in both, and those rewarded for simultaneous decreases showed sizable decreases in both. Similarly, the effectiveness of the operant procedures is even more striking given the finding that differentiation of heart rate and systolic blood pressure (both moving in opposite directions) is also possible, though constrained by a number of biologic factors.

Confirmation of this view of the issue of dissociation was obtained in research on operant control of diastolic blood pressure in man.[56] In this case, in preexperimental conditions, heart rate and diastolic pressure are relatively integrated, although not completely; that is, they tend to move in the same direction at each beat of the heart. Subjects could learn to raise or lower their diastolic pressure in a 35-min training session, with differences between the two conditions of about 10–15% of base line. In this case, heart rate was also influenced to some degree in the same direction. That is, coincidental learning and not dissociation of the two functions was obtained even though only one was reinforced. Furthermore, in a later study when patterns of diastolic blood pressure and heart rate were directly reinforced, integration of the functions was readily learned, although little evidence for learned dissociation was obtained.[49a]

RATIONALE OF CLINICAL APPLICATION

It became clear to researchers, as positive results continued to emerge, that feedback-operant procedures might be of value in medical research and treatment. The major focus of interest has been on psychosomatic disorders and on the possibility of using instrumental learning concepts and techniques in understanding the etiology, maintenance, and modification of psychosomatic symptoms. Miller summarized it as follows: "... evidence of the instrumental learning of visceral responses removes the main basis for assuming that the psychosomatic symptoms that involve the autonomic nervous system are fundamentally different from those functional symptoms, such as hysterical ones, that involve the cerebrospinal system."[39] Thus, a child may be reinforced by the mother for certain symptoms, depending on her particular concern, whether about gastric distress, fever, muscle tension, vasomotor changes, heart rate, or other cardiovascular phenomena. Miller gives the example of a child who is reinforced by being kept home from school on the basis of a particular symptom and thus avoids an important examination for which he is not prepared.[39]

Other related evidence has been obtained by Benson et al., who reported that operant conditioning schedules that exert strong control over a monkey's bar-pressing behavior can also induce marked, persistent elevations in systemic mean arterial blood pressure.[4,5,28] Elevations in blood pressure can be sustained by making shock avoidance contingent on the blood pressure elevation itself, and pressure can be subsequently lowered by reinforcing decreases in pressure. The degree to which the operant learning of abnormal visceral responses can also result in organic damage is a problem under current study but no data have been published as yet.[39]

In considering the possibility of alleviating psychosomatic symptoms directly by reducing their frequency through operant feedback procedures, the assumption is made that the symptom may be dealt with directly through visceral retraining, a kind of autonomic behavior therapy. It remains to be determined whether these techniques are of real therapeutic value, and the material that follows describes some initial exploratory work that leads us to feel optimistic about the possibilities.

ESSENTIAL HYPERTENSION

High blood pressure is the primary symptom in essential hypertension. If pressure levels can be reduced by operant conditioning techniques, then an alternative to pharmacologic and surgical means of lowering pressure can be made available. High blood pressure is also associated with increased risk of atherosclerosis, nephrosclerosis, coronary artery disease, and cerebral vascular accidents,[26] and lowering blood pressure decreases this risk, providing another incentive for developing behavioral methods of controlling pressure. The experimental foundation for this work is derived in part from animal research showing that blood pressure changes can be controlled by means of operant conditioning.[4,17] Normotensive human subjects can also learn to increase or decrease systolic and diastolic pressure by operant-feedback methods.[11,47,49,52,54,55,56,61]

The first study applied the methods developed by Shapiro et al. at the Massachusetts Mental Health Center to seven patients with essential hypertension who were attending the Hypertension Clinic at the Boston City Hospital.[6] Medications were not altered during the experimental procedures. The patients were first adapted to the laboratory by attending 5-16 control sessions during which pressures were recorded but no feedback was given. In the following 1-hr sessions, feedback was given to patients on each heart beat that their systolic pressure fell below criterion levels. Patients were given 8-34 daily conditioning sessions.

Reductions in pressure from the final five control sessions to the final five conditioning sessions ranged from 16 to 34 mm Hg in five of the seven patients showing positive response to the procedure. These results are very promising, although no attempt has been made as yet to evaluate the persistence of lowered pressure outside the laboratory or its maintenance after conditioning. Inasmuch as diastolic pressure is more critical in essential hypertension because of its closer relationship to peripheral resistance, further research is needed to determine whether diastolic levels can be reduced in patients by these techniques.

CARDIAC ARRHYTHMIAS

Bernard Engel et al. have pioneered in research on the operant control of cardiac arrhythmias, the first abstract appearing in 1968.[18] Extensive clinical studies on eight patients with premature ventricular contractions (PVCs) have been reported by Weiss and Engel.[63] The patients were given feedback information about beat-to-beat changes in heart rate. A number of different training conditions were employed: heart rate speeding, heart rate slowing, alternation of speeding and slowing, maintenance of heart rate within a certain range. Whenever the patient's heart rate was doing the right thing according to the particular schedule, a light was turned on to indicate success and a meter accumulated the total time of success. Variations in procedure were determined by individual patient requirements.

Weiss and Engel reported that all of the eight patients showed some degree of heart rate control, and five were able to decrease PVC frequency in association with learned changes in heart rate.[63] Four patients persisted in low PVC frequency in follow-up studies, the longest period being 21 mo. One of the patients had no myocardial infractions during this 21-mo period, as compared with three in the 11 mo prior to the study, and Weiss and Engel conjecture that this result may be related to a decrease in PVCs.[63] Two different mechanisms were isolated on the basis of drug studies that may explain the effects of heart rate training: diminished sympathetic tone and increased vagal tone.

As to the imagery reported by patients in relation to heart-rate control, no consistent pattern was apparent. The investigators noted that one patient had to learn to be comfortable with infrequent PVCs. This patient reported early in the study that frequent PVCs felt comfortable. Another significant technique employed by Weiss and Engel was to phase out feedback for longer and longer periods of time during training in order to wean the patient from dependence on the feedback, and hopefully to have him become more closely aware of his own internal sensations associated with PVCs.[63]

TENSION HEADACHES

Tension headache is related to sustained contraction of the scalp and neck muscles. This activity of specific striate muscles becomes habitual and more or less automatic in persons with the disorder, and functionally resembles the involuntary visceral and neural responses to which operant techniques have been successfully applied.

Budzynski and Stoyva developed an instrument for producing deep muscle relaxation by means of analog information feedback,[13] and this technique was applied to five patients with tension headache, as described in Budzynski et al.[14] In their procedure, electromyogram (EMG) electrodes are placed over the frontalis muscle, and the feedback is in the form of high-pitched tone for high levels of EMG activity. As the muscle tension decreases, the tone lowers in pitch. To keep the tone low, the individual has to relax the muscle. As he improves, the gain or sensitivity of the feedback system is increased, and he is required to maintain still lower tension in order to keep the tone low in frequency.

Patients were given two or three 30-min feedback training sessions a week for 4 wk to 2 mo, depending on the severity of headache. Silent or no-feedback trials were introduced during the training to help the patients maintain muscular relaxation without the use of the feedback. Patients were asked to fill out daily headache charts and encouraged to practice relaxation at home at least once a day.

Average muscle tension and headache ratings for five patients over a 5-wk period decreased as training progressed. As patients advanced through the program, they reported a heightened awareness of maladaptive rising tension; an increasing ability to reduce such tension; and a decreasing tendency to overreact to stress. The authors attribute the first two changes to EMG feedback and the third to a general increase in relaxation outside the laboratory. Follow-up results over a 3-mo period apparently indicate that headaches remained at a low level if patients continued relaxing for a short time each day. These are very promising results on a small number of cases, although control procedures are needed to rule out alternative explanations.

OTHER PSYCHOPHYSIOLOGIC DISORDERS

The foregoing work represents the most comprehensive clinical applications to date. A variety of case studies and preliminary accounts of experiments have also been reported. An impressive case of the use of operant techniques was reported by Lang and Melamed[34] and summarized in Lang as follows:

> Recently I was asked to treat a nine-month-old boy who regularly vomited his entire meal after eating. When I first saw him he was in an advanced state of dehydration and malnutrition. He weighed less than 12 pounds, was being fed through a stomach pump, and was not expected to live.
> Extensive medical tests showed that his condition was not organic. I'm still not sure how he learned this response, but we know that what is learned usually can be unlearned. I measured the muscle potentials along the infant's esophagus and found that on the graph paper I could detect the first wave of reverse peristalsis that just preceded regurgitation. I arranged an apparatus to give aversive electric shocks to his leg whenever his esophagus started to back up, which continued until vomiting had ceased. After only a few meals with this therapy the infant ceased to vomit. He is now a healthy toddler."[33]

This infant had not been helped by any other medical procedure. Noxious conditioning as a treatment for rumination has also been reported by White and Taylor.[64] Block has described an attempt to operantly condition alleviation of pathologic nystagamus.[9] Both auditory and visual feedback procedures were based on eye-movement recordings. There was some indication of clinical improvement in several cases. Feedback for changes in peripheral blood flow and skin temperature have been used in applications to migraine and Raynaud's disease, but no systematic data have been published as yet. In the case of migraine, it has been reported in the press that Elmer Green of the Menninger Clinic has evidence that decreasing skin temperature in the head relative to the hands is beneficial. In our own laboratory we have tried to help two patients with Raynaud's disease by providing feedback for increases in blood volume using a photoplethysmograph, recording from the toe in the first patient and from the finger in the second. The first patient showed a positive response—measured increases in blood flow, less reported pain, and increased feelings of warmth in his toes and feet. This patient used imagery of the sun and heat to help him in the learning process. About 15 mo after the 20 training sessions were concluded, he reported a return of cold feet, and additional feedback sessions are planned. The second patient did not show any marked changes in blood flow, skin temperature, or subjective warmth in her fingers. A critical problem is whether to measure skin temperature directly or to take particular measures of blood flow as psysiological indices appropriate to Raynaud's disease. The possibility of applying feedback-operant techniques to abnormal EEG phenomena has also been mentioned by Lang and by Miller.[33,39]

APPLICATIONS IN PSYCHIATRY AND PSYCHOLOGY

Application of biofeedback training to other problems in psychology and psychiatry have received less attention. Such applications are generally less direct than those in psychosomatic medicine where control of the symptom itself is the desired end, as in tension headaches. As will become apparent, the following applications often involve theoretical conceptions for which the physiology assumes a mediating and/or indirect role in producing the desired psychological changes.

A number of authors have considered the application of biofeedback techniques to systematic desensitization.[66] One obvious application is in the learning of deep muscle relaxation. As described in the foregoing, Budzynski and Stoyva have developed an instrument for producing deep muscle relaxation by means of muscle information feedback.[13] Their belief was that, through EMG feedback, patients could learn to produce deeper states of relaxation, and to do so more quickly. At about the same time, Green et al. presented preliminary observations using a similar EMG-feedback instrument in which 7 of 21 subjects were able to achieve either zero firing or single-motor-unit firing in less than 20 min of a single session.[25] According to these authors, only one person could achieve such control without feedback, and he had practiced yogic meditation for a number of years. However, whether or not such muscular control has clinical value depends in large measure on the role of muscular relaxation in the desensitization of fear, a question that has not been fully answered.[38] Furthermore, even if relaxation itself proves to be of little importance, other factors, such as cognitive development of a sense of self-control, may be clinically significant in the reduction of anxiety and fear.[65]

Only in the past couple of years have researchers become aware of the role of reinforcement administered by either the therapist or patient, or both, in systematic desensitization. Success in desensitization involves progress in moving up the hierarchy, and this requires anxiety (and accompanying physiologic) reduction to a comfortable level before advancing. Since this is the goal that both the therapist and the patient desire, it is not surprising that each move up the hierarchy becomes a rewarding event. Leitenberg et al. have demonstrated that if therapist praise for small improvements is deliberately eliminated from the treatment, fear reduction is retarded.[37] Further analysis reveals that, in this situation, subjects are being rewarded not only for making the verbal or motor response signaling no anxiety, but also for not responding physiologically to the imagined scene. Consequently, it is possible to view desensitization in part as a process involving systematic reinforcement for a particular psychophysiologic state.

Recently we presented preliminary data concerning the possibility of directly controlling physiologic responses to fear stimuli through feedback and reward.[57] Based on research showing increased GSR activity to phobic stimuli, snake-phobic subjects were rewarded for either reducing or increasing their galvanic skin responses (GSRs) to pictures of snakes arranged to provoke increasing fear.[23] The data suggested that subjects could gain some control over their physiologic responses to fearful stimuli in a single session, and that this was accompanied by a relative decrease in fear as indicated by a postexperimental questionnaire. To the extent that the physiologic responses (or patterns) selected for a given subject are significantly related to his fears, and to the extent that subjects can learn to control these functions, it seems possible that feedback control of fear reactions may be possible. However, controlled clinical trials have yet to be performed. Combining this approach with desensitization therapy (e.g., using the physiology to regulate progress through the hierarchy) may prove to be a more powerful technique.

In addition to muscle relaxation training, self-regulation of other physiologic responses has been suggested for producing general relaxation and a feeling of well being. The most prominent has been EEG alpha feedback, which many researchers have reported to be associated with a pleasant, relaxed inner state.[12] Since occipital alpha is generally incompatible with tension, activity, and visual concentration, learning to produce alpha necessarily requires that subjects be in a quiet, nonactive visually passive state (with eyes closed, possibly rolled into the head). Like the EMG work, there are as yet no published studies systematically comparing alpha training with other forms of relaxation training, so the status of alpha control in terms of clinical application remains to be demonstrated.

Based on the observation that certain yogis tend to emit extensive alpha with no alpha blocking and show decreased reactivity to pain while meditating,[1] Gannon and Sternbach have attempted to treat a patient with recurrent headaches through alpha training (the patient was studied for 70 sessions).[21] Given the limitation of interpreting data from a single case, their results suggest that some reduction of pain did occur. These data are encouraging and consistent with scattered reports of the effects of yoga and meditation but controlled studies of the effects of alpha training—or meditation itself—are lacking.

It has been suggested that since alpha, and in some cases theta, often accompanies meditation, it is possible that alpha training per se will speed up the learning process in reaching this desirable state.[25] On the other hand, certain types of meditation, such as transcendental meditation, claim rapid learning (in four brief sessions of individual instruction), and argue against the notion of controlling one's physiology. Whether or not transcendental meditation leads to more rapid alpha, and whether such training proves to be clinically beneficial, is still unknown. However, preliminary work by Benson and Wallace suggests that meditation may be one substitute for drug-taking behavior.[7]

Like meditation, there are many reports concerning the clinical value of autogenic training, a technique that often requires months of training. Green et al., by giving subjects feedback for three physiologic functions at once (EMG, skin temperature changes, and alpha), are attempting to speed up the learning process through autogenic feedback training.[25]

Application of biofeedback training to processes germane to education and learning has been suggested. Kimmel et al. have found that if retarded children are rewarded with candy for producing GSR orienting responses to stimuli of different visual forms, their performance on a later form-board task is improved.[32] By tying into the physiology of attention, the goal here is to use the feedback training to increase attention and therefore increase transfer of learning. Mulholland has suggested similar applications through EEG training and clinical trials are under way.[41]

Another intriguing application related to learning is psychophysiologic training in creativity.[25] Green et al. suggest that creativity often occurs during states of consciousness similar to deep reverie, when hypnogogic-like imagery occurs. They hypothesize that through alpha and theta training subjects will enter a state of consciousness particularly conducive to creative thinking. Indirect support for this idea comes from diverse findings showing that creativity and hypnotic suggestibility are correlated[10] and that biofeedback training in alpha leads to an increase in hypnotic suggestibility.[19] However, research showing a relationship between creativity and alpha is needed, as are data demonstrating practical application to personal experience and growth.

Mention should be made of the possibility of learning to control sexual responses through feedback and reward. Two recently published studies show voluntary inhibition of penile erections without feedback[27,35] and one unpublished study[44] has demonstrated that feedback for penile circumference enhances this control. Altogether, it is apparent that at the present time an abundance of ideas coexists with a paucity of empirical data. Our evaluation of these applications must await the outcome of controlled clinical investigation.

BASIC CLINICAL PROBLEMS AND RESEARCH ISSUES

For all practical purposes the issues concerning the clinical application of biofeedback training are not specific to biofeedback per se but occur in all forms of psychotherapy. If biofeedback training is viewed as a special form of conditioning, then it clearly falls under the heading of a behavior therapy. However, whether or not one wants to classify it as a behavior therapy is not as important as is the realization that it is a therapy. It is our belief that, by considering the clinical problems and research issues in the field of psychotherapy in general, more rapid advancement in the development and evaluation of clinical biofeedback training will occur.

Central to all forms of psychotherapy is patient motivation and involvement. The question, at the simplest level, is how to motivate patients to spend the necessary time practicing the desired behavior. With severely retarded and/or emotionally disturbed children, reinforcers such as food, candy, or other desired objects (e.g., watching a fan spin) are often necessary for modifying behavior. However, with less severe problems, therapist praise, praise from significant others in the patient's environment, or self-reinforcement (a sense of competence) may be more than sufficient. One predictor of success in psychotherapy is the extent to which the patient's problems are causing him pain or suffering.[22] With regard to biofeedback, Miller has suggested that feedback for controlling the function in and of itself may be all that is necessary. However, as progress is made, and the novelty of the situation wears off, what will keep the patient working at the task? In our experience with patients suffering from essential hypertension,[6] a potpourri of rewards was successful, including money for participating and for succeeding (most of these patients were on welfare and thus the money they earned was quite significant), slides of scenes around the world, and general, nonspecific praise from the physician and research assistant running the sessions. However, this was an experiment, and the patients were receiving a sizable monetary reward. One wonders whether they would have spent the same number of hours trying to control their blood pressure in the absence of such incentives, which are impractical on a large-scale basis. In this respect, learning to control physiologic processes is like any other self-control procedure (e.g., dieting) since sacrifice on the patient's part may be necessary for the sake of health.

To the extent that physiologic learning is actually pleasurable, as is suggested to be the case for alpha, producing the state may be reward enough. However, it is not known to what extent such pleasant states are due to suggestion, or placebo effects, related to the novelty and uniqueness of brain-wave feedback.

In addition to motivation, other patient factors such as socioeconomic status, intelligence, and overall personal adjustment may be important predictors of the effectiveness of biofeedback training, as they are in psychotherapy in general.[22] Up to now most of the clinical, and experimental, work on humans has been with highly educated, motivated individuals. Whether or not feedback training can be successfully applied to other segments of the population remains to be demonstrated.

The medical suitability of patients needs to be ascertained since it seems likely that the effectiveness of feedback learning will be limited to the extent that the organ in question is already heavily diseased and physically restricted; e.g., peripheral constriction caused by a buildup of cholesterol may not be amenable to biofeedback learning. Early detection and preventive biofeedback training may be a more reasonable approach. The interaction of standard drugs used in treatment and biofeedback training is a further complication. However, it is conceivable that inducing physiologic changes through drugs in the presence of feedback may further aid the patient in associating the two. Then, by gradually reducing the amount of drug per session, it may be possible for the patient to learn to elicit comparable control himself.

The mention of drugs immediately brings to mind questions of placebo effects, and it would seem that research on biofeedback may be particularly sensitive to this variable. The potential role of suggestion in psychotherapy has been illustrated by Franks, where even death itself (voodoo death) is considered not to be immune from psychological influence.[20] Although many of the biofeedback studies using normal subjects have eliminated suggestion effects by using identical instructions for different groups of subjects and by administering the instructions before the experimental condition is determined so that unintentional nonverbal experimenter effects are eliminated,[45] most of the clinical studies have not.[66]

Although it was initially argued that cognitive factors could not account for the degree of specificity of learning in the autonomic nervous system obtained with feedback training,[15] this conclusion is tempered by data demonstrating that factors such as expectancy and suggestion can elicit specific physiologic changes. For example, Sternbach found that when subjects were given a pill that they believed would cause increased stomach activity, decreased activity, or have no effect (actually the "pill" was a simple magnet used in the measurement of stomach motility) their stomachs behaved in accordance with the anticipated effect of the drug.[59] Earlier work by the Grahams has shown that patients suffering from different psychosomatic disorders have different attitudes, and that when these specific attitudes are suggested to healthy people under hypnosis, physiologic changes occur that partially mimic the associated psychosomatic disorder.[24] Barber has argued that cases such as these are not due to hypnosis per se, and he goes on to review many case studies reporting highly specific elicitation and amelioration of psychosomatic disorders through suggestion alone.[2] Such data point to the importance of controlling for placebo effects in order to determine the role of feedback in visceral training. At our present state of knowledge, it could be argued that the first stage in clinical research should be to demonstrate significant changes with the realization that other potential variables are confounded in the design, and then evaluate the separate components.

Another approach is to view biofeedback as one additional technique; thus one research strategy might be to give a portion of patients already receiving other forms of therapy additional help with biofeedback. Then, if feedback is indeed valuable, it will be shown in terms of increased effectiveness above and beyond the combined effects of the other techniques. Not only is this design feasible and appropriate, it may prove to be the most powerful approach since the interaction of a combination of techniques for a given patient may yield the greatest success.

The concept of treating the "total person" may be important for a number of reasons, not simply for maintaining patient interest and motivation but also for heading off potential problems such as symptom substitution. It is reasonable to anticipate that if a given physiologic symptom is serving some function in the patient's environment, yielding significant secondary (or even primary) gain, then eliminating the symptom eliminates the reward, and the patient therefore has an unfulfilled need that could lead to alternate maladaptive behavior. Such an approach has been clearly expressed by Lazarus concerning behavior therapy, for he suggests that we must not just treat the symptom, but also provide alternative behaviors.[36] In other words, we must consider the patient's needs and help him learn new ways of coping and living. Teaching voluntary control may not be enough since the patient not only must want to control the function, but he must be capable of dealing effectively with his environment outside of the laboratory.

Whether or not patients can learn significant voluntary control of these functions is still a serious question. Little is known yet concerning the best techniques for providing feedback, and in what form; nor is there much information available concerning transfer of learning from external feedback control to internal control in the absence of feedback. One promising approach involves slowly weaning the patient from feedback.[63] Training the patient in self-control may require different instructions and different emphases, such as discrimination training in autonomic perception, to aid him in applying what he has learned in the laboratory to the environment outside the laboratory.

Biofeedback training in a clinical setting takes time and it is expensive and inconvenient for patients to travel to a psychophysiology laboratory to obtain such training. Clearly, the use of home feedback devices, together with regular checkups by the attending physician, may be the most practical approach. Certain types of biofeedback (e.g., alpha) devices are already being marketed on a national scale, and portable devices for determining muscle tension, heart rate, skin temperature, and skin resistance have been developed. One promising direction for research may be to use such devices under controlled conditions, thus making it possible to obtain data from many more subjects under relevant conditions in the environment.[50]

REFERENCES

1. Anand, B. K., Chhina, G. S., and Singh, B.: Some aspects of electroencephalographic studies in Yogis. Electroenceph. Clin. Neurophysiol. 13:452, 1961.

2. Barber, T. X.: LSD, Marihuana, Yoga and Hypnosis. Chicago, Aldine, 1970.

3. Kamiya, J., DiCara, L. V., Barber, T., Miller, N. E., Shapiro, D., and Stoyva, J.: Biofeedback and Self Control: An Aldine Reader on the Regulation of Bodily Processes and Consciousness. Chicago, Aldine, 1971.

4. Benson, H., Herd, J. A., Morse, W. H., and Kelleher, R. T.: Behavioral induction of arterial hypertension and its reversal. Amer. J. Physiol. 217:30, 1969.

5. —, —, —, and —: Behaviorally induced hypertension in the squirrel monkey. Circ. Res. 26, 27:21, (Suppl. 1) 1970.

6. —, Shapiro, D., Tursky, B., and Schwartz, G. E.: Decreased systolic blood pressure through operant conditioning techniques in patients with essential hypertension. Science 173:740, 1971.

7. —, and Wallace, R. K. Decreased drug abuse with Transcendental Meditation: A study of 186 subjects. In Zarafonetis, C. (Ed.): Proceedings of the International Symposium on Drug Abuse. Philadelphia, Lea & Febiger, 1972, p. 369.

8. Black, A. H.: The direct control of neural processes by reward and punishment. Amer. Sci. 59:236, 1971.

9. Block, J. D.: Operant conditioned alleviation of psychological nystagmus. Psychophysiology 5:562, 1969.

10. Bowers, K. S.: Sex and susceptibility as moderator variables in the relationship of creativity and hypnotic susceptibility. J. Abnorm. Psychol. 78:93, 1971.

11. Brener, J., and Kleinman, R. A. Learned control of decreases in systolic blood pressure. Nature (London) 226:1063, 1970.

12. Brown, B. B.: Recognition of aspects of consciousness through association with EEG alpha activity represented by a light signal. Psychophysiology 6:442, 1970.

13. Budzynski, T. H., and Stoyva, J. M.: An instrument for producing deep muscle relaxation by means of analog information feedback. J. Appl. Behav. Anal. 2:231, 1969.

14. —, —, and Adler, C.: Feedback-induced muscle relaxation: Application to tension headache. J. Behav. Ther. Exp. Psychiat. 1:205, 1970.

15. Crider, A., Schwartz, G. E., and Shnidman, S. R.: On the criteria for instrumental autonomic conditioning: A reply to Katkin and Murray. Psychol. Bull. 71:455, 1969.

16. DiCara, L. V.: Learning in the autonomic nervous system. Sci. Amer. 222:30, 1970.

17. —, and Miller, N. E.: Instrumental learning of systolic blood pressure responses by curarized rats: Dissociation of cardiac and vascular changes. Psychosom. Med. 30:489, 1968.

18. Engel, B. T., and Melmon, K. L.: Operant conditioning of heart rate in patients with cardiac arrhythmias. Cond. Reflex. 3:130, 1968.

19. Engstrom, D. R., London, P., and Hart, J. T.: Hypnotic susceptibility increased by EEG alpha training. Nature (London) 227:1261, 1970.

20. Franks, J. D.: Persuasion and Healing. Baltimore, Johns Hopkins, 1961.

21. Gannon, L., and Sternbach, R. A.: Alpha enhancement and a treatment for pain: A case study. J. Behav. Ther. Exp. Psychiat. 2:209, 1971.

22. Garfield, S. L.: Research on client variables in psychotherapy. Psychosom. Med. 24:159, 1962.

23. Geer, J. H.: Fear and autonomic arousal. J. Abnorm. Psychol. 71:253, 1966.

24. Graham, D. T., Kabler, J. D., and Graham, F. K.: Physiological response to the sug-

gestion of attitudes specific for hives and hypertension. Psychosom. Med. 24:159, 1962.

25. Green, E. E., Green, A. M., and Walters, E. D.: Voluntary control of internal states: Psychological and physiological. J. Transpers. Psychol. 2:26, 1970.

26. Gutmann, M. C., and Benson, H.: Interaction of environmental factors and systemic arterial blood pressure: A review. Medicine 50:543, 1971.

27. Henson, D. E., and Rubin, H. B. Voluntary control of eroticism. J. Appl. Behav. Anal. 4:37, 1971.

28. Herd, J. A., Morse, W. H., Kelleher, R. T., and Jones, J. G.: Arterial hypertension in the squirrel monkey during behavioral experiments. Amer. J. Physiol. 217:24, 1969.

29. Kamiya, J., DiCara, L. V., Barber, T., Miller, N. E., Shapiro, D., and Stoyva, J. (Eds.): Biofeedback and Self-Control: An Aldine Reader on the Regulation of Bodily Processes and Consciousness. Chicago, Aldine, 1971.

30. Katkin, E. S., and Murray, E. N.: Instrumental conditoning of autonomically mediated behavior: Theoretical and methodological issues. Psychol. Bull. 70:52, 1968.

31. Kimmel, H. D.: Instrumental conditioning of autonomically mediated behavior. Psychol. Bull. 67:337, 1967.

32. —, Pendergrass, V. E., and Kimmel, E. B.: Modifying children's orienting reactions instrumentally. Cond. Reflex 2:227, 1967.

33. Lang, P. J.: Autonomic control or learning to play the internal organs. Psychology Today, October, 1970.

34. —, and Melamed, B. C.: Case report: Avoidance conditioning therapy of an infant with chronic ruminative vomiting. J. Abnorm. Psychol. 74:1, 1969.

35. Laws, D. R., and Rubin, H. B.: Instructional control of an autonomic sexual response. J. Appl. Behav. Anal. 2:93, 1969.

36. Lazarus, A.: Behavior Therapy and Beyond. New York, McGraw-Hill, 1971.

37. Leitenberg, H., Agras, W. S., Barlow, D. H., and Oliveau, D. G.: Contribution of selective positive reinforcement and therapeutic instructions to systematic desensitization therapy. J. Abnorm. Psychol. 74:133, 1969.

38. Matthews, A. M.: Psychophysiological approaches to the investigation of desensitization and related procedures. Psychol. Bull. 76:73, 1971.

39. Miller, N. E.: Learning of visceral and glandular responses. Science 163:434, 1969.

40. —, DiCara, L. V., Solomon, H., Weiss, J. M., and Dworkin, B.: Learned modifications of autonomic functions: A review and some new data. Circ. Res. 26, 27:3 (Suppl. 1), 1970.

41. Mulholland, T.: The automatic control of visual displays by the attention of the human viewers. In Proceedings of 1969 Conference on Visual Literacy. New York, Pitman, 1970, in press.

42. Nowlis, D. P., and Kamiya, J.: The control of electroencephalographic alpha rhythms through auditory feedback and the associated mental activity. Psychophysiology 6:476, 1970.

43. Razran, G.: The observable unconscious and the inferable conscious in current Soviet psychophysiology: Interoceptive conditioning, semantic conditioning, and the orienting refles. Psychol. Rev. 68:81, 1961.

44. Rosen, R. C.: Voluntary control of penile volume. Unpublished.

45. Rosenthal, R.: Experimenter Effects in Behavioral Research. New York, Appleton-Century-Crofts, 1966.

46. Schwartz, G. E.: Cardiac responses to self-induced thoughts. Psychophysiology 8:462, 1971.

47. —: Voluntary control of human cardiovascular integration and differentiation through feedback and reward. Science 175:90, 1972.

48. —, and Higgins, J. D.: Cardiac activity preparatory to overt and covert behavior. Science 173:1144, 1971.

49. —, Shapiro, D., and Tursky, B.: Learned control of cardiovascular integration in man through operant conditioning. Psychosom. Med. 33:57, 1971.

49a. —, —, and —: Self control of patterns of human diastolic blood pressure and heart rate through feedback and reward. Presented at Society for Psychophysiological Research Meeting, St. Louis, Mo., 1971.

50. —, Shaw, G., and Shapiro, D.: Specificity of alpha and heart rate control through feedback. Presented at Society for Psychophysiological Research Meeting, St. Louis, Mo., 1971.

51. Shapiro, D., and Crider, A.: Psychophysiological approaches in social psychology. In Lindzey, G., and Aronson, E. (Eds.): The Handbook of Social Psychology, Vol. III (ed. 2). Reading, Mass., Addison-Wesley, 1969.

52. —, Tursky, B., Gerson, E., and Stern, M.: Effects of feedback and reinforcement on the control of human systolic blood pressure. Science 163:588, 1969.

53. —, and Schwartz, G. E.: Psychophysiological contributions to social psychology. Ann. Rev. Psychol. 21:87, 1970.

54. —, Tursky, B., and Schwartz, G. E.: Control of blood pressure in man by operant conditioning. Circ. Res. 26, 27:27 (Suppl. 1), 1970.

55. —, —, and —: Differentiation of heart rate and blood pressure in man by operant conditioning. Psychosom. Med. 32:417, 1970.

56. —, Schwartz, G. E., and Tursky, B.: Control of diastolic blood pressure in man by feedback and reinforcement. Psychophysiology, in press.

57. —, —, Shnidman, S. R., Nelson, S., and Silverman, S.: Operant control of fear-related electrodermal responses in snake-phobic subjects. Presented at Society for Psychophysiological Research Meetings, St. Louis, Mo., 1971.

58. Skinner, B. F.: Behavior of Organisms. New York, Appleton-Century-Crofts, 1938.

59. Sternbach, R. A.: The effects of instructional sets on autonomic responsivity. Psychophysiology 1:67, 1964.

60. Stoyva, J., Barber, T., DiCara, L. V., Kamiya, J., Miller, N. E., Shapiro, D. (Eds.): Biofeedback and Self-Control: An Aldine Annual on the Regulation of Bodily Processes and Consciousness. Chicago, Aldine, in press.

61. Tursky, B., Shapiro, D., and Schwartz, G. E.: Automated constant cuff pressure system to measure average systolic and diastolic blood pressure in man. IEEE Trans. Biomed. Engin., in press.

62. Wallace, R. K.: Physiological effects of transcendental meditation. Science 167:1751, 1970.

63. Weiss, T., and Engel, B. T.: Operant conditioning of heart rate in patients with premature ventricular contractions. Psychosom. Med. 33:301, 1971.

64. White, J. D., and Taylor, D.: Noxious conditioning as a treatment for rumination. Ment. Retard. 5:30, 1967.

65. Wilkins, W.: Desensitization: Social and cognitive factors underlying the effectiveness of Wolpe's procedure. Psychol. Bull. 76:311, 1971.

66. Wolpe, J.: Psychotherapy by Reciprocal Inhibition. Stanford, Stanford University Press, 1958.

XI

SELECTED ABSTRACTS

Abstracts of Papers Presented at the Eleventh Annual Meeting of The Society for Psychophysiological Research

3. Lynch, J. J. (University of Maryland School of Medicine) **Cardiac-somatic relationships in operant and Pavlovian conditioning.** Comparisons of heart rate and foot flexion responses in dogs during classical and avoidance shock conditioning revealed no difference in heart rate conditional responses (HR-CRs) during avoidance or classical conditioning. In Pavlovian conditioning, if the foot flexion CR did not occur, or if this response was blocked, the HR response to the shock unconditional stimulus (UCS) was greater than the HR-CR normally observed during shock if the classical flexion response does occur. In addition, if the flexion CR does not occur during the conditional signal (CS) then the HR-CR is significantly diminished. One group of dogs given extensive training with prolonged tone-shock intervals (40 sec) developed clear inhibition of delay in the flexion CR (after 150 trials) so that flexion did not occur until the last 10 sec during the CS. Since the HR-CR concomitantly monitored was diminished on those trials in which the dog was not going to conditionally flex, it was therefore possible to predict whether the flexion was going to occur, some 30 sec in advance of the actual flexion CR.

The classical flexion CR therefore appeared to have instrumental qualities, and the cardiac CR was at least partly coupled to the occurrence or non-occurrence of the flexion response. In these studies the flexion CR appeared to be under greater inhibitory control than the HR-CR, leading to the possibility of predicting the flexion response from the prior HR response.

4. Jones, G. B., & Fenz, W. D. (University of Waterloo) **Relationships between cardiac conditionability in the laboratory and autonomic control in real life stress.** Experienced parachutists, when tested during a jump sequence, have shown a reliable early increase in HR, followed by a steady decline; the more pronounced and orderly pattern was related to better overall performance. The conditioned cardiac response, in an aversive conditioning situation, is one of cardiac acceleration, followed by deceleration during the CS-UCS interval. A similar pattern is also found when the UCS is substituted by a signal to respond in a RT task. There is a clear analogy between the observations from real life and the laboratory; is there a relationship between the two? Continuous HR recordings of 30 experienced sport parachutists were obtained throughout 30 trials in a RT situation, which included an incentive to respond quickly; termination of an 8 sec tone served as the cue to respond. The same Ss were also tested throughout a jump sequence, and their performance was evaluated by independent raters. There was a clear relationship between laboratory and real life in (1) cardiac conditionability during the 8 sec anticipatory interval, an orderliness in the inverted V-shaped HR pattern during the jump sequence, and (2) the magnitude of the deceleratory component of the cardiac response averaged over trials, and the amount of continuous decline in HR before the jump. Both measures were directly related to performance. The discussion explores the analogies between laboratory and real life conditioning. (Supported by The Canada Council)

Reprinted with permission of the publisher from *Psychophysiology*, Vol. 9, 266-280. Copyright 1972, The Society for Psychophysiological Research.

11. **Travis, T. A., Kondo, C. Y., & Knott, J. R.** (University of Iowa) **A controlled study of alpha enhancement.** The effects of no-non-contiguous, and contiguous reinforcement on alpha enhancement were investigated. Three groups were employed: (1) experimental, N = 8 (contiguous reinforcement); yoked control, N = 8 (non-contiguous reinforcement); 3) naive control, N = 8 (no reinforcement). All Ss experienced 6 10-min sessions on each of 2 days. The feedback stimulus was a blue light and the dependent variable was percent alpha-on time. Experimental Ss produced significantly more alpha on day 1 than Ss in either of the control groups. On day 2, both experimental and yoked control groups (who had been switched to relevant feedback) produced significantly more alpha than the naive control group. Significant increases over trials were noted for experimental Ss on day 1 and for both experimental and yoked control Ss on day 2. Naive controls evidenced no changes on either day. Increases in alpha output in our situation appeared to be related to a crucial contiguity between alpha production and the delivery of the feedback stimulus.

12. **Paskewitz, D. A., & Orne, M. T.** (Institute of the Pennsylvania Hospital aand University of Pennsylvania) **Visual effects during alpha feedback training.** In a study of alpha feedback training, no increases in alpha density were observed across 6 days (60 trials) when auditory feedback was presented to 9 Ss in total darkness. For theoretical reasons, it was predicted that the mere addition of low level ambient light would establish lawful increases in alpha densities. A second sample of 7 Ss, given 10 trials a day for 2 days, confirmed the prediction. The original group of Ss, who failed to show increases in total darkness, also increased densities in the presence of light. Furthermore, with light both groups showed significantly higher densities during feedback than during intervening rest periods. These findings may help reconcile some of the conflicting results in the literature. The results support the view that increases in alpha densities can take place only if inhibitory mechanisms are initially present to depress density. Increases in alpha activity may reflect an increasing ability of the S to ignore otherwise distracting influences, and this process may account for some of the reported subjective concomitants of alpha feedback training. (This research was supported in part by the Advanced Research Projects Agency of the Department of Defense and was monitored by the Office of Naval Research under Contract N00014-70-C-0350 to the San Diego State College Foundation.)

13. **Silverman, S.** (University of Denver), &

Sherwood, M. (Massachusetts Mental Health Center) **Operant conditioning of the amplitude component of the EEG.** Feedback contingent on the amplitude component of the EEG irrespective of frequency changes was presented to Ss over 4 1-hr sessions. The feedback signal (soft tone and dim light) was presented each time the S's EEG equalled or exceeded a baseline amplitude obtained with eyes closed during the rest peroid of each session. Four "on" trials alternated with 4 "off" trials in each session. "Off" trials consisted of inhibiting the feedback signal. "On" trials consisted of producing the signal. A 5th no-feedback session concluded the experiment. There was no difference between EEG control in the 4th training session and the no-feedback session. There was a significant amplitude difference between "on" and "off" trials for sessions 4 and 5 combined. In all cases, high amplitude was associated with alpha frequencies, low amplitude with higher frequencies. It was concluded that learned control of EEG amplitude can transfer to a no-feedback condition.

14. **Schwartz, G. E., Shaw, G., & Shapiro, D.** (Harvard University) **Specificity of alpha and heart rate control through feedback.** To explore the effects of repeated alpha wave training on cardiovascular functioning, 12 Ss were pre and post tested in the laboratory on their ability to control their alpha *and* heart rate after 4 daily 1-hr sessions of feedback training to increase and decrease alpha. All Ss received full instructions concerning the nature of the feedback and the desired responses controlled with the following pre and post within S design: 3 min base period, 5 min increase alpha with feedback (100 msec tone per criterion alpha burst), 5 min decrease alpha with feedback; a second 3 min base period, 5 min decrease heart rate with feedback (100 msec tone per heart beat), 5 min increase heart rate with feedback. The pre-test data indicated that Ss were initially unable to differentially increase or decrease alpha, but they were able to produce 5 bpm differences in heart rate which were unaccompanied by alpha changes. However, after the 4 sessions of alpha training, Ss evidenced significant alpha control unaccompanied by heart rate changes, while their ability to control their heart rate was unaffected (identical to the pre-test). Analysis of the subjective report data supports the notion that the mechanisms of alpha and heart rate control are different. Implications for the direct feedback control of *patterns* of alpha and heart rate are discussed.

15. **Honorton, C., Davidson, R., & Bindler, P.** (Maimonides Medical Center) **Shifts**

in subjective state associated with feedback-augmented EEG alpha. Twenty-three Ss completed 1 session with auditory feedback to EEG alpha activity using a closed feedback loop. Ss alternated 2-min trials of alpha generation, rest periods (no feedback), and suppression. Ten trials of each type were given with order counterbalanced by S. In addition to monopolar EEG (occiput-ear lobe), EOG (supraorbital-canthus) and EMG (frontalis) were recorded. Two verbal report measures were used to assess changes in subjective activity: An 'on-line' state report scale in which Ss called numbers between 0–4 to indicate degree of relaxation and attention to external stimuli (elicited during alternate feedback trials) and a post-experimental interview. Successive trials of generation, suppression, and resting alpha were compared: gen. M alpha = 40%, suppr. = 16%, rest = 36%. The 3 conditions differed significantly ($p < .01$) for the group and each S individually had significant gen./suppr. differences ($p < .01$). Alpha gen. was associated with M state report = 2.01, suppr. M = 0.63 ($p < .005$). Ss with strong *shifts* in state reports between conditions had significantly larger alpha shifts than Ss with little shift in state reports ($p < .05$). High state reports were associated with significantly less REM activity and EMG activation than low state reports ($p < .005$). Blind coded interview responses related significantly to state reports ($p < .05$). These converging measures support the hypothesis that relatively high alpha is associated with narrowing of perceptual awareness, relaxation, and with some Ss, 'ASCs.'

16. Schwartz, G. E., Shapiro, D., & Tursky, B. (Harvard Medical School, Massachusetts Mental Health Center) **Self control of patterns of human diastolic blood pressure and heart rate through feedback and reward.** Previous research has shown that Ss can learn to directly integrate their systolic blood pressure (BP) and heart rate (HR) (raise or lower both), or to a lesser degree differentiate these functions (raise one and simultaneously lower the other) when given feedback and reward for the desired BP-HR pattern. The present experiment applied this procedure to the *diastolic* BP-HR relationship in order to determine the extent to which the pattern of two highly integrated functions could be modified through feedback and reward. Forty Ss were given 5 resting, 5 random reinforcement, and 35 conditioning trials, each trial 50 beats in length. Ten Ss were run in each of the 4 possible BP-HR pattern conditions: a 100 msec light and tone served as feedback for each correct BP-HR pattern per beat; 3 sec slides of landscapes, nude females, and money bonuses per 20 feedbacks served as rewards. The data indicated that Ss readily learned to raise or lower both their diastolic BP and HR. However, Ss were less able to raise BP and lower HR at the same time while no evidence for the opposite pattern of differentiation was obtained. The results are explained by a behavioral-physiological model of learned patterning in the ANS. Implications for specific reinforcement of peripheral resistance in hypertension are outlined.

17. Obrist, P. A., LeGuyader, D. D., Howard, J. L., Lawler, J. E., & Galosy, R. A. (University of North Carolina School of Medicine) **Operant conditioning of heart rate: Somatic correlates.** The operant modification of heart rate was attempted using shock avoidance as reinforcement in 5 groups of either 12 or 15 human Ss each. The groups varied with regard to the extent of experimental control of somatic activity. Effective discriminative control of heart rate was achieved only where 1) either no control or minimal control of somatic activity was attempted, and 2) when only heart rate increases were reinforced. Discriminative control of heart rate was either minimal or absent when 1) maximum control of somatic activity was used, both for reinforcement of increases and decreases of heart rate, and 2) where no control of somatic activity was attempted and heart rate was reinforced for decreasing. Consistent with these heart rate results was the observation that 3 of the parameters of somatic activity assessed on each S (general activity, chin EMG, and respiration) showed similar influences of the operant contingency. There was observed, however, a modification of heart rate which was independent of both somatic indices as well as the reinforcement contingencies. (Supported by USPHS Grant MH07995)

18. Ray, W. J. (Langley Porter Neuropsychiatric Institute), **and Strupp, H. H.** (Vanderbilt University) **Locus of control and the voluntary control of heart rate.** Twenty internal locus of control and 20 external locus of control Ss were instructed to increase heart rate on the initial part of the trial and to decrease heart rate on the second part of the trial pair for 8 trial pairs. This order was counterbalanced. During the first 4 trial pairs, no feedback concerning heart rate was given. During the second 4 trial pairs, the Ss received analog feedback via a light panel of 16 lights concerning the IBI interval. Across groups there was a significant increase in the magnitude of heart rate changes on the feedback trials as compared to those without. Across all trials, the internal locus of control Ss were better able to increase their heart rate as compared with externals and the external locus of control Ss were better able to decrease heart rate as compared with internals. Self-report measures concerned with the strategies utilized for controlling heart rate demonstrated locus of control differences in each group's approach to the task. These strategy differences relate to previous heart rate studies not involving the direct control of heart rate.

19. Volow, M. R., & Hein, P. L. (Duke University Medical Center) **Bidirectional operant conditioning of peripheral vasomotor responses with augmented feedback and prolonged training.** A program of training in voluntary control of finger pulse volume (FPV), using a discrete trials design, required dilatation and constriction from each of 8, male, undergraduate Ss. Each trial consisted of a 10 sec passive baseline period; followed by a 20 sec active baseline period (S maintained some intermediate FPV level); followed by a 70 sec trial period, for which S_D was either a green slide (dilatation) or a red slide (constriction). S received continuous visual, analog feedback; and also intermittent binary feedback (tone) whenever FPV exceeded its mean active baseline level in the appropriate direction. Monetary rewards were based on Ss' performance. Approximately 70 trials were given in each direction; in alternating groups of 7 each at first, and in randomized order during later test sessions. There were 11 1-hr sessions. MANOVAs of individual difference scores (covariance-adjusted for passive baseline variability) showed: 2 Ss could both dilate and constrict reliably; 4 Ss could only constrict reliably; 1 S could only dilate reliably; 1 S learned neither task. All Ss showed a non-specific constrictive effect, superimposed upon the first 15 sec of both dilatation and constriction trials.

20. Shapiro, D., Schwartz, G. E., Shnidman, S., Nelson, S., & Silverman, S. (Harvard Medical School, Massachusetts Mental Health Center) **Operant control of fear-related electrodermal responses in snake-phobic subjects.** Twenty volunteer women with snake phobias were shown 80 slides of snakes, arranged in order of increasing fear (5 sec each, 2 per min). Slides 1–10 were used for baseline. On slides 11–70, 10 Ss were reinforced with a tone (monetary bonus) if the GSR elicited by the slide was larger than the GSR to the preceding slide. Ten Ss were similarly reinforced but only when the GSR was smaller in amplitude. A shaping procedure was used to determine criterion GSR amplitudes. Slides 71–80 were used for extinction, with no reinforcement. Although both groups showed smaller GSRs over the slides, the change was larger in the decrease group, this difference being significant during extinction. Subjective reports obtained by a snake and spider fear questionnaire given before and after the session suggested that a greater reduction of fear of snakes, but not spiders, occurred in the decrease group. Implications for desensitization and behavior control will be discussed.

21. Ballard, P., Doerr, H., & Varni, J. (University of Washington) **Arrest of a disabling eye disorder using biofeedback.** Essential Blepharospasm is a long-recognized disorder consisting of involuntary clonic and tonic spasms of the eyelids and accessory musculature. The clinical course is one of progressive disability terminating in severe or total curtailment of sight-dependent activities. The commonly used surgical treatments have produced universally disappointing, and either brief, or disfiguring results. Original referral was made for depression secondary to the psychosocial consequences and anticipated surgery. Confirming diagnoses were independently made in two opthalmology clinics, and base rates established by randomly selecting sampling periods daily over 2 mos. Both biofeedback and conditioned avoidance training were given. Auditory feedback was generated from the electro-oculograph eye-blink potentials. Shock was used as the avoidance stimulus. Simultaneous recordings during feedback and avoidance training showed greater effect for the former, and reflected the progressive voluntary control. The surprising absence of troublesome spasms at 9 mos follow-up suggests the possible long-term effectiveness of biofeedback techniques in painlessly treating a number of neuromuscular disorders. (Supported in part by a Special Research Fellowship No. 1 F03 MH 47938-01 from the National Institute of Mental Health)

29. Libby, W. L. (University of Windsor) **Awareness of own cardiac and pupillary response to pictures.** According to several studies people are able to report changes in heart rate occurring upon presentation of stimuli which vary in *acceleratory* effect. The present 54 male undergraduate Ss, using 7-point scales, reported changes in both heart rate and pupil size in response to pictorial stimuli which vary in known *deceleratory* effect upon the heart and dilatant effect upon the pupil. The stimuli consisted of black and white pictures selected, primarily upon the basis of their rated attention-value, from photography magazines and other sources. Results indicated that: 1) accuracy in reporting dilations approached, but did not reach significance, 2) the greater the actual average cardiac deceleration produced by a picture, the greater the acceleration reported by Ss ($p < .01$). Pictures for which greater accelerations were reported were rated by Ss as more exciting and attention-getting. Evidently Ss' reports were based upon picture content rather than upon awareness of their own autonomic responses. Implications for learning self-control of autonomic responses are drawn.

46. Hord, D., Naitoh, P., & Johnson, L. C. (Navy Medical Neuropsychiatric Research Unit) **EEG spectral features of self-regulated high alpha states.** Intensity ($\mu V^2/Hz$) and coherence (relation between pairs of electrodes) from frontal, temporal, and occipital scalp regions were obtained from 6 Ss who had demonstrated an ability to self-regulate alpha in an operant conditioning experiment. The positive reinforcer was an alpha-contingent auditory signal. Spectral data was transformed to three-dimensional contour maps that represent intensity and coherence as functions of the frequency-time plane. The contours completed so far indicate that: 1) slow frontal activity, which is due primarily to eye movements, is less intense during successful alpha regulation that during baselines; 2) there is little evidence for increased intensity in frequency ranges other than the alpha range during alpha regulation; 3) occipital alpha tends to be more coherent with frontal than with temporal alpha during occipital alpha regulation. (Supported by Department of Defense ARPA Order No. 1596)

53. Watanabe, T., Shapiro, D., & Schwartz, G. E. (Harvard Medical School) **Meditation as an anoxic state: A critical review and theory.** Existing physiological data describe meditation as a state of synchronized EEG and reduced respiratory activity. Such physiological states, and possibly correlated altered states of consciousness, reached through meditation techniques, can be assumed to be attributable to a unique combination of physical, motivational, and environmental factors. Chief among them is a reduction in oxygen consumption. In all forms of meditation, oxygen consumption is reported to be reduced by 20% to 30%. This paper discusses the critical role of anoxia in achieving a meditative state. Several lines of reasoning are suggested from an extensive, critical review of the literature. EEG changes, not only in their sequence but also in their localization, during early stages of anoxia are very close to those found during meditation. A biochemical process due to a lack of oxygen is possibly suggested by reduced blood pH in meditation and delay of recovery in normal EEG after meditation. Emotional relaxation is seen to follow exposure to moderate oxygen deficiency in CO_2 therapy and high altitude studies. Impairment of visual functioning occurs in both anoxia and meditation. Possible mechanisms underlying meditation are discussed in reference to the anoxic state.

Abstracts of Papers Presented at the Biofeedback Research Society Annual Meeting, 1972

PERFORMANCE OF TRAINED HR SLOWING: EFFECTS OF INSTRUCTIONS, FEEDBACK, AND SHOCK
Philip W. Dunbar and Paul E. Baer
Baylor College of Medicine

Although it has been shown that subjects can be trained to slow HR, there is no evidence for generalization of this ability to stressful situations which produce an accelerating rate.

For two identical sessions, twelve normal subjects were given eight 120 beat training periods, with instructions to slow HR and continuous visual feedback. Each training period was preceded by a 45 beat rest period. There were seven 120 beat test periods following training, each preceded by a 60 beat rest period. The test periods consisted of first, three non-shock periods comprised of instructions to slow without feedback, feedback without instructions, and a training period. The other four test periods constituted a shock phase and were all conducted while subjects were periodically shocked. In addition to a shock baseline period excluding both instructions and feedback, the shock phase also included periods identical to those comprising the non-shock test phase. Finally, there were two additional control periods, one with and one without shock, during which subjects tracked the feedback meter without instructions to slow. The subjects lowered HR throughout training beginning with the first 10 beats of the first training period. For the non-shock test phase there was slowing for instructions alone, but not for feedback alone. The shock baseline period yielded acceleration. There was slowing under shock with instructions alone, feedback

Reprinted with permission. Copyright 1973 by the Biofeedback Research Society.

alone, and with both instructions and feedback. Slowing with instructions exceeded slowing obtained with feedback alone. Slowing was also obtained during tracking trials, both with and without shock.

The results indicated that following training to slow HR subjects could maintain a slowed HR despite the administration of shock which otherwise produced HR acceleration. With respect to the relative importance of instructions and feedback for the performance of HR slowing, the results support previous reports of the superior effectiveness of instructions.

SIMULTANEOUS FEEDBACK AND CONTROL OF HEART RATE AND RESPIRATION RATE
Robert W. Levenson and Hans H. Strupp
Vanderbilt University

Experimental subjects were required to either increase or decrease their heart rate (HR), relative to a pre-trial baseline period, while keeping their respiration rate (RR) at its baseline level. Subjects were provided with external feedback of their HR and RR independently, via a digital display device which was used to present numerical representations of their current HR and RR relative to baseline rates. The amount of feedback was varied on different trials to consist of either no feedback, HR feedback, or HR and RR feedback. HR and RR data were quantified on-line by a digital computer and were analyzed in terms of (1) HR inter-beat-interval (IBI) changes; (2) RR inter-cycle-interval (ICI) changes; (3) the number of heart beat intervals which were in the instructed (i.e., increase or decrease) direction; and (4) the number of heart beat intervals which were in the instructed direction and were accompanied by respiration intervals which were at baseline rate.

Data analysis revealed that subjects were reliably able to produce HR changes regardless of the amount of feedback available, but that parallel changes in RR accompanied these HR changes. These RR changed occurred in spite of specific instructions not to use breathing changes to effect HR change and regardless of whether feedback of RR was available. On the basis of the combined measure of HR and RR intervals which met the instructed criteria, it was concluded that subjects were unable to produce HR change without accompanying RR changes. Implications of this finding for the general question of mediation in autonomic control research and for future studies using human subjects were discussed.

HEART RATE CONTROL UNDER FALSE FEEDBACK CONDITIONS
Dieter Vaitl
Department of Clinical Psychology
University of Muenster, Germany

The hypothesis was tested that Ss are able to stabilize their HR under the conditions of false feedback, consisting of different degrees of HR variability. Four groups (N=56) were instructed to stabilize their HR in 6 training periods, but the visual process-informations-feedback was only presented to three experimental groups. The Ss of group I were informed of their own HR whereas a non-contingent feedback was administered to the other two groups: a highly stabilized HR to group II and the feedback of normal variability to group III. Control Ss (group IV) did not get any visual feedback, but only instructions about the influence of certain control techniques (respiration, relaxation) on the HR variability. The $U(x)$ scores as measures of the uncertainty of HR processes indicated, that there was a significant increase of the HR control (frequency and variability) only under the conditions of highly stabilized non-contingent feedback. This effect might be influenced by the experimental reduction of irritations usually emanating from contingent analog feedback procedures. Self-report measures concerned with the strategies utilized for controlling HR, demonstrated different approaches in each group to the adequate somatic and emotional state for this task. These individual differences are explained by the model of labelling. No significant correlation between HR control and respiration could be observed. Implications for the use of 'false' feedback in HR control experiments are discussed.

COGNITIVE AND SOMATIC MECHANISMS IN VOLUNTARY CONTROL OF HEART RATE

Iris R. Bell
Stanford University
Neuro and Biobehavioral
Sciences Program

Gary E. Schwartz
Harvard University
Department of Psychology
and Social Relations

In a previous study, Bell and Schwartz found that when subjects were simply instructed to "control and raise" their heart rate, they could immediately increase their heart rate by 7.6 bpm without feedback, and further increase their heart rate to 9.6 or more bpm when given continuous meter feedback for beat by beat changes in heart rate. Analysis of subjective reports and somatic indices suggested that both cognitive and somatic processes were involved.

To explore the role of self-induced imagery in cardiac rate control, three experimental conditions were studied. One group of 10 subjects were told to "control and raise" their heart rate, and were tested for four 1 minute periods without (pre) feedback, eight 1 minute periods with meter feedback, and four 1 minute periods without (post) feedback. A second group of subjects were given imagery-arousal instructions, specifically told to "make themselves aroused" by thinking sexual, angry, or other arousing thoughts, while during the feedback period instructed to use the meter information to help find the imagery most effective in producing arousal. A third group of subjects were told to rest during prefeedback, then "control an internal response by making the needle go to the right." Hence, these subjects were given "control" commands but were not instructed as to direction of control (thereby short circuiting imagery immediately elicited by the instruction "raise").

Whereas the control-and-raise group showed a 6.3 bpm increase in heart rate during prefeedback, the imagery-arousal subjects only increased their heart rate by 2.9 bpm. Then, with the addition of feedback, the control-and-raise group immediately produced a further increase to 8 bpm; the imagery-arousal subjects also increased by a couple of bpm at feedback onset but then declined back to their pre-feedback level. Only the control-alone group showed a "learning curve" during feedback, reaching the high levels obtained by the control-and-raise group at postfeedback.

These data indicate that arousing imagery per se is not _the_ mechanism for self produced heart rate increases. It is hypothesized that (1) "control" instructions initiate the self generation of somatic control processes affecting heart rate through central and peripheral mechanisms, and (2) focusing attention on the self generation of imagery _per se_ can inhibit the generation of motor activity, thereby minimizing the magnitude of self regulated heart rate increases.

AUTOREGULATION OF SKIN TEMPERATURE USING A VARIABLE INTENSITY FEEDBACK LIGHT
Edward Taub and Cleeve Emurian
Institute for Behavioral Research
Silver Spring, Maryland

A technique is described enabling most humans to establish rapid operant control of their own skin temperature. It involves operant shaping of small changes in skin temperature by means of a variable intensity feedback light. Nineteen of the last 20 consecutive subjects have been able to learn autoregulatory control of this parameter. Training to a level of unequivocal acquisition rarely required more than four 15-minute periods. After that time the mean change per 15 minute session for all subjects was approximately 2.5° F, ranging up to 6.5° F. After original training some subjects were asked to reverse the direction of their autoregulatory control. The 4 subjects trained longest displayed an ability to alter temperature in opposite directions during successive periods on the same day, and routinely displayed ranges of 8 - 15° F within 15 minutes. Retention of the task was virtually perfect over an interval of 4 - 5 months in the four cases tested. It was further found that after sufficient training autoregulation of skin temperature was as good without feedback as with feedback. Transfer of control from one portion of the body to another was also found to be easily accomplished late in training. (Supported by ARPA of DOD under ONR Contract N00014-70-C-0350 to the San Diego State College Foundation.)

VOLUNTARY CONTROL OF SKIN TEMPERATURE:
UNILATERAL CHANGES USING HYPNOSIS AND AUDITORY FEEDBACK
Alan H. Roberts, University of Minnesota
Donald G. Kewman and Hugh Macdonald
Stanford University

To demonstrate the ability of human subjects to achieve a high degree of control over a specific autonomic function under conditions providing evidence of voluntary control, a select group of four female and two male subjects received five to nine individual one-hour training sessions using both auditory feedback and hypnosis. Three one-hour experimental sessions followed with each subject asked to raise the temperature of a designated hand relative to the other for eight minutes, then reverse the hand temperatures for two additional eight minute trials during each session. Two subjects participated in two additional experimental sessions, without auditory feedback. Statistical analyses showed significant skin temperature changes in pre-designated directions for all subjects. One subject attained a maximum temperature difference of 9.2°C. on a single trial, and an average maximum change of 4.92°C. across six trials. Two subjects evaluated without feedback for six additional trials over two sessions performed even better than with feedback. Response patterns used by subjects were different including making both hands colder (or warmer) but at different rates, holding one hand constant while changing the temperature of the other, or diverging the temperatures of the two hands. Personality test data and subjective reports and observations will be discussed as time allows. It is clearly demonstrated that some individuals can achieve a high degree of voluntary control of peripheral skin temperature of sufficient magnitude to suggest therapeutic applications.

INTRA-HEMISPHERIC PHASE RELATIONSHIPS DURING
SELF-REGULATED ALPHA ACTIVITY
D.J. Hord, P. Naitoh, L.C. Johnson and M.L. Tracy
Navy Medical Neuropsychiatric Research Unit
San Diego, California

Scalp EEG from six practiced "alpha-regulators" was monitored continuously from Fp2, T4, and O2 leads during 1-hr. alpha feedback sessions. Data was obtained for each of three separate days. Spectral analysis was performed on 1-min. samples derived from the pre-feedback, feedback, and post-feedback periods. Three-dimensional linear contours of intensity, coherence, and phase for the frequency range from 0.0 to 16.0 Hz provided descriptions of the electrophysiological correlates of the self-regulated alpha state. During the feedback period a stable phase angle developed between the frontal and occipital leads in which frontal lagged occipital by approximately 150 degrees (42 msec) at the frequency of maximal intensity in the alpha band. Other frequencies and lead combinations showed no such stable phase relationships. In addition, there was a significant increase in the phase angle toward the end of the feedback period. The results are interpreted theoretically in terms of shifting loci in the thalamus. It is also suggested that biofeedback offers a useful technique in the stabilization of physiological processes for the purpose of detailed study.

THE CONTROL OF THE EEG THETA RHYTHM
Pola Sittenfeld
University of Colorado Medical Center and University of Duesseldorf

The control of the 4 - 7 Hz EEG theta rhythm has proved to be valuable for the self induction of sleep in 4 out of 7 pilot subjects who suffered from sleep onset insomnia. These subjects fell asleep in the laboratory twice within 20 minutes, as measured by EEG criteria, after training with theta feedback.

Basically, the feedback technique involves providing the subject with continuous information on how much theta he is producing. The feedback, which he hears through his headphones, consists of variable clicks--the more theta, the more clicks.

Since the feedback control of theta poses certain difficulties (it is little evident in the normal, alert EEG and is generally associated with a disconnected mental state, which makes it's control difficult), we tested a two phase procedure: subjects were first trained in deep muscle relaxation by EMG feedback. This was designed to increase the amount of theta, on the assumption that deep muscle relaxation is associated with a generalized low arousal state. Subjects were then shifted to the feedback of theta itself.

EEG, forehead EMG, forearm EMG and heart rate were recorded in 20 male, paid volunteers, age 35-50, for 3 baseline sessions, 8 feedback sessions, and 2 post-baseline sessions. The time of day was kept constant for each subject. Subjects were divided into 4 groups according to their baseline EMG levels (high or low frontalis muscle tension) and the type of feedback they received (theta feedback only or the two phase procedure).

The data were treated by separate analyses of variance. The main results are as follows:

1. Subjects baseline EMG levels influence their ability to control theta. Subjects with high EMG baselines reach control over theta only when given relaxation training as a first step. Subjects with low EMG baselines on the other hand did as well when given theta feedback only.

2. Heart rate decreased significantly during the training.

3. There is a marked and sudden drop in frontalis EMG and heart rate with the onset of higher theta activity.

The implications of these results for the induction of low arousal states will be discussed.

READ BY TITLE ONLY

INTERACTION EFFECTS OF SEX OF SUBJECT AND
EXPERIMENTER IN BIOFEEDBACK TRAINING
Terry A. Travis, Univ. of So. Ill., School of Medicine
C. Y. Kondo and John Knott
Dept. of EEG and Neurophysiology and Psychiatry
University of Iowa School of Medicine

The task of enhancing alpha in a feedback setting seems on <u>a priori</u> basis to be particularly susceptible to motivational variables, and since the sex interaction phenomenon can be related to motivation, the interaction of sex of experimenter vs. sex of subject was investigated. Twenty-eight subjects (male, n=14 and female, n=14) were run by two experimenters, one male and one female, each of whom ran seven male and seven female subjects. All subjects received five, ten-minute eyes-open training sessions. Feedback was presented when criterion alpha was reached. <u>Criterion alpha</u> was defined as follows: 0.07 to 0.13 seconds 8-13 Hz. activity which was 50% or more of the amplitude of the maximum eyes-closed alpha observed during an initial calibration period. Subjects received the blue light feedback only when they equalled or exceeded this criterion. The results indicate that no sex of Experimenter-Subject interaction exists in this eyes-open feedback situation.

MEASUREMENT AND QUANTIFICATION OF SURFACE EMG SIGNALS IN STATES OF RELAXATION
Rolf R. Engel
Department of Psychology
University of Duesseldorf, West Germany

With the use of low-noise amplifiers and constant reset level integration techniques the accuracy of quantifying the mean amplitude of EMG signals (as well as EEG-Alpha or other signals) has greatly improved.

The quantification of surface EMG signals during biofeedback relaxation training, however, bears particular problems which are introduced by two facts.

(a) Since the signal-to-noise ratio is very low with relaxed subjects, the comparability of amplitude scores obtained in different therapeutical sessions depends to a great deal on the electrode resistances in each particular session. Instead of allowing the resistances to change anywhere under a certain level- which is the usual technique-, it is suggested to measure exactly the particular noise level introduced by the varying electrode resistances in different sessions and to correct the over-all integration count by this special value. This may be done either off-line or on-line by means of operational amplifier circuits.

(b) Similar to other variables which have a fixed limit on only one tail, the scores for the mean EMG amplitude are not normally distributed but rather right skewed. To adjust this kind of distribution, a hyperbolic transformation along the equation $y=a/x$ is suggested. As with the correction for noise, this may be done either by off-line calculations or on-line by means of a multiplier/divide unit or an analog computer. In the case of off-line corrections the scores have to be transformed prior to any averaging procedures.

The effects of this transformation are demonstrated using the data of one of our EMG feedback studies. It is shown that the interpretation of the raw scores alone may lead to misinterpretations.

EFFECTS OF FEEDBACK CONTROL OF HEART RATE
ON JUDGMENTS OF ELECTRIC SHOCK INTENSITY
Alan Sirota, Department of Psychology,
    Pennsylvania State University
Gary E. Schwartz, Department of Psychology and
    Social Relations, Harvard University
David Shapiro, Department of Psychology
Harvard Medical School

Twenty female subjects (age 21-27) were instructed either to increase (N=10) or to decrease (N=10) their heart rate while anticipating electric shock. They were also provided with visual and auditory feedback and monetary incentives for the appropriate heart rate changes. The experiment consisted of 72 trials of 15 sec. duration each, half followed by electric shock delivered to the left forearm on a random schedule. By the end of conditioning, heart rate in the Increase group was significantly greater by 13 bpm than in the Decrease group during the anticipatory periods. Phasic heart rate changes showed similar differences, with shock trials producing elevations in heart rate over no-shock trials for the Increase but not for the Decrease group. Shock ratings were significantly higher for increase subjects, but this difference appeared on the earliest trials. However, when subjects in each group were subdivided into Cardiac-Reactors and Non-cardiac-Reactors on the basis of a post-experimental questionnaire designed to measure awareness of physiological states during fear or anxiety, a highly significant subgroup by trial interaction evolved. Increase Cardiac-Reactors on the basis of a post-experimental questionnaire designed to measure awareness of physiological states during fear or anxiety, a highly significant subgroup by trial interaction evolved. Increase Cardiac-Reactors rated the shocks as increasing in intensity over the trials whereas Decrease Cardiac-Reactors showed decreases in their ratings. Non-cardiac-Reactors showed no such differentiation. These data suggest that a conditioned and instructionally mediated change in an autonomic response can affect subjective reaction to painful stimuli, particularly in subjects for whom such a change is relevant in terms of their normal awareness of autonomic functioning in fearful situation.

A TENTATIVE PROCEDURE FOR THE CONTROL OF PAIN: MIGRAINE AND TENSION HEADACHES
S. Alexander Weinstock, Ph.D.
Department of Psychology, Columbia University

The purpose of this paper is to present a method that shows promise in relieving headache pain. The method is based on a combined technique of self-hypnosis, EMG, thermal biofeedback training, and the learning of alternative responses to stressful situations. The four procedures are designed to give the patient potential alternatives, both physiological and behavioral, to headache pain as a means of coping with stress. In addition, self-hypnosis, EMG, and thermal biofeedback emphasize the lowering of a person's state of physiological tension and arousal, which, according to Harold G. Wolff's theory of stress, should make it easier for the patient to learn more appropriate patterns of response to stressful phenomena. This paper does not deal with a theory about the causes of the different types of headaches and does not differentiate between tension and migraine patients. The data are based on seven cases, with headache histories ranging from two years to thirty-five.

During the first therapy session, a brief psychological history was taken, focusing on those situations the patient found especially stressful or difficult to cope with, followed by a rather detailed explanation of the nature of self-control. Then the patient was given an example of self-induced hypnotic relaxation. This part of the first session was recorded, and the patient was instructed to practice on his own at least twice a day. All reported relief in their state of tension and anxiety after the first exposure to the hypnotic-relaxation tape.

At the second session, EMG biofeedback was started, with relaxation of the frontalis muscle as an overall measure of tension. When a patient's tension level reached a low of between 5 and 7 microamps, usually within 10 to 14 sessions, thermal biofeedback training was started. With practice, usually within 4 to 10 sessions, each patient was able to increase his differential in temperature between the head and right index finger by 2.5 to 15 degrees Fahrenheit. Throughout therapy the patient was questioned about any antecedent external variables relating to his headaches, in order to uncover any environmental supports maintaining the headache behavior. Whenever such were found, role-playing and training in alternative response modes were instituted.

At the time of writing, all seven patients seem to be functioning without headaches, though numerous changes in their lives have occurred at the same time. This is not to suggest that headaches of the tension or migraine type are functional in nature, but rather reflects the author's belief that any aberrant habit that is sufficiently strong and long-standing will be incorporated into a person's life style and will be used as an adaptive mechanism in certain situations unless this behavior is changed simultaneously. The patient needs to be taught more adaptive ways of coping. Unquestionably, biofeedback training combined with a sound understanding of behavioral, biological and medical approaches is one of the most promising developments in the field of psychotherapy. Because of the short time I have been working with this particular approach,

4 to 5 months, more follow-up will be needed at a later date to further evaluate the method.

TRAITEMENT DE L'EPILEPSIE PAR RETRO-ACTION SONORE
(Treatment of epilepsy by auditory feedback)
Fernand Poirier, Montreal

Quarante-quatre sujets, dont l'épilepsie a été prouvée de façon formelle, font l'objet de cette étude. Leur répartition quant au sexe est égale. Leur âge varie entre 2 et 44 ans. La majorité d'entre eux présente une épilepsie dite "sous-corticale"; un bon nombre sont photo-sensibles.

On leur fait entendre leurs ondes cérébrales insonorisées. Pour ce faire, on fait passer celles-ci par un circuit qui comprend des filtres (avec des bandes de fréquences pré-établies), un amplificateur de voltage, un sélecteur de seuil, un "covertisseur" en sons "unitaires", et "blancs", ou modulés. Le sujet est ainsi informé de la fréquence, de l'amplitude et de l'allure de ses ondes cérébrales. On l'instruit également des sons produits par ses décharges épileptiques.

On développe des méchanismes d'occlusion et d'anticipation de celles-ci; on parvient même à inhiber les décharges provoquées, par exemple, par la stimulation lumineuse intermittente, par renforcement de l'activité alpha normale.

Cet apprentissage à générer des ondes cérébrales normales, vigoureuses et résistantes aux stimuli nocifs, semble avoir une certaine perdurabilité. C'est ainsi qu'un enfant apprend à marcher, à aller à bicyclette, sans tomber!

Ce traitement s'inscrit dans le cadre des techniques de rétro-action ("feedback") et de conditionnement opérant, qui trouvent des applications de plus en plus nombreuses en médecine, et permet d'entrevoir la possibilité, entre autres, de traiter éventuellement les épileptiques sans médication.

d-LSD-25 IN THE TREATMENT OF EPILEPSY
Larry O. Rouse
California State University at Fresno

d-LSD is an anticonvulsant at high doses that interfer with neural transmission. It has not been shown to be a convulsant at any dosage. Learning and memory are facilitated by d-LSD in minute doses far below the direct anticonvulsant threshold. A variety of procedures now exist for establishing conditioned inhibition of epileptic seizures. d-LSD at 3.65 micrograms per mouse strengthens habituation and conditioned inhibition of audiogenic seizures in BALB/cCrgl without affecting seizure susceptibility directly. In humans certain components of the seizure's stimulus antecedents (auras, dreams, prodromes) have been frequently found to occur in response to the drug in 50—250 microgram doses in the absence of the motor seizure proper. d-LSD should facilitate conditioned inhibition of epileptic behavior in humans. Such an effect would suggest its use as an adjunct to current behavioral and electrophysiological feedback procedures used in treating a variety of illnesses.

PROBLEMS IN THERAPEUTIC APPLICATION OF EMG FEEDBACK
W. A. Love, Jr.
Department of Psychology, Nova University

The data presented is derived from clinical rather than experimental situations; however, two experimental studies are discussed: One deals with optimal scheduling of EMG feedback, the other with EMG feedback for hypertensive subjects.

Standard procedures involved in using feedback EMG therapeutically are considered. Topics include electrode placement and the scheduling of both treatment and home practice.

The question of who is a good candidate for therapeutic EMG feedback is discussed. Subject types reviewed are: 1. Psychotics, 2. Obsessive compulsive neurotics, 4. Insomniacs, 5. Those with low I.Q.'s,

6. Those short on motivation, 7. Neurotics with marked anxiety, 8. Neurotics with minimal anxiety, 9. Psychopaths, 10. Crisis intervention situations, 11. Visceral internalizers, 12. Depressives, and 13. Post ECT patients.

The psychological mechanisms operating in EMG feedback therapy are discussed, and the situations where it stands alone as opposed to those situations where other treatment procedures such as chemotherapy required are covered. The paper concludes with a review of EMG procedures useful in Behavior Therapy.

A SIMPLE ON-LINE TECHNIQUE FOR REMOVING EYE MOVEMENT ARTIFACT FROM THE EEG
Dexter G. Girton and Joe Kamiya
Langley Porter Neuropsychiatric Institute

A simple on-line technique for removing potentials generated by eye movements from the EEG is presented. A circuit is described which subtracts a fraction of both the vertical and horizontal EOG signals from the EEG signal to be corrected. EEG potentials derived from vertex-right mastoid electrodes were used to demonstrate the technique. By visual inspection of the corrected EEG trace it appears that the eye movement artifact is almost completely removed. These results hold for both small and large eye movements and are not affected by light and dark adaptation. Computer averaging techniques were used to verify the results.

This is the abstract of a paper entitled, "A Simple On-Line Technique for Removing Artifacts from the EEG."*

*Electroencephology and Clinical Neurophysiology, 1973, 34, 212-216.

SOME PHYSIOLOGICAL CORRELATES OF ANALGESIA IN VARIOUS STAGES OF HYPNOTIC DEPTH
S. Alexander Weinstock, Ph.D.
Department of Psychology, Columbia University

A woman with low self-esteem, periods of depression and suicidal tendencies was referred to me for hypnotic treatment. In order to better understand what was happening to her physiologically, the patient agreed to be hooked up to a number of physiological instruments during the course of treatment. Measures were taken of rau EEG, separating alpha and theta bands, moment-to-moment heart rate, heart rate per minute, respiration rate, EMG in the frontalis and in the forearm, and surface temperature. The data reported below were collected over the first two sessions.

In the first session, the patient proved to be a very good "hypnotic subject." At the second, she went into a self-induced trance very easily, her heart rate and respiration rate dropping dramatically. Once in a hypnotic trance, it was suggested to her that she could take herself down through four levels of hypnosis and that she should go down to the fourth, the lowest, stage by stage. This deep stage would be characterized by a tingling sensation starting in her toes, flowing through her body and reaching the top of her head, accompanied by an increasing self-confidence. At this point there were no changes whatsoever from the previous basal multigraph patterns. Then a form of hypnotic analgesia was induced and repeated at a lighter stage of hypnosis.

While still in the fourth stage, the patient was told that she would be completely oblivious to any kind of pain at all. She was then told that I would prick her left hand with a needle. When I did so five times, she reported that she felt no pain nor did she exhibit any physiological changes. She was then told to go to stage three, a lighter level of hypnosis, and the procedure was repeated. This time, in response to the pin pricks, sharp physiological changes occurred on all measures but one. Her frontalis contractions increased, her breathing was disrupted, there was blockage in the production of alpha waves, and the patient reported that she was very uncomfortable and fearful. The only physiological indicator which did not change was the EMG reading on her right hand (I had pricked her left). It is interesting to note that whereas the EMG recording of her hand remained stable, that of her frontalis showed a change in contractions in response to the pricks, pointing to the frontalis as potentially a good measure of overall tension. Also interesting is the fact that in both the first and second hypnotic sessions, surface temperature on her middle finger went consistently down, from 81 degrees to 78, a phenomenon difficult to put into any existing theoretical framework.

INTEGRATED ALPHA: COMPARISON WITH CRITERION
OR REWARDED ALPHA
C. Y. Kondo, Dept. of EEG and Neurophysiology,
and Psychiatry, Univ. of Iowa School of Medicine
Terry A. Travis, Univ. of So. Ill., School of Medicine
John A. Knott, Dept. of EEG and Neurophysiology
and Psychiatry

It is suggested that a common or standardized dependent variable be adopted to eliminate the possibility that unique dependent variables may produce what would appear to be disparate results in EEG feedback research. This study examines integrated alpha and presents a method for voltage integration (modified from Knott, McAdam and Grass, 1964). Sixteen subjects were run, eight of these received eyes-open (Male, n=4; female, n=4) and eight (male, n=4; female, n=4) eyes-closed alpha enhancement training. Integrated alpha was defined as all occipital EEG which passed the active twin-T filter (8-13 Hz) used in our set-up. Criterion alpha was defined as percent time alpha occurring for a minimum of seconds and which was 50% or more of the amplitude of the maximum voltage of eye-closed alpha observed during an initial calibration period. Both measures were recorded simultaneously for each subject during all training trials. "Reward" (blue light) was given only when criterion was exceeding. Eyes-open and eyes-closed training data demonstrated that integrated alpha bears a strong morphological similarity to criterion alpha.

EFFECTS OF INSTRUCTIONAL SET, REINFORCEMENT, AND INDIVIDUAL DIFFERENCES IN EEG ALPHA FEEDBACK TRAINING

David H. Walsh*
Psychophysiology Laboratory
Veterans Administration Hospital
Bedford, Massachusetts

The effects of EEG alpha feedback training, instructional set (suggestion), reinforcement, and certain individual differences (alpha baseline, field orientation, hypnotic susceptibility, intuition, introversion, and attitude toward "mind altering" techniques) were studied in forty college students. Percent time alpha activity and reported subjective experiences during feedback were examined experimentally with a four way factorial "drug-drug set" design. Alpha and neutral instructional sets were each paired with alpha and no-alpha training, in order to observe the effects of set and reinforcement. Individual differences were studied correlationally. The results showed that for an "alpha experience" to occur, both alpha activity and alpha set are necessary; neither alone is sufficient. No effects were attributable to reinforcement. Alpha activity never increased above baseline in accord with earlier findings, and was positively correlated with hypnotic susceptibility, intuition and attitude. Theoretical considerations based on the Schachter and Singer (1962) drug model are discussed along with implications for future work in alpha feedback training.

*Present address:

David H. Walsh, Ph.D.
Department of Psychology
Merrimack College
North Andover, Mass. 01845

NAME INDEX

Ackner, B. A., 300
Adler, C., 92n
Adrian, E. D., 273
Adrian, E. E., 168
Akutsu, K., 355
Anand, B. K., 172, 175, 353, 355
Antrobus, J. S., 61
Aristotle, 273
Arnold, M. B., 376, 377
Asmussen, E., 283
Assagioli, 166
Axelrod, 473

Baer, Paul E., 500–501
Bagachi, G., 175
Bagchi, B. K., 353, 355
Ballard, P., 498
Banuazizi, A., 91, 329, 338
Barber, T. X., 147, 365n, 366, 368, 373, 374, 375, 377, 378, 379, 381, 382, 398n–99n, 479, 488
Barcroft, J., 46
Basmajian, J. V., 67, 73
Bateman, D., 148
Batenchuk, C., 90
Beatty, J. T., 254
Bell, I. R., 503–4
Bennett, T. L., 122, 123
Benson, H., 173, 179, 481, 485
Berger, 168
Bergman, J.S., 249
Bertini, M., 146–47, 148, 149, 150
Billig, 288
Bindler, P., 496–97
Black, A. H., 90, 309, 321, 327, 328n
Bleecker, Eugene, 193, 207
Bliznitchenko, F., 148
Block, J. D., 483
Blunton, 3

Borsook, 288
Botkin, 4, 17
Bowers, K. S., 388
Bradley, R. J., 131
Braginsky, B. M., 387
Braginsky, D. D., 387
Braun, J. J., 91
Braunewald, E., 216n
Brener, J., 235
Brenman, M., 379, 381, 388
Brenner, J., 215n, 229, 249
Bronk, D. W., 273
Bross, Thérèse, 353
Brown, B. B., 149, 170, 172, 175, 455
Browning, C. W., 413
Budzynski, T. H., 92n, 444, 452, 482, 484

Calverley, D. S., 366, 375, 378, 379, 381, 398n–399n
Cameron, D. E., 449
Cameron, H. G., 34, 38, 41, 42, 43
Cannon, Walter B., 363
Carlsöö, S., 281
Carmona, A., 131
Carmona, S., 90
Carpenter, T. M., 4, 35, 41, 42
Carter, E. P., 23
Cason, H., 32
Castle, A., 168
Cavonius, C. R., 127–28
Cayce, Edgar, 165
Chapman, R. M., 127–28
Chavat, J., 456
Chaves, J. F., 387, 407
Chhina, G. S., 172, 355
Chiene, 31
Chism, R. A., 248
Claude-Bernard, 4
Cleveland, S. E., 451

519

Index

Coe, W. C., 406
Cohn, D. H., 193, 203
Cohen, Sidney, 174
Collins, J. K., 393–94
Conger, J. J., 465
Coolidge, F., 148
Costello, J., 128
Crasnilneck, H. B., 413
Crider, A., 215n, 309, 311, 312, 316, 328n
Cronin, D. M., 368
Cross, D. V., 67
Cumming, W. W., 61
Czermak, 4

Dalal, A., 381
Das, N. N., 355
Davidson, R., 496–97
daVinci, Leonardo, 273
Davis, H., 30
Davis, R. C., 65, 66, 67
Davison, G. C., 413, 426
Deikman, A. J., 180
Delabare, 238
Delgado, J. M. R., 131
Dell, P., 466
Dewan, E. M., 136–37, 169
Diamond, M. J., 178–79, 368
DiCara, L. V., 88, 90, 91, 215n, 263, 321, 322–23, 327, 455, 479
Doerr, H., 498
Dollard, J., 460, 463
Donders, 5, 7
Dunbar, Philip W., 500–501
Dune, Marvin, 162
Durup, G., 168, 170
Dworkin, Barry, 92
Dykman, R. A., 193

Eccles, J. C., 30
Edelberg, R., 299
Edfeldt, A., 281
Edmonston, W. E., 398n
Eldridge, L., 55
Elliot, A. H., 43
Emmons, W., 148
Emurian, Cleeve, 504
Engel, Bernard T., 92, 235, 248, 456, 482
Engel, Rolf, 452, 508–509
Engstrom, D. R., 178, 183
Epstein, S. E., 216n
Erickson, M. H., 177, 371, 372, 404–5, 413
Ernest, J. T., 127–28
Estes, W. K., 54
Euler, U. v., 43
Evans, F. J., 388
Eysenck, H. J., 75

Fairchild, M. D., 130

Farmer, R. G., 451, 452
Fatheree, T. J., 43
Fehmi, L. G., 264
Felton, G. S., 282
Fenwick, P. B., 168–69
Fenz, W. D., 495
Fere, 3
Ferguson, Daniel, 160
Ferster, C. B., 316
Fessard, A., 168, 170
Fetz, E. E., 124–25, 216n
Fields, C., 90
Finnochio, D. V., 124–25, 216n
Fisher, S., 451
Fixott, R. S., 413
Flynn, J. P., 193
Folkow, B., 329, 456
Forsyth, R. P., 200
Fothergill, Millner, 4
Foulkes, D., 145, 146, 150
Fox, S. S., 109, 115, 132, 133
Frank, 4
Franks, J. D., 487
Freud, Sigmund, 152, 180
Fromer, 239

Galen, 273
Galin, D., 265, 268
Galosy, R. A., 497
Galvani, 273, 284
Gannon, L., 485
Gannt, W. H., 84, 193, 204
Gastaut, Henri, 146, 355
Gavalas, R. J., 91n, 316
Gellhorn, E., 441, 449
Germana, J., 441, 449
Gershon, E., 215n
Gershorn, E., 456
Gill, M. M., 379, 381, 388
Gindes, B. C., 370, 371, 376
Girton, Dexter G., 514
Glick, G., 216n
Goesling, W. J., 249
Goldstein, I. B., 276
Goldstein, L., 177
Golseth, 288
Goodman, Paul, 55, 56n
Gordon, W., 43
Gorski, R. A., 468
Gottleib, S. H., 199
Graham, D. T., 488
Graham, F. K., 488
Gray, J. A., 122, 123
Green, A. M., 113, 148, 153, 260, 455, 484, 486
Green, D. M., 69n
Green, E. E., 113, 148, 260, 434n, 455, 483, 484, 486

Name Index 521

Green, W. A., 311
Greenlaw, R. K., 283
Grinker, R. R., 463, 464
Grollman, 43, 46

Hadfield, J. A., 376–77
Ham, M. L., 374
Hamano, K., 239–40. 241, 242
Hansen, S. D., 466
Hansen, S. P., 235
Harano, K., 377
Hardyk, L. D., 282
Harlow, H. F., 205
Harrower, Molly, 56n
Hart, J. T., 178
Hartland, J., 367, 368, 371
Hein, P. L., 498
Helson, H., 427
Henderson, R. L., 280
Heron, W., 147
Hess, W. R., 441, 449
Hilgard, E. R., 388, 398n
Hill, F. A., 91
Hines, E. A., 43
Hirai, T., 172–73, 355
Hitchcock, F. A., 35, 41, 42
Honorton, C., 496–97
Hooker, Donald R., 22
Hord, D. J., 497, 505–6
Hoskins, R. G., 34, 35, 41, 42
Hothersall, D., 235
Howard, J. L., 321, 497
Howe, R. C., 129–30
Hudgins, C. V., 32
Hull, C. L., 376
Hunter, W. S., 32
Hyoda, M., 244

Inman, V. T., 273

Jacobs, A., 282
Jacobson, E., 74–75, 376, 437, 438, 442
Jenney, C. C., 177
Johnson, H. J., 249
Johnson, L. C., 499, 505–6
Johnston, V. S., 131
Jonas, Saran, 92
Jones, G. B., 495

Kabat, H., 57
Kamiya, J., 61–62, 90, 107, 110, 128, 169, 170–71, 172, 175, 177, 178, 263, 438, 479, 514
Kasamatsu, A., 172–73, 355
Katkin, E. S., 215n, 328n, 479
Kelley, C. R., 413, 419
Kewman, Donald G., 505
Kilkowitsch, 17

Kimmel, H. D., 91, 215n, 239, 316, 328n, 479, 486
Kissin, M., 43
Kitano, 243–44
Klausen, K., 283
Kleinman, R. A., 215n, 249
Klemm, W. R., 124
Kline, M. V., 413
Klinger, B. I., 375
Knott, John R., 496, 507–8, 516
Köhler, Wolfgang, 56n
Kondo, C. Y., 496, 507–8, 516
Kostjurm, 15
Kotake, Y., 238
Krutch, J. W., 113
Kuhn, T. S., 386, 404
Kurschner, 4

Lacey, B. C., 85, 87
Lacey, J., 85, 87
Lachman, R., 328n
Lacy, J. I., 317
Lader, M. H., 450
Lambert, E. F., 30
Lane, H. L., 67
Lang, P. J., 442, 455, 483, 484
Lapides, J., 91, 469
Lawler, J. E., 497
Lazarus, A. A., 438, 450, 455, 488
LeCron, L. M., 413
Legewie, H., 256
LeGuyaden, D. D., 497
Leitenberg, H., 484
Lesh, T. V., 180–81, 183
Leuba, C., 148
Levenson, Robert W., 501–2
Levitt, G., 43
Levy, M., 148
Lewis, H. B., 146–47
Libby, W. L., 498–99
Liljestrand, G., 43
Lindsley, D. B., 278–79
Lisina, M. I., 317
London, P., 178
Lopresti, R. W., 130
Love, W. A., Jr., 513–14
Luria, A. R., 429
Luthe, W., 179, 442
Lynch, J. J., 127, 128, 495

McClure, J. N., Jr., 361
MacConaill, M. A., 276
MacDonald, Hugh, 505
MacDonald, L. R., 129–30
Macvaugh, G. S., 368
Magladery, J. W., 30
Malmo, R. B., 445
Masserman, 84

Index

Mathews, B., 168
Matthews, A. M., 450
Maupin, E., 179
Mearles, A., 388
Melamed, B. C., 483
Menzies, R., 376–77
Miller, Neal E., 88, 90, 91, 170, 215n, 263, 309, 321, 322, 327, 328n, 329, 338, 430, 433, 434, 455, 463n, 479, 480–81, 484, 487
Miyake, S., 238, 241, 242
Miyata, Y., 238, 239–40
Monod, H., 277
Mordey, T. R., 376
Morgagni, 4
Moss, C. F., 365
Mossa, 13
Mowrer, O. H., 238
Mulholland, T., 76, 109, 115, 128, 168, 169, 170, 172, 176, 486
Mundy, 168
Murphree, A. A., 177
Murphy, Gardner, 72–73, 153–54
Murray, E. N., 215n, 247, 328n, 479
Myers, A. K., 470

Naitoh, P., 499, 505–6
Naruse, G., 377
Nelson, S., 498
Neumann, 3
Noble, M., 91n
Nowlis, D. P., 107, 172
Nuzum, F. R., 43

Obrist, P. A., 193, 321, 327, 449, 497
Ogawa, K., 377
Okita, T., 241, 244–45
Olds, J., 131
Orne, M. T., 111, 128, 148, 388, 496
Ornstein, R., 265, 268
Oswald, I., 169

Pappas, B. A., 90, 91
Paskewitz, D. A., 111, 127, 128, 496
Pattie, F. A., 375
Pavlov, Ivan P., 64, 81, 85
Pease, E. A., 38
Peper, E., 107, 109, 169, 170, 172, 175–76
Perez-Reyes, M., 193
Perls, Fritz, 55, 56n
Pfeiffer, G. B., 177
Phillips, C. G., 73
Pickering, G. W., 43
Pitts, Ferris N., Jr., 361
Pitts, L. H., 193
Plumlee, L., 90
Pohorecky, Larissa, 473
Poirier, Fernand, 512

Pollen, 169
Porges, S., 317
Post, J., 170
Powers, W. R., 281
Probst, W., 256

Rachman, S., 450
Randt, Clark T., 92n
Raskin, D., 317
Ray, W. J., 497–98
Rempel, B., 30
Rice, D. G., 91n
Rieckert, H., 359–60
Ring, K., 387
Roberts, Alan H., 505
Robertson, T. G., 398n
Rosenblueth, A. A., 30
Rosenfeld, J. P., 104, 132
Ross, J., 216
Roszak, T., 56n
Rouse, Larry O., 513
Rubin, F., 148
Rubenstein, E. H., 329
Rudell, A. D., 109, 115, 132, 133
Rushmer, R. F., 215n

Salamon, I., 170
Salome, Eugene, 6–7, 16, 17–18
Sarbin, T. R., 375, 390, 406
Sargent, J. D., 434n
Savage, W. F., 21, 25
Scher, A. M., 215n
Scherrer, J., 277
Schildkraut, J. J., 473
Schoenfeld, W. N., 61
Schultz, Johann, 152, 153, 160, 179, 376, 442
Schumova, 10
Schwartz, Gary E., 121n, 198, 215n, 328n, 480, 496, 497, 498, 503–4, 509–10
Schwarz, Jack, 157, 166
Scully, H. E., 281
Sells, S. B., 413
Shapiro, David, 91n, 215n, 309, 311, 312, 316, 456, 481, 497, 498, 509–10
Shapiro, J., 178–79
Shaw, G., 496
Shaw, George B., 64
Shearn, D. W., 91
Sherrington, C. S., 277
Sherwood, M., 496
Shnidman, S. R., 215n, 328n, 498
Shock, N. W., 42
Shor, R. E., 388
Short, P. L., 168
Silverman, S., 496, 498
Simeons, A. T. W., 456
Simon, C., 148
Singh, Baldeu, 172, 355

Singleton, L., 413, 426
Sirota, Alan, 509–10
Sittenfeld, Pola, 452, 506–7
Skinner, B. F., 58, 205n, 238, 316, 465
Smith, O. C., 278–79
Snyder, C., 91n
Soley, M., 42
Sonnenblick, E. H., 216n
Spanos, N. P., 368, 373, 374, 377
Sperry, R. W., 73–74
Spiegel, H., 371
Spiegel, J. P., 463, 464
Staab, L., 148
Stampfl, T., 148
Sterman, M. B., 98, 129–30
Stern, M., 215n, 456
Sternbach, R. A., 65, 485, 488
Stolov, W. C., 276
Stone, E. A., 90
Stoyva, J., 92n, 438, 479, 482, 484
Strupp, Hans H., 501–2
Stumpf, Ch., 99
Sugarman, H., 177
Sugi, Y., 355
Sutcliffe, J. D., 389
Sutterer, J. R., 321
Svyadoshch, A., 148
Swami, Rama, 154, 157, 160–66
Swets, J., 69n

Tanaka, R., 241
Tarchanoff, J. R., 38, 47
Tart, C. T., 178, 398n
Taub, Edward, 504
Taylor, D., 483
Taylor, N. B., 38, 41, 42, 43
Terrace, H. S., 61
Thorndike, E. L., 465
Tohei, K., 183
Townsend, 5
Tracy, M. L., 505–6
Travis, Terry A., 496, 507–8, 516
Trosman, H., 146
Tuke, Daniel H., 4
Tursky, B., 215n, 312, 456, 497
Tuttle, R., 284

Uyeshiba, M., 183

Vaitl, Dieter, 502
Vanderwolf, C. H., 123, 214
Van de Velde, T. H., 46
Varni, J., 498
Venables, P. H., 444
Vesalius, 273
Vogel, G., 145, 146
Volow, M. R., 498

Waitzkin, 107
Walker, S., 168–69
Wallace, J. D., 131, 165
Wallace, R. K., 173, 179, 260, 485
Walter, Grey, 463
Walter, W. G., 75, 168
Walters, Dale, 148, 164
Walters, E. D., 113, 260, 434n, 455
Webb, R. A., 321
Weber, 4, 7
Wechsler, N. H., 216n
Wedenski, 20
Weinstock, S. Alexander, 510–12, 515–16
Weiss, J. M., 321, 322, 470, 473
Weiss, Theodore, 196–97, 207, 482
Weitzenhoffer, A. M., 365, 367, 371, 377, 413
Wenger, M. A., 175, 353, 355
West, H. F., 21, 25
Whatmore, George B., 74–75, 76
White, J. D., 483
White, R. W., 375
Wilkins, B. R., 280
Wilson, D. L., 372, 375
Wing, J. K., 444
Winnick, W. A., 55
Witkin, H. A., 146–47
Wolberg, L. R., 367, 368, 372
Wolf, G., 88
Wolff, H. G., 456
Wolpe, J., 75, 437, 438, 441, 449, 450, 455
Wood, D. M., 193
Wright, J. M. C., 451, 452
Wyrwicka, W., 98

Yogi, Maharishi Mahesh, 357
Young, G. A., 90

Zimbardo, P. G., 430

SUBJECT INDEX

Abductive tensions, 62–63, 66–67, 70
Acetylcholine, 83
Adductive tensions, 62–63, 66–67, 70
Age progression and regression, 407
Aikido, 182, 183
Alcohol, effects of, 464–65
Aldomet, 93
All-India Institute of Medical Science, 353
Alpha rhythm sleep, 145, 146
   compared to theta rhythm sleep, 147
Alpha/theta feedback, 147, 148, 156–57
   see also, Alpha wave production
Alpha wave production, 107, 111, 131n, 145, 147, 149, 150, 155, 163, 173, 256, 361, 496, 505–6, 516
   amount of, 169
   analyses of variances in producing (ANOVAs), 257, 259
   and arousal, 178
   benefits from, 179, 183–84
   blocking, 131, 168
   courses in controlling, 167
   in electroencephalogram, 61–62
   electroencephalographic feedback and, 262–69
   emotional effects during, 172, 176, 255
   and empathy, 181
   experiments in producing, 111, 126–29, 154–55, 264, 266–68, 496, 499, 505–6
   and extrasensory perception, 175
   feedback, 148, 438–41, 479
   good subjects for, 177, 179
   and headache control, 485
   and learning, 175–76
   localized, 266–68
   and mediation, 254, 355, 361
   occipital, 168, 265
   physical changes during, 169, 171
   during presleep, 145–50
   for schizophrenia treatment, 181–82
   and trust, 178
   voluntary control of, 111, 126–29, 154–55, 162–63, 167–84, 254–57, 455, 496, 499
   during Zen meditation, 172–75
Alternative for hypnosis. *See* Paradigm, alternative for hypnosis
Aminophylline, 50
Amnesia, 407
   combat, 462–63
Amphetamine, 50
Amygdala electrical activity, 131
Analgesia, 407, 515–16
Anoxia, self-induced, 128
Anxiety, 145, 361, 449–50
   neurotic, 452
   pervasive, 452
Arch support, 277
Arousal levels
   and alpha training, 184
   producing low, 145–50
Arrectores pilorum muscles, voluntary control of, 26–33
   and electronic potentials of the brain, 31, 33
   experiments in, 26–31
   physical changes during, 26, 28–31, 32
"Arrest of a Disabling Eye Disorder Using Biofeedback," 498
Arsenic, and heart rate control, 16–17
Arthritis, rheumatoid, 84
Astigmatism, 414
Atherosclerosis, 481
Atrial fibrillation, 193, 207
Atrial flutter, 162
Atrial-ventricular node, 193–96
Atropine, 46, 50, 194, 197
Attitude. *See* Response bias
Audioneuromuscular re-education. *See* Neuromuscular re-education

Subject Index 525

Aurobindo's Integral Yoga, 158
Autogenic Feedback Training, 154, 179, 441
Autogenic training, 152, 153, 160, 485–86
Autonomic nervous system
    and cardiac conditioning, 193
    central aspects of, 83–84
    divisions, 81
    experiments in controlling, 26–31
    Japanese study of, 239–45
    and meditation, 355, 360–61
    and muscle relaxation, 449
    neurotransmitters in, 83
    parasympathetic divisions, 82, 83, 449
    physical changes while controlling, 26, 28–31
    reactivity, 275–76
    sympathetic divisions, 82, 83
    and ventricular rate, 192
    visceral regulation, 82
    voluntary control of, 3, 26–33, 81–84, 238–45, 297–309, 311–18, 328, 355, 455, 495
    see also, Galvanic skin responses
Autonomic neurotransmitters, 83
Autonomic-somatic integrations, 441
"Autoregulation of Skin Temperature Using a Variable Intensity Feedback Light," 504
Avoidance behavior, 54–55, 244–45
    in rats, 328–38
"Awareness of Own Cardiac and Pupillary Response to Pictures," 498

Barber Suggestibility Scale, 366, 369, 371, 372, 381, 394, 414
Barbiturates, 464–65
Barrier test, 451
Basal metabolism rate, 165
Beckman Biopotential Skin Electrodes, 231
Behavior
    avoidance, 54–55
    biofeedback techniques for, 437–56
    computer use in controlling, 454–55
    conditioning, 58–60
    experiments, 438–41, 442–49
    Gestalt therapy, 55–58
    hypnotic, 386–408
    prospects for controlling, 454–56
    and psychophysiological feedback loop, 437
    systematic desensitization in, 438, 442–46
    therapy, 437–56, 514
    voluntary control of, 53–76
Behavioral clock, 315–16, 317
Belladonna, 25
Bell's palsy, 288
Beta waves, 165, 255
    operant feedback for, 479

    producing, 149, 155, 156, 172, 253, 254–59
"Bidirectional Operant Conditioning of Peripheral Vasomotor Responses with Augment Feedback and Prolonged Training," 498
Bidirectional procedure, 97–100, 337
Biofeedback
    applied to anxiety problems, 145
    applied to psychosomatic disorders, 145
    clinical potential of, 438
    home systems for, 150, 489
    interaction effects of sex during, 507
    and meditation, 54–66
    operant, 478–79
    presleep produced by, 145–50
Biofeedback Research Society, 72
Blood
    lactate level, 361
    mobilization of, 43–46
Blood pressure, 59, 198, 481
    during arrectores pilorum muscle control, 27, 30–31, 32
    behavioral relationship to heart rate, 210–15, 219–20, 223
    experiments in controlling, 34–48, 92, 218–24, 479–80
    during heart rate control, 5, 9–11, 12, 13, 14, 35, 42, 209, 467, 469, 479
    during hypnosis, 362
    during meditation, 357–58
    in monkeys, 218
    and operant feedback, 479
    in rats, 90–91, 218
    voluntary control of, 34–48, 90, 91, 191, 209–16, 456, 467, 481–82
    see also, Diastolic blood pressure; Systolic blood pressure
Blood vessel constriction, 361
Body hair, voluntary control of, 26–33
    see also, Arrectores pilorum muscles, voluntary control of
Boston City Hospital, 481
Bradycardia, conditioned, 193
Brain
    animal, 84
    conscious domains, 158
    disorders, 264, 265
    electrical activity of, 167–84
    electronic potentials of, 31, 33
    hemispheres of, 264–65
    paleo-cortex of, 158
    subcortical, 158
    unconscious domains, 157
    visceral, 158
    voluntary control of electrical activity in, 167–84
Brain-behavior relationships, 114

Carbon dioxide elimination, 46
Cardiac Arrest, 49–50
    see also, Heart rate
Cardiac arrhythmia, 193, 482
Cardiac function
    cardiodynamics, 192–97
    conditioning, 191–208
    hemodynamics, 197–201
    neural regulations, 202–4
    psychological regulations, 204–8
    see also, Heart
Cardiac output, 43
"Cardiac-somatic Relationships in Operant and Pavlovian Conditioning," 495
Cardiodynamics, 191, 192–97
    chronotropic action, 192
    introtropic action, 192
Cardiovascular control. See Heart rate, voluntary control of
Cassette tape systems, 150
Cats
    curarization of, 339–50
    evoked potential to light in, 126, 132
    sensorimotor cortex rhythm of, 129, 339–47
    experiments with, 339–47
    voluntary control in, 339–50
Central nervous system
    changes in, 97–100
    constraints on conditioning, 111–12
    controlling electrical activity in, 96–139
    efficiency of conditioning, 107–11
    goals of conditioning, 112–17
    and muscle relaxation, 452
    operant conditioning of, 115–16, 118–30, 202–4
Central-peripheral interactions, 74, 75
Cerebral cortex, 158
Cerebral thrombosis, 287
Cerebral vascular accidents, 481
Cerebrospinal system, 481
CNS. See Central nervous system
"Cognitive and Somatic Mechanisms in Voluntary Control of Heart Rate," 503
Colitis, ulcerative, 84
Compulsive ideation, 455
Computer-assisted teaching programs, 150
Concomitant measures of electrical activity, 118–27
Conditioning
    classical, 59, 86, 91, 110, 193, 203–4, 297, 335, 465–66, 467
    and control, 90–94, 204–5
    dissociative, 121, 138
    independent, 347–48
    instrumental, 59, 60, 85–86, 91, 170, 465–66
    interospective, 81
    operant, 96–139, 170, 203–4, 281, 297, 311–18, 321–27, 343–45, 363, 430,
465–66, 496, 497, 498
Pavlovian, 86, 478
    in Russia, 81, 85, 86, 91
    semantic, 81
    theories of neurosis and personality, 81
"Contingent negative variation," 75
"Control of the EEG Theta Rhythm, The," 506–7
Control, voluntary, see Voluntary control
"Controlled Study of Alpha Enhancement, A," 496
Coracohumeral ligament, 277
Coronary artery disease, 481
Cortical activity
    and convulsion-producing drugs, 134
    manipulating, 66
Cortical desynchronization, 119, 127
Craniospinal nervous system, 188
Creativity
    feedback enhancement of, 147, 150
    psychophysiologic training for, 486
    related to reverie-and-hypnagogic imagery, 147
Curarization
    of animals, 86–87, 317, 433, 478–79
    of cats, 339–50
    of dogs, 105–6
    of humans, 309, 466, 469
    of rats, 90, 263, 321–27, 328–38
Cycloplegic drugs, 426

Deautomatization, in alpha, 180
"Defense alarm" reaction, 363–64, 456
Delta wave production, 155, 156, 164–65
    experiments in, 164–65
Desensitation, systematic. See Behavior
Diabetes insipidus
    in rats, 88
    relaxation as treatment of, 179
Diastolic blood pressure, 43
    experiments in, 218–24, 480, 497
    and heart rate, 15, 35, 46, 47, 224
    during meditation, 358
    medium, 219
    in monkeys, 90
    voluntary control of, 34–38, 93, 218–24, 497
    see also, Blood pressure
Discrimination training, 67
Dissociation, 392
Dissociative conditioning, 121
"d-LSD-25 in the Treatment of Epilepsy," 513
Dogs
    conditioning experiments on, 495
    curarization of, 105–6
    hippocampel electrical activity in, 97–100, 105–6, 118–24
    voluntary heart rate control in, 90

Subject Index 527

Drugs
  and fear, 464–65
  d-tubocurarine chloride, 105
Dyspoesis, 74, 75

Ear muscles, voluntary control of, 19
Edema, 286, 287
Edrophonium, 194
Education
  during alpha state, 175–76
  during presleep, 148, 150
  and stress, 469–75
"EEG Spectral Features of Self-regulated High Alpha States," 499
"Effects of Feedback Control of Heart Rate on Judgments of Electric Shock Intensity," 509–10
Eidetic imagery, 429
Electrical activity
  of amygdala, 131
  of brain, 113
  in central nervous system, 96–139
  experiments in, 97–100, 105–6, 118–24
  hippocampal, 97–100, 105–6, 118–24
  significance of, 114–15
Electrical stimulation, as reward, 90
Electrodermal activity, 59
Electrodermal modification, operant, 311–18
  behavioral clock for, 315–16, 317
  experiments in, 312–16
Electrodes
  bipolar, 274–75
  intramuscular, 279
Electroencephalogram patterns
  abnormal readings of, 263
  alpha rhythms in, 61–62, 127, 355
  amplitude component of, 496
  autoregulation of, 253–54
  experiments in controlling, 254–57, 264, 499
  and eye movements, 514
  feedback, 108, 156, 253–61, 348–49, 438–41, 479, 485
  during presleep, 145, 146
  during sleep, 362
  theta waves in, 506–7
  see also, Electroencephalographic control
Electroencephalographic control
  desynchronization of, 253
  experiments in, 254–57, 264, 266–68
  localized, 262
  and physiological language development, 262, 263
  for producing alpha feedback, 262–69
  see also, Electroencephalogram patterns
Electromyogram, 65, 66–67, 275, 287
  during avoidance learning, 330
  in behavior therapy, 441–54, 455
  feedback, 149, 441–54, 442–46, 455, 513–14
  in neuromuscular control, 288–93
  during presleep, 146
  during relaxation, 508–9
  sensitivity, 446–49
  signal storing, 275
  see also, Electromyography
Electromyography, 273–84, 482–83
  in apes, 284
  kinesiology of, 282–83
  types of electrodes used in, 274–75, 279
  see also, Electromyogram
Electrophysiology, 53–76, 153
  see also, Psychophysiology
Emotional tachycardia, 25
Ephedrine, 50
Epilepsy, 264
  controlling, 262, 269, 512, 513
Epinephrine, 43, 47
Erb's paralysis, 291
Ergotropic-trophotropic distinction, 441
Esalen Institute, 178
Escape-avoidance responding, 200
ESP. See Extrasensory perception
Essential blepharospasm, 498
Evolutionary theory, 284
Experiments and research
  alpha wave production control, 111, 126–29, 154–55, 264, 266–68, 438–41, 496, 497, 499, 505–6, 516
  alternative paradigm to hypnosis, 394–403
  amplitude component of the EEG, 496
  arrectores pilorum muscle control, 26–31
  atrial-ventricular node, 193–96
  autonomic nervous system control, 26–31
  avoidance behavior, 54–55, 244–45, 328–38
  behavior therapy, 438–41, 442–49
  behavioral relationship between blood pressure and heart rate, 210–15
  blood pressure control, 34–48, 92, 218–24, 479–80, 497
  body hair control, 26–31
  brain wave patterns, 162–64
  cardiac-somatic relationships, 495
  with cats, 339–47
  delta wave production control, 164–65
  diastolic blood pressure control, 218–24
  electroencephalograph control, 254–57, 264, 499
  electromyogram feedback, 513–14
  epilepsy treatment, 512, 513
  evoked potentials to light in cats, 126
  evolutionary theory, 284
  eye disorders, 498
  false feedback, 502
  fear, 460–63, 498
  finger pulse control, 498
  galvanic skin responses, 240–41, 298–99, 300–307

headache control, 510–12
heart rate control, 6–11, 17–18, 19–20,
    22–23, 34–48, 49–50, 161, 175,
    199–201, 202–4, 219–20, 229–36,
    241–42, 247–49, 467, 479–80, 496, 497,
    498–99, 500–501, 503
  hippocampal electrical activity control,
    97–100, 105–6, 118–24
  homeostasis maintenance, 88
  hypnosis, 177–79, 373, 414–24
  intestinal control, 328–38, 467, 505
  motor unit control, 280–81
  muscle control, 284
  neuromuscular control, 73
  operant conditioning, 495
  Pavlovian conditioning, 495
  perceived muscular tension, 60–64, 154–55
  pupil dilation, 498–99
  with rats, 322–26, 328–38
  response control, 60–64, 66–72
  schizophrenia, 176–77
  sensorimotor activity, 129–30
  single-cell activity in monkeys, 124–26,
    127–29
  shock avoidance, 111, 509–10
  skin temperature control, 154–55, 160,
    430–33
  somatic change, 312–16, 322–26
  somatic control, 497
  somatic psychology, 56–57
  theta wave production, 506–7
  transcendental meditation, 357–61
  urination control, 469
  visceral learning, 479–80
  visual improvement, 414–24
External clock, 315–16, 317
Exteroceptive feedback loops, 66
Extrasensory perception, 174–75
Eye refraction
  and alpha production, 169
  experiments, 414–24
  voluntary control of, 412–27

Fantasy, 146
  goal-directed, 373, 375
Fear, 460–63
  and alcohol, 464–65
  combat, 470
  and drugs, 464–65
  experiments in, 498
Feedback
  added by experimenter, 108
  alpha/theta, 147
  by electromyographic responses, 108n,
    513–14
  exteroceptive, 247–49, 297–309
  false, 183, 502
  galvanic skin response control, 300–307

heart rate control, 206–8, 215, 227–36,
    247–49, 502
  hypertensia control, 217–18, 223
  for insomnia control, 149
  meters, 157
  motor skills training, 109
  naturally occurring, 108
  operant, 478–79, 498
  during presleep, 145–50
  for psychological uses, 484–86
  for psychosomatic disorders, 154–66, 484–86
  response state and reinforcement, 107–11
  sensory, 76n
  tonal, 171
  training, 66–67
Finger joints, voluntary control of, 19
Foundation for the Study of Consciousness,
    160
Fowler's arsenic, 16
Free association, 146–47
Freudian psychology, 75–76

Gallamine, 105
Galvanic skin response
  amplitude data, 305–6
  controlling spontaneous, 238, 240–41,
    297–309
  experiments in, 240–41, 298–99, 300–307
  and feedback, 300–307
  heart rate data, 307
  measuring, 234–35
  respiration data, 306
  tonic data, 306
  see also, Automatic nervous system
Gamma motoneurons, 280
Generalization testing, 67
Gestalt, 56n
Gestalt Theorie, 56n
*Gestalt Therapy: Excitement and Growth in
    the Human Personality,* 55–58, 60, 64
Glandular responses, controlling, 90–94
Glenohumeral joint, 277
Glycosuria, 43
Goose pimples
  and arrectores pilorum muscle control, 26,
    32
  during heart rate control, 6
Group therapy, 76

Hallucination, 149
  conditioning, 61, 407
Harvard Group Scale of Hypnotic
    Susceptibility, 423
Harvard Medical Unit, 357
Hatha Yogis, 175
Headaches, relieving, 92n, 434, 482–83, 485,
    510–12

Subject Index 529

Heart
  conditionability of, 495
  function, 191–208
  ventricular rate, 192
  see also, Heart rate; Heart rate, voluntary control of
Heart rate, 62, 119, 120, 316
  during arrectores pilorum muscle control, 28–30, 31
  behavioral relationship to blood pressure, 210–15, 219–20, 224
  during diastolic blood pressure control, 217, 224
  electrocardiogram traces of, 23
  emotional influences on, 4
  during galvanic skin response control, 307
  influence of heat upon, 15–16
  during hypnosis, 362
  measuring, 153, 230–33, 498–99
  median, 219
  during meditation, 357
  see also, Heart; Heart rate, voluntary control of
"Heart Rate Control Under False Feedback Conditions," 502
Heart rate, voluntary control of, 3–20, 21–25, 26–33, 34–47, 59, 90–91, 161, 175, 198, 199–201, 205, 209–16, 241–42, 247–49, 456, 477, 496, 497, 500–501, 503
  autonomic response profile of, 3
  blood pressure change during, 209, 469, 479
  cardiac arrest during, 49–50
  under curare, 322–23
  data, 9–11
  in dogs, 90
  and exhaustion, 16
  experiments in, 6–11, 17–18, 19–20, 22–23, 34–48, 49–50, 161, 175, 199–201, 202–4, 219–20, 229–36, 241–42, 247–49, 479–80, 496, 497, 500–501, 503
  feedback for, 206–8, 215, 227–36, 247–49, 479
  and heart output, 18, 37, 43, 198
  history of patients and, 25, 34–35, 49
  by meditation, 353–58
  by monkeys, 90, 199
  and nitrous acid, 16, 17, 47
  operant conditioning study of, 197–201
  physical changes during, 6–7, 12, 21, 22, 38–46, 47, 49, 199, 205
  by rats, 90–91, 321–27
  and respiration rate, 8, 38–41, 50, 321
  see also, Heart; Heart rate
Hemodynamics, 191, 192, 197–201
Hexobarbitol, 464
Hippocampal electrical activity, 97–100, 105
  in dogs, 97–100, 118–24
  experiments in, 97–100, 105–6, 118–24

  and high-frequency RSA, 118–24
  in rats, 109, 118–24
Histamine, 83
Homeostasis maintenance, 87–88
Hunger-motivating approach, 465
Hydrodiuril, 93
Hypercirculation, voluntary control of 34–48
  see also, Blood pressure, voluntary control of; Heart rate, voluntary control of
Hyperglycemia, 43
Hyperponesis, 74
  see also, Muscle
Hypertension, 84–85, 92, 364
  controlling, 92, 217, 481–82, 487
  cortico-visceral theory, 85
  studies with animals, 85
  see also, Blood pressure
Hypertension Clinic, 481
Hypertensive pathology, 179
Hyperthyroidism, 84
Hyperventilation, during voluntary heart rate control, 38, 41–42
Hypnagogic state, 146
  see also, Presleep state
Hypnosis, 152, 362, 365–84, 386–408, 479–505
  age regression in, 413
  alternative paradigm to, 390–93
  and analgesia, 515–16
  assumptions about, 367, 380–81, 388–89
  cooperation during, 368, 375
  drug, 372
  education, 367–68
  effect of cues during, 381–82
  experiments during, 177–79, 373, 414–24, 505
  eye-fixation technique, 371–72
  eyes closed during, 368–69
  focused thoughts during, 376
  good subjects for, 177, 179
  hand levitation technique, 372
  human plank feat in, 393
  imagining during, 372–74
  induction procedures for, 365–84
  labeling as, 366–67
  mediating variables in, 374–77
  metronome technique, 372
  and operant feedback, 479
  physical changes during, 362–63, 365, 377–80, 388–84
  and positive expectancy, 375
  postural sway technique, 372
  research, 177–79
  and skin temperature control, 429–34
  stage, 394, 403
  suggestions during, 370–72, 374, 375–77, 382
  support for, 403–4
  susceptibility during, 177–79

Index

test suggestions without, 393–403
trance paradigm, 387–89
and trust, 178
variables in, 377–83
visual improvement under, 412–27
Hypometabolic state, 361, 364

Iatrogenic anxiety, 207
Insomnia, feedback training for, 149
"Integrated Alpha: Comparison with Criterion or Rewarded Alpha," 516
"Interaction Effects of Sex on Subject and Experimenter in Biofeedback Training," 507
"Intra-hemispheric Phase Relationships During Self-regulated Alpha Activity," 505–6
Intrathoracic pressure, 5
Iso-bias contour, 70–71
Isoproterenol, 194, 196
Iso-sensitivity contour, 71

Jungfrau pattern, 72

Kansas University Medical Center, 162
Karate, 182

Laboratory of Dermatology Research, 431
Laser scintillation technique, 422–23
L-dopa, 93
Learning
    during alpha state, 175–76
    during presleep, 148, 150
    and stress, 469–75
Lindsay Municipal Hospital, 49
"Locus of Control and the Voluntary Control of Heart Rate," 497
Low arousal levels, 145
Lumber lordosis, increasing, 283

Magnitude estimation, 62
Massachusetts Mental Health Center, 481
"Measurement and Quantification of Surface EMG Signals in States of Relaxation," 508–9
Mediation, 204, 309, 467
    criteria for, 101–2
    of neural events, 100–107
    physiological correlates of, 179–80
    relation to alpha wave activity, 254
Meditation, 180–81
    and alpha wave control, 172–73, 355
    experiments in, 357–61
    and heart rate control, 353–58
    physiological effects of, 173–74, 355–64
    transcendental, 357–61, 485
    see also, Yoga meditation; Zen meditation
Menninger Foundation, 154, 161, 455
Mental retardation, 486

Metabolism
    changes during meditation, 355, 357
    changes during voluntary heart rate control, 38–41, 46
Metaphysics, 283
Millicent Foundation, 160
Milton Fund, 216n
"Mobilizing the Self," 56
Monkeys
    avoidance learning by, 90
    blood pressure in, 218
    hemodynamic study of, 200–201
    motor cortex single-cell activity in, 124–26
    voluntary control of heart rate by, 90, 199
Monomethylhydrazine, 130
Motor unit, 286
    applications of training in control of, 281–82
    experiments in controlling, 280–81
    and operant feedback, 479
    potentials, 274–75
    training, 278–79, 280–81, 286–93
    transient, 290
Muscles
    abductive contraction of, 63–64, 66–67, 70
    abductor pollicis brevis, 279
    abductor digiti minimi, 279
    adductive contractions of, 63–64, 66–67, 70
    alpha motor neuron discharge in, 276
    and anxiety, 449–50
    atrophy in, 289
    and autonomic nervous system, 449
    axon of, 274
    biceps braichii, 279
    brachialis, 277, 279
    caracobrachialis, 287
    and central nervous system, 452
    contraction perception, 60–64
    coordinating, 277–78
    electrical activity in, 273–84
    experiments in controlling, 284
    eye, 420–21
    fatigue in, 275–77
    fiber of, 273–14
    fibrosis in, 291
    frontalis, 445
    hypertonia of, 288
    nerve impulses in, 289
    pectoralis, 287
    plantar ligaments in, 277
    pronator quadratus, 277
    psoas major, 283
    relaxation of, 441, 442, 449–50, 452, 482–83, 484
    shoulder, 273, 277
    striated, 174
    supraspinatus, 277, 287
    sustained contraction of, 62–64, 66–67
    tension in 42–53, 60–64, 154–55, 441

Subject Index 531

teres major, 287
tibialis anterior, 279
tone of, 275–77
trapezius, 287
trunk, 283
voluntary control of, 273–84, 286–93, 442–46
Mydriatic agents, 426
Myoelectric artificial aids, 281
Myopic improvement, 419
see also, Visual improvement

Nearsightedness, 412
see also, Visual improvement
Nephron necrosis, 84
Nephrosclerosis, 481
Neural regulations, 113–14, 191
Neuromuscular re-education, 286–93
experiments in, 73, 498
and nerve impulses, 287, 289–90
objectives, 293
and reflexes, 290
sprouting process in, 292–93
voluntary control in, 73, 290–91, 498
Neurophysiologic engineering, 75
Neuroses, treating, 180, 460–75
and coping, 473–75
experiments in, 460–63
and fear, 460–63
and learning, 469–75
New York University Medical Center, 92
Nitrous acid, and heart rate control, 16, 17, 47
Norepinephrine, 83, 361, 473–75

"Operant Conditioning of the Amplitude Component of the EEG," 496
"Operant Conditioning of Heart Rate: Somatic Correlates," 497
"Operant Control of Fear-related Electrodermal Responses in Snake-Phobic Subjects," 498
Optometric training techniques, 413
Osaka Prefectural Institute of Public Health, 239

Parachutists, research on, 495
Paradigm, alternative for hypnosis, 386–408
experiments in, 394–403
human plank feat in, 393–94
negative attitudes in, 392
positive attitudes in, 392, 406
prospects for, 404–8
response to, 391–403
task-motivational instructions in, 398
test suggestions for, 393–403
Paradigm, trance, 387–89
assumptions about 388–89
see also, Hypnosis

Paralysis, 286, 287
functional, 288
Paraplegia, spasticity in, 278
Parasympathetic activation, 83, 441, 449, 453–54
Pathological blocks, 286–87
Pathological nystagamus, 483
Pathophysiology, visceral, 84–88
Perception, increased, 183
"Performance of Trained HR Slowing: Effects of Instructions, Feedback, and Shock," 500–501
Personality. *See* Behavior
Pflüger's Archives, 3
Phenylephrine, 197
Phobias, 462
Phoria, 415
Physical therapy, 282
Physiological unconscious, 205
Pigeons, conditioned tachycardia in, 193
Pilomotors. *See* Arrectores pilorum muscles
Placebo effects, 487
Poliomyelitis, anterior, 288, 292
Polykrotismus, 15
Postreinforcement synchronization, 131
Posture, 282–83
"Power law," 62
Premature ventrical contractions, 196
Presleep state, 145–50
education during, 150
physical changes during, 145
producing, 147
psychotherapy during, 150
suggestibility during, 147
Pressor hyperactivity, modifying, 191
Private events, 58
see also, Behavior
"Problems in Therapeutic Application of EMG Feedback," 513–14
Propranolol, 194
Proprioception, 278–79
Proprioceptive facilitation, 57
Psychoanalytic trauma theory, 53
Psychological regulations of cardiac function, 191
Psychology
Freudian, 75–76
Gestalt therapy, 55–58
group therapy, 76
neurophysiologic engineering, 75
somatic, 55–58
Psychophysiology, 53–76, 284
to analyze private events, 58–12
and biofeedback, 72–76
feedback loop in, 437
see also, Behavior
Psychosomatic disorders, 60, 83, 145, 480–81
training programs for, 153, 263, 308, 434

see also, Neuroses, treating
Psycho synthesis, 166
Psychotherapy, 75, 76
   behavioral, 462
   Freudian, 75–76, 462
Pulse rate. *See* Heart rate; Heart rate, voluntary control of
Punishment, of rats, 54
Pupil, dilation of, 498–99
   during voluntary control of arrectoris pilorum muscles, 28
   voluntary control of, 32

Rats
   avoidance behavior in, 54–55, 111, 328–38
   blood pressure control in, 90, 91, 218
   control of vasomotor dilation in ear of, 263, 456
   curarized, 90, 263, 321–27, 328–38
   experiments with, 54–55, 322–26, 328–38
   fear in, 464
   heart rate control in, 90, 91, 321–27, 467
   hippocampal electricity activity in, 109, 118–24
   RSA response in, 122
   urination control in, 467
   uterus contraction control in, 90, 467
   visceral learning in, 87
Raynaud's disease, 483
Reciprocal inhibition, 277–78
Refractive error, 414n
Relational approach to electrical activity, 114–15
"Relationship between Cardiac Conditionability in the Laboratory and Autonomic Control in Real Life Stress," 495
Research. *See* Experiments and research
Respiration rate, 5, 62
   during hypnosis, 362
   measuring, 230, 233–34
   and meditation, 355, 356, 361
   of rats, 55
   during sleep, 362
   and voluntary heart rate control, 8, 38–41, 50, 321, 501–2
Respiratory quotient, 39n
Response Bias, 64–72
   signal detection theory, 65
Reverie-and-hypnagogic imagery, 147
Reverie state, *see* Presleep state
Reversible physiological block, 286, 287
Rheumatic heart disease, 50
Rockefeller University, 430

Salivation, 59
   and heart rate, 6
   operant feedback for, 479

Schizophrenia, 184n
   research, 176–77
   treating, 181–82
"Self Control of Patterns of Human Diastolic Blood Pressure and Heart Rate through Feedback and Reward," 497
Self-healing, 166
Self-perception, 64
Self-regulation, for psychosomatic disorders, 154–66
Sensorimotor rhythms, 262
   in cats, 129–30, 131, 135
Sensory aphasia, 287
Sensory deprivation, 147
Serotonin, 83
Sexual response control, 486
"Shifts in Subjective State Associated with Feedback-Augmented EEG Alpha," 496–97
Shock, electric, 500–501, 509–10
   avoidance of, 111, 328–38
   offset, 337
   response to, 202, 247–49
Signal detection concepts, 69n
"Simple On-line Technique for Removing Eye Movement Artifact from the EEC, A," 514
"Simultaneous Feedback and Control of Heart Rate and Respiration Rate," 501–2
Sinus arrhythmia, 25
Skin potential response, 3, 311–12, 316–18
Skin resistance levels
   during hypnosis, 362
   measuring, 234
   during meditation, 355, 357, 360, 361
   during sleep, 362
Skin responses, galvanic. *See* Galvanic skin responses
Skin temperature
   controlling, 154–55, 157, 160, 429–34, 483, 504, 505
   experiments in, 430–33, 504, 505
   feedback, 455, 479
   and heart rate control, 512
Sleep
   alpha rhythm, 145
   compared to presleep, 145
   control of in Russia, 148
   descending, 145
   learning, 147–49
   physical changes during, 362
   suggestibility during, 148
Sleep onset. *See* Presleep state
Sodium amytal, 464
Somatic processes, 321–-27
Somatocardio vascular relations, 321, 327, 497
"Some Physiological Correlates of Analgesia

in Various Stages of Hypnotic Depth," 515–16
Somnambule, 387, 406, 407
Soviet science
  classical conditioning, 81, 85
  pathological model systems, 84–85
  sleep control, 148
Specificity, 204
"Specificity of Alpha and Heart Rate Control Through Feedback," 496
Spinal motoneurons, 278, 280
Spine, changes in curve of, 283
Spiritualism, 152
Stanford Hypnotic Susceptibility Scale, 372, 382
Stanford Medical Center, 431
Stirling Homex, 167
Stomach contractions, 62
  voluntary control of, 456
Succinylcholine, 469
Suggestion, as visual improver, 412–27
Suppression amblyopia, 413
Sympathetic activation, 83, 441
Sympathetic cardioaccelerator nerves, 21
Sympathetic causalgia, 287
Sympathetic nervous system, 364
Sympathomimetic drugs, 50
Synergist, 278
Synthesis of Yoga, The, 158
Systems approaches, to electrical activity, 114–15
Systolic blood pressure, 42, 46, 47, 198
  and heart rate, 15, 35, 46, 47, 209, 223
  during meditation, 358
  voluntary control of, 34–48, 209–16, 480
  see also, Blood pressure
Systolic murmur, 18

Tachycardia, conditioned, 193
Teleology, 282–83
Tension patterns, 62
Tensions
  abductive, 62–63, 66–67, 70
  adductive, 62–63, 66–67, 70
  perceiving, 60–64
  prestimulus, 65
"Tentative Procedure for the Control of Pain: Migraine and Tension Headaches," 510–12
Theta rhythms
  compared to alpha rhythms, 147, 355
  controlling, 155, 163–64, 452, 455, 506–7
  enhancement during research, 149
  during meditation, 355, 361
  operant feedback for, 479
  during presleep, 145, 146, 150
  training program, 156
Thorndike Memorial Laboratory, 357

Thumb twitching, 60–64
"Traitement de L'Epilepsie Par Retro-Action Sonore" (Treatment of Epilepsy by Auditory Feedback), 512
Transfer procedure studies, 129–30
"Twilight learning," 148
Twilight state. *See* Presleep state

Ulcer, peptic, 84
Ulcerated colitis, 84
University of California, 353, 357
University of Florida, 148
University of Michigan Medical School, 353
University of Tokyo, 355
Urination, voluntary control of, 91, 456, 467, 469
Uterus, voluntary control of, in rats, 90

Vagus cardioinhibitory nerves, 21
Valsalva maneuver, 353–58
Vasoconstriction, 43, 59
Vasomotor system
  during heart rate control, 12–13
  operant feedback for, 479, 498
Vedas, 158
Ventricular contractions, premature
  voluntary control of, 92
Ventricular rate, voluntary control of, 194–96
Veterans' Administration Hospital, 160
Videotape self-confrontation, 76
Visceral learning, 477–80
  experiments, 479–80
  feedback-operant model of, 478
  see also, Visceral responses
Visceral responses
  controlling, 90–94, 433, 469
  modifying, 85–87
  in rats, 87
  recognizing, 94
  and visceral pathophysiology, 84–88
"Visual Effects during Alpha Feedback Training," 496
Visual evoked potential, 131
  in cats, 132
Visual improvement, 412–27
  experiments, 414–24
  by hypnosis, 412–27
  laser scintillation technique for, 422–23
Visual motor systems, 127
Volition and cardiac arrest, 49–50
Voluntary control
  of alpha wave production, 111, 126–29, 154–55, 162–63, 167–84, 254–57, 355, 361, 455, 496, 497, 505–6, 516
  of arrectores pilorum muscles, 26–33
  of autonomic nervous system, 3, 26–33, 81–84, 238–45, 297–309, 311–18, 455
  of behavior, 53–76, 437–56

of blood pressure, 34–48, 90–91, 191, 209–16, 218–24, 456, 467
of body hair, 26–33
of brain electrical activity, 113, 167–84
in cats, 339–47
of central nervous system, 96–139
of delta wave production, 164–65
of diastolic blood pressure, 34–48, 93, 218–24
of ear muscles, 19, 456
of eye muscles, 498
of eye refractive powers, 412–27
of finger joints, 19
of finger pulse, 498
of galvanic skin responses, 238, 240–41, 297–309
of glandular responses, 90–94
of headaches, 510–12
of heart rate, 3–20, 21–25, 26–33, 34–47, 59, 91, 161, 175, 198, 199–201, 205, 209–16, 241–42, 247–49, 321–27, 456, 467, 496, 500–501, 503
of hippocampal electrical activity, 97–100, 105–6, 122
of hypercirculation, 34–48
by hypnosis, 414–24, 429–34
of internal processes, 113–14, 136, 328–38
of intestines, 328–38, 467
by meditation, 353–58
of muscles, 273–84, 286–93, 442–46
of neuromuscular units, 73, 286–93
of pain, 510–12
of penile erection, 486
for psychosomatic disorders, 154–66
of psychological processes, 113–14
of pupil, 32
recognizing, 136
in rats, 321–27, 328
of respiration rate, 355, 356
of skin temperature, 154–55, 160, 429–34, 504, 505
of stomach contractions, 456
of theta wave production, 149, 163–64, 165, 452–53, 455, 506–7
of urination, 91, 456, 467, 469
of ventricular rate, 72, 194–96
of visceral responses, 90–94
"Voluntary Control of Skin Temperature: Unilateral Changes Using Hypnosis and Auditory Feedback," 505
Voluntary Controls Project, 154
Voluntary muscular system, 158

Washington University School of Medicine, 361

Yoga meditation, 131, 152, 167, 180, 205, 353, 356, 437
and alpha wave production, 167, 173
Aurobindo's Integral, 158
and physiological control, 174
physiological correlates of, 172
Raja, 158
and respiration rate, 355
and voluntary heart rate control, 175, 355–58
see also, Meditation

Zen meditation, 131, 167, 180, 184, 353, 356, 437
and alpha wave production, 172–73
and heart rate control, 355–58
and physiological control, 174
see also, Meditation

DATE DUE